Molecular Environmental Soil Science

Progress in Soil Science

Aims and Scope

Progress in Soil Science series aims to publish books that contain novel approaches in soil science in its broadest sense – books should focus on true progress in a particular area of the soil science discipline. The scope of the series is to publish books that enhance the understanding of the functioning and diversity of soils in all parts of the globe. The series includes multidisciplinary approaches to soil studies and welcomes contributions of all soil science subdisciplines such as: soil genesis, geography and classification, soil chemistry, soil physics, soil biology, soil mineralogy, soil fertility and plant nutrition, soil and water conservation, pedometrics, digital soil mapping, proximal soil sensing, soils and land use change, global soil change, natural resources and the environment.

For further volumes:
http://www.springer.com/series/8746

Jianming Xu • Donald L. Sparks
Editors

Molecular Environmental Soil Science

Springer

Editors
Jianming Xu
Institute of Soil and Water Resources
and Environmental Science
Zhejiang University
Hangzhou
People's Republic of China

Donald L. Sparks
Department of Plant and Soil Sciences
Delaware Environmental Institute
University of Delaware
Newark Delaware
USA

ISBN 978-94-007-4176-8 ISBN 978-94-007-4177-5 (eBook)
DOI 10.1007/978-94-007-4177-5
Springer Dordrecht Heidelberg New York London

Library of Congress Control Number: 2012948663

Printed on acid-free paper

Springer is part of Springer Science+Business Media (www.springer.com)

Preface

Chemical, biological, physical, and geological processes operate together to support life at the Earth's surface for sustainability. Advances on the frontiers of knowledge on this subject matter require scientists to cross disciplines and scales to integrate understanding of the processes ranging in scale from the environmental mineral-organism-humus-water-air interfaces to the impact on the globe and humankind. However, there are great knowledge gaps on how and to what extent the processes, especially in the rhizosphere, are affected by the interfacial interactions among environmental nanoparticles, mineral colloids, humus, water, and organisms (flora, fauna, and roots), at the molecular level. Therefore, an international symposium on the above-mentioned subject area was proposed and held in Zhejiang University, Hangzhou, China, on October 10–14, 2009 to explore the contribution of chemistry into this new field. The conference provided a forum for the exchange of information, interactions, and collaboration among soil chemists, mineralogists, microbiologists, and physicists with allied scientists including pure chemists, mineralogists, biologists, environmental scientists, ecologists, and eco-toxicologists to address the current state of the art on "Molecular Environmental Soil Science" which is fundamental to understanding and regulating soil processes at the molecular level.

The International Symposium of Molecular Environmental Soil Science at the Interfaces in the Earth's Critical Zone (ISMESS2009) was the first international conference focused on the above-mentioned subject area with special significance for understanding environmental pollution and global change processes. The conference was cosponsored by the International Union of Soil Sciences (IUSS) and the International Union of Pure and Applied Chemistry (IUPAC). Drs. J.M. Xu (China), P.M. Huang (Canada), A. Violante (Italy), H.H. Cheng (USA), J. Berthelin (France), J.M. Zhou (China), K. Wada (Japan), R.F Shen (China), R.G. Burns (Australia), and W. E. H. Blum (Austria) served in the Symposium Organizing Committee. The conference attracted 266 delegates from 21 countries in the world including Australia, Austria, Belgium, Brazil, Canada, China, D.P.R. Korea, Egypt, France, Iran, Italy, Kenya, New Zealand, Pakistan, R.O. Korea, Russia, South Africa, Spain, the UK, USA, and Vietnam. The theme of ISMESS2009 was "Biophysico-Chemical

Processes in Soil Environments at the Molecular Level and the Impact on the Globe and Humankind." The conference program was divided into the following five sessions: (1) The Role of Mineral Colloids in Carbon Turnover and Sequestration and the Impact on Climate Change; (2) Biogeochemical Interfacial Reactions and the Transformation, Transport, and Fate of Vital and Toxic Elements; (3) Anthropogenic Organics, Crop Protection, and Ecotoxicology; (4) Environmental Nanoparticles: Distribution, Formation, Transformation, Structural and Surface Chemistry, and Biogeochemical and Ecological Impacts; and (5) Environmental Processes and Ecosystem Health. There were 2 IUPAC plenary lectures, 18 invited speakers, 29 oral presentations, and 56 posters. Dr. D. L. Sparks from University of Delaware, USA, presented one plenary lecture entitled "Advances in the Use of Synchrotron Radiation to Elucidate Environmental Interfacial Reaction Processes and Mechanisms in the Earth's Critical Zone." Dr. G. M. Gadd from University of Dundee, UK, presented another plenary lecture entitled "Microbial Role in Global Biogeochemical Cycling of Metals and Metalloids at the Interfaces in the Earth's Critical Zone."

The 15 chapters in this book are mainly the papers from the plenary and invited speakers of ISMESS2009. They address the current state of the art on chemical-biological-physical-geological processes and the interfacial interactions at the molecular level. The book presents a variety of issues on the fundamental processes and interactions among nanoparticles, soil minerals, native and anthropogenic organic compounds/pollutants, and microorganisms and their impacts on the globe and humankind. This book is divided into four parts.

Part 1 addresses current research into fundamental processes in the soil-microbial ecosystem. The latest research is presented on a wide range of soil-based microbial processes, including organic pollutant degradation and the role of exocellular enzymes in soil processes. Other work addresses the structure and biological activity and environmental fate of both natural Bt proteins and those produced by genetically modified plants.

In Part 2, the latest research findings concerning the fundamentals of elemental and mineral transformations and metal cycling processes and reaction and transformation mechanisms of metals and metalloids are discussed. In addition, new mechanisms to account for the interaction between heavy metals and variable charge surfaces, and ion diffusion/adsorption as affected by microbes, biomolecules, and the electrostatic field from surface charges are presented. Advances in the use of synchrotron radiation to elucidate these various environmental interfacial reaction processes and mechanisms are also reviewed in depth.

Part 3 presents some of the latest developments in the field of surface charge chemistry and ion diffusion/adsorption by soil. The effects of "ageing" on the retention, bioreactive chemical transformations and transport in soil are reviewed extensively. Further work is presented on the sorption characteristics of pharmaceuticals and hydrophobic organic chemicals in the environment and the role of black carbon in the adsorption of organics from aqueous solutions.

Finally, Part 4 is concerned with soil fertility and soil quality as affected by agricultural management and microbial activities. This is a broad field and the focus here is on subsoil acidification, including causes and solutions, use of microbial indicators in soil quality evaluation and, finally, the impacts of agricultural management practices on soil properties.

These chapters address current knowledge gaps and provide future research directions at the molecular level. We believe that the publication of this ISMESS2009 special book will promote in-depth studies in molecular environmental soil science and the book will provide an invaluable reference for research scientists, professors, graduates, and consultants working in soil, agricultural, microbial, biological, ecological and environmental sciences.

We wish to extend our special thanks to the many sponsors including Organization for the Prohibition of Chemical Weapons (OPCW), National Natural Science Foundation of China (NSFC), Soil Science Society of China (SSSC), Zhejiang University, and Zhejiang Provincial Key Laboratory of Subtropical Soils and Plant Nutrition.

In addition to this book, a book of proceedings composed of extended abstracts that were subject to peer review by external referees, by International Scientific Committee members of the symposium, and by the editors of the proceedings was published by Zhejiang University Press and Springer. Volunteered papers presented at ISMESS2009 and accepted after rigorous external review were also published as a special issue by the *Journal of Soils and Sediments*. The book of proceedings and the special issue serve as companion volumes of this IUSS- and IUPAC-sponsored book published by Springer-Verlag.

Contents

Part I
Soil Microbial Processes

Chapter 1
The Sorption Behavior and Bioavailability of Persistent Organic Pollutants (POPs) in Soils

Jianming Xu, Tahir Hayat, and Na Ding

Abstract Persistent organic pollutants (POPs) pose a threat to the environment because of their potential for long-range atmospheric transport, bioaccumulation and toxicity. The POPs behave dynamically in the environment according to their different processes e.g. volatilization, sorption, desorption and transportation. Environmental variables like temperature, soil pH and moisture have serious effects on POPs behavior in soil. Organic as well as inorganic compounds may react with the xenobiotics and play an important role in their transformation in soil. Manganese and iron oxides and clay minerals (e.g. smectites containing Fe(III)) had well-documented properties to promote the oxidation of a number of organic pollutants. Organic matter is considered the most important factor limiting the availability and mobility of POPs in soil and a substantial percentage of the organic contaminants applied to a soil might become associated with soil humic fraction. Organic pollutants strongly adsorbed to carbonaceous sorbents such as black carbon. In particular, activated charcoal (AC) had a strong adsorptive capacity due to its high specific surface area. Adsorption to activated charcoal could render hazardous organic pollutants in soils and sediments less available to organisms and hindered their dispersal into unaffected environment. Some studies also showed that some sorbents from natural organic materials, such as peat, soybean stalk and pine needle under superheated temperature/pressure conditions, significantly affected the sorption of polycyclic aromatic hydrocarbons (PAHs) in contaminated soils. Natural materials significantly decreased the extractability and bioavailability of POPs from contaminated soils. Organic matter also plays a vital role in controlling the contamination of POPs in soil. The rhizosphere effect is showing promising to control POPs contamination in the environment. This review evaluated the work on the environmental behavior and bioavailability of POPs in soils as affected by various environmental variables.

J. Xu (✉) • T. Hayat • N. Ding
Institute of Soil and Water Resources and Environmental Science, Zhejiang University, Hangzhou 310058, China
e-mail: jmxu@zju.edu.cn

J. Xu and D.L. Sparks (eds.), *Molecular Environmental Soil Science*, Progress in Soil Science, DOI 10.1007/978-94-007-4177-5_1,

Keywords Persistent organic pollutants • Polychlorinated biphenyls • Polycyclic aromatic hydrocarbons • Pentachlorophenol • Sorption • Desorption • Bioavailability • Organic matter

1.1 Introduction

The term persistent organic pollutants (POPs) refers to the organic chemical compounds that persist and bio-accumulate in the environment through the food web, and are a risk to human and environmental health. People may be exposed to POPs in occupational settings through the inhalation of contaminated workplace air and dermal contact or with wood products treated with the chemicals. Also, general population exposure may occur through contact with contaminated environmental media, particularly in the vicinity of wood treatment facilities and hazardous wastes sites. In addition, some other important routes of exposure seem to be the inhalation of contaminated air, ingestion of contaminated ground water used as a source of drinking water, ingestion of contaminated food, and dermal contact with soils or products treated with the chemicals. POPs are highly dangerous because they are highly toxic, persistent, lasting for years or even decades before degrading into less dangerous forms. They may evaporate and travel long distances through air and water and accumulate in living fatty tissues. These properties make them, arguably, the most problematic chemicals that natural systems can be exposed to. POPs can be divided into three main categories: (1) pesticides e.g. DDT, (2) industrial chemicals e.g. polychlorinated biphenyls (PCBs) & polycyclic aromatic hydrocarbons (PAHs), pentachlorophenol (PCP), and (3) unintentional by-products e.g. dioxins, furans.

POPs keep accumulating through the food web as one animal eats another, so that the highest levels are found in predator animals at the top level of the food web. Research from numerous countries has demonstrated that measurable quantities of POPs are present in human adipose tissue, blood and breast milk, because they are soluble in fats and are not easily broken down in the body (http://www.epa.ie/whatwedo/resource/hazardous/pops). POPs are of environmental concern because of their potential for long-range atmospheric transport, bioaccumulation and toxicity (Klecka et al. 2000). They display high affinities for soils where they can persist over years to decades. Persistence of organic pollutants in soils has been attributed to sequestration mechanisms including intra-organic matter and intra-particle diffusion of the parent compounds and transformation products; contaminant humification to soil organic matter (SOM); low aqueous solubility; and/or microbial recalcitrance (Northcott and Jones 2000). Quite a number of organic pollutants exist in soil environment and these comprise a very significant source of pollution. We can not review all POPs in this article. Instead we emphasize only some organic pollutants and their entry into the environment through industrial and agricultural processes or as unintentional by-products such as polychlorinated biphenyls (PCBs), pentachlorophenol (PCP) and polycyclic aromatic hydrocarbons (PAHs).

The commercial production of PCBs began in 1930. Industrial manufacturing of PCBs was based on the chlorination of biphenyl in the presence of suitable catalysts. Over one million tons of PCBs have been produced in industrialized countries. The commercial use of PCBs began in 1965 in China, most commercially used PCBs were PCB_3 and PCB_5. In 1972, the production and use of PCBs were banned by law in Switzerland, UK and USA. Now, only the congeners with low degrees of chlorination can be sold and used as dielectrics in closed systems (WHO 1979). After that, many countries restricted the sale of PCBs. PCBs were also listed as priority pollutants by US Environmental Protection Agencies (USEPA) since the 1970s (Hu and Wang 1998; Keith and Telliard 1979). By the end of 1974, most factories stopped the production of PCBs and in the early 1980s, and the commercial production of PCBs was completely banned in China. Within this period of time, approximately 10^7 kg PCBs were produced and most of the products, about 9,000 t, were used as dielectric fluid in electric capacitors, except that a small part were used as additives in paints. Most of the electric capacitors containing PCBs have been exhausted. Furthermore, some electric capacitors and transformers containing PCBs were imported into China without pre-notification from some developed countries from 1950 to 1970 or even later. Although the commercial production of PCBs has been banned for almost two decades, improper disposal of the PCBs containing appliances will cause and has caused serious pollution in many parts of the world. It is worth mentioning that PCBs can also be used in many industrial processes such as chloranil and dicloran production. Therefore, PCBs will enter the environment when they are produced and used. PCBs entering the environment will go into the global cycling of materials and energy. These PCBs become major pollution sources in soil and water and ultimately through food cycle enter in the human body and become a serious cause of diseases such as liver cancer and biliary track. PCBs also have been shown to mimic the action of estrogen in breast cancer cells and can enhance breast carcinogenesis. Due to toxic effect as well as classified as persistent organic pollutants the production of PCBs was banned by the United States Congress in 1979 and by Stockholm Convention on Persistent Organic Pollutants in 2001.

Pentachlorophenol (PCP) is a synthetic chlorinated phenol that was first produced in the 1930s. It can be found in two forms: PCP itself or as the sodium salt of PCP, which dissolves easily in water. In the past, it has been used as a herbicide, insecticide, fungicide, algaecide, disinfectant and as an ingredient in antifouling paint. Some applications were in agricultural seeds (for nonfood uses), leather, and masonry, wood preservation, and cooling tower water, rope and paper mill systems. PCP can be produced by the chlorination of phenol in the presence of catalysts (anhydrous aluminum or ferric chloride) and a temperature of up to approximately 191°C. However this process was incomplete and resulted in the production of commercial grade PCP of between 84 and 90% purity. During the process several contaminants including other polychlorinated phenols, polychlorinated dibenzo-p-dioxins and polychlorinated dibenzofurans were also produced, which are more toxic than the original PCP. After entering into soils, PCP could be irreversibly bound to soil particles (Chen et al. 2004).

PAHs are ubiquitous environmental pollutants, with well-documented toxic, mutagenic, and carcinogenic properties. PAHs can be formed naturally by low-temperature and high-pressure reactions of natural organic matter. These compounds are produced by natural processes, including forest and grass fires, volcanoes, and seepage of oil into the environment. However, the majority of PAHs are released from a wide variety of anthropogenic sources, including petroleum spills, incomplete combustion of fossil fuels in electric power generation, refuse incineration, home heating, internal combustion engines, as well as the production of coke, carbon black, coal tar and asphalt (Eisler 1987; Burgess et al. 2003). The petroleum industry is an important source of PAHs, especially in the vicinity of petroleum refinery facilities. Waste water and soil in these areas are often characterized by PAHs concentration 100 times higher than in other areas (Douben 2003). PAHs are found in air, soil, vegetation, water and ice (Holoubek et al. 2000). PAHs are also found in remote areas far from the emission sources (e.g. the arctic) (Sehili and Lammel 2007). Sediments in aquatic environment function as a reservoir for PAHs (Ankley et al. 1994). PAHs are highly hydrophobic compounds, with physicochemical properties that generally favor association with particulate matter. These compounds are usually associated with airborne particulate matter (Fox and Olive 1979; Legzdins et al. 1995; Oda et al. 1998), soil (Chen et al. 2004; Lors and Mossmann 2005; Lampi et al. 2006), sediments in the water column (Lampi et al. 2001, 2006; Kilemade et al. 2004; Olajire et al. 2005), as well as organic surface films in urban environments (Gingrich et al. 2001; Butt et al. 2004).

1.2 General Behavior of POPs

In recent decades, a vast number of POPs have been produced and used worldwide (http://www.greenpeace.org/international/campaigns/toxics/toxic-hotspots/what-are-persistent-organic-pop). Many are still in production and are used in everyday products. POPs behave dynamically in the environment according to their nature of action as they volatilize in the air and cause pollution. These chemicals have become widespread environmental pollutants. POPs contaminated areas close to sites where they were released into the environment from industry and agriculture. However, volatile and semi-volatile POPs also contaminated regions remote from their source, because they could be transported for thousands of kilometers via air currents. POPs might also be transported for long distances by rivers, ocean currents and as contaminants in wildlife. Their sorption and desorption processes were very important in order to understand their toxicity levels. PCBs have a high degree of physical and chemical stability under normal conditions. They are very resistant to acids, bases and oxidants. They are, in practice, fire resistant with rather high flash points. The liquid phase vapor pressures of PCBs at ambient temperature are not very high but the change of temperature affects them significantly (Dai et al. 1997). PCBs have low aqueous solubility and most of their values of K_{OW} are greater than 10^4 (Wan and Mackay 1986).

Environmental pollution with PCBs is caused by their release into the atmosphere via leakage and improper disposal in industry. Other pathways of dispersal into the environment include combustion of industrial and urban wastes and storage and dumping of wastes in landfill sites. Although the amount emitted by these ways is much less, however, they can enter the food chains (WHO 1979). PCBs volatilized into the atmosphere during their use and disposal and returned to the terrestrial and aquatic environment as dry material or precipitation deposition. In most areas of North America, PCBs in the soil mainly originated from the atmosphere. There were from 1×10^3 to 2×10^3 ton PCBs deposited on the soil from the atmosphere, especially in urban areas (Meng et al. 2001). Other than deposition, PCBs in soil derived from the use of sewage sludge as a fertilizer, leaching from landfill sites and the use of pesticides containing PCBs. They entered water mainly from discharge of industrial and urban wastes into rivers, lakes and ocean. When discharged into open water, the water soluble PCBs would concentrate onto dissolved particles, and then sinked into the sediment at different rates according to the particle size and accumulated there. The sinking rate of PCBs from the surface to deeper layers in the open ocean was relatively slower in tropical waters than in high latitude waters (WHO 1993). PCBs in sediments are the main source to enter in the food chains (Dai et al. 1997; WHO 1979).

Sorption is a phase distribution process that accumulates solutes at surfaces and interphases (i.e., adsorption) or from one phase to another (i.e., partitioning). This process affected the transport and reduced the chemical and biological reactivity of relatively hydrophobic organic chemicals (HOCs) such as PAHs and chlorinated aliphatic and aromatic compounds in surface aquatic and groundwater systems (Weilin et al. 2003). Sewage treatment could remove the solid combined PCBs, but it could not remove the dissolved fraction. It has been widely recognized that the ecotoxicity or biodegradability of POPs e.g. PAHs, PCP and PCBs in soils was controlled by the rate and extent of desorption from the solid phase. The sorption ability decreased with increasing residence time of the compounds in soil (Alexander 2000; Reid et al. 2000) and depended on the nature and quality of SOM. This was reflected by the wide variation of organic carbon water partition coefficient (KOC) ranging up to three orders of magnitude among different soils (Chiou et al. 1998; Krauss and Wilcke 2001).

Because of the high hydrophobicity of PCBs, their environmental behavior would significantly depend on sorption phenomena (Voice et al. 1983). PCBs in the aquatic environment were mainly adsorbed onto sediments and suspended particles and then transported to areas far away from the source of pollution. Due to the binding of PCBs, the concentration of PCBs in the sediment near estuaries was as high as $2–5 \times 10^3$ $\mu g\ kg^{-1}$ (Wang 1991; WHO 1993). In highly polluted waters, the concentration of PCBs was several times higher than their solubility due to the interaction between PCBs and solid particles in water (Dai et al. 1997; WHO 1979).

PCBs in water could not be hydrolyzed and their retention period in water was very short because of their high hydrophobicity and chemical stability. PCBs could not be decomposed chemically other than by photolysis. However, biological degradation was important in decreasing PCBs. When PCBs entered food chains, the congeners with low chlorination would be biodegraded stepwise hence they could be totally decomposed in this way (Dai et al. 1997; Wang 1991; WHO 1979).

1.3 Sorption Behavior of POPs in Soil

Soils and sediments are heterogeneous organic and inorganic materials. The heterogeneity has an impact on the linearity of the sorption isotherm and the sorption mechanism. Soil particles can be divided into three size-classes: first, the exposed inorganic minerals, which had little impact on the sorption of HOCs; second, the amorphous organic matter, which included humic and fulvic acids; third, the condensed organic matter, the existence of which was proven by X-ray diffraction spectra (Piatt and Brusseau 1998). Sorption of HOCs by condensed soil organic matter (SOM) appeared to be energetically more favorable and more nonlinear than that by amorphous SOM, most likely because of the more heterogeneous composition and less polar character of condensed organic matter (Weber and Huang 1996; Chiou et al. 1979). The change in the composition of organic matter resulted in a decrease or increase of the relative amounts of oxygen-containing functional groups. High amounts of oxygen containing functional groups might result in an increase in the polarity of organic polymers composed of natural organic matter, and thus in a lower affinity for nonionic compounds, and thereby decreased the sorption capacity. Therefore, a relationship existed between the elemental composition and the partition coefficient.

Organic carbon content and particle size had a significant impact on the sorption of PCBs. Chiou et al. (1983) found that the sorption mechanisms of HOCs like PAHs and PCBs were different between coarse particles and fine particles. The sorption coefficient of PCBs was inversely proportional to their solubility. When Aroclors transported to the soil, they would be fractionated by soil similarly to chromatography and they had a poor mobility (Chu et al. 1995; McGinley et al. 1993). The distributed coefficients changed with time (Girvin and Scott 1997). Different kinds of soil had different abilities to adsorb POPs.

He et al. (2006a) studied the sorption isotherms of PCP in ten different soils with various physico-chemical properties. They also investigated the sorption-desorption hysteresis phenomena and its possible mechanism. They applied a Freundlich and Langmuir-derived dual mode model (DMM) to study the sorption onto and into fractal sorbents (Fig. 1.1). They produced typical sorption isotherms of PCP on soil spanning two orders of magnitude in concentration. The isotherms showed an inflection point which was solute dependent within the solution concentration of 0–5 mg L^{-1}. Both DMM and Freundlich model had a good fit to sorption data, but the DMM model appeared to be superior according to the actual R^2 values. Examples of fits of the DMM and Freundlich models to the full isotherms of soil 2, 6, and 10 were shown in Fig. 1.2. With the increase of soil pH, the nonlinearity of the isotherms gradually disappeared (Fig. 1.2). In this experiment, He et al.

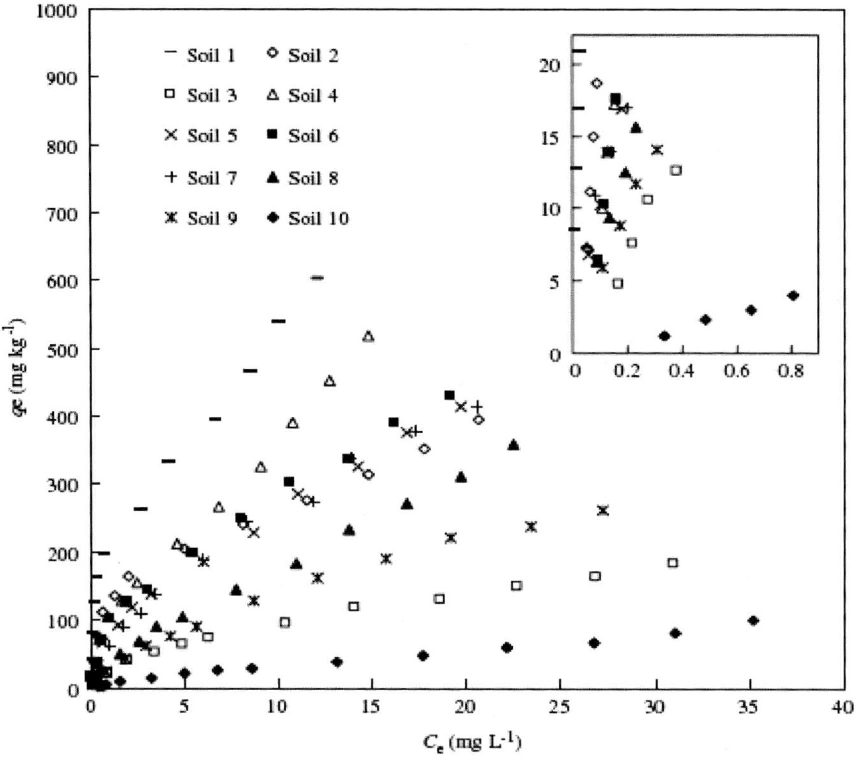

Fig. 1.1 Sorption isotherms of pentachlorophenol (PCP) on 10 soils. q_e is the amount of PCP sorbed per mass of soils, C_e is the equilibrium PCP concentration. The *symbols* denote experimental data points (He et al. 2006a)

(2006a) reported that total nitrogen in soils might contribute in these processes. Correlations of stepwise regression equations confirmed that pH, soil organic carbon and fractions of organic carbon, and particle size distribution were the main factors responsible for the sorption and desorption hysteresis (Table 1.1). These factors acted in concert in influencing the fate of PCP sorption-desorption in soil.

The retention of HOCs in the soil-water system was strongly correlated with soil organic matter (SOM) content, and the soil mineral fractions played a comparatively minor role except in the absence of SOM (Xing and Pignatello 1997; Xia and Ball 1999). He et al. (2006b) reported that the presence of SOM showed a significant positive effect on the sorption of PCPs in six soils. By removing 80% of the organic carbon from soil with H_2O_2, sorption of PCP decreased by an average of 50%. They reported the hysteresis index (HI) for pentachlorophenol sorption on the six soils at five representative concentrations before and after treatment of soil with H_2O_2. Figure 1.3 showed that the value of HI generally decreased after treatment with H_2O_2, especially in soils 1, 2, 4 and 5. In soil 4, PCP desorption was completely reversible (zero or negative HI value indicated that sorption-desorption hysteresis

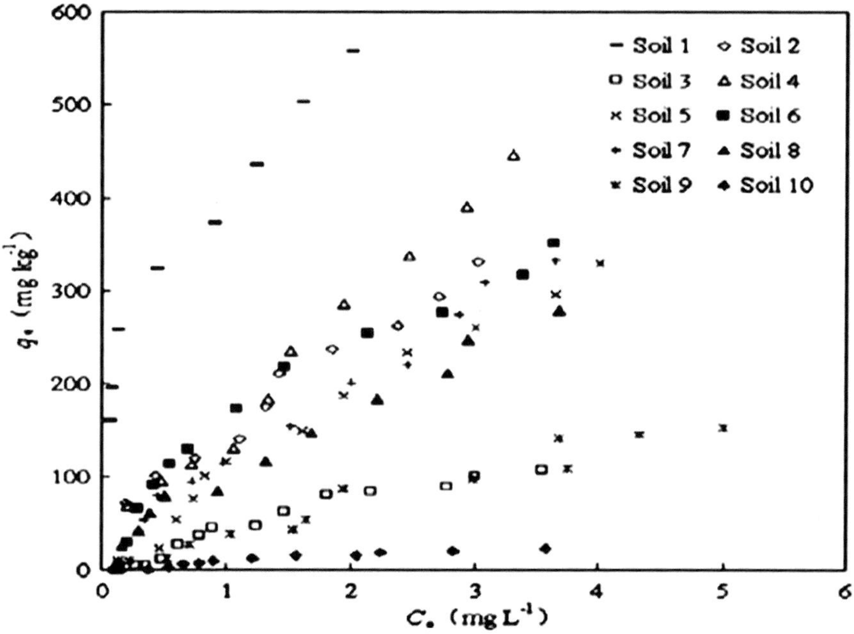

Fig. 1.2 Single-dilution desorption isotherms of pentachlorophenol (PCP) on 10 soils. q_e is the amount of PCP sorbed per mass of soils, C_e is the equilibrium PCP concentration. The *symbols* denote experimental data points (He et al. 2006a)

was statistically insignificant). These results also indicated that the H_2O_2 treatment had great effects on the TOC and pH values of different soils. However, soils 3 and 6 showed unexpected increases in the value of HI with the majority of concentrations of PCP tested. This might happen due to minor surface changes in clay and oxide minerals. He et al. (2006b) also reported that the removal extent of TOC was low. Thus the increment of HI was possibly related to the change of pH value.

The impact of clay minerals and DOM on the competitive sorption-desorption of PAHs was studied by Hwang and Cutright (2002). They reported that the amount of desorbed pyrene was similar in subsoil where SOM had been chemically removed compared to bulk soil. However, when clay minerals were removed, a significant amount of pyrene was desorbed. For instance, 3.97 and 12.76 mg kg^{-1} pyrene was desorbed when it was added at the rate of 3 and 15 mg L^{-1}, respectively. These values indicated that in the clay-depleted subsoil containing SOM the amount of pyrene desorbed was 1.5 (3 mg L^{-1}) to 3.4 (15 mg L^{-1}) times greater than the bulk soil. The contribution of clay minerals to phenanthrene (PHE) desorption was more significant at lower initial concentrations. Desorbed amounts of PHE increased by 3.2 (3 mg L^{-1}) to 1.7 (15 mg L^{-1}) times in clay depleted soil, desorption of pyrene decreased by 26.7 and 64.8% at 3 and 15 mg L^{-1}, respectively, when only clay minerals were present as sorbent. They also reported that the clay mineral contribution to PHE sorption was 2.6% greater than SOM when the level of added

Table 1.1 Stepwise regression models for predicting parameters of PCP sorption on soils based on soil properties (He et al. 2006a)

			T value and R^2 of the partial regression coefficient		
Stepwise regression models[a]	R^2	F value[b]		T value[b]	R^2
$Y_1 = 261.791 - 21.810X_1 - 56.379X_2 + 3.009X_3 + 30.523X_5$	0.939	19.406**	X_1	2.553*	0.566
			X_2	5.482**	0.857
			X_3	4.593**	0.808
			X_5	2.717*	0.596
$Y_2 = 14.960 - 1.722X_1 - 0.102X_6 - 0.004X_8 + 0.059X_{10}$	0.843	6.685*	X_1	3.987**	0.760
			X_6	2.864**	0.621
			X_8	3.250*	0.679
			X_{10}	1.984	0.441
$Y_3 = 62.296 - 12.416X_1 + 0.431X_7 - 0.012X_8 + 0.362X_{10}$	0.897	10.844*	X_1	6.139**	0.884
			X_7	3.808**	0.743
			X_8	2.280	0.510
			X_{10}	2.591*	0.573
$Y_4 = 62.311 - 5.349X_1 - 13.623X_2 + 0.464X_3 + 0.398X_9$	0.901	11.284*	X_1	3.465*	0.706
			X_2	5.320**	0.850
			X_3	3.601*	0.723
			X_9	2.208	0.494
$Y_5 = 4.102 - 0.295X_1 - 0.459X_2 + 0.039X_4$	0.960	47.884**	X_1	9.487	0.937
			X_2	6.070	0.859
			X_4	4.028	0.729
$Y_6 = 1.288 - 0.623X_2 + 0.078X_4$	0.856	20.760**	X_2	6.259	0.674
			X_4	6.165	0.845

[a] Y_1, S^0, mgkg^{-1}; Y_2, b, L mg^{-1}; Y_3, K_D, L kg^{-1}; Y_4, $(S_0b)/K_D$; Y_5, LogK_{oc}, L kg^{-1}, Y_6, HI (T=20°C, C_e=0.8 mg L^{-1}); X_1, pH; X_2, TN, g kg^{-1}; X_3, CEC, cmol (+) kg^{-1}; X_4, TOC, g kg^{-1}; X_5, FA-C, g kg^{-1}; X_6, HA-C, g kg^{-1}; X_7, Humin-C, g kg^{-1}; X_8, C_{mic}, mgkg^{-1}; X_9, clay content, %; X_{10}, the silt content, %.
[b] *Correlation is significant at 0.05 probability level; **correlation is significant at 0.01 probability level.

PHE was 3 mg L^{-1} at a clay concentration of 12.22%. Therefore, the desorption resistant fraction was attributed to the PAHs binding to the clay minerals. The PAHs bound on the SOM surface were easily desorbed due to relatively weak binding forces compared to PAHs bound to clay minerals. Therefore the PAHs desorption in soil containing SOM was greater than in subsoil or bulk soil containing clay minerals.

From the above discussion it could be concluded that the dynamic behavior of POPs sorption was the main risk for global contamination. The dynamic behavior of POPs sorption in the environment was directly related to the types of minerals and organic matter present in the soil. The sorption ability decreased with increasing residence time of the compounds in soil and depended on the nature and quality of the SOM. Soil texture also played a significant role in sorption-desorption reactions of POPs in soil. For example, the presence of clay had a pronounced positive effect on the availability of POPs in soil with low content of organic matter.

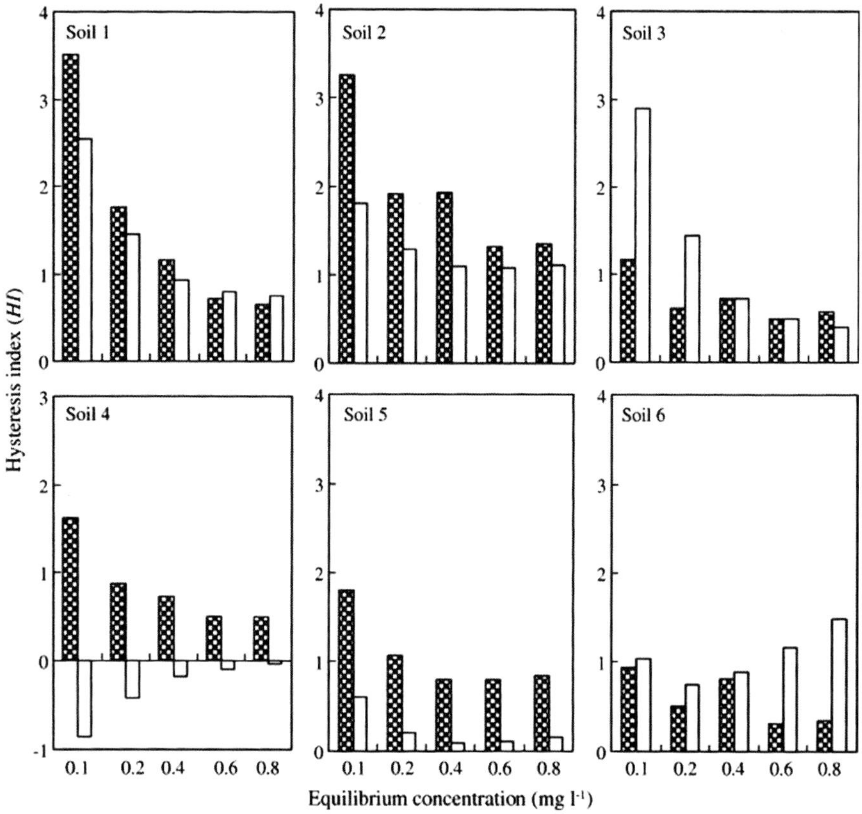

Fig. 1.3 Hystersis index (*HI*) fir pentachlorophenol sorption on six soils at five representative concentrations before and after removal of the organic carbon by H_2O_2. *Symbols*: (■) before H_2O_2 – treatment; (□) after H_2O_2 – treatment (He et al. 2006b)

1.4 POPs Bioavailability in the Presence of Organic Matter

Organic matter is a very important factor that can limit the availability and mobility of POPs in soil (Northcott and Jones 2000). A substantial percentage ranging from 20 to 90% of the total organic contaminant applied to a soil might become associated with the humic fraction (Xie et al. 1997; Burauel and Fuhr 2000). Humic substances have been traditionally divided into three operationally defined fractions: (i) fulvic acid, (ii) humic acid (HA), and (iii) humin which is defined as that fraction which is insoluble in an aqueous solution at any pH (Kohl and Rice 1998). According to Fabio and Alessandro (2001), microcosms that received humic substance (HS) at 1.5% showed a higher persistence of specialized bacteria and yields of PCB biodegradation and dechlorination products about 150 and 100%, larger, respectively, than those in HS-free microcosms. Therefore, humic substances and similar materials might enhance PCB bioavailability and biodegradation during soil bioremediation. Kastner and Mahro (1996) also reported that the presence of the solid organic matrix of compost seemed to be essential for the enhanced degradation of PAHs in soil.

PCBs binding to humus in the soil environment and the adsorption of PCBs have been related to soil organic matter content. The toxicity of Aroclor 1254 decreased in the presence of organic matter (Strek and Weber 1982a). Moza et al. (1979) speculated that PCBs in soil were metabolized to phenols which polymerized to form bound residues. However, it was more commonly accepted that these xenobiotics were bound to humic substances through charge-transfer complexing, electrostatic attractions, and hydrophobic bonding (Strek and Weber 1982b). Adsorption increased as the number of chlorine atoms in the PCB molecule increased and was influenced by the position of the chlorine substituent on the biphenyl ring. The 2-,4-monochloro isomers and 3,4-dichlorobiphenyl were more likely to bind to organic matter than other biphenyls (Bollag and Loll 1983).

Organic pollutants strongly adsorbed to carbonaceous sorbents such as black carbon (Bucheli and Gustafsson 2000; Cornelissen et al. 2006; Koelmans et al. 2006; Ghosh 2007). In particular, activated charcoal had a strong adsorptive capacity due to its high specific surface area (Cornelissen et al. 2005). Adsorption to activated charcoal could render hazardous organic pollutants in soils and sediments less available to organisms (Burgess et al. 2009) and hindered their dispersal into unaffected environments. Tomaszewski et al. (2007) added 3.2% carbon or reactivated carbon to contaminated sediment (containing up to 252 mg DDT kg^{-1}) and thereby reduced aqueous equilibrium concentrations by up to 83%. In a study by Vasilyeva et al. (2001), the addition of 1% activated carbon to a TNT–contaminated soil (2,000 mg kg^{-1}) reduced the concentrations of extractable TNT and its metabolites by 89% within 120 days. The accelerated removal was attributed to reduced toxicity of the contaminants leading to enhanced microbial degradation.

Pan et al. (2007) also studied the effect of organic matter fractions on sorption of phenanthrene, naphthalene and o-xylene. They adopted humic acid and de-ashed humin using batch equilibration techniques. Aliphatic-rich de-ashed humin exhibited the highest sorption capacity for HOCs while mineral associated humin had the strongest nonlinearity. Aliphatic components of SOM contributed greatly to sorption of HOCs in soil and sediments.

Wen et al. (2009) studied the immobilization of pentachlorophenol in soils amended with 2.0% char, humic acid (HA) or peat. The extractability of PCP was significantly decreased (Fig. 1.4). Desorption kinetics indicated that the amendments could lead to strong binding and slow desorption of PCP in soils. The results of both physicochemical and biological tests suggested that the amendments reduced PCP bioavailability quickly and implied that addition of carbonaceous materials, especially char, was a potentially attractive in-situ remediation method for sequestration of PCP in contaminated soil.

Weber et al. (2006) proposed in-situ remediation methods to reduce the bioavailability of organic pollutants in soils. The organic contaminants were partitioned to carbonaceous sorbents that were mixed with soil to enhance the natural process of contaminant stabilization. They developed engineered sorbents from natural

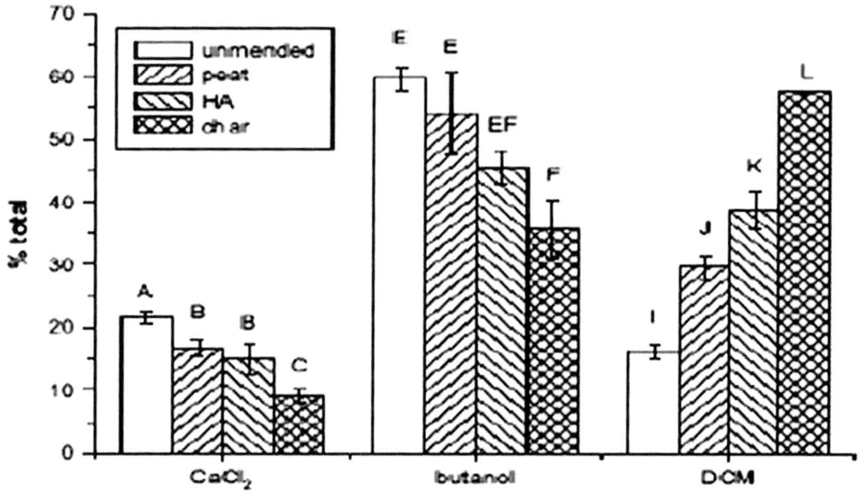

Fig. 1.4 Extractable fractions of PCP in soil amended with carbonaceous materials for 250 days using $CaCl_2$, butanol and DCM as extractants. *Error bars* represent one standard deviation. *Bars* with same letter or *letters* imply no significant difference ($P < 0.05$) (Wen et al. 2009)

organic materials, such as peat, soybean stalk and pine needles under conditions of superheated temperature and high pressure, for sorption of PAHs in contaminated soils. Both natural and engineered natural materials significantly decreased the extractability and bioavailability of PAHs from contaminated soils (Tang and Weber 2006; Weber et al. 2006; Tang et al. 2007). Addition of active carbon to sediments increased the desorption of PCBs and PAHs (Zimmerman et al. 2004). The bioavailability and bioaccumulation of PCBs were reduced markedly by activated carbon amendment in both laboratory (Millward et al. 2005) and field (Cho et al. 2007) tests. This clearly demonstrated the effectiveness of amendments for immobilization of hydrophobic organic contaminants in soils and sediments.

Rhizosphere technology is one of the emerging techniques to control POPs activity in soil. This technology has been applied to bioremediate POPs through plant growth in order to accelerate degradation rates. The presence of plants had positive effects upon the degradation of a wide range of organic compounds (Hayat et al. 2011; Ma et al. 2010). These included the degradation of simple phenolics, aliphatic hydrocarbons, trichloroethylene, polycyclic aromatic hydrocarbons (PAHs), chlorinated aromatics and insecticides. He et al. (2005) reported that the effect of root proximity (i.e. the distance from the root surface) was important in the removal of PCPs. The rhizosphere environment enhanced the ability of soil microorganisms to remove POPs, in addition to enhancing phosphatase and urease activity which were active in the degradation of PCPs in the rhizosphere of ryegrass. As a result, the corresponding degradation gradients of PCP occurred (Fig. 1.5). He et al. (2009) demonstrated that the rhizosphere gradient effect of ryegrass (*Lolium perenne* L.) on the first two principal components of phospholipids fatty acids

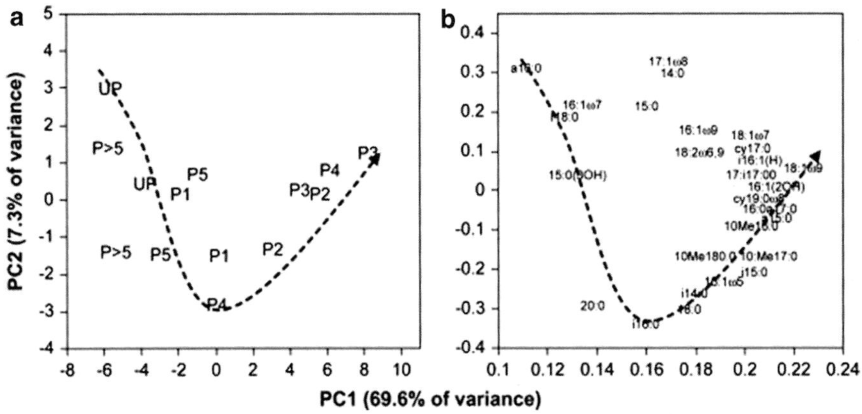

Fig. 1.5 Rhizosphere gradient effect of ryegrass (*Lolium perenne* L.) on the first two principal components of phospholipids fatty acids (PLFAs) data set. (**a**) results present principal components (PC) scores for PC1 and PC2. Plot of samples where '*P*' indicates planted treatments, and the number of 1–5 mm and >5 mm are the distance from the root surface, which denote 1–5 mm near-rhizosphere and >5 mm far-rhizosphere, respectively; '*UP*' indicates unplanted treatments, analyzed using the mixed samples of each soil layer from unplanted treatment, (**b**) results present the component matrix loadings of the individual PLFA for PC1 and PC2 (He et al. 2009)

(PLFAs) in PCP contaminated soil at different distance from the root. The PLFA analyses of samples with and without ryegrass demonstrated changes in microbial composition at different distance from the root. The first principal component (PC_1) and second principal component (PC_2) together accounted for 77% of the variation and differentiated the sample both in the planted soil in different layers of the rhizosphere and in the unplanted treatment. He et al. (2009) reported a distinct selective enrichment of the PLFA biomarkers in the near rhizosphere compared to the far rhizosphere and unplanted treatment, especially for the 2 mm and 3 mm near rhizosphere layers where the most rapid PCP depletion occurred.

Phytoemediation is also a promising technique for the remediation of PAHs in soil (Ma et al. 2010). The profiles of sequentially extracted PAHs varied with plant species and the difference of toxicity equivalency concentrations between rhizosphere and bulk soil was significantly higher than that of total PAHs concentration. Ma et al. (2010) described the sequentially extracted fractions of PAHs in the rhizosphere of four different plant species (Fig. 1.6). Concentrations of sequentially extracted PAHs were lower than in bulk soils, especially in oat rhizospheres. The dichloromethane (DCM) extractable fraction was the primary fractionation among three extractable fractions, while among three non-extractable fractions crude humin-bound PAHs dominated. In bulk soils of different plant species, the profile of sequential extraction varied, particularly the proportions of n-butanol and DCM extractable PAHs. The concentration of n-butanol extractable PAHs ranged from 1.6 mg kg^{-1} under soybean to 5.4 mg kg^{-1} under rape. The DCM extractable PAH concentration in bulk soil under soybeans was 10.1 mg kg^{-1} but reached

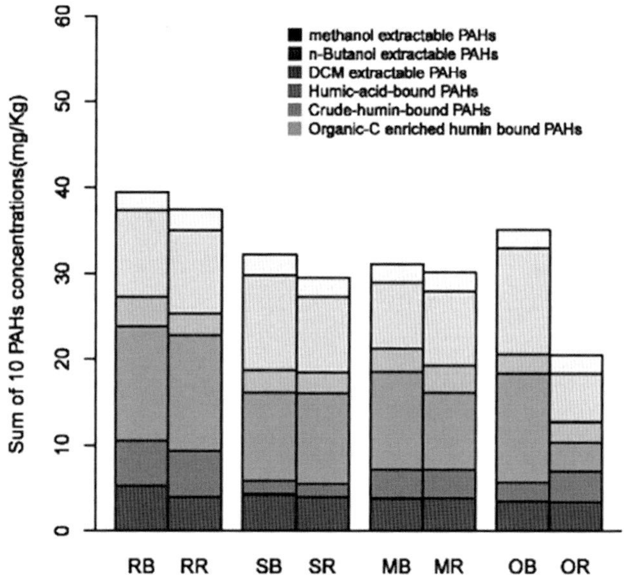

Fig. 1.6 Total concentration of PAHs in rhizosphere and bulk soils based on sequential extraction. *RR* rhizosphere soil of rape, *RB* bulk soil of rape, *SR* rhizosphere soil of soybean, *SB* bulk soil of soybean, *MR* rhizosphere soil of maize, *MB* bulk soil of maize, *OR* rhizosphere soil of oat, *OB* bulk soil of oat (Ma et al. 2010)

13.5 mg kg^{-1} under rape. Comparison of profiles among the plant species showed that profiles of sequential extraction were relatively similar between bulk and rhizosphere soils under the same plant species. However, the concentration of PAHs fractions were significantly lower than in the corresponding bulk soil, including the DCM extractable PAHs in the rhizosphere of maize, methanol extractable PAHs in the rhizosphere of rape, and crude humin-bound PAHs in the rhizosphere of soybeans and oat.

Ding et al. (2009) reported that rhizoremediation played a vital role in the degradation of PCBs in contaminated soils. They measured the degradation of Aroclor 1242 in the rhizosphere of ryegrass in specially designed rhizobox in which they measured Aroclor 1242 at the millimeter scale in the near and far rhizosphere. The total removal rates of Aroclor 1242 in planted soils were much higher than in unplanted soils (Fig. 1.7). Ryegrass removal rates were affected by the proximity to the root. The degradation of Aroclor 1242 near the rhizosphere was higher than in the root compartment and far rhizosphere. However, there were no significant differences in the removal rates with increasing distances from the central compartment.

It can be concluded that organic matter present in soil or came through plant in the form of root exudates has significant interactions with POPs activity in soil. Natural organic matter plays an important role in controlling the movement as well

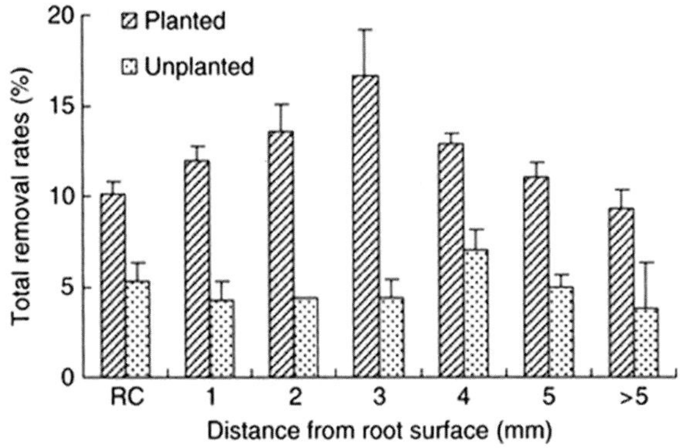

Fig. 1.7 Aroclor 1242 total removal rates (%) in various sampling zones as a function of proximity to ryegrass roots grown in the soil treated with 8.0 mg kg^{-1} of Aroclor 1242. *Bars* are the SE of means of three replicates. *RC*, root compartment (Ding et al. 2009)

as the bioavailability of POPs. Rhizosphere effect can also play a vital role in the remediation of POPs in the soil because roots release significant amounts of organic compounds in the form of root exudates which directly affect POPs bioavailability as well as microbial activity in the rhizosphere, thus increase the degradation of the POPs in rhizosphere vicinity.

1.5 Environmental Factors Affecting Transport, Sorption and Bioavailability of POPs

POPs undergo long range atmospheric transport and reach remote regions where they have never been produced or used. This process could happen as a simple emission-transport-deposition event or following a series of "hops" by deposition onto a surface and subsequent emission (Wania and Mackay 1996). In general, the capacity of an environmental compartment to retain and release chemicals, which depended on several processes, was widely variable both spatially and temporally. The distribution of POPs based on the environmental compartments depended on their properties and on the capacity of the exchange surfaces (vegetation, soil, water) to retain them (Mackay 2001). Therefore, the effects of possible future climatic changes, which can imply, for example, the acceleration of desertification, changes in land use, the reduction of glacier surfaces, can significantly influence global POPs cycling (e.g. Dalla Valle et al. 2004; Macdonald et al. 2005).

The transport distance and the number of air-surface exchange episodes depended on the type of surface (soil, water, vegetation, etc.) and on the physical–chemical properties of the given chemical. According to this principle, persistent chemicals with high volatility would be transported long distances followed by deposition in distant areas (e.g. arctic regions), while those with lower vapor pressure would be deposited preferentially in areas closer to the emission source. Many other mechanisms and factors were able to influence the distribution of POPs in the atmosphere: the capacity of the environmental compartments to accumulate or to degrade POPs; the atmospheric circulation patterns; the spatial and temporal distribution of primary sources and the kinetic of the air-surface exchange. POPs in air could be degraded by photo-degradation and via reactions with hydroxyl radicals, both depending, to a certain extent, on the solar radiation intensity (Dalla Valle et al. 2007) and other reactive agents.

Temperature had a significant effect on the microbial growth and physiological activity including uptake and enzymatic dehalogenation of PCB congeners. The influence of temperature was multifaceted, which included changes in the adsorption and desorption kinetics of PCBs from soil particles. pH affected the equilibrium between PCBs that were dissolved and those that were adsorbed to organic matter and thus influenced the bioavailability of PCBs in soil (Jota and Hassett 1991). Temperature was one of the main factors which influenced the global distribution of POPs, even if several other processes played an important role. Processes directly depending on temperature were: emission rates from primary and secondary sources, the gas-particle partitioning of POPs in air, reaction rates (biodegradation, photolysis and oxidation in air), and the air-surface exchange (Halsall et al. 1999). The degradation rates, for example, were estimated to increase by a factor of two to three for every 10 °C increment (Lammel et al. 2001; Macdonald et al. 2005). In addition, important atmospheric depletion mechanisms for organic contaminants were photochemical degradation and the degradation initiated by the hydroxyl radicals, whose concentration was related to solar activity and cloud cover (Mandalakis et al. 2003). Recently, an attempt to estimate the spatial and temporal variability of POPs distribution has been made for the oceans (Jurado et al. 2004), surface soils (Dalla Valle et al. 2005) and vegetation (Dalla Valle et al. 2004). The spatial variability of this capacity depended on the combination of several factors e.g. organic carbon content (in soil or within plant leaves), green biomass, or more generally the ecosystem type. Those factors even could vary markedly within a limited space (e.g. along a mountain slope), thus influencing POPs distribution according to their properties (Grimalt et al. 2004; Van Droodge et al. 2004; Daly and Wania 2005). The temporal variability was mainly dependant on temperature (e.g. Hornbuckle and Eisenreich 1996; Halsall et al. 1999), therefore showing daily and seasonal oscillations, but also on biomass trends and biogeochemical cycles (e.g. phytoplankton cycles; Dachs et al. 2002), which could be seasonal or exhibiting a long term trend, and on soil use and properties. Another complication was given by the wide range of physical–chemical properties of the different classes of chemicals. Other climate related variables such as the frequency of storm surges, precipitation, the seasonality and the intensity of the atmospheric events, could influence the fate of POPs in the environment. Increased precipitation, for example, could cause an increase of POPs deposition onto the soil. More frequent storm surges might enhance the mobilization of chemicals stored in the soil compartment, which could be transported by land runoffs, making them available to aquatic organisms.

The degradation of POPs might be further limited by a number of environmental parameters, such as temperature, pH, presence of oxygen, nutrients, moisture and salinity. Soil and ground water temperatures varied regionally and seasonally. Optimizing these variables in-situ would enhance and maintain bioavailability of

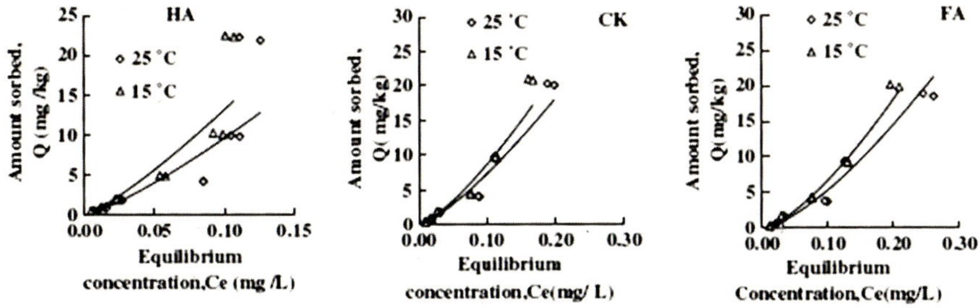

Fig. 1.8 Sorption isotherms for phenantherene (PHE) in soils at different temperature levels (Ping et al. 2006)

POPs. Most biodegradation studies have been performed at temperatures from 20°C and 30°C. Only a limited number of studies have illustrated the biodegradation of PAHs at low temperatures and they generally agreed that PAHs biodegradation rates and efficiencies were lower at lower temperatures (Eriksson et al. 1999; Phillips et al. 2000; Yuan et al. 2000). However, some studies have shown that a decrease in temperature did not affect PAHs biodegradation (Whyte et al. 1997; Mohammed et al. 1998).

Moisture levels determined the oxygen content, redox potential, and type of microflora of a soil. The amount of water in soil then had a profound effect on the interactions between humus and POPs. When soils were flooded and, therefore, became anaerobic, dinitroanilines and nitrobenzenes were converted to anilines, which could react with humus (Moza et al. 1979). Inorganic minerals also had some interactions with the POPs and played an important role in the transformation of POPs. Manganese and iron oxides and some clay minerals promoted the oxidation of a number of organic pollutants (Mortland and Halloran 1976; Pizzigallo et al. 1995).

Ping et al. (2006) reported that the sorption of phenanthrene along with environmental factors such as pH and temperature also affected exogenous humic acids (Figs. 1.8 and 1.9). Humic acid and fulvic acid from the same source had different effects on the adsorption of phenanthrene. Adsorption of phenanthrene was higher in soil treated with fulvic acid compared to humic acid. Added humic substances influenced the adsorption of phenanthrene, which was also affected by environmental conditions. Exogenous humic substances played an important role in the regulation of soil PAHs mobility under appropriate conditions leading to a decrease in PAHs pollutant diffusion.

Ghosh et al. (2001) investigated the kinetics and thermodynamics of PAHs desorption from sediment. They demonstrated that desorption of PAHs associated with clay/silt particles was faster at room temperature and was characterized by low desorption activation energies, whereas desorption of PAHs associated with coal-derived material at room temperature was much slower and was characterized by high desorption activation energies. Particles derived from coal in the sediment, which contained the majority of PAHs, were in the state of non-equilibrium even after decades of aging in the field due to their extremely slow diffusion through

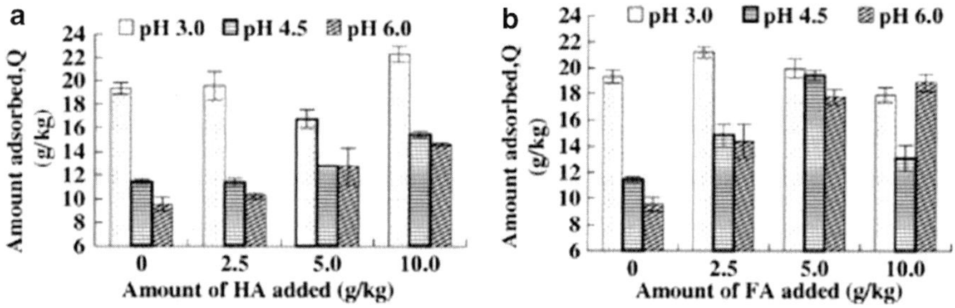

Fig. 1.9 Adsorption of PHE under different pH conditions (Ping et al. 2006)

the polymeric coal-derived materials. Internal non-equilibrium conditions slowed down the long-term desorption from the coal-derived materials. To strongly bond PAHs on coal-derived materials high activation energy required for desorption and relatively unavailable at ambient temperatures. PAHs associated with media having large activation energies might thus comprise the unavailable PAHs fraction in sediments, and these PAHs might pose less environmental risk than PAHs in clay and silt fractions (Luthy et al. 2000).

The potential capacity of soil abiotic components (e.g. oxides, clay minerals, humic substances) to promote the transformation of aromatic molecules, was unexplored for POPs characterised by a very low water solubility. Taking into account that the adsorption process represented the first fundamental step of the oxidation reaction, the interaction between the inorganic catalyst and the POPs could be studied by mixing and grinding the two reactants in solid phase. This method called "mechanochemical" was well known in environmental fields and has attracted attention in relation to the degradation of organo-chlorides (Loiselle et al. 1997; Nasser et al. 2000; Mio et al. 2002). The treatment provoked a solid-state transformation at the mineral surface without any deformation of the structure. The solid contact of the modified surface with the POPs induced the production of free radicals, which were very active in subsequent oxidative coupling reactions. Thus, the "mechanochemical" procedure could start the reaction between the POPs and the minerals without any interference from the organic solvents required to solubilise non-polar compounds. Mechanochemical treatments of soils contaminated with organochlorides molecules were conducted by grinding using a ball mill with different dechlorinating reagents (Loiselle et al. 1997; Mio et al. 2002). Few authors provided information on the mechanochemical technique applied to oxidative reactions. Some researchers (Nasser et al. 2000; Shin et al. 2000) successfully tested the degradation of herbicides such as 2,4-D and atrazine using very gentle grinding of manganese oxides with herbicides together. The authors advocated that the grinding of a reactive mineral in the presence of adsorbed organic molecules could operate as a surface alteration of the mineral structure producing structural defects (Pizzigallo et al., 2004). These defects might produce chemical activation, which promoted the breaking of chemical bonds and consequently the degradation of chlorinated compounds. Therefore, mechanochemical treatments provided a way to initiate a reaction between organic compounds. This technique could represent an important alternative to more drastic oxidative processes in removing recalcitrant organochlorine compounds from heavily contaminated soils.

1.6 Conclusion

Persistent organic pollutants (POPs) pose a serious threat to both human and environmental health. They are of environmental concern because of their potential for long-range atmospheric transport, bioaccumulation and toxicity. POPs undergo long range atmospheric transport, and reach remote regions where they have never been produced or used. Their sorption and desorption processes were very important to study their toxicity levels. Temperature was one of the main environmental factors being able to influence the global distribution of POPs although several other factors (moisture contents, pH, Eh, soil texture) also played an important role. Organic matter is considered to be a very important factor that could limit bioavailability and mobility of POPs in soil, and a substantial percentage of the organic contaminants applied to a soil might become associated with the humic fraction. The bioavailability and bioaccumulation of PCBs, PCP and PAHs were greatly reduced by the addition of activated carbon in soils. Therefore, sorption/desorption processes were very important in relation to POPs bioavailability and their impact on other environmental components. Rhizosphere effect also played a vital role in controlling POPs activity in soil, and in releasing many contaminated lands for productive purpose. Many theories and models have been proposed to account for POPs bioavailability. However, none of them was suitable under all circumstances. It is therefore important to further study the sorption mechanism of POPs, which would provide a scientific basis to predict the fate of POPs in the environment and permit scientific risk assessment and management. Future research should be directed towards microbiological research and rhizosphere effect in order to identify useful microbes to control POPs bioavailability under different environmental conditions.

Acknowledgement This work was supported by the National Natural Science Foundation of China (41090284, 20977077). We would like to thank Professor Philip C. Brookes from Rothamsted Research, UK for his kind help for editing the manuscript.

References

Alexander M (2000) Aging bioavailability and overestimation of risk from environmental pollutants. Environ Sci Technol 34:4259–4265

Ankley GT, Collyard SA, Monson PD, Kosian PA (1994) Influence of ultraviolet-light on the toxicity of sediments contaminated with polycyclic aromatic-hydrocarbons. Environ Toxicol Chem 13:1791–1796

Bollag JM, Loll MJ (1983) Incorporation of xenobiotics into soil humus. Experientia 39:1221–1231

Bucheli TD, Gustafsson O (2000) Quantification of the soot-water distribution coefficient of PAHs provides mechanistic basis for enhanced sorption observations. Environ Sci Technol 34:5144–5151

Burauel P, Fuhr F (2000) Formation and long-term fate of nonextractable residues in outdoor lysimeter studies. Environ Pollut 108:45–52

Burgess RM, Ahrens MJ, Hickey CW (2003) Geochemistry of PAHs in aquatic environments: source, persistence and distribution. In: Douben PT (ed) PAHs: an ecotoxicological perspective. Wiley, Etobicoke

Burgess RM, Perron MM, Friedman CL, Suuberg EM, Pennell KG, Cantwell MG, Pelletier MC, Ho KT, Serbst JR, Ryba SA (2009) Evaluation of the effects of coal fly ash amendments on the toxicity of a contaminated marine sediment. Environ Toxicol Chem 28:26–35

Butt CM, Diamond JT, Truong J, Ikonomou MG, Helm PA, Stern GA (2004) Semivolatile organic compounds in window films from lower Manhattan after the September 11th world trade center attacks. Environ Sci Technol 38:3514–3524

Chen B, Xuan X, Zhu L, Wang J, Gao Y, Yang K, Shen X, Lou B (2004) Distributions of polycyclic aromatic hydrocarbons in surface waters, sediments and soils of Hangzhou city China. Water Res 38:3558–3569

Chiou CT, Peters LJ, Freed ZH (1979) A physical concept of soil-water equilibria for nonionic organic compounds. Science 206:831–832

Chiou CT, Porter PE, Schmedding DW (1983) Partition equilibria of nonionic organic compounds between soil organic matter and water. Environ Sci Technol 17:227–231

Chiou CT, McGroddy SE, Kile DE (1998) Partition character- istics of polycyclic aromatic hydrocarbons on soils and sediments. Environ Sci Technol 32:264–269

Cho YM, Smithenry DW, Ghosh U, Kennedy AJ, Millward RN, Bridges TS, Luthy RG (2007) Field methods for amending marine sediment with activated carbon and assessing treatment effectiveness. Marine Environ Res 64:541–555

Chu SG, Yang C, Xu XB (1995) Polychlorinated biphenyl congener residuals in sediment and soil from pollution area. China Environ Sci 15:199–203 (in Chinese)

Cornelissen G, Gustafsson O, Bucheli TD, Jonker MTO, Koelmans AA, Van Noort PCM (2005) Extensive sorption of organic compounds to black carbon, coal, and kerogen in sediments and soils: mechanisms and consequences for distribution, bioaccumulation, and biodegradation. Environ Sci Technol 39:6881–6895

Cornelissen G, Breedveld GD, Kalaitzidis S, Christanis K, Kibsgaard A, Oen AMP (2006) Strong sorption of native PAHs to pyrogenic and unburned carbonaceous geosorbents in sediments. Environ Sci Technol 40:1197–1203

Dachs J, Lohmann R, Ockenden WA, Mejanelle L, Eisenreich SJ, Jones KC (2002) Oceanic biogeochemical controls on global dynamics of persistent organic pollutants. Environ Sci Technol 36:4229–4237

Dai SG, Yue GC, Wang XR (1997) Environmental chemistry (in Chinese). Higher Education Press, Beijing, pp 269–275

Dalla Valle M, Dachs J, Sweetman AJ, Jones KC (2004) Maximum reservoir capacity of vegetation for persistent organic pollutants: implications for global cycling. Glob Biogeochem Cycles 18:GB4032

Dalla Valle M, Jurado E, Sweetman AJ, Jones KC, Dachs J (2005) The maximum reservoir capacity of soils for persistent organic pollutants: implications for global cycling. Environ Pollut 134:153–164

Dalla Valle M, Codato E, Marcomini A (2007) Climate change influence on POPs distribution and fate: a case study. Chemosphere 67:1287–1295

Daly GL, Wania F (2005) Organic contaminants in mountains. Environ Sci Technol 39:385–398

Ding N, Guo HC, Hayat T, Wu YP, Xu JM (2009) Microbial community structure changes during Aroclor 1242 degradation in the rhizosphere of ryegrass (*Loliummultiflorum* L.). FEMS Microbiol Ecol 70:305–314

Douben PET (2003) PAHs: an ecotoxicological perspective: John Wiley & Sons Ltd. Dioxygenase and polycyclic aromatic hydrocarbon dihydrodiol dehydrogenase in *Pseudomonas putida* Ous82. J Bacteriol 176:2444–2449

Eisler R (1987) Polycyclic aromatic hydrocarbon hazards to fish, wildlife, and invertebrates: a synoptic review. U.S. Fish and Wildlife Service. Report nr 85(1.11)

Eriksson M, Dalhammar G, Borg-Karlson AK (1999) Aerobic degradation of a hydrocarbon mixture in a natural uncontaminated potting soil by indigenous microorganisms at 20°C and 6°C. Appl Microbiol Biotechnol 51:532–535

Fabio F, Alessandro P (2001) Effect of humic substances on the bioavailability and aerobic biodegradation of polychlorinated biphenyls in a model soil. Biotech Bioeng 77:204–221

Fox MA, Olive S (1979) Photooxidation of anthracene on atmospheric particulate matter. Science 205:582–583

Ghosh U (2007) The role of black carbon in influencing availability of PAHs in sediments. Hum Ecol Risk Assess 13:276–285

Ghosh U, Talley JW, Luthy RG (2001) Particle-scale investigation of PAH desorption kinetics and thermodynamics from sediment. Environ Sci Technol 35:3468–3475

Gingrich SE, Diamond ML, Stern GA, McCarry BE (2001) Atmospherically derived organic surface films along an urban-rural gradient. Environ Sci Technol 35:4031–4037

Girvin DC, Scott AJ (1997) Polychlorinated biphenyl sorption by soils: measurement of soil-water partition coefficients at equilibrium. Chemosphere 35(9):2007–2025

Grimalt J, Borghini F, Sanchez-Hernandez JC, Barra R, Torres Garcia CJ, Focardi S (2004) Temperature dependence of the distribution of organochlorine compounds in the mosses of the Andean mountains. Environ Sci Technol 38:5386–5392

Halsall CJ, Gevao B, Howsam M, Lee RGM, Ockenden WA, Jones KC (1999) Temperature dependence of PCBs in the UK atmosphere. Atmos Environ 33:541–552

Hayat T, Ding N, Ma B, He Y, Shi JC, Xu JM (2011) Dissipation of Pentachlorophenol in the aerobic-anaerobic interfaces established by the rhizosphere of rice (Oryza sative L.) root. J Environ Qual 40:1722–1729

He Y, Xu JM, Tang CX, Wu YP (2005) Facilitation of pentachlorophenol degradation in the rhizosphere of ryegrass (*Lolium perenne* L.). Soil Biol Biochem 37:2017–2024

He Y, Xu JM, Wang HZ, Ma ZH, Chen JQ (2006a) Detailed sorption isotherms of pentachlorophenol on soils and its correlation with soil properties. Environ Res 101:362–372

He Y, Xu JM, Wang HZ, Ma ZH, Akmal M (2006b) Potential contribution of clay minerals and organic matter to pentachlorophenol retention in soils. Chemosphere 65:497–505

He Y, Xu JM, Lv XF, Ma ZH, Wu JJ, Shi JC (2009) Does the depletion of pentachlorophenol in root-soil interface follow a simple linear dependence on the distance to root surface? Soil Biol Biochem 41:1807–1813

Holoubek I, Korinek P, Seda Z, Schneiderova E, Holoubkova I, Pacl A, Triska J, Cudlin P, Caslavsky J (2000) The use of mosses and pine needles to detect persistent organic pollutants at local and regional scales. Environ Pollut 109:283–292

Hornbuckle KC, Eisenreich SJ (1996) Dynamic of gaseous semivolatile organic compounds in a terrestrial ecosystem – effects of diurnal and seasonal climate variations. Atmos Environ 30:3935–3945

Hu YM, Wang MJ (1998) Methodology for studying the fate and mobility of organic pollutants in soil. Adv Environ Sci 6(4):44–55 (in Chinese)

Hwang S, Cutright TJ (2002) Impact of clay minerals and DOM on the competitive sorption/desorption of PAHs. Soil Sediment Contam 11(2):269–291

Jota MAT, Hassett JP (1991) Effect of environmental variable on binding of PCB congener by dissolved humic substances. EnvironToxic Chem 10:483–491

Jurado E, Lohmann R, Meijer S, Jones KC, Dachs J (2004) Latitudinal and seasonal capacity of the surface oceans as a reservoir of polychlorinated biphenyls. Environ Pollut 128:149–162

Kastner M, Mahro B (1996) Microbial degradation of polycyclic aromatic hydrocarbonsin soils affected by the organic matrix of compost. Appl Microbiol Biotechnol 44:668–675

Keith LH, Telliard WA (1979) Priority pollutants a perspective view. Environ Sci Technol 13:416–423
Kilemade M, Hartl MGJ, Sheehan D, Mothersil C, van Pelt FNAM, O'Brien NM, O'Halloran J (2004) An assessment of the pollutant status of surficial sediment in cork harbour in the south east of Ireland with particular reference to polycyclic aromatic hydrocarbons. Mar Poll Bull 49:1081–1096
Klecka G, Boethling B, Franklin J, Graham G, Grady L, Howard P, Kannan K, Larson R, Mackay D, Muir D, van der Meent K (2000) Evaluation of persistence and long-range transport of organic chemicals in the environment. SETAC Special Publication Series, Pensacola
Koelmans AA, Jonker MTO, Cornelissen G, Bucheli TD, Van Noort PCM, Gustafsson O (2006) Black carbon: the reverse of its dark side. Chemosphere 63:365–377
Kohl SD, Rice JA (1998) The binding of contaminants to humin:a mass balance. Chemosphere 36:251–261
Krauss M, Wilcke W (2001) Predicting soil-water partitioning of polycyclic aromatic hydrocarbons (PAHs) and polychlorinated biphenyls (PCBs) by desorption with methanol-water mixtures at different temperatures. Environ Sci Technol 35:2319–2325
Lammel G, Feichter J, Leip A, Semeena VS (2001) Long-range transport and environmental fate of modern pesticides – model development and application. In: Lammel G (ed) Slowly degradable organics in the atmospheric environment and air–sea exchange Report Max Planck Institute for Meteorology No. 335, Hamburg, pp 51–56
Lampi MA, Huang XD, El-Alawi YS, McConkey BJ, Dixon DG, Greenberg BM (2001) Occurrence and toxicity of photomodified polycyclic aromatic hydrocarbon mixtures present in contaminated sediments. In: Greenberg BM, Hull RN, Roberts MH, Gensemer RW (eds) Environmental toxicology and risk assessment: science, policy, and standardization – implications for environmental decisions: tenth volume STP 1403. American Society for Testing and Materials, West Conshohocken
Lampi MA, Gurska J, McDonald KIC, Xie FL, Huang XD, Dixon DG, Greenberg BM (2006) Photoinduced toxicity of polycyclic aromatic hydrocarbons to daphnia magna: ultraviolet-mediated effects and the toxicity of polycyclic aromatic hydrocarbon photoproducts. Environ Toxicol Chem 25:1079–1087
Legzdins AE, McCarry BE, Marvin CH, Bryant DW (1995) Methodology for bioassay-directed fractionation studies of air particulate material and other complex environmental matrices. Int J Environ Anal Chem 60:79–94
Loiselle S, Branca M, Mulas G, Cocco G (1997) Selective mechanochemical dehalogenation of chlorobenzenes over calcium hydride. Environ Sci Technol 31:261–265
Lors C, Mossmann JR (2005) Characteristics of PAHs intrinsic degradation in Two coke factory soils. Polcycl Aromat Comp 25:67–85
Luthy RG, Ghosh U, Gillette JS, Zare RN, Talley JW, Tucker S (2000) Chemical speciation and reactivity in water chemistry and water technology. In: A symposium in honor of James J. Morgan, American Chemical Society, Washington, DC, pp 497–500
Ma B, Chen H, He Y, Wang H, Xu J (2010) Evaluation of toxicity risk of polycyclic aromatic hydrocarbons (PAHs) in crops rhizosphere of contaminated field with sequential extraction. J Soils Sediments 10:955–963
Macdonald RW, Harner T, Fyfe J (2005) Recent climate change in the arctic and its impact on contaminant pathways and interpretation of temporal trend data. Sci Total Environ 342:5–86
Mackay D (2001) Multimedia environmental models: The Fugacity Approach, 2nd edn. Lewis Publishers, Boca Raton, p 261
Mandalakis M, Berresheim H, Stephanou EG (2003) Direct evidence for destruction of polychlorobiphenyls by OH radicals in the subtropical troposphere. Environ Sci Technol 37:542–547
McGinley PM, Katz LE, Weber WJ Jr (1993) A distributed reactivity model for sorption by soils and sediments: 2. Multicomponent systems and competitive effects. Environ Sci Technol 27:1524–1531
Meng QY, Chu SG, Xu XB (2001) Sorption phenomena of PCBs in environment. Chin Sci Bull 46(2):89–97

Millward RN, Bridges TS, Ghosh U, Zimmerman JR, Luthy RG (2005) Addition of activated carbon to sediments to reduce PCB bioaccumulation by a polychaete (*Neanthes arenaceodentata*) and an amphipod (*Leptocheirus plumulosus*). Environ Sci Technol 39:2880–2887

Mio H, Saeki S, Kano J, Saito F (2002) Estimation of mechanochemical dechlorinated rate of poly (vinyl chloride). Environ Sci Technol 36:1344–1348

Mohammed SA, Sorensen DL, Sims RC, Sims JL (1998) Pentachlorophenol and phenanthrene biodegradation in creosote contaminated aquifer material. Chemosphere 37:103–111

Mortland MM, Halloran LJ (1976) Polymerization of aromatic molecules on smectite. Soil Sci Soc Am J 40:367–370

Moza P, Schneunert I, Klein W, Korte F (1979) Studies with 2,4′,5-trichlorobiphenyl-14C and 2,2′,4,4′,6-pentachlorobiphenyl-14C in carrots, sugar beets, and soil. J Agric Food Chem 27:1120–1124

Nasser A, Sposito G, Cheney MA (2000) Mechanochemical degradation of 2,4-D adsorbed on synthetic birnessite. Colloid Surf 163:117–123

Northcott GL, Jones KC (2000) Experimental approaches and analytical techniques for determining organic compound residues in soils and sediment. Environ Pollut 108:19–43

Oda J, Maeda I, Mori T, Yasuhara A, Saito Y (1998) The relative proportions of polycyclic aromatic hydrocarbons and oxygenated derivatives in accumulated organic particulates as affected by air pollution sources. Environ Technol 19:961–976

Olajire AA, Altenburger R, Küster E, Brack W (2005) Chemical and ecotoxicological assessment of polycyclic aromatic hydrocarbon-contaminated sediments of the Niger delta, southern Nigeria. Sci Total Environ 340:123–136

Pan B, Baoshan X, Shu T, Wenxin L, Xiumei L, Yang X, Hancheng D, Xianming Z, Yanxv Z, Huishi Y (2007) Effect of physical forms of soil organic matter on phenanthrene sorption. Chemosphere 68:1262–1269

Phillips TM, Seech AG, Liu D, Lee H, Trevors JT (2000) Monitor ing biodegradation of creosote in soils using radiolabels, toxicity tests, and chemical analysis. Environ Toxicol 15:99–106

Piatt JJ; Brusseau ML (1998) Rate-limited sorption of hydrophobic organic compounds by soils with well-characterized organic matter. Environ Sci Technol 32: 1604–1608

Ping L, Yongming L, Longhua W, Wei Q, Jing S, Peter C (2006) Phenanthrene adsorption by soils treated with humic substances under different pH and temperature conditions. Environ Geochem Health 28:189–195

Pizzigallo MDR, Napola A, Spagnuolo M, Ruggiero, P (2004) Mechanochemical removal of organo-chlorinated compounds by inorganic components of soil. Chemosphere 55:1485–1492

Pizzigallo MDR, Ruggiero P, Crecchio C, Mininni R (1995) Manganese and iron oxides as reactants for oxidation of chlorophenols. Soil Sci Soc Am J 59:444–452

Reid BJ, Jones KC, Semple KT (2000) Bioavailability of persistent organic pollutants in soils and sediments—a perspective on mechanisms, consequences and assessment. Environ Pollut 108:103–112

Sehili AM, Lammel G (2007) Global fate and distribution of polycyclic aromatic hydrocarbons emitted from Europe and Russia. Atmos Environ 41:8301–8315

Shin JY, Buzgo CM, Cheney MA (2000) Mechanochemical degradation of atrazine adsorbed on four synthetic manganese oxides. Colloid Surf 172:113–123

Strek HJ, Weber JB (1982a) Adsorption and reduction in bioactivity of polychlorinated biphenyl (Aroclor 1254) to redroot pigweed by soil organic matter and montmorillonite clay. Soil Sci Soc Am J 46:318–322

Strek HJ, Weber JB (1982b) Behavior of polychlorinated biphenyls (PCBs) in soils and plants. Environ Pollut A 28:291–312

Tang J, Weber WJ (2006) Development of engineered natural organic sorbents for environmental applications. 2. Sorption characteristics and capacities with respect to phenanthrene. Environ Sci Technol 40:1657–1663

Tang J, Petersen EJ, Huang Q, Weber WJ Jr (2007) Development of engineered natural organic sorbents for environmental applications: 3. Reducing PAH mobility and bioavailability in contaminated soil and sediment systems. Environ Sci Technol 41:2901–2907

Tomaszewski JE, Werner D, Luthy RG (2007) Activated carbon amendment as a treatment for residual DDT in sediment froma superfund site in San Francisco Bay, Richmond, California, USA. Environ Toxicol Chem 26:2143–2150

Van Droodge B, Grimalt J, Camarero L, Catalan J, Stuchli E, Torres Garcia CJ (2004) Atmospheric semivolatile organochlorine compounds in European high-mountain areas (central Pyrenees and high Tatras). Environ Sci Technol 38:3525–3532

Vasilyeva GK, Kreslavski VD, Oh BT, Shea PJ (2001) Potential of activated carbon to decrease 2,4,6-trinitrotoluene toxicity and accelerate soil decontamination. Environ Toxicol Chem 20:965–971

Voice TC, Rice CP, Weber WJ (1983) Effect of solids concentration on the sorptive partitioning of hydrophobic pollutants in aquatic systems. Environ Sci Technol 17:513–518

Wan YY, Mackay D (1986) A critical review of solubilities, vapor pressures, Henry's law constants, and octanol-water partition coefficients of the polychlorinated biphenyl. J Phys Chem Ref Data 15(2):911–929

Wang LS (1991) Organic pollutant chemistry (in Chinese). Science Press, Beijing, pp 233–244

Wania F, Mackay D (1996) Tracking the distribution of persistent organic pollutants. Environ Sci Technol 37:1352–1359

Weber WJ, Huang WL (1996) A distributed reactivity model for sorption by soils and sediments: 4. Intraparticle heterogeneity and phase-distribution relationships under nonequilibrium conditions. Environ Sci Technol 30(3):881–888

Weber WJ, Tang J, Huang Q (2006) Development of engineered natural organic sorbents for environmental applications. 1. Materials, approaches, and characterizations. Environ Sci Technol 40:1650–1656

Weilin H, Ping'an P, Zhiqiang Y, Jiamo F (2003) Effects of organic matter heterogeneity on sorption and desorption of organic contaminants by soils and sediments. Appl Geochem 18:955–972

Wen B, Li R, Zhang S, Shan X, Fang J, Xiao K, Khan SU (2009) Immobilization of pentachlorophenol in soil using carbonaceous material amendments. Environ Pollut 157:968–974

Whyte LG, Bourbonniere L, Greer CG (1997) Biodegradation of petroleum hydrocarbons by psychrotrophic Pseudomonas strains possessing both alkane (alk) and naphthalene (nah) catabolic pathways. Appl Environ Microbiol 63:3719–3723

World Health Organization (1979) Polychlorinated biphenyls and terphenyls (in Chinese). Chinese Environmental Science Publishing House, Beijing, pp 3–65

World Health Organization (1993) Polychlorinated biphenyls and terphenyls, 2nd edn. WHO, Geneva, pp 92–98

Xia G, Ball WP (1999) Adsorption-partitioning uptake of nine low polarity organic chemicals on a naturalsorbent. Environ Sci Technol 33:262–269

Xie H, Guetzloff TF, Rice JA (1997) Fractionation of pesticideresidues bound to humin. Soil Sci 162:421–429

Xing B, Pignatello JJ (1997) Dual-mode sorption of low-polarity compounds in glassy poly (vinyl chloride) and soil organic matter. Environ Sci Technol 31:792–799

Yuan SY, Wei SH, Chang BV (2000) Biodegradation of polycyclic aromatic hydrocarbons by a mixed culture. Chemosphere 41:1463–1468

Zimmerman JR, Ghosh U, Millward RN, Bridges TS, Luthy RG (2004) Addition of carbon sorbents to reduce PCB and PAH bioavailability in marine sediments: physicochemical tests. Environ Sci Technol 39:1199–1200

Chapter 2
Microbial Extracellular Enzymes and the Degradation of Natural and Synthetic Polymers in Soil

Richard G. Burns

Abstract Bacteria and fungi encounter a large number of complex organic materials in soil, many of which represent a source of the energy, carbon and the nutrients that are required for cell maintenance and growth. Cellulose and lignin are the two most abundant plant polymers but contaminated soils may also contain recalcitrant xenobiotics including polycyclic aromatic hydrocarbons and polychlorinated biphenyls. However, all of these natural and anthropogenic structures are only *potential* substrates because bacteria and fungi cannot absorb insoluble macromolecules. Instead, they depend on the activities of extracellular enzymes to generate the soluble low molecular mass structures that are recognized by cell membrane receptors and transported into the cell. Some of the organic pollutants in soil are not only polymeric and poorly soluble but are also toxic and may require transformation prior to uptake. In this chapter the diversity of extracellular enzymes and some of the polymers that they degrade are reviewed as are the many locations and multiple fates of these enzymes once they have left the cytoplasm. We summarise the diversity of soil enzymes and describe some of the natural and synthetic macromolecules that they degrade. Aspects of the regulation of extracellular enzyme synthesis and secretion and the many locations and multiple fates of microbial enzymes after they are externalized will be outlined. The chemical, physical and biological properties of soil all affect enzyme diffusion, survival and substrate turnover, as well as the proportion of the product that is available to and assimilated by the producer cells. The ways in which microbes and their extracellular enzymes attempt to overcome the generally destructive, inhibitory and competitive properties of the soil matrix and the various strategies they adopt for effective substrate detection and utilization will be described.

R.G. Burns (✉)
School of Agriculture and Food Sciences, The University of Queensland,
Brisbane, QLD 4072, Australia
e-mail: r.burns@uq.edu.au

J. Xu and D.L. Sparks (eds.), *Molecular Environmental Soil Science*,
Progress in Soil Science, DOI 10.1007/978-94-007-4177-5_2,

Keywords Soil enzymes • Cellulose • Lignin • Xenobiotics • PAH • PCB • Bioremediation • Microbial ecology

2.1 Introduction

Microbial enzyme synthesis and secretion is an energetically expensive process and many regulatory mechanisms and ecological strategies have evolved to ensure the efficient capture of the products of exogenous substrate catalysis. In other words, the allocation of resources to generating and externalizing enzymes must be compensated by the gain in energy and nutrients for the processes of microbial growth and cell division. However, soil is an hostile environment and once extracellular enzymes leave the cell they are subject to denaturation, degradation, adsorption and dilution. The locations and functions of enzymes in soil have been researched and discussed for decades (Burns 1978, 1982; Burns and Dick 2002; Caldwell 2005; Nannipieri et al. 2002; Wallenstein and Weintraub 2008) and are currently the subjects of an international collaborative research effort (http://enzymes.nrel.colostate.edu). Recent advances in molecular, microscopic and analytical techniques (Bouws et al. 2008; Wallenstein and Weintraub 2008) and some original thinking (Allison 2005; Burns 2003; Burns and Stach 2002; Theuerl and Buscot 2010) coupled with the need to understand how the activities of enzymes contribute to a large number of industrial, medical and environmental processes, has lead to great advances. For example, those investigating composting (Crecchio et al. 2004; Raut et al. 2008), waste water (Shackle et al. 2006) and sludge treatment (Whiteley and Lee 2006), paper production (Ravalason et al. 2008; Witayakran and Ragauskas 2009) and the conversion of plant materials to fermentable sugars for bioethanol production (Wackett 2008; Wilson 2008, 2009) need to understand the functions and efficiency of the many enzymes responsible for lignocellulolysis. Combating the invasive activities of phytopathogens (Kikot et al. 2009) and the complex enzymology of ruminant digestion (Morrison et al. 2009) also depend on a detailed knowledge of organic polymer solubilization and mineralization.

The possible consequences of climate change on soil processes have stimulated a great deal of experimental research, modeling and theorizing, some of it focused on soil organic matter (SOM) formation and decomposition (Bradford et al. 2008; Davidson et al. 2000; Eliasson et al. 2005; Fang et al. 2005; Jones et al. 2005; Kirschbaum 2004; Knorr et al. 2005). The SOM is a recalcitrant polyphenolic and polysaccharidic complex and is a long-term repository for sequestered carbon as well as being a major contributor to crumb structure, soil stability, plant nutrient and water retention, microbial diversity and activity, and an host of properties that contribute to soil fertility and plant productivity. Increased soil (and water) temperatures, elevated atmospheric carbon dioxide (Finzi et al. 2006), and more frequent wetting and drying cycles (Allison and Treseder 2008; Henry et al. 2005) and increasing salinity (Chowdhury et al. 2011) will change microbial community composition and probably accelerate growth and enzyme activities (Allison and

Treseder 2008; Henry et al. 2005) either directly or following stimulation of plant growth and root exudation. The long debated and controversial process of soil ‘priming’ has again become topical because of the impact of increased plant growth on the size and activities of the microbial biomass (Kuzyakov 2010) and consequential changes in SOM stability (Jones et al. 2003; Melillo et al. 2002) as have the ongoing arguments concerning the significance of abiological SOM degradation (Brookes et al. 2009; Kemmitt et al. 2008; Kuzyakov et al. 2009). Those attempting to quantify the outcomes of global warming (Gillabel et al. 2010; Kleber 2010; Luo 2007; Macias and Arbestain 2010) must take into account any predicted increases in microbial processes, including soil enzyme activities, and any associated declines in the hitherto recalcitrant humic matter (Davidson and Janssens 2006). Many potential organic pollutants, such as the polycyclic aromatic hydrocarbons and the polychlorinated biphenyls, are both chemically complex and poorly water soluble and require extracellular catalysis and sometimes detoxification prior to microbial uptake and catabolism. Inevitably, the rational and successful bioremediation of contaminated soil will depend on a thorough understanding of the relevant enzymatic processes (Asgher et al. 2008; Wackett 2004).

2.2 Substrate and Enzyme Diversity in Soils

Extracellular enzymes in soils catalyse the degradation of plant, animal and microbial macromolecules in addition to many xenobiotics: this they do primarily through hydrolytic and oxidative reactions. Hydrolytic enzymes are substrate-specific: they catalyze reactions that cleave specific bonds (e.g. C–O and C–N bonds) between monomers. Oxidative enzymes, on the other hand, act on a much broader range of substrates that share similar bonds (e.g. C–C and C–O–C) and use either oxygen (oxygenases) or hydrogen peroxide (peroxidases) as their electron acceptors. The physical and chemical complexity of the many potential C and energy sources demands the combined catalytic activities of many extracellular enzymes. In addition, the mineral components (especially clays) and such variable properties as pH, temperature, cation exchange capacity and oxygen levels suggest that the catabolism of complex organics is likely to be a function of a large number of enzymes and isozymes generated by a community of interacting microorganisms.

2.2.1 Cellulose and Lignin Degradation

Degradation of the most common plant polymers, cellulose, hemicelluloses, pectin, starch and lignin, depends on the simultaneous and/or sequential activities of a number of extracellular enzymes produced by a diverse community of bacteria and fungi; it is probable that more than 50 different enzymes are involved in the extracellular depolymerisation of a plant leaf. In other words, the processes necessary for successful biopolymer degradation depend on cascades of enzymes secreted in an organized sequence and surviving long enough to generate pentoses, hexoses, phenols, amino acids, amino sugars, etc. This complexity is well-illustrated by the microbiology and biochemistry of cellulose and lignin decay.

Cellulose, the most abundant form of plant carbon, is a chemically simple yet structurally complex insoluble polymer that is composed of linear chains of 5,000 or more glucose units. These chains are held together with H-bonds to form rigid microfibrils which are themselves linked to hemicelluloses, pectins, glycoproteins and lignins. The enzymology of cellulolysis and some of the strategies for degrading plant cell walls have been reviewed recently (Wilson 2008). Basidiomycete and ascomycete fungi and many bacteria are major degraders of cellulose, employing three groups of extracellular hydrolytic enzymes, primarily endo-1,4-β-glucanases (EC 3.2.1.4), cellobiohydrolases (EC 3.2.1.91) and β-glucosidases (EC 3.2.1.21) (Baldrian and Valášková 2008). The plurality of these enzymes is a necessity given the complexity of the substrate and the many and variable soil microenvironments. Many aerobic fungi secrete all three although some bacteria retain β-glucosidases within the cytoplasm or periplamic space, presumably because they are able to transport the dimer cellobiose through the cell wall. In addition to the principal cellulases, the overall decomposition of cellulose involves cellobiose dehydrogenase (EC1.1.99.18). Other ill-defined extracellular enzymes (e.g. 'expansins', 'swollenins') help to loosen the structure of cell walls prior to cellulase penetration (Kim et al. 2009; Tsumuraya 1996). Some anaerobic bacteria use cellobiose phosphorylase (EC 2.4.1.20) to convert the dimer to glucose and glucose-1-phosphate. The best known cellulose degrader, *Trichoderma reesei* (*Hypocrea jecorina*) has 30 or more glycosyl hydrolases (including seven endo-glucanases) and a secretome containing greater than 100 proteins (Martinez et al. 2008) whereas its close relative, *Trichoderma harzianum*, produces more than 250 extracellular proteins (Suarez et al. 2005). Cellulolysis is less common in bacteria but there are many important strains, especially ruminant anaerobes (Morrison et al. 2009) and those found in high temperature environments (Yang et al. 2009).

Not all cellulolytic microorganisms secrete a full complement of endo- and ecto-cellulases and must rely on other microbes in order to successfully degrade cellulose. This reinforces the concept of a community-driven process and it was reported a long time ago that endo- and ecto-cellulases from unrelated microorganisms can act synergistically sometimes with specific activities up to 15× those shown by the individual enzymes (Irwin et al. 1993). An important question is why do fungi and bacteria express so many cellulases? There are a number of possible answers to this but certainly the physical diversity of the plant cell wall with its amorphous and crystalline cellulose regions, the many differing contributions of, and associations with, other structural polymers, and the variable chemical and physical properties of soil and climatic factors all play a part. The microbial community needs cellulases for every occasion and in every situation in order to exploit the huge carbon and energy resource offered by plant residues.

Lignin, with which cellulose is usually associated, is the second abundant polymer in plant debris. It is a chemically complex phenylpropanoid that is degraded by an array of non-specific oxidative enzymes (Baldrian 2006; Cullen and Kersten 2004; Osono 2007; Sinsabaugh 2010). Prominent among these are laccases (E.C. 1.10.3.2), manganese peroxidases (EC 1.11.1.13) and lignin peroxidases (EC 1.11.1.14). Degradation also demands the input of enzymes generating hydrogen peroxide and Fe^{2+} and Mn^{2+}. Lignin breakdown is carried out mainly by basidiomycetes (Hatakka 1994; Osono 2007), some ascomycetes, such as *Xylaria* spp. (Kellner et al. 2007) and certain actinobacteria (Kirby 2006). Both saprotrophic (Valaskova et al. 2007) and ectomycorrhizal fungi (Chen et al. 2001) can produce lignin-degrading enzymes. Interestingly, the laccase gene is widespread among bacteria (Alexandre and Zhulin 2000; Claus 2003), and their importance in lignin degradation may be underestimated (Kellner et al. 2008). Laccase and lignin breakdown is a story in itself (Theuerl and Buscot 2010). Lignin peroxidase (EC 1.11.1.14), cellobiose dehydrogenases (EC1.1.99.18), versatile peroxidase (EC 1.11.1.16) and pyranose-2-oxidases (EC 1.1.3.10) are also implicated in lignin degradation (Baldrian 2006; Nyanhongo et al. 2007). Clearly, as is the situation with cellulose, lignin degradation is a complex process involving a raft of enzymes. Much of our knowledge arises from studies with lignin-like compounds such as veratryl alcohol and the applicability of this to plant tissue penetration is far from clear; enzymes will find it difficult to diffuse into intact wood (Blanchette et al. 1997) and may need to be carried there by aggressively-growing fungal hyphae.

There are three general classes of fungi involved to a greater or lesser extent in lignin degradation: white-rot, soft-rot, and brown-rot fungi (Osono 2007). White-rot fungi are the only group able to completely mineralize lignin which they accomplish through a combination of hydroxylation and demethylation followed by oxidative degradation of the aromatic rings. In contrast, brown-rot fungi modify lignin by removing methoxyl groups but do not oxidise the aromatic rings. Soft-rot fungi, including ascomycetes and deuteromycetes, break down the middle lamella of the cell wall, which acts to soften the lignin structure. Soft-rot fungi appear to produce peroxidases enzymes that are specific to lignin in hardwoods. The saprophytic white rot fungus, *Phanerochaete chrysosporium*, has been sequenced and has more than 85 genes for glycosyl hydrolases, in excess of 100 for 'ligninases' and a secretome of almost 800 proteins (Vanden Wymelenberg et al. 2006) – including lignin peroxidases (8), manganese peroxidases (8), other oxidoreductases (87) and glycoside hydrolases (90). The hydrogen peroxide required for peroxidase activity is produced by a collection of enzymes: glyoxal oxidases (EC1.1.3 -), superoxide dismutases (EC1.15.1.1) and aryl alcohol oxidases (EC 1.1.3.7). The lignolytic basidiomycete *Coprinopsis cinerea* is predicted to secrete 1,769 proteins and 76 of these have been identified as enzymes (Bouws et al. 2008).

The physical structure of plant detritus is an important determinant of the order in which the polymers are degraded. Soluble carbohydrates may be located on the outer fiber structures of litter or diffuse out of the complex, making them susceptible to the early colonizers in the degradation sequence. On the other hand, some plant components that are otherwise relatively easily degraded can persist because they are physically protected from degradation. Lignin acts as a barrier around cellulose and needs to be degraded to increase the access of cellulases to their substrate. It is probable that lignin does not function as a sole carbon source and therefore its degradation has to be coupled to cellulolysis (or the metabolism of other carbohydrates) in order to generate the carbon and energy needed for microbial invasion and enzyme attack. Furthermore, the complexity of the process probably demands that fungi and bacteria act in concert.

In addition to their lignocellulosic activities, 'ligninases' may have other functions during the life-cycle of the fungus and can certainly degrade soil humic matter composed of an array of condensed aromatic compounds (vanillin, ferulic acid, guiacol, etc.). Importantly, a wide range of organic pollutants including polycyclic aromatic hydrocarbons, polychlorinated biphenyls, pesticides, halogenated aromatics, cyanides, pentachlorophenol and trinitroglycerine, are subject to attack by the lignin degrading enzymes.

2.3 Xenobiotic Pollutants and Their Breakdown

Xenobiotics are compounds that are structurally different from naturally occurring organics and are sometimes characterized by long half-lives in the environment. Most of the persistent structures are organochlorines, introduced into the soil by human activities and include industrial chemicals, agrochemicals and the byproducts of commercial processes. Examples include DDT (dichlorodiphenyltrichlorethane), PCBs (polychlorinated biphenyls), heptachlor, dioxins, toxaphene and chlordane. Other anthropogenic products have their counterparts in nature but are nonetheless recalcitrant and may possess mutagenic, teratogenic, or carcinogenic properties and have deleterious consequences for the environment at large. This second group includes PAHs (polycyclic aromatic hydrocarbons) and many pesticides. The feasibility of using fungi and bacteria (and their enzymes) for the degradation of these and other persistent organics has been studied extensively (e.g. Harms et al. 2011; Strong and Claus 2011; Husain et al. 2009; Rubilar et al. 2008).

PAHs are a group of hydrocarbons made up of multiple interconnected 5- or 6-carbon benzene rings. They arise from both natural and anthropogenic sources: forest fires, volcanic eruptions, vehicular emissions, wood burning, petroleum catalytic cracking, and the combustion of fossil fuels. Several have been identified by the US EPA as priority pollutants (Gan et al. 2009) and include: phenanthrene and anthracene (three rings); benzo(a)anthrene, fluoranthene and pyrene (four rings); benzo(a) pyrene (five rings); and dibenzo(a,h)anthracene (six rings). Soils have been reported as containing total PAH concentrations ranging between 1 $\mu g\ kg^{-1}$ and 300 $g\ kg^{-1}$ (Bamforth and Singleton 2005; Kanaly and Harayama 2000) sometimes in association with other contaminants such as benzene, toluene, ethyl-benzene, xylene (collectively BTEX) and heavy metals.

In addition to their structural stability, PAHs are hydrophobic, easily sorb onto and within the soil organic matter and, as a consequence, become non-bioavailable and persistent. Nonetheless, low-molecular-mass PAHs, such as naphthalene, phenanthrene and anthracene, are degraded by some bacteria and used as a sole carbon source both under laboratory conditions and in the soil (Cerniglia 1993; Irwin et al. 1993; Kandeler 1990; Smith et al. 1997, 1999). The genera *Sphingomonas* and *Mycobacterium* seem particularly proficient at degrading many low ring number PAHs (Guo et al. 2010; Kim et al. 2008) but there have also been reports describing the microbial oxidation of four-ring PAHs, such as pyrene and chrysene, sometimes as sole carbon and energy sources (Willison 2004; Zheng and Obbard 2002, 2003). However, microbial metabolism and growth on PAHs containing five or more rings is uncommon.

The aerobic biodegradation of PAHs has been reviewed recently (Gan et al. 2009; Loick et al. 2009; Peng et al. 2008). The low substrate specificity of the white rot fungi extracellular enzymes, including lignin peroxidases, manganese peroxidases and laccases, facilitates the breakdown of not just the PAHs but many xenobiotics (Habe and Omori 2003). Once the products enter the cell a number of dioxygenases cleave the rings (e.g. naphthalene dioxygenase has been reported in many soil bacteria) using either the ortho or meta pathway. As with ligno-cellulose decay, the complexity of the extracellular enzymatic processes required to completely degrade many of the priority PAHs is illustrated by the requirement for a microbial consortia often composed of both fungi and bacteria. In flooded, compacted or sub-soils (and anaerobic microsites in aerobic soils), where oxygen is limited, PAH biodegradation can also occur anaerobically. Some anaerobes use nitrate, sulphate, iron, manganese and carbon dioxide as their electron acceptors whilst others use the actual organic contaminants in fermentation (Ambrosoli et al. 2005; Chang et al. 2008).

The second important group of soil pollutants are the PCBs, which for many years were components of hydraulic fluids, plasticizers, adhesives, paints and dielectric fluids in transformers, capacitors and coolants. PCBs are composed of up to ten chlorine atoms attached to two benzene rings and, in general, the higher the degree of chlorination the greater the persistence. They are typically mixtures of dozens of congeners and exhibit chirality: the challenge to the microbial world is, therefore, immense. Nonetheless, many PCBs are metabolized or co-metabolised readily in *in vitro* experiments (Borja et al. 2005; Rein et al. 2007) but in soil, as the number of substituents increases, so does adsorption and this imparts further resistance to degradation. On the other hand, although adsorption reduces bioavailability it also restricts movement and thus leaching to ground water may not be an issue. Of course, some industrial sites may also have organic solvents in the soil and this must inevitably inform risk assessments regarding the likelihood of PCB leaching. The manufacture and use of PCBs is very much restricted nowadays but their persistence in soils and sediments (and their potential to accumulate in animal tissues) means that contaminated sites are the target for chemical, physical as well as biological cleanup.

Diverse microorganisms are able of degrading PCBs but there are dozens of different dehalogenation routes and the process is often incomplete and chlorobenzoates (CBA) accumulate. In their turn, CBAs may inhibit soil microbes (Adebusoye et al. 2008). Once again the white rot fungi are involved in the extracellular degradation of PCBs. However, fungi (including *Penicillium* and *Fusarium* species) which do not secrete lignolytic enzymes, are also implicated in PCB breakdown

(Tigini et al. 2009) although they may be opportunistic and utilize products of the extracellular enzyme reactions performed by other species. Cometabolism is implicated in the overall process of the aerobic degradation of polychlorinated dioxins by the white rot fungi (Field and Sierra-Alvarez 2008) and probably PCBs. Anaerobic dehalogenation by bacteria such as *Dehalococcoides* is important in aquatic environments (Pieper and Seeger 2008) and probably in sub-soils. Little is known about the distinctions between extracellular and intracellular catalysis of PCBs but fungal peroxidases are usually secreted during secondary metabolism and are generally induced by nutrient (especially N) depletion. The expression of ligninolytic enzymes under nutrient-limited conditions is probably an adaptation of the white rot fungi to the low nutrient levels (and high C to N ratios) found in wood (Fernando and Aust 1994) but how this relates to PCBs and PAHs is far from clear. The regulation and control of ligninases is a complex issue (Elisashvili et al. 2010; Giardina et al. 2010) linked to substrate and product levels, C:N ratios, and a plethora of soil chemical and physical properties that induce stress responses in the microflora.

2.3.1 Soil Bioremediation

Bioremediation is a developing technology that attempts to apply our knowledge of microbial activities in soil to problems of soil pollution. Treatment of both PAH (Gan et al. 2009) and PCB (Vasilyeva and Strijakova 2007) contaminated soils has involved *in situ* techniques, including land farming and composting, and *ex situ* methods using bioreactors. On site methods involving biostimulation of the indigenous biomass and bioaugmentation with selected cultures have been used. In the latter case, inoculants have included spent mushroom compost containing *Agaricus bisporus*, white rot fungi (e.g. *Irpex lacteus, Pleurotus ostreatus, Phanerochaete chrysosporium*) and sometimes ill-defined mixtures of bacteria and fungi. However, even the most effective treatments have resulted in limited degradation of the five- and six-ringed structures. The white rot fungi are commonly used for aerobic remediation because they have the capability of degrading the wide range of aromatic compounds often found in polluted soils. Introducing microorganisms to soil can be a hit-and-miss process since, even if the inoculants survive, they may not express the desired catabolic processes or even move away from the target compounds. In spite of these difficulties, there are a number of commercially available microbial mixtures that appear not only to survive but also to degrade PCBs in soil (Sobiecka et al. 2009) and prior oxidation treatments (e.g. using Fenton reagents) may generate more easily degradable products. It is important to note that many of the microbial introductions have been accompanied by intentional changes to soil chemical and physical properties and have involved aeration, irrigation, carbon amendments, and fertilization with N and P. All these are targeted at creating a more amenable environment for microbial activity. Developments in PAH and PCB bioremediation technology include introducing encapsulated microbes or their enzymes into soil and formulating the carrier such that it slowly releases the degraders (Gianfreda and Bollag 2002; Karabika et al. 2009; Su et al. 2008) and thereby reduces their exposure to the hostile soil environment.

Microbes may accelerate xenobiotic breakdown indirectly by improving the bioavailability of the target compounds. In this context it is important to recall that the 'ligninases' can degrade humic matter and release complexed PCBs and PAHs. Microbial extracellular hydrolases may serve the same purpose by depolymerising and solubilising the polysaccharide adsorbents (Wicke and Reemtsma 2010). Some microbes and treatment sequences may use surfactants to render the target organics more accessible (Tsai et al. 2009) and many other mobilizing agents have been assessed in this context (Leonardi et al. 2008). Other ways in which microbes increase pollutant bioavailability may be to change the microenvironment pH and bring about desorption. All these possibilities should be considered in the design of microbial remediation programs although nowadays molecular biology presents many more opportunities to improve the success of soil inoculants. These include the use of genetic constructs (gene bioaugmentation) and transgenic bacteria have been used to mineralize a range of PCB congeners, and under certain circumstances, these survive and proliferate once introduced into the contaminated soil (Saavedra et al. 2010). The use of transgenic plants containing bacterial biphenyl dioxygenases, coupled with a PCB metabolising root community, are exciting developments in what is known as phytoremediation (Aken et al. 2010). Recent papers report the significant reduction of PCBs in contaminated soils taking place in the rhizosphere of horseradish (*Armoracia rusticana*) (Mackova et al. 2009) and mycorrhizal alfalfa inoculated with rhizobia (Teng et al. 2010); what is due to the stimulation of many indigenous microbes and what is the combined action of the plant-microbe synergists (Sinsabaugh et al. 2002a) awaits discovery. To go a step further, engineering the microbial remediation genes directly into a plant and controlling the expression and exudation of enzymes from the roots renders unnecessary the use of microbial inoculants. Of course, the release of genetically manipulated microbes and plants into the environment continues to be a controversial issue.

2.4 Ecology of Extracellular Enzymes

Microbes and their extracellular enzymes must be capable of transforming organic debris into soluble monomers or low number oligomers that can be transported into the cytoplasm. In addition, enzymes must overcome the physical and chemical protection of otherwise bioavailable molecules complexed within or sorbed to clay and humic aggregates. Furthermore, most enzymes work effectively within a narrow range of physical conditions and so such factors as temperature, pH and redox potential must be considered when considering the ecology of microbe-enzyme-substrate interactions. Ultimately, the successful depolymerization of complex carbon and energy sources and the subsequent uptake and metabolism of low molecular mass products will be governed by the growth requirements of the microorganisms and, in that context, C:N:P ratios may often be rate limiting. Leaf litter has a C:N ratio in the range 20–80 and a C:P ratio of around 3,000:1. Microbial biomass, on the other hand, has C:N ratios of 5–10:1 and C:P ratios of 50–100:1 (Cleveland and Liptzin 2007). Therefore, microbes degrading plant residues need not only the right compliment of enzymes but also ways to access additional N and P (and all the other elements that make up a microbial cell). It therefore follows that polysaccharase production is increased when there is plentiful supply of soluble nitrogen (Sinsabaugh et al. 2002b) and excess carbon may increase protease synthesis. One often overlooked source of N and P are dead microbial cells – although these are also composed of many complex polymers and, no doubt, a great deal of enzymic mineralization occurs after cell lysis. All this suggests that additional extracellular enzyme synthesis and secretion must be coordinated with those releasing soluble C in order to provide the appropriate elemental ratio for microbial growth.

In some cases, microbes release small amounts of extracellular enzymes semi-continuously as a mechanism to detect substrate. If the substrate is present, these constitutive enzymes generate products or signals that induce additional enzyme manufacture. Of course in a soil, when substrates are discontinuous both in time and space, this is a high risk strategy and there is a limit to how long a cell can continue to generate enzymes without reward before it is energetically compromised. However, the multiple functional locations of extracellular enzymes, combined with the capacity of the microbial community to detect and migrate towards potential substrates, suggest that there are many more efficient ways in which macromolecular organics are transformed into soluble matter.

2.4.1 Extracellular Enzyme Location and Efficiency

Extracellular enzymes may be associated with the cell's plasma membrane, contained within or attached to the walls of the periplasmic space, cell wall and glycocalyx, or released into the soil aqueous phase (Sinsabaugh 1994). The periplasm may provide Gram-negative bacteria with a reservoir of enzyme activity that is retained until an external signal for secretion is received: in these circumstances synthesis and secretion are separate processes. The proteins and enzymes may be modified *en route* to the extracellular environment or even co-secreted with protective proteases and antibiotics in order to prepare them for the hostile 'real world'. Of course, enzymes that degrade soluble sugar dimers, such as cellobiose and mannose, and some phosphatases and nucleases are often (but not always) restricted to the periplasmic space and further transform their substrates prior to transport into the cytoplasm.

Enzymes that are retained on the external faces of the cell wall (known as mural or peripheral enzymes) are likely to be configured such that their active sites are exposed and the components that are susceptible to attack by microbes and their proteases are shielded. Although this location gives extracellular enzymes some

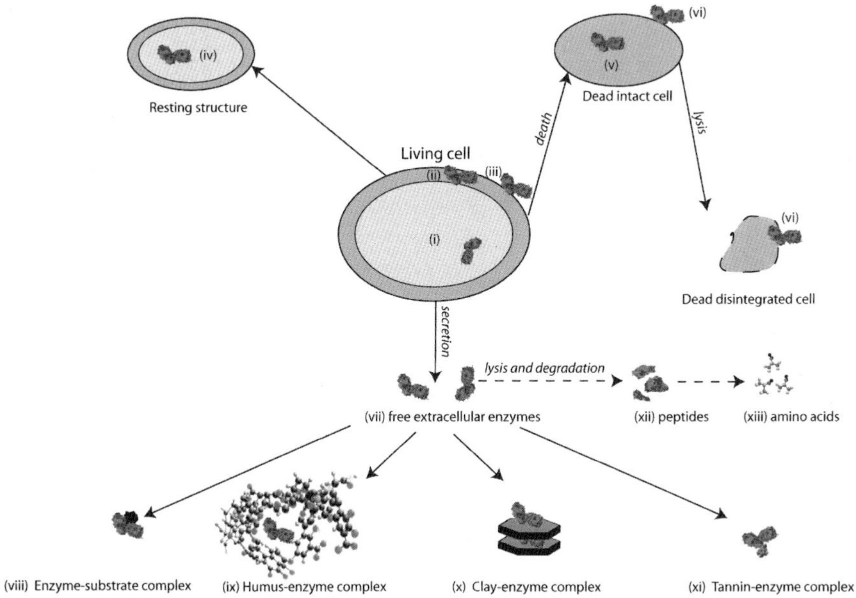

Fig. 2.1 Locations of enzymes in soil (Adapted from Klose (2003) and Wallenstein and Burns 2011) (*i*) Enzymes functioning within the cytoplasm of microbial cells; (*ii*) enzymes restricted to the periplasmatic space of Gram-negative bacteria; (*iii*) enzymes attached to the outer surface of cells with their active sites extending into the soil environment, contained within cellulosomes, or retained by biofilms; (*iv*) enzymes within resting cells including fungal spores, protozoal cysts and bacterial endospores; (*v*) enzymes attached to entire dead cells and cell debris; (*vi*) enzymes leaking from intact cells or released from lysed cells; (*vii*) free extracellular enzymes in soil water; (*viii*) enzymes associated with enzyme-substrate complexes; (*ix*) enzymes complexed with humic colloids by absorption, entrapment or co-polymerization; (*x*) enzymes sorbed to the external and internal surfaces of clay minerals; and (*xi*) enzymes bound to condensed tannins. Extracellular enzymes can be degraded abiotically or through the activity of proteases into (*xii*) peptides and then degraded by peptidases into their constituent amino acids (*xii*)

protection from the external environment, inevitably their scavenging ability is much reduced. This suggests that cells with mural enzymes need signals to initiate movement towards carbon and energy sources (chemotaxis) or must rely on much less efficient phenomena such as Brownian motion and random movement in order to make contact with their substrates (Fig. 2.1).

Other cell-bound extracellular enzymes include those contained within a multicellular polysaccharidic biofilm and those that are organized within microscopically visible structures attached to but protruding from the cell wall. The former may develop on the surface of organic debris and be composed of many different but interactive genotypes (Flemming and Wingender 2002; Romani et al. 2008) and multispecies biofilm communities have been shown to degrade PAHs more effectively that conventionally enriched mixed cultures (Stach and Burns 2002). Pollutant degrading microbial biofilms have been evaluated frequently in bioreactors (Cao et al. 2009). The latter are the cellulosomes and contain cellulases (as well as other hydrolases) arranged on a scaffold that facilitates efficient cleavage of polysaccharides. Cellulosomes were first described for the anaerobic thermophile *Clostridium thermocellum* and have been much researched since (Bayer et al. 2004, 2008; Fontes and Gilbert 2010; Gold and Martin 2007; Peer et al. 2009). *C. thermocellum,* produces a large number extracellular enzymes, including endoglucanases, exoglucanses, β-glucosidases, xylanases, lichenases, laminarinases, xylosidases, galactosidases, mannosidases, pectin lyases, pectin methylesterases, polygalacturonate hydrolases, cellobiose phosphorylases and cellodextrin phophorylases (Demain et al. 2005; Lamed et al. 1983) and many of them are housed within cellulosomes. Interestingly, there is another group of extracellular enzymes that are effective without passing into the external environment. These have been reported in some pathogenic bacteria which transfer their extracellular enzymes directly into the target cell using what is termed a secretory needle (Hauser 2009; Pastor et al. 2005). It is not know if such a mechanism functions in soil organic matter degradation.

Truly extracellular enzymes diffuse away from their parent cell and are then subject to sorption, (Sinsabaugh 1994) denaturation and degradation. Some of these enzymes are more persistent than their intracellular counterparts because they are glycosylated and have disulfide bonds: structural modifications that confer thermostability, a broad pH range for activity, and resistance to proteases. Once in contact with their substrate, cellulases have a variety of ways in which they optimize their activity. One mechanism involves the all-important cellulose binding moieties (CBM) and their impressive substrate binding affinities (Hilden and Johansson 2004). Fungal and bacterial CBMs belong to many different families (Wilson 2008) and serve to anchor the enzyme to its substrate at appropriate sites for the catalytic domain (CD) to cleave the β-1, 4-linkages (Boraston et al. 2003). CBMs may also detach and slide across the fibrillar surface and thereby move the associated CD along the cellulose chain. This relocates the CD and processively hydrolyzes the substrate (Jervis et al. 1997).

Many extracellular enzymes become further stabilized through interactions with clay minerals and humic acids and retain significant levels of activity (Burns et al. 1972; Quiquampoix and Burns 2007). Enzymes stabilized as components of humic matter and associated with organomineral complexes also retain some activity (Tate 2002). Colloid-immobilized soil enzymes represent a reservoir of potential activity and may be the first catalytic response to changes in substrate availability in soils as well as serving as the generator of signal molecules for the microbial community. Soil-bound enzymes may also be a source of potential activity in soils during periods when microbial biomass is low or shut down due to stressed conditions (Stursova and Sinsabaugh 2008). In fact, much of the activity of some enzymes is associated with organic and inorganic colloids rather than being free in solution (Kandeler 1990). It should be noted that some of these bound enzymes (e.g. urease, phosphatases) are not extracellular *sensu stricta* but rather they have become externalized as a result of the death and lysis of their parent cell. Stabilized enzymes often have reduced *in-situ* activity, as complexation can restrict substrate accessibility, occlude active sites, and cause conformational changes (Allison and Jastrow 2006; Nannipieri 2006; Quiquampoix and Burns 2007). The turnover rates of immobilized enzymes in soils and their contribution to overall catalysis of specific substrates have not yet been measured.

The extracellular degradation of complex organics can occur if: (i) diffusing enzymes survive; (ii) substrate is present and accessible; (iii) substrate is located; (iv) correct combination of enzymes is assembled; and (v) appropriate chemical and physical conditions exist. However, assuming that this sequence of events is in place and a more soluble diffusible product is generated then other microbes may intercept the molecules before the originating cell can benefit. These opportunistic 'cheaters' may not have invested any resources in extracellular enzyme generation yet will reduce the efficiency of the entire process as far as the actual enzyme producers are concerned (Allison 2005). Some microbes may employ antibiotics and enzymes to counteract the 'highjacking' of products or rely on the activities of predators to control their rivals: in the rumen stomach protozoa may have this function (Modak et al. 2007). Of course, what might appear as cheating may be part of a complex microbial community synergy: the so-called cheaters provide some direct or indirect benefit to the cheated. Or it may be that the benefits of extracellular depolymerization to the overall microbial community far outweigh the disadvantages to the principal enzyme generators. A mass balance that considers the bioenergetics of enzyme synthesis and secretion, substrate degradation, product assimilation, and microbial growth is much needed.

One way of overcoming at least some of these constraints is suggested by a mechanism that involves microbes 'sensing' both the substrate and their own population numbers. In this way gene function is connected to cell density and enzymes are only synthesized and/or secreted when cell numbers are high enough to have a major impact. This is a process known as quorum sensing (Shank and Kolter 2009) and has been well-described for many phytopathogens especially *Erwinia* species (Barnard and Salmond 2007). Gene products in this situation include pectin methyl esterase, pectic lyase and polygalacturonase, all of which depolymerise the protective coat of the target seed or fruit and facilitate microbial penetration and pathogenesis. Many other effector proteins pass into the host plant tissue and reduce its resistance to attack. Quorum sensing in the rhizosphere is believed to be an important controlling process for all sorts of microbial interactions and is a subject of intensive research and speculation (Babalola 2010; Badri et al. 2009; Mathesius 2009; Pang et al. 2009) but its role in the turnover of natural and anthropogenic substrates in soil is yet to be evaluated.

2.5 Conclusions

Identifying the functional locations of extracellular enzymes in soils and quantifying their individual contribution to the catalysis of a particular substrate are challenges yet to be overcome by soil biochemists and microbiologists. The majority of research on soil enzymes has focused on quantifying potential enzyme degradation rates in soil slurries where substrate supply is non-limiting. In undisturbed soils, microbe, enzyme and substrate diffusion are restricted by the microscale spatial structure of the soil, which limits solvent and solute movement and isolates microsites. This chapter has emphasized the role of extracellular enzymes in the degradation of lignocelluloses and two groups of soil pollutants. It must be remembered that the overall turnover of these compounds will inevitably involve the activities of invertebrates including earthworms (Nozaki et al. 2009), ants (Ikeda-Ohtsubo and Brune 2009) and termites (Douglas 2009) and the symbiotic activities of the cellulolytic microbes they cultivate or harbor in their digestive tracts. These topics are outside the purview of this chapter.

An enhanced knowledge of extracellular enzyme function will have many practical applications in soil biology not only for bioremediation as described here, but also biocontrol, plant nutrient generation and availability, aggregate stability, and carbon cycling and sequestration. One of the greatest challenges in soil biology is to link the functional and ecological aspects of microbial extracellular enzyme activities to organic matter degradation. We are now equipped with the analytical (electrophoretic, chromatographic, mass spectrometric), microscopical (fluorescence, scanning probe, atomic and ultrasonic force, confocal laser, differential interference, digital image analysis), molecular (genomic, proteomic, metabolomic, secretomic, metagenomic) and bioinformatics tools to achieve many of these objectives. The coordination of extracellular enzyme production, the spatial distribution of enzymes and the microbial ecology of substrate utilization is the function of a complex microbial community in which a close synergy exists between different taxa. In this context it may prove rewarding to consider game theory (Aken et al. 2010; Babalola 2010; Bouws et al. 2008; Davidson and Surette 2008; Velicer 2003) and the consequences of intra-species variation, individual sacrifice and social networks within the microbial world (Kraemer et al. 2010; Krug et al. 2008). Finally, new ideas about order and efficiency arising from the self organization of interconnected networks may be relevant to microbial life in soil (Macklem and Seely 2010).

References

Adebusoye SA, Picardal FW, Ilori MO, Amund OO (2008) Influence of chlorobenzoic acids on the growth and degradation potentials of PCB-degrading microorganisms. World J Microbiol Biotechnol 24:1203–1208

Aken BV, Correa PA, Schnoor JL (2010) Phytoremediation of polychlorinated biphenyls: new trends and promises. Environ Sci Technol 44:2767–2776

Alexandre G, Zhulin IB (2000) Laccases are widespread in bacteria. Trends Biotechnol 18:41–42

Allison SD (2005) Cheaters, diffusion and nutrients constrain decomposition by microbial enzymes in spatially structured environments. Ecol Lett 8:626–635

Allison SD, Jastrow JD (2006) Activities of extracellular enzymes in physically isolated fractions of restored grassland soils. Soil Biol Biochem 38:3245–3256

Allison SD, Treseder KK (2008) Warming and drying suppress microbial activity and carbon cycling in boreal forest soils. Glob Chang Biol 14:2898–2909

Ambrosoli R, Petruzzelli L, Minati JL, Marsan FA (2005) Anaerobic PAH degradation in soil by a mixed bacterial consortium under denitrifying conditions. Chemosphere 60:1231–1236

Asgher M, Bhatti HN, Ashraf M, Legge RL (2008) Recent developments in biodegradation of industrial pollutants by white rot fungi and their enzyme system. Biodegradation 19:771–783

Babalola OO (2010) Beneficial bacteria of agricultural importance. Biotechnol Lett 32:1559–1570

Badri DV, Weir TL, van der Lelie D, Vivanco JM (2009) Rhizosphere chemical dialogues: plant-microbe interactions. Curr Opin Biotechnol 20:642–650

Baldrian P (2006) Fungal laccases – occurrence and properties. FEMS Microbiol Rev 30:215–242

Baldrian P, Valášková V (2008) Degradation of cellulose by basidiomycetous fungi. FEMS Microbiol Rev 32:501–521

Bamforth SM, Singleton I (2005) Bioremediation of polycyclic aromatic hydrocarbons: current knowledge and future directions. J Chem Technol Biotechnol 80:723–736

Barnard AML, Salmond GPC (2007) Quorum sensing in Erwinia species. Anal Bioanal Chem 387:415–423

Bayer EA, Belaich JP, Shoham Y, Lamed R (2004) The cellulosomes: multienzyme machines for degradation of plant cell wall polysaccharides. Annu Rev Microbiol 58:521–554

Bayer EA, Lamed R, White BA, Flint HJ (2008) From cellulosomes to cellulosomics. Chem Rec 8:364–377

Blanchette RA, Krueger EW, Haight JE, Akhtar M, Akin DE (1997) Cell wall alterations in loblolly pine wood decayed by the white-rot fungus, Ceriporiopsis subvermispora. J Biotechnol 53:203–213

Boraston AB, Notenboom V, Warren RAJ, Kilburn DG, Rose DR, Davies G (2003) Structure and ligand binding of carbohydrate-binding module CsCBM6-3 reveals similarities with fucose-specific lectins and "galactose-binding" domains. J Mol Biol 327:659–669

Borja J, Taleon DM, Auresenia J, Gallardo S (2005) Polychlorinated biphenyls and their biodegradation. Process Biochem 40:1999–2013

Bouws H, Wattenberg A, Zorn H (2008) Fungal secretomes – nature's toolbox for white biotechnology. Appl Microbiol Biotechnol 80:381–388

Bradford MA, Davies CA, Frey SD, Maddox TR, Melillo JM, Mohan JE, Reynolds JF, Treseder KK, Wallenstein MD (2008) Thermal adaptation of soil microbial respiration to elevated temperature. Ecol Lett 11:1316–1327

Brookes PC, Kemmitt SJ, Addiscott TM, Bird N (2009) Comments on the paper by Kemmitt et al (2008) Soil Biol Biochem 40:61–73

Burns RG (1978) Soil enzymes. Academic, New York

Burns RG (1982) Enzyme-activity in soil – location and a possible role in microbial ecology. Soil Biol Biochem 14:423–427

Burns RG (2003) Soil bioremediation: bioavailability, biofilms and complexity. British Crop Protection Council, Farnham

Burns RG, Dick RP (2002) Enzymes in the environment. Activity, ecology, and applications. Marcel Dekker, New York

Burns RG, Stach JEM (2002) Microbial ecology of soil biofilms: substrate bioavailability, bioremediation and complexity. In: Violante A et al (eds) Soil mineral-organic matter-microorganism interactions and ecosystem health, volume 28b – ecological significance of the interactions among clay minerals, organic matter and soil biota, vol 28B. Elsevier Science Bv, Amsterdam, pp 17–42

Burns RG, Pukite AH, McLaren AD (1972) Concerning the location and persistence of soil urease. Soil Sci Soc Am Proc 36:308–311

Caldwell BA (2005) Enzyme activities as a component of soil biodiversity: a review. Pedobiologia 49:637

Cao B, Nagarajan K, Loh K-C (2009) Biodegradation of aromatic compounds: current status and opportunities for biomolecular approaches. Appl Microbiol Biotechnol 85:207–228

Cerniglia CE (1993) Biodegradation of polycyclic aromatic hydrocarbons. Curr Opin Biotechnol 4:331–338
Chang B, Chang IT, Yuan SY (2008) Anaerobic degradation of phenanthrene and pyrene in mangrove sediment. Bull Environ Contam Toxicol 80:145–149
Chen DM, Taylor AFS, Burke RM, Cairney JWG (2001) Identification of genes for lignin peroxidases and manganese peroxidases in ectomycorrhizal fungi. New Phytol 152:151–158
Chowdhury N, Marschner P, Burns RG (2011) Soil microbial activity and community composition as affected by interaction of matric and osmotic potential. Soil Biol Biochem 43:1229–1236
Claus H (2003) Laccases and their occurrence in prokaryotes. Arch Microbiol 179:145–150
Cleveland CC, Liptzin D (2007) C:N:P stoichiometry in soil: is there a "Redfield ratio" for the microbial biomass? Biogeochemistry 85:235–252
Crecchio C, Curci M, Pizzigallo MDR, Ricciuti P, Ruggiero P (2004) Effects of municipal solid waste compost amendments on soil enzyme activities and bacterial genetic diversity. Soil Biol Biochem 36:1595–1605
Cullen D, Kersten PJ (2004) Enzymology and molecular biology of lignin degradation. In: Brambl R, Marzluf GA (eds) Mycota III: biochemistry and molecular biology, vol 3, 2nd edn. Springer, Berlin, pp 249–273
Davidson EA, Janssens IA (2006) Temperature sensitivity of soil carbon decomposition and feedbacks to climate change. Nature 440:165
Davidson CJ, Surette MG (2008) Individuality in bacteria. Annu Rev Genet 42:253–268
Davidson EA, Trumbore SE, Amundson R (2000) Biogeochemistry – soil warming and organic carbon content. Nature 408:789–790
Demain AL, Newcomb M, Wu JHD (2005) Cellulase, clostridia, and ethanol. Microbiol Mol Biol Rev 69:124–154
Douglas AE (2009) The microbial dimension in insect nutritional ecology. Funct Ecol 23:38–47
Eliasson PE, McMurtrie RE, Pepper DA, Stromgren M, Linder S, Agren GI (2005) The response of heterotrophic CO_2 flux to soil warming. Glob Chang Biol 11:167–181
Elisashvili V, Kachlishvili E, Khardziani T, Agathos SN (2010) Effect of aromatic compounds on the production of laccase and manganese peroxidase by white-rot basidiomycetes. J Ind Microbiol Biotechnol 37:1091–1096
Fang CM, Smith P, Moncrieff JB, Smith JU (2005) Similar response of labile and resistant soil organic matter pools to changes in temperature. Nature 433:57–59
Fernando T, Aust SD (1994) Biodegradation of toxic chemicals by white rot fungi. In: Chaudhry GR (ed) Biological degradation and bioremediation of toxic chemicals. Dioscorides Press, Portland, pp 386–402
Field JA, Sierra-Alvarez R (2008) Microbial degradation of chlorinated dioxins. Chemosphere 71:1005–1018
Finzi AC, Sinsabaugh RL, Long TM, Osgood MP (2006) Microbial community responses to atmospheric carbon dioxide enrichment in a warm-temperate forest. Ecosystems 9:215–226
Flemming HC, Wingender J (2002) What biofilms contain – proteins, polysaccharides, etc. Chem Unserer Zeit 36:30–42
Fontes C, Gilbert HJ (2010) Cellulosomes: highly efficient nanomachines designed to designed to deconstruct plant cell wall complex carbohydrates. In: Kornberg RD (ed) Annual review of biochemistry, vol 79. Annual Reviews, Palo Alto, pp 655–681
Gan S, Lau EV, Ng HK (2009) Remediation of soils contaminated with polycyclic aromatic hydrocarbons (PAHs). J Hazard Mater 172:532–549
Gianfreda L, Bollag JM (2002) Isolated enzymes for the transformation and detoxification of organic pollutants. In: Burns RG, Dick RP (eds) Enzymes in the environment: activity, ecology, and applications. Marcel Dekker, New York, pp 495–538
Giardina P, Faraco V, Pezzella C, Piscitelli A, Vanhulle S, Sannia G (2010) Laccases: a never-ending story. Cell Mol Life Sci 67:369–385
Gillabel J, Cebrian-Lopez B, Six J, Merckx R (2010) Experimental evidence for the attenuating effect of SOM protection on temperature sensitivity of SOM decomposition. Glob Chang Biol 16:2789–2798

Gold ND, Martin VJJ (2007) Global view of the Clostridium thermocellum cellulosome revealed by quantitative proteomic analysis. J Bacteriol 189:6787–6795
Guo CL, Dang Z, Wong YS, Tam NF (2010) Biodegradation ability and dioxgenase genes of PAH-degrading sphingomonas and mycobacterium strains isolated from mangrove sediments. Int Biodeter Biodegr 64:419–426
Habe H, Omori T (2003) Genetics of polycyclic aromatic hydrocarbon metabolism in diverse aerobic bacteria. Biosci Biotechnol Biochem 67:225–243
Harms H, Schlosser D, Claus H (2011) Untapped potential: exploiting fungi in bioremediation of hazardous chemicals. Nature Rev Microbiol 9:177–192
Hatakka A (1994) Lignin-modifying enzymes from selected white-rot fungi – production and role in lignin degradation. FEMS Microbiol Rev 13:125–135
Hauser AR (2009) The type III secretion system of Pseudomonas aeruginosa: infection by injection. Nat Rev Microbiol 7:654–665
Henry HAL, Juarez JD, Field CB, Vitousek PM (2005) Interactive effects of elevated CO2, N deposition and climate change on extracellular enzyme activity and soil density fractionation in a California annual grassland. Glob Chang Biol 11:1808–1815
Hilden L, Johansson G (2004) Recent developments on cellulases and carbohydrate-binding modules with cellulose affinity. Biotechnol Lett 26:1683–1693
Husain Q, Husain M, Kulshrestha Y (2009) Remediation and treatment of organopollutants mediated by peroxidases: a review. Crit Rev Biotechnol 29:94–119
Ikeda-Ohtsubo W, Brune A (2009) Cospeciation of termite gut flagellates and their bacterial endosymbionts: trichonympha species and 'Candidatus Endomicrobium trichonymphae'. Mol Ecol 18:332–342
Irwin DC, Spezio M, Walker LP, Wilson DB (1993) Activity studies of 8 purified cellulases – specificity, synergism, and binding domains. Biotechnol Bioeng 42:1002–1013
Jervis EJ, Haynes CA, Kilburn DG (1997) Surface diffusion of cellulases and their isolated binding domains on cellulose. J Biol Chem 272:24016–24023
Jones CD, Cox P, Huntingford C (2003) Uncertainty in climate-carbon-cycle projections associated with the sensitivity of soil respiration to temperature. Tellus B Chem Phys Meteorol 55:642–648
Jones C, McConnell C, Coleman K, Cox P, Falloon P, Jenkinson D, Powlson D (2005) Global climate change and soil carbon stocks; predictions from two contrasting models for the turnover of organic carbon in soil. Glob Chang Biol 11:154–166
Kanaly RA, Harayama S (2000) Biodegradation of high-molecular-weight polycyclic aromatic hydrocarbons by bacteria. J Bacteriol 182:2059–2067
Kandeler E (1990) Characterization of free and adsorbed phophatases in soils. Biol Fertil Soils 9:199–202
Karabika E, Kallimanis A, Dados A, Pilidis G, Drainas C, Koukkou AI (2009) Taxonomic identification and use of free and entrapped cells of a new mycobacterium sp., strain spyr1 for degradation of polycyclic aromatic hydrocarbons (PAHs). Appl Biochem Biotechnol 159:155–167
Kellner H, Luis P, Buscot F (2007) Diversity of laccase-like multicopper oxidase genes in Morchellaceae: identification of genes potentially involved in extracellular activities related to plant litter decay. FEMS Microbiol Ecol 61:153–163
Kellner H, Luis P, Zimdars B, Kiesel B, Buscot F (2008) Diversity of bacterial laccase-like multicopper oxidase genes in forest and grassland cambisol soil samples. Soil Biol Biochem 40:638–648
Kemmitt SJ, Lanyon CV, Waite IS, Wen Q, Addiscott TM, Bird NRA, O'Donnell AG, Brookes PC (2008) Mineralization of native organic matter is not regulated by the size, activity or composition of the soil microbial biomass – a new perspective. Soil Biol Biochem 40:61–73
Kikot GE, Hours RA, Alconada TM (2009) Contribution of cell wall degrading enzymes to pathogenesis of Fusarium graminearum: a review. J Basic Microbiol 49:231–241

Kim SJ, Kweon OG, Jones RC, Edmondson RD, Cerniglia CE (2008) Genomic analysis of polycyclic aromatic hydrocarbon degradation in mycobacterium vanbaalenii PYR-1. Biodegradation 19:859–881

Kim ES, Lee HJ, Bang WG, Choi IG, Kim KH (2009) Functional characterization of a bacterial expansin from Bacillus subtilis for enhanced enzymatic hydrolysis of cellulose. Biotechnol Bioeng 102:1342–1353

Kirby R (2006) Actinomycetes and lignin degradation. Adv Appl Microbiol 58:125

Kirschbaum MUF (2004) Soil respiration under prolonged soil warming: are rate reductions caused by acclimation or substrate loss? Glob Chang Biol 10:1870–1877

Kleber M (2010) What is recalcitrant soil organic matter? Environ Chem 7:320–332

Klose S (2003) Enzyme mediated reactions and microbial biomass in agricultural and fly ash influenced forest ecosystems. Habilitation dissertation Dresden University of Technology, Germany

Knorr W, Prentice IC, House JI, Holland EA (2005) Long-term sensitivity of soil carbon turnover to warming. Nature 433:298

Kraemer SA, Toups MA, Velicer GJ (2010) Natural variation in developmental life-history traits of the bacterium Myxococcus xanthus. FEMS Microbiol Ecol 73:226–233

Krug D, Zurek G, Revermann O, Vos M, Velicer GJ, Muller R (2008) Discovering the hidden secondary metabolome of Myxococcus xanthus: a study of intraspecific diversity. Appl Environ Microbiol 74:3058–3068

Kuzyakov Y (2010) Priming effects: interactions between living and dead organic matter. Soil Biol Biochem 42:1363–1371

Kuzyakov Y, Blagodatskaya E, Blagodatsky S (2008) Comments on the paper by Kemmitt et al (2008). Soil Biol Biochem 41:435–439

Lamed R, Setter E, Bayer EA (1983) Characterization of a cellulose-binding, cellulase-containing complex in clostridium thermocellum. J Bacteriol 156:828–836

Leonardi V, Giubilei MA, Federici E, Spaccapelo R, Sasek V, Novotny C, Petruccioli M, D'Annibale A (2008) Mobilizing agents enhance fungal degradation of polycyclic aromatic hydrocarbons and affect diversity of indigenous bacteria in soil. Biotechnol Bioeng 101:273–285

Loick N, Hobbs PJ, Hale MDC, Jones DL (2009) Bioremediation of Poly-Aromatic hydrocarbon (PAH)-contaminated soil by composting. Crit Rev Environ Sci Technol 39:271–332

Luo YQ (2007) Terrestrial carbon-cycle feedback to climate warming. Annu Rev Ecol Evol Syst 38:683–712

Macias F, Arbestain MC (2010) Soil carbon sequestration in a changing global environment. Mitig Adapt Strat Glob Chang 15:511–529

Macklem PT, Seely A (2010) Towards a definition of life. Perspect Biol Med 53:330–340

Mackova M, Prouzova P, Stursa P, Ryslava E, Uhlik O, Beranova K, Rezek J, Kurzawova V, Demnerova K, Macek T (2009) Phyto/rhizoremediation studies using long-term PCB-contaminated soil. Environ Sci Pollut Res 16:817–829

Martinez D, Berka RM, Henrissat B, Saloheimo M, Arvas M, Baker SE, Chapman J, Chertkov O, Coutinho PM, Cullen D, Danchin EGJ, Grigoriev IV, Harris P, Jackson M, Kubicek CP, Han CS, Ho I, Larrondo LF, de Leon AL, Magnuson JK, Merino S, Misra M, Nelson B, Putnam N, Robbertse B, Salamov AA, Schmoll M, Terry A, Thayer N, Westerholm-Parvinen A, Schoch CL, Yao J, Barbote R, Nelson MA, Detter C, Bruce D, Kuske CR, Xie G, Richardson P, Rokhsar DS, Lucas SM, Rubin EM, Dunn-Coleman N, Ward M, Brettin TS (2008) Genome sequencing and analysis of the biomass-degrading fungus Trichoderma reesei (syn. hypocrea jecorina). Nat Biotechnol 26:553–560

Mathesius U (2009) Comparative proteomic studies of root-microbe interactions. J Proteomics 72:353–366

Melillo JM, Steudler PA, Aber JD, Newkirk K, Lux H, Bowles FP, Catricala C, Magill A, Ahrens T, Morrisseau S (2002) Soil warming and carbon-cycle feedbacks to the climate system. Science 298:2173–2176

Modak T, Pradhan S, Watve M (2007) Sociobiology of biodegradation and the role of predatory protozoa in biodegrading communities. J Biosci 32:775–780

Morrison M, Pope PB, Denman SE, McSweeney CS (2009) Plant biomass degradation by gut microbiomes: more of the same or something new? Curr Opin Biotechnol 20:358–363

Nannipieri P (2006) Role of stabilised enzymes in microbial ecology and enzyme extraction from soil with potential applications in soil proteomics. Nucleic Acid Proteins Soil 8:75–94

Nannipieri P, Kandeler E, Ruggiero P (2002) Enzyme activities and microbiological and biochemical processes in soil. In: Burns RG, Dick RP (eds) Enzymes in the environment. Marcel Dekker, New York, pp 1–33

Nozaki M, Miura C, Tozawa Y, Miura T (2009) The contribution of endogenous cellulase to the cellulose digestion in the gut of earthworm (Pheretima hilgendorfi: megascolecidae). Soil Biol Biochem 41:762–769

Nyanhongo GS, Gubitz G, Sukyai P, Leitner C, Haltrich D, Ludwig R (2007) Oxidoreductases from trametes spp. In biotechnology: a wealth of catalytic activity. Food Technol Biotechnol 45:250–268

Osono T (2007) Ecology of ligninolytic fungi associated with leaf litter decomposition. Ecol Res 22:955–974

Pang Y, Liu X, Ma Y, Chernin L, Berg G, Gao K (2009) Induction of systemic resistance, root colonisation and biocontrol activities of the rhizospheric strain of serratia plymuthica are dependent on N-acyl homoserine lactones. Eur J Plant Pathol 124:261–268

Pastor A, Chabert J, Louwagie M, Garin J, Attree I (2005) PscF is a major component of the pseudomonas aeruginosa type III secretion needle. FEMS Microbiol Lett 253:95–101

Peer A, Smith SP, Bayer EA, Lamed R, Borovok I (2009) Noncellulosomal cohesin- and dockerin-like modules in the three domains of life. FEMS Microbiol Lett 291:1–16

Peng R, Xiong A, Xue Y, Fu X, Gao F, Zhao W, Tian Y, Yao Q (2008) Microbial biodegradation of polyaromatic hydrocarbons. FEMS Microbiol Rev 32:927–955

Pieper DH, Seeger M (2008) Bacterial metabolism of polychlorinated biphenyls. J Mol Microbiol Biotechnol 15:121–138

Quiquampoix H, Burns RG (2007) Interactions between proteins and soil mineral surfaces: environmental and health consequences. Elem 3:401–406

Raut MP, William S, Bhattacharyya JK, Chakrabarti T, Devotta S (2008) Microbial dynamics and enzyme activities during rapid composting of municipal solid waste – a compost maturity analysis perspective. Bioresour Technol 99:6512–6519

Ravalason H, Jan G, Molle D, Pasco M, Coutinho PM, Lapierre C, Pollet B, Bertaud F, Petit-Conil M, Grisel S, Sigoillot JC, Asther M, Herpoel-Gimbert I (2008) Secretome analysis of phanerochaete chrysosporium strain CIRM-BRFM41 grown on softwood. Appl Microbiol Biotechnol 80:719–733

Rein A, Fernqvist MM, Mayer P, Trapp S, Bittens M, Karlson UG (2007) Degradation of PCB congeners by bacterial strains. Appl Microbiol Biotechnol 77:469–481

Romani AM, Fund K, Artigas J, Schwartz T, Sabater S, Obst U (2008) Relevance of polymeric matrix enzymes during biofilm formation. Microb Ecol 56:427–436

Rubilar O, Diez MC, Gianfreda L (2008) Transformation of chlorinated phenolic compounds by white rot fungi. Crit Rev Environ Sci Technol 38:227–268

Saavedra JM, Acevedo F, Gonzalez M, Seeger M (2010) Mineralization of PCBs by the genetically modified strain Cupriavidus necator JMS34 and its application for bioremediation of PCBs in soil. Appl Microbiol Biotechnol 87:1543–1554

Shackle V, Freeman C, Reynolds B (2006) Exogenous enzyme supplements to promote treatment efficiency in constructed wetlands. Sci Total Environ 361:18–24

Shank EA, Kolter R (2009) New developments in microbial interspecies signaling. Curr Opin Microbiol 12:205–214

Sinsabaugh RL (1994) Enzymatic analysis of microbial pattern and process. Biol Fertil Soils 17:69–74

Sinsabaugh RL (2010) Phenol oxidase, peroxidase and organic matter dynamics of soil. Soil Biol Biochem 42:391–404

Sinsabaugh RL, Carreiro MM, Alvarez S (2002a) Enzyme and microbial dynamics of litter decomposition. In: Burns RG, Dick RP (eds) Enzymes in the environment: activity, ecology, and applications. Marcel Dekker, New York, pp 249–265
Sinsabaugh RL, Carreiro MM, Repert DA (2002b) Allocation of extracellular enzymatic activity in relation to litter composition, N deposition, and mass loss. Biogeochemistry 60:1–24
Smith MJ, Lethbridge G, Burns RG (1997) Bioavailability and biodegradation of polycyclic aromatic hydrocarbons in soils. FEMS Microbiol Lett 152:141–147
Smith MJ, Lethbridge G, Burns RG (1999) Fate of phenanthrene, pyrene and benzo[a]pyrene during biodegradation of crude oil added to two soils. FEMS Microbiol Lett 173:445–452
Sobiecka E, Cedzynska K, Bielski C, Antizar-Ladislao B (2009) Biological treatment of transformer oil using commercial mixtures of microorganisms. Int Biodeter Biodegr 63:328–333
Stach JEM, Burns RG (2002) Enrichment versus biofilm culture: a functional and phylogenetic comparison of polycyclic aromatic hydrocarbon-degrading microbial communities. Environ Microbiol 4:169–182
Strong P, Claus H (2011) Laccase: a review of its past and its future in bioremediation. Crit Rev Environ Sci Technol 41:373–434
Stursova M, Sinsabaugh RL (2008) Stabilization of oxidative enzymes in desert soil may limit organic matter accumulation. Soil Biol Biochem 40:550–553
Su D, Li PJ, Wang X, Stagnitti F, Xiong XZ (2008) Biodegradation of benzo[a]pyrene in soil by immobilized fungus. Environ Eng Sci 25:1181–1188
Suarez MB, Sanz L, Chamorro MI, Rey M, Gonzalez FJ, Llobell A, Monte E (2005) Proteomic analysis of secreted proteins from Trichoderma harzianum – Identification of a fungal cell wall-induced aspartic protease. Fungal Genet Biol 42:924–934
Tate RL (2002) Microbiology and enzymology of carbon and nitrogen cycling. In: Burns R, Dick R (eds) Enzymes in the environment: activity, ecology, and applications. Marcel Dekker, Inc, New York, pp 227–248
Teng Y, Luo YM, Sun XH, Tu C, Xu L, Liu WX, Li ZG, Christie P (2010) Influence of arbuscular mycorrhiza and rhizobium on phytoremediation by alfalfa of an agricultural soil contaminated with weathered PCBs: a field study. Int J Phytoremediation 12:516–533
Theuerl S, Buscot F (2010) Laccases: toward disentangling their diversity and functions in relation to soil organic matter cycling. Biol Fertil Soils 46:215–225
Tigini V, Prigione V, Di Toro S, Fava F, Varese GC (2009) Isolation and characterisation of polychlorinated biphenyl (PCB) degrading fungi from a historically contaminated soil. Microb Cell Fact 8:14
Tsai TT, Kao CM, Yeh TY, Liang SH, Chien HY (2009) Remediation of fuel oil Contaminated soils by a three-stage treatment system. Environ Eng Sci 26:651–659
Tsumuraya Y (1996) Expansins, nonenzymatic proteins promoting cell wall expansion in higher plants. Trends Glycosci Glycotechnol 8:443–444
Valaskova V, Snajdr J, Bittner B, Cajthaml T, Merhautova V, Hoffichter M, Baldrian P (2007) Production of lignocellulose-degrading enzymes and degradation of leaf litter by saprotrophic basidiomycetes isolated from a Quercus petraea forest. Soil Biol Biochem 39:2651–2660
Vanden Wymelenberg A, Sabat G, Mozuch M, Kersten PJ, Cullen D, Blanchette RA (2006) Structure, organization, and transcriptional regulation of a family of copper radical oxidase genes in the lignin-degrading basidiomycete Phanerochaete chrysosporium. Appl Environ Microbiol 72:4871–4877
Vasilyeva GK, Strijakova ER (2007) Bioremediation of soils and sediments contaminated by polychlorinated biphenyls. Microbiology 76:639–653
Velicer GJ (2003) Social strife in the microbial world. Trends Microbiol 11:330–337
Wackett LP (2004) Evolution of new enzymes and pathways: soil microbes adapt to s-triazine herbicides. Pesticide Decontam Detoxification 863:37–48
Wackett LP (2008) Biomass to fuels via microbial transformations. Curr Opin Chem Biol 12:187–193

Wallenstein MD, Burns RG (2011) Extracellular enzyme activities and organic matter degradation in soil: a complex community driven process. In: Dick RP (ed) Methods in soil enzymology. Soil Science Society of America Inc, Madison Wisconsin USA, pp 35–55

Wallenstein MD, Weintraub MN (2008) Emerging tools for measuring and modeling the in situ activity of soil extracellular enzymes. Soil Biol Biochem 40:2098–2106

Whiteley CG, Lee DJ (2006) Enzymology of sludge-floc degradation using confocal laser scanning microscopy: an enzyme-floc model. Nova Science Publishers, Inc, Hauppauge

Wicke D, Reemtsma T (2010) Mobilization of hydrophobic contaminants from soils by enzymatic depolymerization of soil organic matter. Chemosphere 78:996–1003

Willison JC (2004) Isolation and characterization of a novel sphingomonad capable of growth with chrysene as sole carbon and energy source. FEMS Microbiol Lett 241:143–150

Wilson DB (2008) Three microbial strategies for plant cell wall degradation. In: Wiegel J et al (eds) Incredible anaerobes: from physiology to genomics to fuels, vol 1125. Blackwell Publishing, Oxford, pp 289–297

Wilson DB (2009) Cellulases and biofuels. Curr Opin Biotechnol 20:295–299

Witayakran S, Ragauskas AJ (2009) Modification of high-lignin softwood kraft pulp with laccase and amino acids. Enzyme Microb Technol 44:176–181

Yang SJ, Kataeva I, Hamilton-Brehm SD, Engle NL, Tschaplinski TJ, Doeppke C, Davis M, Westpheling J, Adams MWW (2009) Efficient degradation of lignocellulosic plant biomass, without pretreatment, by the thermophilic anaerobe "Anaerocellum thermophilum" DSM 6725. Appl Environ Microbiol 75:4762–4769

Zheng ZM, Obbard JP (2002) Oxidation of polycyclic aromatic hydrocarbons (PAH) by the white rot fungus, Phanerochaete chrysosporium. Enzyme Microb Technol 31:3–9

Zheng ZM, Obbard JP (2003) Oxidation of polycyclic aromatic hydrocarbons by fungal isolates from an oil contaminated refinery soil. Environ Sci Pollut Res 10:173–176

Chapter 3
Structure, Biological Activity and Environmental Fate of Insecticidal Bt (*Bacillus thuringiensis*) Cry Proteins of Bacterial and Genetically Modified Plant Origin

Nordine Helassa, Hervé Quiquampoix, and Siobhán Staunton

Abstract The properties and environmental fate of insecticidal proteins produced by the bacterium *Bacillus thuringiensis* are reviewed. These proteins, either produced directly by the bacterium and prepared in formulations of biopesticides or synthesised by genetically modified plants, have important roles to play in the control of agricultural pests and insect vectors of disease. Each strain of the bacterium produces a distinctive cocktail of crystal proteins, known as Cry proteins. The insecticidal properties of each protein are highly target species-specific. The mechanism of action of the proteins and differences between the proteins produced by the bacterium and by GM plants are discussed. Protein nomenclature and structure are briefly reviewed. The published literature on the interactions of Cry proteins with soils, soil minerals and reference clay minerals is reviewed. There is currently no link between protein structure and the extent of interaction with soil components. Hydrophobic interactions may play some part in these interactions. Both field studies and controlled laboratory studies of the persistence of Cry proteins in soil are analysed. There is ongoing debate as to the soil properties that determine the persistence of the proteins. There is very little information available on the environmental relevance of ongoing detection of the toxins in soil, since studies usually assay the protein after chemical extraction with no check on the insecticidal properties of the protein. New research directions are suggested on the basis of both the existing information on the environmental fate of Cry proteins in soil and the knowledge gaps identified.

Keywords *Bacillus thuringiensis* • Soil • Biopesticides • Genetically modified plants • Protein adsorption • Persistence

---

N. Helassa (✉) • H. Quiquampoix • S. Staunton (✉)
INRA-Eco&Sols (Ecologie Fonctionnelle et Biogéochimie des Sols et des Agrosystèmes), place Viala, 34060 Montpellier, France
e-mail: staunton@montpellier.inra.fr

J. Xu and D.L. Sparks (eds.), *Molecular Environmental Soil Science*, Progress in Soil Science, DOI 10.1007/978-94-007-4177-5_3,

3.1 Introduction

A large fraction of crop production worldwide is lost because of pests, despite the development of chemical pesticides and various management practices. Insecticidal proteins produced by the bacterium *Bacillus thuringiensis* have important roles to play in the protection of crops from pests and thus in the improvement of crop yield and quality. These proteins are currently used for pest control in two distinct modes; firstly and historically as biopesticides, used in both traditional and organic agriculture and to control insects in non agricultural contexts and secondly in genetically modified plants, commercialised since 1996 that account for rapidly increasing acreage worldwide. We shall review the environmental fate and action of these insecticides, emphasising the similarity and difference depending on the origin of the protein (biopesticide or biotechnology).

Bacillus thuringiensis (Bt) is a gram-positive spore-forming bacterium. It is ubiquitous, having been found in soil, insect remains and grain, although its precise ecological role is not fully understood. Since its discovery, various strains of Bt have been isolated from habitats including soil (Carozzi et al. 1991; Delucca et al. 1979; Hastowo et al. 1992; Martin and Travers 1989; Smith and Couche 1991), insects (Carozzi et al. 1991), butterfly scales or dust (Burges and Hurst 1977; Delucca et al. 1979; Meadows et al. 1992), and conifer needles (Kaelin et al. 1994; Smith and Couche 1991). Its entomopathogenic properties have made it the object of intense scientific study. During sporulation Bt produces parasporal crystals containing a large amount of insecticidal proteins, up to 30% by weight (Agaisse and Lereclus 1995). These proteins, known as Cry proteins, are highly specific to a narrow range of targets species. Each strain of Bt produces a small number of insecticidal proteins, usually between one and five. It is thought that the remains of insects killed by the Cry proteins provide a nutritious environment favouring germination of spores. *Bacillus thuringiensis* was first isolated in Japan by Ishiwata (Ishiwata 1901) in larvae from sick silk worms (*Bombyx mori*). It was later described by Berliner in 1911 in Thuringe (Germany), whence its name. *Bacillus thuringiensis* is one of a group of six bacilli, *Bacillus cereus*, along with *B. anthracis*, *B. cereus*, *B. mycoides*, *B. pseudomycoides* and *B. weihenstephanensis*. The similarity between *B. thuringiensis*, *B. cereus* and *B. anthracis* is such that it has even been suggested that they are a single species (Helgason et al. 2000). However the pathogenic properties of these spore-forming bacilli are very different (Crickmore 2006). There is no reported evidence of mammal pathogenicity of Bt, whereas *B. cereus* is a common human pathogen and *B. anthracis* is the cause of the often lethal disease, anthrax. The number of strains of Bt and of Cry proteins identified has increased over the decades and will probably continue to rise. A large collection of bacilli is available from the Bacillus Genetic Stock Center (www.bgsc.org). To date a limited number of strains have been used commercially as biopesticides and a very small of Cry proteins has been inserted in genetically modified plants.

Bt has been used in biopesticides for decades with no evidence of harmful side-effects and little concern for negative environmental impact. The toxins produced by genetically modified plants are considered to be substantially equivalent to those produced by bacteria, and indeed most risk assessment evaluations have been carried out using proteins of bacterial origin. However there is both scientific and public concern about use of genetically modified plants producing insecticidal proteins. This concern is fuelled by subtle differences in the toxin of bacterial and biotechnological origin, but more importantly by differences in the mode of introduction to the environment and hence exposure to target and non target organisms. Public concern also reflects varying degrees of suspicion or refusal of new technologies produced by powerful multinational companies. The level of concern is exacerbated by the rapid increase in agricultural land worldwide used for the production of genetically modified crops, leaving insufficient time for an appreciation of long-term, unexpected side effects.

Much of the research on the environmental fate of Cry proteins is directed towards the development of resistance and the effect on non target organisms. As for chemical pesticides, it is thought to be inevitable that target insects develop resistance to Bt toxins. There were early predictions of rapid and massive resistance, which has not been observed *in situ*. A high degree of resistance within a target population would decrease the efficiency of the toxin making the treatment ineffective, hence a loss of investment for the producers of the biopesticide or plant variety. Another consequence could be the transfer of resistance by horizontal gene flow to other organisms, thus exacerbating problems of crop management. Non target organisms can be directly affected by Bt toxin only if they are exposed and sensitive to the toxin. The absence of any observable effect of Bt toxins on non target organisms may thus reflect the absence of exposure, which could be a result of experimental design or could show that the organism is not naturally exposed to the toxin. Similarly, an organism that is exposed to the toxin, for example by consuming plant tissue containing the toxin, would not exhibit symptoms of toxicity if specific receptors are not present in its digestive system. Indirect effects are possible and indeed probable. The most obvious being that the removal of a pest from an ecosystem modifies the equilibrium of that ecosystem, freeing a niche for secondary pests, depriving predators of a food source and disrupting a complex food chain. Other effects may result from secondary differences in genetically modified plants. For example, it has been suggested that the lignin content of Bt maize is greater than that of non transformed varieties (Saxena and Stotzky 2001b). This modifies its digestibility by detrivorous species and hence possibly their growth rate, mortality and reproductive capacity and also the dynamics of soil microorganisms involved in the cycling of crop residues. An observable effect on a non target organism is thus not proof of toxicity.

A smaller volume of environmental research has been directed towards the monitoring of persistence of Bt, either the spores and bacteria applied as biopesticides or the toxin released into soil from genetically modified crops. The underlying scientific question may be the long-term advantage or hazard of the treatment, the probability of undesired effects (which may be assumed to increase with time of possible exposure) or more simply to establish monitoring programmes. A major hurdle that remains to be resolved for such research is the method of detection of Bt, particularly Bt toxin in soils.

In this review we aim to give a broad overview of the major scientific questions concerning the insecticidal properties of *Bacillus thuringiensis* as found in biopesticides and genetically modified plants. We shall place particular emphasis on the properties of these delta-endotoxins. Aspects of the structure, function, nomenclature and environmental fate of the bacillus and the toxin will be discussed. References will be given for more specialised reviews and studies.

3.2 Insecticidal Properties of Cry Proteins (*Bacillus thuringiensis*)

Cry proteins are active against insects of different orders, Lepidoptera, Coleoptera, Hymenoptera, Diptera and some nematodes, but all have the same mode of action with distinct steps (Gill et al. 1992; Gazit and Shai 1995; Schnepf 1995; Schnepf et al. 1998; de Maagd et al. 2001). First the crystal proteins must be ingested by the target insect and the crystals solubilised by the alkaline pH of the insect midgut. The proteins at this stage are protoxins, with molecular mass in the range 70–130 kDa. They are not toxic. Secondly the soluble proteins are enzymatically cleaved by proteases in the insect gut to produce toxins that have molecular masses of about 60 kDa. The principal proteases of Lepidoptera and Diptera are serine-like proteins, whereas those of Coleoptera are mainly cathepsin-like proteins. This insect-specific difference in digestive proteases contributes to the specificity of Cry proteins. The toxins are relatively resistant to further protease cleavage. These monomeric proteins undergo conformational changes to bind with specific receptors on membrane lipids and form oligomers, possibly tetramers. Pores are thus formed in the membranes leading to cell lysis, osmotic shock and death by septicaemia. Each of the three steps is highly species-specific and each step contributes to the strong species specificity of Bt toxins. The proteins produced by genetically modified Bt plants are the toxins, not the crystal protoxins, therefore two of the three steps are excluded, but the proteins are nevertheless found to express strong species specificity, as will be discussed later.

The C-terminus of the protoxin, cleaved to form the toxin, is thought to be important for the formation of crystals within the bacillus. Cry proteins are globular proteins made of three structural domains. Although the relationship between structure and function has not been firmly established, the essential features are generally agreed upon. Domain I is made up of α-helices and is important for pore formation. It has been pointed out that there are structural similarities between this domain and other pore-forming bacterial toxins, possibly suggesting a common origin (de Maagd et al. 2003). Domain II consists of anti-parallel β-sheets and is thought to interact with the membrane receptor. Domain III is a β-sandwich of two anti-parallel β-sheets and has roles in both pore formation and receptor binding. The intra- and inter-molecular salt bridges in domains II and III contribute to the stability of the protein conformation in the globular state.

3.3 Structure and Nomenclature of Cry Proteins

Two systems of classification and nomenclature of Cry proteins exist. The earlier classification is based on the primary structure of the proteins and their spectrum of target organisms (Höfte and Whiteley 1989). The proteins were assigned to four major classes to which a fifth was soon added

- CryI active against Lepidoptera
- CryII active against Lepidoptera and Diptera
- CryIII active against Coleoptera
- CryIV active against Diptera
- CryV active against Lepidoptera and Coleoptera

However the rapid increase in the number of genes coding for Cry proteins and the number of proteins identified, made this system of classification inadequate. There were an increasing number of inconsistencies between observed insect spectra and the degree of homology. Crickmore proposed a different classification based on the similarity of amino acid sequences (Crickmore et al. 1998). The code for the proteins is alpha-numerical with four ranks. A new (Arabic) numeral is assigned when there is <45% identity with a known protein, a new capital letter when identity is between 45 and 75% and a new lower case letter when identity is between 75 and 95%. The optional quaternary rank (a number) may distinguish proteins that are in fact identical. At the time of writing, 68 major classes and over 4000 proteins are listed in the full list of Cry proteins (Crickmore et al. 2010). Only a very small number of three-dimensional crystal structures have been determined by X-ray crystallography (Cry1Aa, Cry2Aa, Cry3Aa, Cry3Bb, Cry4Aa and Cry4Ba) (Bravo et al. 2007).

Although there is considerable variation of genetic sequences between the various classes of Cry toxin, the comparison of the three-dimensional structures of Cry1, Cry2 and Cry3 shows a high degree of similarity of structure and thus probably of functionality (de Maagd et al. 2001). The structural similarities are between of between 50 and 90% within the same rank and in the range 20–30% between ranks (Crickmore et al. 1998; Höfte and Whiteley 1989). The structures of Cry3A and Cry1Aa are very similar although they only share 36% similarity of protein sequences. The principal differences are at the level of domain II, which is implicated in receptor recognition. A comparison of the crystal structures of Cry1Aa and Cry1Ac shows major differences in domains II and III (Li et al. 2001). These differences probably explain the contrasting target specificity of the toxins. Cry2A has only 20% amino-acid sequence similarity with Cry3A and Cry1Aa, because of a very different N-terminal domain (propeptide). However, their three-dimensional structure remains very similar, suggesting that this N-terminal field does not play an important role in protein conformation (Parker and Feil 2005).

3.4 Detection of Bt and Bt Toxins in Soil

The persistence of *Bacillus thuringiensis* in soil may be monitored by germinating spores and using traditional microbiological techniques to culture and count bacteria. More recently DNA has been extracted from soil and analysed to monitor the presence of Bt, particularly genes inserted in insect resistant plants. The most common technique adopted to monitor the fate of Cry proteins released from genetically modified plants is chemical extraction of the protein from soil followed by immunochemical detection. The presence of Cry toxins and protoxins may be measured by biotests, using target insects. Each of these approaches has advantages and disadvantages.

The quantification of Bt spores and bacteria, as for all microbiological quantification assays, is limited by the culture of the bacillus in controlled conditions. *In situ*, spore counts are usually reported to decrease rapidly, however spores may remain dormant for long periods and germinate when environmental conditions become favourable.

The detection and quantification of Bt proteins in soil are subject to the same limitations as other biologically active proteins. Two approaches are possible. Biological activity may be measured and the result is a combined function of the concentration of protein and the expression of its activity in the presence of soil. Alternatively, the proteins may be desorbed after which various analytical techniques used for quantification. To date, no method of *in situ* detection exists (in the presence of soil) which is at the same time reliable, specific and reproducible.

Biotests require no sample pre-treatment and detect the active toxin which may not be the case for chemical and biochemical tests that may detect inactive fragments of the protein (Sims and Holden 1996; Sims et al. 1996). However, they are expensive, labour intensive, time consuming and subject to biological variability. Experimental artefacts may arise if the insects are not correctly and uniformly exposed to the toxin in the sample (they do not ingest the sample, or individuals consume differing amounts of the sample) or if the target sample is not initially uniform in size and stage of maturity. Furthermore there is no uniformly accepted method with reported endpoints being growth, mortality or both (Sims and Holden 1996; Sims and Ream 1997; Tapp and Stotzky 1998; Herman et al. 2002). Moreover, intrinsic biological variability with the larvae of insects led to data that are difficult to reproduce in intra-and interlaboratory trials. Furthermore, soil contaminated *in situ* may contain other toxins which could distort the results.

The alternative methods to detect toxin Bt in the grounds require a preliminary chemical extraction. In common with other soil analyses, a major limiting factor of these analyses is the first step of non destructive extraction of proteins or DNA before quantitative assaying. The extraction method described by (Palm et al. 1994) combines a high pH, a strong concentration of salts and a surfactant. These conditions mimic the conditions in the intestines of target insects and an extractant specificity designed to mimic gut conditions has been proposed (Shan et al. 2005). Extraction yields vary in the range 27–60% depending on the soil.

The soil dependent yield limits the comparison of Cry dynamics in different soils. Zwitterionic and nonionic surfactants such as CHAPS, Triton-X-100, and Tween 20 have been found to improve extraction yield more than anionic surfactants such as Sarkosyl, and SDS which may denature the protein and decrease its subsequent immunochemical recognition (Helassa et al. 2009). Extraction yields of trace amounts of Bt Cry protein have been further improved and differences observed between contrasting soils decreased when another protein, bovine serum albumin, was added to the extraction cocktail (Helassa et al. 2011b).

DNA can be assayed by various PCR reactions (Polymer Chain Reactions) after extraction from soil, to establish the persistence of the *Bacillus* and spores in soil. Cry proteins can be separated using by HPLC (High Pressure Liquid phase Chromatography) followed by UV-spectroscopic detection, however this technique is now rarely used because of high detection limit and analytical interferences (Clark et al. 2005). Flow cytometry has also been used to measure Cry protein concentrations, but the detection limit of this technique is too high for it to be used quantitatively for environmental samples (Tapp and Stotzky 1995a; 1997). Most authors now prefer immunological detection of type western blot (Sims et al. 1996) and now almost exclusively, ELISA (Enzyme-Linked ImmunoSorbent Assay) (Palm et al. 1994, 1996; Sims and Berberich 1996). Various ELISA-based kits are now commercially available for environmentally relevant proteins with low detection limits ($<$ng g^{-1} soil). ELISA assays are very sensitive and selective for Cry toxins. They are not subject to interferences, but further dilutions may be required to bring sample solutions to the linear range of quantification and the use of denaturing detergents that could reduce the immunochemical recognition of the protein must be avoided. Kits may detect various proteins from the same rank, for example, Cry1Aa, Cry 1Ab and Cry1Ac.

3.5 Agricultural Use of Bt Cry Toxins: Biopesticides

The interest for the biological pesticides began in the mid 1970s because of increasing insect resistance to chemical pesticides such as pyrethrinoid (Watkinson 1994). For over 20 years, one of the most successful applications of Bt has been the control of Lepidoptera in the Canadian and N. American forests, using the strain HD-1 that produces the toxins Cry1Aa, Cry1Ab, Cry1Ac and Cry2A, thereby significantly reducing the use of chemical insecticides (Lacey et al. 2001; Bauce et al. 2004; Crickmore 2006). Several million hectares of forest are pulverized every year by products derived from Btk (*kurstaki*) for the treatment against the Gypsy moth (*Lymantria dispar*) and the Spruce Budworm (*Choristoneura fumiferana*). A significant advantage of the use of Bt products is that insect acquisition of resistance to chemical pesticides has been considerably slowed because of the more limited application. For example, in Australia, the United States and Far East, they have been used against pyrethrinoid-resistant *Heliothis* ssp. and night butterflies. Chemical insecticides are thus used less often, the risk of resistance is decreased and these products can continue to be used effectively when required (Watkinson 1994). However, some cases of resistance to Bt toxin have been reported. Bti (*israelensis*) used for 80% of the control of Dipterous such as Black flies (*Simulium*) and the mosquito, which are vectors of many diseases in the African countries because of their resistance to the organophosphates (Bravo et al. 2007). This native strain is very active since the various toxins of its crystal inclusions function in a synergistic manner (Angsuthanasombat et al. 1992; Chang et al. 1993; Crickmore et al. 1995; Delecluse et al. 1993; Poncet et al. 1995; Wu et al. 1994). Similarly there are synergistic actions between certain toxins (Ravoahangimalala et al. 1993; Vanfrankenhuyzen et al. 1991), between toxins and bacteria (Dubois and Dean 1995) and between certain toxins and spores (Delecluse et al. 1993; Dubois and Dean 1995; Johnson and McGaughey 1996; Miyasono et al. 1994; Moar et al. 1995, 1989; Tang et al. 1996).

Commercial products containing the spores are thought to be more effective since spores can induce septicaemia, thus increasing the insect mortality rate (Watkinson 1994). In 1982, a new strain of Bt was developed, Btt (*tenebrionis*) by the Mycogen companies and Ecogen for the control of the Colorado beetle in the United States and in Europe (*Leptinotarsa decemlineata*). In 1992, Bta (*aizawai*) was developed by the Abbott laboratories for the control of the night butterfly (*Plutella xylostella*), which was resistant to chemical pesticides (Watkinson 1994). The formulations of Bt are varied: organic powders, granules, briquettes or suspensions, giving farmers more flexibility in the use of the product, particularly for mixtures of products (Watkinson 1994). The best known are Dipel®, Thuricide® and Delfin® but their use remains limited because of their short lifetime in soil (they are rapidly degraded by ultra-violet radiation and micro-organisms) and of their high cost (Roush 1994).

Biotechnology laboratories have developed various strains of bacteria transformed to express genes of insecticidal Cry. The first bacterium in which protein Bt genes were cloned is *Escherichia coli*, a bacterium that is easy to manipulate genetically. However, the expression yield was poor due to various regulatory factors (Broadwell 1994). Later, the Mycogen company in the United States developed a strain of Pseudomonas in which a gene of Btk was inserted and found to be stable (Watkinson 1994). This strain preserves its cellular membrane intact after death thus protecting the crystals from solar radiation. The performance in soil is thereby increased (Watkinson 1994). Another approach, developed by Crop Genetics International, is to transform an endophyte bacterium, *Clavibacter xyli* ssp. cynodontis with a gene of Bt. The impregnation of corn seeds by this bacterium allows the colonization of the vascular system of the adult plant, then protecting it from the European corn borer (*Ostrinia nubilalis*) (Watkinson 1994). Genes of Bt have also been inserted in algae for the control of the mosquitoes and in viruses for insect control taking advantage of the high level of infectivity of the viruses (Watkinson 1994). However, the best hosts for these genes remain *Bacillus*. Therefore, by cloning these genes into other species of *Bacillus* such as *Bacillus sphaericus*, the spectrum of hosts can be increased with possible synergistic actions between toxins and the spores (Broadwell 1994). In addition, new approaches of formulation are proposed, in particular the use of toxin in an adsorbed state, which is more effective against filter-feeding larvae of mosquitoes (Lee et al. 2003).

Bt, used as alternative or in addition to chemical pesticides in agriculture, is now the most widely used biopesticide on a global scale. More than 250 biopesticides in the world (representing 1% of the pesticide market are commercialised) and Bt accounts for approximately 80% of the sold biopesticides (Whalon and Wingerd 2003). Although the use of the biological agents is much lower than that of the chemical pesticides, the perceived environmental benefit of Bt products gives impetus to an increasing share of the market. The sales strategy of Bt products takes three major parameters into account (Bryant 1994):

- Physicochemical characteristics of Bt: the toxin becomes active only in the intestine of the target insects, thus any process interfering with the nutrition of the insects decreases effectiveness. Effective life-span is strongly decreased by the action of ultra-violet radiation and washing out by rainfall (Crecchio and Stotzky 1998; Koskella and Stotzky 1997). To improve the effectiveness of Bt, compounds were developed to allow a better penetration of the leaf surface, stabilising deposition by limiting rain washing and containing "UV-protectors" such as starch or other polymers (Brar et al. 2006; Cökmüs et al. 2000; McGuire et al. 1990).
- Mode of application: the amount of Bt applied should either be lethal or afford weaker chronic exposure. In addition, the zone of contact must be targeted according to the feeding pattern of the target insect.
- The behaviour of the target species: The amount to be applied is a function of the size of the larvae, a large larva requiring a larger exposure to the toxin (McGaughey 1978).

Since insects and the *B. thuringiensis* bacterium evolved together, resistance to Bt toxins is expected to be less than for synthetic chemical pesticides. However, studies as early as 1980 demonstrated the development of various levels of resistance by several species of insects exposed to crystal proteins of *Bacillus thuringiensis* under laboratory conditions (Ferré et al. 1995; Tabashnik 1994). In theory, the mechanisms of resistance to Bt can be found at each stage of the action of toxin: solubilisation, proteolytic activation, passage through the peritrophic membrane, binding with the receptor, insertion in the membrane, formation of the pore and osmotic lysis of the intestinal cells (Ferré and Van Rie 2002; Crickmore 2005). The most commonly observed imply a deterioration of the connection of toxin to the receptor (prevalent), a deterioration of proteolytic activation or a rapid regeneration of the intestinal epithelium (Ferré and Van Rie 2002). The first case of resistance was observed for Dipel® (Abbott Laboratories), a formulation of spores and crystals commercial of Btk HD-1, against *Plodia interpunctella*, Lepidoptera (McGaughey 1985). When the insect feeds on food containing Dipel® at a concentration that can produce 70–90% larval mortality, the insect develops a resistance to toxin. In only two generations, resistance increases by a factor 30 and at the end of 15 generations by a factor of up to 100. This resistance remains stable in absence of selective pressure. However, the insect remains sensitive to the other stocks of Bt suggesting that it acquires only the resistance of Cry contained in Dipel®.

3.6 Agricultural Use of Bt Cry Toxins: GM Crops

The incorporation of Bt genes in plants to protect against insects offers many advantages over both biopesticides and chemical pesticides (Peferoen 1997; Betz et al. 2000; Shelton et al. 2002). Since the toxin is produced by the plant to be protected, the drawbacks associated with application and limited life-span of biopesticides are avoided. In comparison with chemical pesticides, GM-produced Bt toxin is more target specific, is found directly in the plant to be protected thus limiting environmental contamination and does not produce toxic secondary products. These advantages should lead to agricultural cost reduction and improvements in crop quality and quantity. It is also claimed that the use of GM crops allows savings in energy consumption due to the reduced number of pesticide applications, but most of such data refer to herbicide resistant and not insect tolerant GM crops.

Several *cry* genes have been introduced into plants, starting with tobacco (Barton et al. 1987) followed by other species including potato, rice, corn and cotton (Adang et al. 1993; Fujimoto et al. 1993; Koziel et al. 1993; Nayak et al. 1997; Perlak et al. 1990, 1993). The first genetically modified plant was tobacco produced in 1983 using *Agrobacterium tumefasciens*, which is the causal agent of crown gall disease in over 140 dicot species. This was thus the first virus-resistant, herbicide-tolerant and insect-resistant GM crops (Llewellyn et al. 1994; Zambryski et al. 1983). Later techniques such as microinjection or cellular bombardment were developed, but the use of *Agrobacterium* remained preferred for the introduction of genes coding for Cry proteins (Finer and McMullen 1990). The *cry* genes inserted produce the protein directly in its toxic form because this proved more effective than whole proteins (Barton et al. 1987). However, further research was required to attain a high level of expression of Bt in the transgenic plants (Llewellyn et al. 1994). In common with corn, both cotton and potato are of great economic importance, the phytophage insect larvae that feed on these cultures were the subject of great attention.

Currently the most widespread GM plants produced by MONSANTO express the toxins Cry1A and Cry3A (Shelton et al. 2002). The most widespread varieties are the corn (MON810, Bt11, Event176) with 85% of Bt corn planted in the world being MON810; cotton (531, GK) and the potato (Russet Burbank). Bt corn is the most important culture of Bt plants in the world with 11.3 million hectares (James 2007). Corn MON810 expressing Cry1Ab, which targets the European corn borer, is the most widely grown Bt plant to date (Benedict and Ring 2004). The levels of Cry1Ab produced in Bt11 and MON810 vary from 3 to 10 $\mu g\ g^{-1}$ in the leaves and from 0.2 to 1.4 $\mu g\ g^{-1}$ in seeds, with the highest values for Bt11 (Icoz and Stotzky 2008). With regard to Bt cotton expressing Cry1Ac, the principal targets are *Helicoverpa zea*, *Pectinophora gossypiella* and *Heliothis virescens* (Benedict and Ring 2004). The quantity of Cry1Ac produced by Bt cotton is in the range 1.56–12.6 $\mu g\ g^{-1}$ in the leaves and from 0.86 to 12.7 $\mu g\ g^{-1}$ in seeds, depending on the transformation events (Icoz and Stotzky 2008).

In 1996, the first seeds of transgenic potato, cotton and corn were sold to farmers. The principal perceived advantages were continuous production of toxin and a longer persistence in plant material (Sims and Holden 1996; Sims and Ream 1997). These GM plants allowed a greater diversity of insects to be targeted, including sucking insects and nematodes, which were not targeted by simple biopesticide treatments. The applications of chemical insecticides could thus be considerably reduced and used only as complements. Adequate crop protection may thus be achieved with reduced production costs and less harmful effects on human health and the environment (Shelton et al. 2002). For example, the use of Bt cotton led to an 80% decrease in the average number of foliar applications of insecticides in the United States and 60–80% decreases in China between 1995 and 2001 (Romeis et al. 2006a, b). This success has led to spectacular increases in the agricultural land on which GM crops are grown. Between 1996 and 2009, this area has increased from less than one million hectares to 134 million hectares. Of this area, about a third is Bt transformed, of which cotton accounts for 53% and corn 46%.

Bt-crops, as for all GM crops were first commercially grown in the USA and the USA still accounts for the largest proportion of GM hectares worldwide (64 Mha of the 134 Mha planted to GM crops) (James 2009). The top five countries (USA, Brazil, Argentina, India and Canada) account for over 90% of planted area. The initial rapid increase in their use was mostly in both North and South America, and more recently rapid expansion has been observed in India, China and S. Africa. The take-up rate has been much slower in Europe. Only Spain has at least 0.1 Mha planted in GM crops (Bt-corn, representing 80% of the european GM Bt maize) (James 2009). The reasons for this are both social and commercial. There is considerable public distrust of genetically modified organisms and so both the European Commission and individual countries have limited or refused the introduction of GM crops. In addition, the target pests for currently available Bt crops are not responsible for major crop damage in all of the European countries and so the potential advantage of Bt crops (particularly corn) is more limited than in other agricultural zones. This may evolve as western corn rootworn, *Diabrotica virgifera*, a common N. American pest of corn has been introduced into Europe and changing climatic conditions will favour its development (Ciosi et al. 2008; Aragón et al. 2010).

In some studies, the use of Bt crops has been found to lead to marked improvements in crop quality. The reason is the decreased mycotoxin content of Bt crops (Betz et al. 2000; Abbas et al. 2008; Hammond et al. 2008; Barros et al. 2009; Ostry et al. 2010). Mycotoxins are produced by fungi that may colonise plants physically damaged by insects (Bennett and Klich 2003). When climatic conditions are favourable to fungal growth, the accumulation of mycotoxins may render crops unfit for human, and even animal, consumption (Doohan et al. 2003). Analytical techniques have only recently improved to allow the quantification of toxic levels of mycotoxins.

As for chemical pesticides, a major concern for the long-term effectiveness of Bt crops, is the development of insect resistance (Ferré and Van Rie 2002; Bourguet 2004; Tabashnik et al. 2004). The development of resistance was predicted to be enhanced by the continuous exposure of pests to the toxin in GM plants.

Although the early predictions of very rapid development of resistance have not been observed, efforts have been made to limit the appearance of this phenomenon.

Various strategies have been proposed to decrease the appearance of resistant insects. The combination of toxins with different modes of action, crop rotation, crops with very high expression levels and the use of spatial or temporal refuge should limit the emergence of resistant insect homozygotes (McGaughey and Whalon 1992; Roush 1994; Tabashnik 1994, 2001; Tabashnik et al. 2008). The refuge strategy is currently the most widely adopted management strategy. For this strategy to be effective, the frequency of the resistant alleles and the survival of the heterozygotes must be weak. Refuges allow sensitive alleles to be preserved within the population. However, the necessary size of these refuges remains subject for debate since they limit the economic benefit of the sales of transgenic seeds (Shelton et al. 2000, 2002). Theoretical models suggest that the use of two different toxins in the same plant would more delay the appearance of phenomena of resistance compared to the production of a single toxin, and would require smaller refuges (Roush 1998). This model has been confirmed for transgenic cotton and tobacco (Greenplate et al. 2000; Zhao et al. 2000, 1999). The second generation of GM plants contain more than one gene, being for example both herbicide resistant and insecticidal. One commercial variety, SmartStax™, already contains eight different genes and synthesises six insecticidal proteins: Cry1A.105, Cry1F, Cry2Ab, Cry3Bb1, Cry34 and Cry35Ab1 (James 2009). In Bt plants, genes will be stacked to express more than one toxin, with different membrane receptors, thus increasing mortality and decreasing the development of resistance (Moar 2003). Already stacked traits are commercially planted in 11 countries of which 8 are developing countries (James 2009).

3.7 Environmental Impact of Bt Crops

One of the most serious negative effects associated with the use of broad spectrum chemical pesticides is their toxicity towards the mammals. The large scale use of Bt cultures should reduce this risk (Hossain et al. 2004). Given exposure route of Bt toxin in genetically modified plants and the specificity of the various steps in the mechanism of action of Bt toxin it is thought that the risk of adverse effects on non target organisms is low. Furthermore the fact that the insect receptor sites do not exist in mammals means that they should not be sensitive to the toxin. When 14 insect species (Coleopteras, Dipteras, Homopteras, Hymenopteras, Lepidopteras, Neuropteras, Orthopteras) were exposed to Cry1Ac at a level 100 times that found in transgenic cotton, only Lepidopteras showed more than 50% mortality (Sims 1995). Although some reports have suggested negative impact on non target species, including the monarch butterfly (Losey et al. 1999) most reviews agree that the introduction of Bt crops has not caused a significant risk for non target insects (Gatehouse et al. 2002; Shelton et al. 2009). Nevertheless a large number of studies have been undertaken to evaluate the possible harmful effects of Bt toxins in

GM plants. These studies, in addition to the regulatory safety studies prior to the commercial introduction of GM crops, include large scale field studies and smaller scale controlled studies in the laboratory or green-house. The European Food Safety Authority (EFSA) has recently published guidelines on the evaluation of GM crops (Andersson et al. 2010). That report emphasises the necessity to consider the natural variability of alternative non-GM commercial varieties and points out the value of statistical analysis of many characteristics simultaneously (multi-variant analysis). Another fear is the risk of gene transfer to other plants (Vacher et al. 2003), but to date no evidence of this has been obtained for commercially grown GM crops (Dale et al. 2002). The environmental impact of Bt crops has been the object of several excellent reviews (O'Callaghan et al. 2005; Clark et al. 2005; Icoz and Stotzky 2008).

Earthworms are important soil organisms that play essential roles in the decomposition of organic matter and bioturbation of soil thereby transporting matter and solutes and contributing to physical structure and aeration. Icoz and Stotzky have summarized the studies of the effect of Cry toxins on earthworms. It is seen that Cry proteins added directly to soil or contained in GM plant material have little or no significant effect on earthworms (Icoz and Stotzky 2008). Other recent studies draw similar conclusions (Saxena and Stotzky 2001a; Clark and Coats 2006; Vercesi et al. 2006; Zwahlen et al. 2003b; Schrader et al. 2008). Such conclusions are however subject to caution since in feeding experiments it is difficult to know the exact exposure of animals to the toxin. If the contaminated material is not consumed, either because it is not in a form that enters its usual diet or because subtle changes in the material make it less delectable or less digestible than material not containing the toxin, then there was no exposure. It has been suggested that transgenic maize has a higher lignin content than its non transgenic equivalent making it less easily degradable and this could lead to an indirect adverse effect on earthworm growth (Saxena and Stotzky, 2001a).

Microarthropodes are key indicators of soil health and fertility. They are often present in the rhizosphere and would thus be very exposed to Cry proteins exuded by the roots of Bt plants. There is no published evidence of negative effects of Cry proteins on the reproduction and survival of collemboles *Folsomia candida* (Al-Deeb and Wilde 2003; Clark and Coats 2006; Sims and Ream 1997; Yu et al. 1997). Soil mites, *Oppia nitens*, do not appear to be negatively affected by Cry proteins (Al-Deeb and Wilde 2003; Yu et al. 1997). The species of macroarthropodes usually taken as a model species is *Porcellio scaber* which is active in decomposition of plant debris. Another study found no effect of Cry protein on this species (Escher et al. 2000).

Nematodes play roles in the metabolic activity of soil, including the mineralisation of nutrients and are valuable indicators of soil quality. Impact studies showed no significant effect on the biodiversity and the number of nematodes in the rhizosphere (Manachini and Lozzia 2002; Saxena and Stotzky 2001a). However, the proportions of bacteriophage, mycophage and phytophage nematodes vary on soils planted with Bt crops or not (Icoz and Stotzky 2008). This probably reflects changes in population dynamics that intervene in the food chain of nematodes because of successful pest control. In addition, the few studies relating to only one species of nematodes generally show negative effects of the Cry proteins on their growth and their reproduction (Icoz and Stotzky 2008).

Micro-organisms are essential soil organisms both in terms of biomass and activity. They are implicated in various processes including the decomposition of the organic matter and mineralisation of the elements. They account for more than 80% of the total biomass of soil (Kowalchuk et al. 2003) and are mainly contained in the rhizosphere (Ten times more than in the non-rhizospheric soil) (Icoz and Stotzky 2008). Any change in the nature and quantity of compounds exuded by the plants could induce a change of biodiversity and activity of micro-organisms. However, it is more difficult to establish a link between soil ecology, soil function and an ecological pressure, such as that which could be imposed by Bt plants (Lilley et al. 2006). The majority of the studies show only little or no significant effects of plants Bt on the structure of the microbial communities and their growth (Blackwood and Buyer 2004; Brusetti et al. 2004; Devare et al. 2004, 2007; Donegan et al. 1995; Ferreira et al. 2003; Flores et al. 2005; Griffiths et al. 2006; Koskella and Stotzky 2002; Naef et al. 2006; Saxena and Stotzky 2001a; Shen et al. 2006; Wu et al. 2004a). It appears that plant age and variety, as well as seasonal variations and environmental factors have greater effects on bacterial diversity that do Bt plants (Baumgarte and Tebbe 2005; Blackwood and Buyer 2004; Donegan et al. 1995; Fang et al. 2005; Griffiths et al. 2005, 2006; Naef et al. 2006). However, other authors show that the exudates from Bt plants change the structure of the bacterial communities (Brusetti et al. 2004; Castaldini et al. 2005; Rui et al. 2005; Xue et al. 2005) and significantly reduce the growth of an endomycorrhizal fungus, *Glomus mosseae*, (Turrini et al. 2004). Some studies point to some change in enzyme activities in soil, including phosphatases, deshydrogenases, ureases, invertases, cellulases and arylsulfatases, in the presence of Bt toxins (Sun et al. 2007; Wu et al. 2004a, b). However, no significant effect of Bt toxins on many other enzymes, microbial biomass as on various processes such as the potential mineralisation of nitrogen, nitrification and respiration were observed by others (Cortet et al. 2006; Devare et al. 2004, 2007, Flores et al. 2005; Shen et al. 2006).

On the basis of the data currently available, the Environmental Protection Agency (EPA) concludes that Bt toxins do not have any effect on total biomass, bacteria, nematodes, fungi, mammals, nontarget organisms and the environment. Furthermore, to date no specific effects of the transgenic varieties currently in commercial use have been identified (Cry1Ab, Cry1Ac). Risk assessment considers unintended effects of novel foodstuffs (Cellini et al. 2004). It has been claimed that differing regulatory frameworks leads to confusion in data analysis and comparison (Ramessar et al. 2009, 2010) GM plants may also have positive effects on the environment. The adoption of GM plants can considerably reduce the amount of pesticides used for equivalent crop quality and yield while preserving environment. Combining GM plants with chemical pesticides has been shown to be a good strategy to reduce emergence of resistance and emission of CO_2. Moreover, use of GM crops could be a good strategy to develop biofuels, that are more environmentally friendly than fossil-based fuels. Both biofuels and the reduction in the use of chemical pesticides can contribute to the reduction of greenhouse gases and thus mitigate against climate change (James 2009).

3.8 Fate of Bt Toxins in Soil: Interactions and Persistence

The fate of Cry proteins in soil may be determinant for the appearance of undesirable effects, and a good understanding of the environmental fate is essential for environmental monitoring. Reviews of the environmental fate of Cry toxins have been published recently. As one of these reviews points out existing data are subject to caution because analytical methods are not ideal (Clark et al. 2005). Since effects are not large, meta-analyses may be required for their detection (Marvier et al. 2007). Bt plants release toxins into soil by root exudation (Saxena et al. 1999; Saxena and Stotzky 2000; Saxena et al. 2004), by pollen deposition (Losey et al. 1999; Obrycki et al. 2001) and by the decomposition of plant residues (Stotzky 2004; Zwahlen et al. 2003a). These releases may lead to a potential accumulation of toxin in soil. If all the biomass of a transgenic plant was incorporated in soil, the resulting soil concentration would be about 1.6 μg Cry1Ac kg^{-1} for transgenic cotton, 280 μg Cry1Ab kg^{-1} for Bt11 corn and 210 μg Cry1Ab kg^{-1} for MON810 corn (Clark et al. 2005). In addition, these toxins have been found in the intestinal tract and the excrement of cows (Einspanier et al. 2004), of the decomposer *Porcellio scaber* (Pont and Nentwig 2005) and in earthworms (Saxena and Stotzky 2001a). Cry toxin has also been detected in both the tissues and faeces of the garden snail, *Helix aspersa*, although no accompanying effect on growth was noted (Kramarz et al. 2007; de Vaufleury et al. 2007). Bt corn decomposes more slowly in soil than an isogenic variety because of its higher lignin content (Saxena and Stotzky 2001b; Flores et al. 2005). This could lead to a longer residence period of the toxin in soil and hence a longer exposure period of non target organisms, with possible adverse effects, and exposure of target insects to sub lethal concentrations thereby increasing the selection pressure for the acquisition of resistance. However not all studies have concluded that Bt corn varieties have markedly different lignin contents, and consequently slower composition times, than non Bt-corn (Yanni et al. 2010).

Various studies of the fate of Bt toxin in soil show that after a lag period there is a period of rapid loss of up 90% detectable toxin, followed by a much slower decrease (Herman et al. 2002; Palm et al. 1994; Tapp and Stotzky 1998; West et al. 1984). Cry1A toxins can persist in soil for more than 200 days (Crecchio and Stotzky 2001; Saxena and Stotzky 2001a; Saxena et al. 2002; Sims and Holden 1996; Tapp and Stotzky 1998; Vettori et al. 2003) with half-lives usually in the range 1.6–22 days but half-lives of up to 46 days have been reported (Shelton et al. 2002). The initial rapid decline has been attributed to the metabolisation of the toxin by the micro-organisms, and the second stage determined by the protection of the toxin by clays and organic matter (Palm et al. 1996). Any free toxin would be rapidly used as source of carbon and energy by soil micro-organisms, in contrast to adsorbed toxin.

Cry1Ab has been reported to have a shorter half-life in water than in soil as would be expected (Douville et al. 2005). The persistence of protein is determined by its rapid and quasi-irreversible adsorption to soil components. Adsorption is thought to confer physical protection since protein residues are rendered inaccessible for cleavage by proteases (Chevallier et al. 2003; Crecchio and Stotzky 2001; Koskella and Stotzky 1997; Tapp and Stotzky 1995a, b). In contrast, other studies have concluded that Bt toxins from maize are very bioavailable and easily degraded by micro organisms, thus they do not accumulate in soil (Dubelman et al. 2005; Hopkins and Gregorich 2003).

Stotzky and coworkers report faster loss of detectable protein at alkaline pH due to the pH dependency of both microbial activity and adsorption (Crecchio and Stotzky 2001; Lee et al. 2003; Tapp et al. 1994; Tapp and Stotzky 1998; Venkateswerlu and Stotzky 1992). Few studies have investigated the effect of soil pH with environmentally relevant levels of toxin. A shorter half-time of Cry1Ab from Bt rice has been reported in alkaline soil (11.5 days) compared to acid soil (34.4 days) (Wang et al. 2006). It should be noted that the soils were not identical apart from their pH and so the observed difference could have resulted from their microbial populations or their texture, mineralogical and chemical properties. No significant effect on the dynamics of Cry1Aa in 4 contrasting soils was observed when the pH was adjusted (Helassa et al. 2011b).

The variability in the reported residence periods of plant derived Bt toxin in soil certainly arises from differences in the level of toxin in plants, the nutritional status of soil and the physico-chemical and mineralogical composition (pH, clay content and type, organic matter content and type) (Icoz and Stotzky 2008). Persistence also depends on the initial concentration, on the rate of inactivation and degradation by UV radiation as well as soil microbes and the rate of consumption by the insect larvae (Crecchio and Stotzky 1998; Koskella and Stotzky 1997; Stotzky 2000). Nevertheless, the current understanding of the dynamics of Cry toxins, and indeed the variation between different Cry proteins, makes it impossible to predict the kinetics of release of Cry protein from GM plants and the subsequent residence in soil.

3.9 Interaction with Reference Clays and Soil Extracted Clays

Proteins are macromolecules with pH dependant charge that usually have a great affinity for surfaces. Interaction with surfaces depends on protein structure and conformation and varies with pH. Adsorption of proteins is often at its maximum near the isoelectric point on charged surfaces such as clays. The isoelectric point of Bt proteins is often between 4.4 and 5.5 (Bietlot et al. 1989; Venkateswerlu and Stotzky 1992). For pH values above the isoelectric point, the repulsive forces between like-charged clay and protein tend to reduce adsorption. At pH below the isoelectric point the attractive interaction between the proteins and the surface may lead to an increase in the points of contact and changes in conformation leading to spreading of the protein on the surface and a decrease in the amount of protein adsorbed (Quiquampoix 2008; Quiquampoix and Ratcliffe 1992; Quiquampoix et al. 1993). Proteins may be classified as hard (not likely to undergo conformational changes) or soft (likely to undergo conformational changes) (Haynes 1994). Only a small number of studies have attempted to monitor conformational changes of Cry proteins after adsorption on surfaces.

The adsorption of Bt insecticidal toxins on reference clays such as homoionic montmorillonite and kaolinite (Tapp et al. 1994; Venkateswerlu and Stotzky 1992; Helassa et al. 2009), on clay-sized fractions of soils, and soils (Chevallier et al. 2003; Muchaonyerwa et al. 2002; Sundaram 1996; Tapp and Stotzky 1995a), is rapid (less than 1 h), not strongly temperature dependent and generally of weak affinity (L-type). Adsorption is significantly lower on kaolinite than on montmorillonite because of its lower specific surface and cation exchange capacity and may vary with the nature of the charge compensating cation (Tapp et al. 1994; Tapp and Stotzky 1995b; Helassa et al. 2009). Adsorption capacity is estimated or measured from the shape of the adsorption isotherm and so large concentrations of the protein in solution are usually required. This may led to experimental artefacts, since Cry proteins have a strong tendency to oligomerise and polymerise, unless alkaline pH and large salt concentrations are maintained in solution. The loss of protein from solution due to time dependant precipitation may be confused with adsorption. Affinity may be measured at lower solution concentrations since it may be assessed from the slope of an adsorption isotherm, the fraction adsorbed, the initial rate of adsorption kinetics or the solid–liquid distribution coefficient.

Adsorption capacity decreases with increasing pH above the isoelectric point (Helassa et al. 2009). In that study, lower pH values were not investigated to avoid polydispersivity and insolubility of the protein. There is also evidence that adsorption affinity for various minerals decreases with increasing pH (Zhou et al. 2007). At very high pH, above ten, decreased affinity due to the formation of oligomers formed with di-sulphide bonds between monomers has been reported (Zhao et al. 2000). Decreased affinity of trace amounts of Cry1Ab on montmorillonite has been observed when pH was increased beyond six (Pagel-Wieder et al. 2007). Adsorption of Cry1Ab on quartz was seen to decrease with increasing pH above the isoelectric point (Sander et al. 2010) and the authors concluded that electrostatic interactions governed adsorption. A greater affinity of Cry1Aa on mica acid-washed hydrophilic glass at pH near the isoelectric point than at alkaline pH has also been reported, showing the importance of electrostatic interactions (Janot et al. 2010). However these authors also found stronger adsorption on hydrophobic glass than hydrophilic glass, indicating that hydrophobic interactions may be important in some circumstances.

Soil clays are coated by organic matter and mineral oxides. The chemical removal of mineral oxides from soil clays has been found to increase the adsorption of Cry toxin (Crecchio and Stotzky 2001; Muchaonyerwa et al. 2006). In contrast, the chemical removal of organic matter has been found to decrease adsorption to an extent that depends on the quantity and quality of the organic matter (Muchaonyerwa et al. 2002, 2006). A trend for affinity of Cry1Ab for different soils to decrease with increasing organic matter content has been observed (Pagel-Wieder et al. 2007) report a general. The interaction of Bt toxin with montmorillonite coated with organic matter also follows an isotherm of weak affinity type (Giles et al. 1974). Adsorption is rapid, and 70% complete within 1 h (Crecchio and Stotzky 2001). As for reference minerals, the greatest adsorption is observed around the isoelectric point, pH 5–6, and decreases with increasing pH above this value. Remarkably similar conclusions were drawn from studies of the interaction of Cry toxin with humic acids (Crecchio and Stotzky 1998). Adsorption was found to be rapid. Washing with water removed between 20 and 55% of adsorbed toxin, depending on the nature of the humic acid. Toxin adsorbed on humic acid had roughly the same toxicity, as assessed by the lethal concentration, LC50 as soluble protein and appears to be protected against microbial breakdown.

Protein adsorption is often observed to be completely or partially irreversible, and this irreversibility increases with time of contact between the protein and the surface. The probable mechanism is changing conformation of the adsorbed protein leading to increasing points of contact between the protein and the surface and an increased specific interfacial area. The currently available data indicate different degrees of reversibility of Cry proteins. This difference may arise from the nature of the surface investigated and the experimental conditions. It should also be borne in main that experimental conditions may lead to artefacts and misinterpretation, for example, the quantification of protein remaining in solution in contact with moist mineral after the adsorption step and non linear detection of the protein in solution. Considerable reversibility of adsorption for Cry protein on pure clays (30%) has been reported (Tapp et al. 1994). More recently, adsorption of a Cry protein on quartz was reported to be completely reversible (Sander et al. 2010). Other studies report less desorption by washing with water or buffer solution (Lee et al. 2003; Muchaonyerwa et al. 2006; Tapp and Stotzky 1995b; Venkateswerlu and Stotzky 1992; Crecchio and Stotzky 2001; Zhou et al. 2005). Desorption from montmorillonite was increased by the addition of surfactant, sodium dodecyl sulphate (SDS) or β-mercaptoethanol (Venkateswerlu and Stotzky 1992). Solutions of alkaline pH containing a surfactant are used to desorb Cry proteins prior to quantification (Palm et al. 1996). It has been shown that detergents differ in their improvement of extraction yield of Cry1Aa from clays (Helassa et al. 2009) and soils (Helassa et al. unpublished). More recently, the addition of another protein to the desorption solution has been shown to improve the extraction yield from soils with contrasting compositions (Helassa et al. 2011b).

Cry proteins have a strong tendency to oligomerise. Both dynamic light scattering and infrared spectroscopy have been used to follow oligomerisation in solution (Masson et al. 2002). The tendency to oligomerise may be related to their mode of action in contact with membrane receptors where oligomers must be formed to create pores leading to cell lysis in the target insect. Some authors report the precautions taken to remove polymers formed in concentrated stock solutions prior to experiments, but few have considered the polydispersivity of clear solutions and the different properties of monomers and soluble oligomers. Centrifugation or filtration of turbid solutions prior to use will have removed all or most of the large polymers but certainly left in solution a mixture of monomers and various oligomers. Environmentally relevant protein concentrations are low and so it is likely that they remain in a monomeric state. Only after contact with the membrane receptors in intestinal cells of the target insects would they oligomerise. When more concentrated solutions are studied, as may be required by analytical techniques, high pH and large salt concentration are required to prevent oligomerisation (Masson et al. 2002; Helassa et al. 2009). Masson and coworkers also reported that CAPES is the ideal buffer to store Cry toxins, not only because of the high pH but also because it stabilises the conformation of the protein in the monomeric form thereby preventing oligomerisation. We have recently demonstrated that adsorption of Cry1Aa on montmorillonite inhibits the process of oligomerisation that is rapid in solution (Helassa et al. 2011a). The adsorption properties of oligomeric and monomeric Cry1Aa protein have been shown to differ (Janot et al. 2010). The oligomer showed greater affinity for glass and mica surfaces. Furthermore, conformational changes during oligomerisation favoured interaction with hydrophobic surfaces.

Insecticidal activity has been reported to be preserved, either completely or partially, following adsorption on clay minerals (Crecchio and Stotzky 1998, 2001; Koskella and Stotzky 1997; Lee et al. 2003; Stotzky 2000; Tapp and Stotzky 1995b; Venkateswerlu and Stotzky 1992; Zhou et al. 2007). Insecticidal activity has been observed as long as months or years after soil contamination (Vettori et al. 2003; Saxena et al. 2002). This means that soil organisms will be exposed to the toxin long after the protein is released into soil and adsorbed. However more studies are required to investigate the long term dynamics of insecticidal activity in the adsorbed state, and to investigate how they vary as a function of contrasting soil organo-mineral surfaces. The preservation of insecticidal activity in the adsorbed state suggests either that protein is desorbed in the insect digestive tract or that conformation of the whole protein or of the domain responsible for toxicity is not greatly modified by adsorption. A recent study of Cry1Aa adsorbed on montmorillonite by Fourier Transform infrared spectroscopy confirms that conformational changes in the adsorbed state are small (Helassa et al. 2011a) and FRAP experiments demonstrated that the toxin is not mobile once adsorbed (Helassa et al. 2010). A recent study on quartz has concluded that changes in Cry protein conformation following adsorption are negligible (Sander et al. 2010). There are no published studies of the desorption of Cry proteins in media that mimic the conditions of the digestive tract of insects. However the alkaline pH conditions would favour desorption (Helassa et al. 2009; Zhou et al. 2007). As reported for other proteins, adsorption on mineral surfaces is thought to protect the Cry proteins from microbiological degradation (Chevallier et al. 2003; Crecchio and Stotzky 1998, 2001; Koskella and Stotzky 1997). This observation would further contribute to an extended exposure of soil organisms to the toxin.

As stated above, it is generally assumed that while inactivation of Bt in biopesticides containing spores and crystals is determined by competition with soil microflore and UV-inactivation, the decrease in toxins from GM plants is limited

Carozzi NB, Kramer VC, Warren GW et al (1991) Prediction of insecticidal activity of *Bacillus thuringiensis* strains by polymerase phain reaction product profiles. Appl Environ Microbiol 57:3057–3061

Castaldini M, Turrini A, Sbrana C et al (2005) Impact of Bt corn on rhizospheric and on beneficial mycorrhizal symbiosis and soil eubacterial communities iosis in experimental microcosms. Appl Environ Microbiol 71:6719–6729

Cellini F, Chesson A, Colquhoun I et al (2004) Unintended effects and their detection in genetically modified crops. Food Chem Toxicol 42:1089–1125

Chang C, Yu YM, Dai SM et al (1993) High level cryIVD and cytA gene expression in *Bacillus thuringiensis* does not require the 20-kilodalton protein, and the coexpressed gene products are synergistic in their toxicity to mosquitos. Appl Environ Microbiol 59:815–821

Chevallier T, Muchaonyerwa P, Chenu C (2003) Microbial utilisation of two proteins adsorbed to a vertisol clay fraction: toxin from *Bacillus thuringiensis* subsp *tenebrionis* and bovine serum albumin. Soil Biol Biochem 35:1211–1218

Ciosi M, Miller NJ, Kim KS et al (2008) Invasion of Europe by the western corn rootworm, *Diabrotica virgifera virgifera*: multiple transatlantic introductions with various reductions of genetic diversity. Mol Ecol 17:3614–3627

Clark BW, Coats JR (2006) Subacute effects of CrylAb Bt corn litter on the earthworm *Eisenia fetida* and the springtail *Folsomia candida*. Environ Entomol 35:1121–1129

Clark BW, Phillips TA, Coats JR (2005) Environmental fate and effects of *Bacillus thuringiensis* (Bt) proteins from transgenic crops: a review. J Agric Food Chem 53:4643–4653

Cökmüs C, Sayar AH, Sacilik SC et al (2000) Effects of UV-light on *Bacillus sphaericus* and its protection by chemicals. J Basic Microbiol 40:215–221

Cortet J, Andersen MN, Caul S et al (2006) Decomposition processes under Bt (*Bacillus thuringiensis*) maize: results of a multi-site experiment. Soil Biol Biochem 38:195–199

Crecchio C, Stotzky G (1998) Insecticidal activity and biodegradation of the toxin from *Bacillus thuringiensis* subsp. *kurstaki* bound to humic acids from soil. Soil Biol Biochem 30:463–470

Crecchio C, Stotzky G (2001) Biodegradation and insecticidal activity of the toxin from *Bacillus thuringiensis* subsp. *kurstaki* bound on complexes of montmorillonite-humic acids-Al hydroxypolymers. Soil Biol Biochem 33:573–581

Crickmore N (2005) Using worms to better understand how *Bacillus thuringiensis* kills insects. Trends Microbiol 13:347–350

Crickmore N (2006) Beyond the spore – past and future developments of *Bacillus thuringiensis* as a biopesticide. J Appl Microbiol 101:616–619

Crickmore N, Bone EJ, Williams JA et al (1995) Contribution of the individual components of the delta-endotoxin crystal to the mosquitocidal activity of *Bacillus thuringiensis* subsp. *israelensis*. FEMS Microbiol Lett 131:249–254

Crickmore N, Zeigler DR, Feitelson J et al (1998) Revision of the nomenclature for the *Bacillus thuringiensis* pesticidal crystal proteins. Microbiol Mol Biol Rev 62:807–813

Crickmore N, Zeigler DR, Schnefp E et al (2010) *Bacillus thuringiensis* toxin nomenclature. http://www.lifesci.sussex.ac.uk/Home/Neil_Criclmore/Bt/

Dale PJ, Clarke B, Fontes EMG (2002) Potential for the environmental impact of transgenic crops. Nat Biotechnol 20:567–574

de Maagd RA, Bravo A, Crickmore N (2001) How *Bacillus thuringiensis* has evolved specific toxins to colonize the insect world. Trends Genet 17:193–199

de Maagd RA, Bravo A, Berry C et al (2003) Structure, diversity, and evolution of protein toxins from spore-forming entomopathogenic bacteria. Annu Rev Genet 37:409–433

de Vaufleury A, Kramarz PE, Binet P et al (2007) Exposure and effects assessments of Bt-maize on non-target organisms (gastropods, microarthropods, mycorrhizal fungi) in microcosms. Pedobiologia 51:185–194

Delecluse A, Poncet S, Klier A et al (1993) Expression of cryIVA and cryIVB genes, independently or in combination, in a crystal negative strain of *Bacillus thuringiensis* subsp. *israelensis*. Appl Environ Microbiol 59:3922–3927

Delucca AJ, Simonson J, Larson A (1979) Two new serovars of *Bacillus thuringiensis* – serovars Dakota and Indiana (serovars 15 and 16). J Invertebr Pathol 34:323–324

Devare MH, Jones CM, Thies JE (2004) Effect of Cry3Bb transgenic corn and tefluthrin on the soil microbial community: biomass, activity, and diversity. J Environ Qual 33:837–843

Devare M, Londono-R LM, Thies JE (2007) Neither transgenic Bt maize (MON863) nor tefluthrin insecticide adversely affect soil microbial activity or biomass: a 3-year field analysis. Soil Biol Biochem 39:2038–2047

Donegan KK, Palm CJ, Fieland VJ et al (1995) Changes in levels, species and DNA fingerprints of soil microorganisms associated with cotton expressing the *Bacillus thuringiensis* var. *kurstaki* endotoxin. Appl Soil Ecol 2:111–124

Doohan FM, Brennan J, Cooke BM (2003) Influence of climatic factors on Fusarium species pathogenic to cereals. Eur J Plant Pathol 109:755–768

Douville M, Gagne F, Masson L et al (2005) Tracking the source of *Bacillus thuringiensis* Cry1Ab endotoxin in the environment. Biochem Syst Ecol 33:219–232

Dubelman S, Ayden BR, Bader BM et al (2005) Cry1Ab protein does not persist in soil after 3 years of sustained Bt corn use. Environ Entomol 34:915–921

Dubois NR, Dean DH (1995) Synergism between CryIA insecticidal crystal proteins and spores of *Bacillus thuringiensis*, other bacterial spores, and vegetative cells against *Lymantria dispar* (Lepidoptera: Lymantriidae) larvae. Environ Entomol 24:1741–1747

Einspanier R, Lutz B, Rief S et al (2004) Tracing residual recombinant feed molecules during digestion and rumen bacterial diversity in cattle fed transgene maize. Eur Food Res Technol 218:269–273

Escher N, Kach B, Nentwig W (2000) Decomposition of transgenic *Bacillus thuringiensis* maize by microorganisms and woodlice *Porcellio scaber* (Crustacea: Isopoda). Basic Appl Ecol 1:161–169

Fang M, Kremer RJ, Motavalli PP et al (2005) Bacterial diversity in rhizospheres of nontransgenic and transgenic corn. Appl Environ Microbiol 71:4132–4136

Ferré J, Van Rie J (2002) Biochemistry and genetics of insect resistance to *Bacillus thuringiensis*. Annu Rev Entomol 47:501–533

Ferré J, Escriche B, Bel Y et al (1995) Biochemistry and genetics of insect resistance to *Bacillus thuringiensis* insecticidal crystal proteins. FEMS Microbiol Lett 132:1–7

Ferreira L, Molina JC, Brasil C et al (2003) Evaluation of *Bacillus thuringiensis* bioinsecticidal protein effects on soil microorganisms. Plant Soil 256:161–168

Finer JJ, McMullen MD (1990) Transformation of cotton (*Gossypium hirsutum* L.) via particle bombardment. Plant Cell Rep 8:586–589

Flores S, Saxena D, Stotzky G (2005) Transgenic Bt plants decompose less in soil than non-Bt plants. Soil Biol Biochem 37:1073–1082

Fujimoto H, Itoh K, Yamamoto M et al (1993) Insect resistant rice generated by introduction of a modified delta-endotoxin gene of *Bacillus thuringiensis*. Bio/Technology 11:1151–1155

Gatehouse AMR, Ferry N, Raemaekers RJM (2002) The case of the monarch butterfly: a verdict is returned. Trends Genet 18:249–251

Gazit E, Shai Y (1995) The assembly and arganization of the alpha-5 and alpha-7 helices from the pore-forming domain of *Bacillus thuringiensis* delta-endotoxin – relevance to a functional model. J Biol Chem 270:2571–2578

Giles CH, D'Silva AP, Easton IA (1974) A general treatment and classification of the solute adsorption isotherm: part: II. Experimental interpretation. J Colloid Interface Sci 47:766–778

Gill SS, Cowles EA, Pietrantonio PV (1992) The mode of action of *Bacillus thuringiensis* endotoxins. Annu Rev Entomol 37:615–636

Greenplate JT, Penn SR, Shappley Z et al (2000) Bollgard II efficacy: quantification of total lepidopteran activity in a 2-gene product. 2000 Proceedings beltwide cotton conferences, vol 2, pp 1041–1043, San Antonio, 4–8 January 2000

Griffiths BS, Caul S, Thompson J et al (2005) A comparison of soil microbial community structure, protozoa and nematodes in field plots of conventional and genetically modified maize expressing the *Bacillus thuringiens* is CryIAb toxin. Plant Soil 275:135–146

Griffiths BS, Caul S, Thompson J et al (2006) Soil microbial and faunal community responses to Bt maize and insecticide in two soils. J Environ Qual 35:734–741

Hammond B, Campbell K, Cea J et al (2008) Opportunities for mycotoxin reduction in maize using biotechnology. In: Siantar DP et al (eds) Food contaminants: mycotoxins and food allergens, vol 1001. American Chemical Society, Washington, DC, pp 109–124

Hastowo S, Lay BW, Ohba M (1992) Naturally occurring *Bacillus thuringiensis* in Indonesia. J Appl Bacteriol 73:108–113

Haynes CA (1994) Globular proteins at solid/liquid interaces. Colloids Surf B Biointerfaces 2:517–566

Helassa N (2008) Devenir de la protéine insecticide Cry1Aa issue de *Bacillus thuringiensis* (Bt) dans le sol. Thèse de doctorat, Montpellier SupAgro

Helassa N, Daudin G, Quiquampoix H, Noinville S, Janot JM, Déjardin P & Staunton S (2010) Mobility of adsorbed Cry1Aa insecticidal toxin from *Bacillus thuringiensis* (Bt) on montmorillonite measured by Fluorescence Recovery After Photobleaching (FRAP). Phil Mag 90:2365–2371

Helassa N, Revault M, Quiquampoix H, Dejardin P, Staunton S Noinville S (2011a) Adsorption on Montmorillonite prevents oligomerization of Bt Cry1Aa Toxin. J Coll Interface Sci 356: 718–725

Helassa N, M'Charek A, Quiquampoix H, Noinville S, Dejardin P, Frutos R & Staunton S (2011b) Effects of physicochemical interactions and microbial activity on the persistence of Cry1Aa Bt (*Bacillus thuringiensis*) toxin in soil. Soil Biol Biochem 43:1089–1097

Helassa N, Quiquampoix H, Noinville S et al (2009) Adsorption and desorption of monomeric Bt (*Bacillus thuringiensis*) Cry1Aa toxin on montmorillonite and kaolinite. Soil Biol Biochem 41:498–504

Helgason E, Okstad OA, Caugant DA et al (2000) *Bacillus anthracis*, *Bacillus cereus*, and *Bacillus thuringiensis* – one species on the basis of genetic evidence. Appl Environ Microbiol 66:2627–2630

Herman RA, Wolt JD, Halliday WR (2002) Rapid degradation of the Cry1F insecticidal crystal protein in soil. J Agric Food Chem 50:7076–7078

Höfte H, Whiteley HR (1989) Insecticidal crystal proteins of *Bacillus thuringiensis*. Microbiol Rev 53:242–255

Hopkins DW, Gregorich EG (2003) Detection and decay of the Bt endotoxin in soil from a field trial with genetically modified maize. Eur J Soil Sci 54:793–800

Hossain F, Pray CE, Lu YM et al (2004) Genetically modified cotton and farmers ' health in China. Int J Occup Environ Health 10:296–303

Icoz I, Stotzky G (2008) Fate and effects of insect-resistant Bt crops in soil ecosystems. Soil Biol Biochem 40:559–586

Ishiwata S (1901) On a kind of severe flacherie (sotto disease). Dainihon Sanshi Kaiho 114:1–5

James C (2007) Global status of commercialized biotech/GM crops: 2007. ISAAA Briefs, Ithaca

James C (2009) Global status of commercialized biotech/GM crops: 2009. ISAAA Briefs, Ithaca

Janot JM, Boissiere M, Thami T et al (2010) Adsorption of alexa-labeled Bt toxin on mica, glass, and hydrophobized glass: study by normal scanning confocal fluorescence. Biomacromolecules 11:1661–1666

Johnson DE, McGaughey WH (1996) Contribution of *Bacillus thuringiensis* spores to toxicity of purified cry proteins towards Indianmeal moth larvae. Curr Microbiol 33:54–59

Kaelin P, Morel P, Gadani F (1994) Isolation of *Bacillus thuringiensis* from stored tobacco and *Lasioderma serricorne* (F). Appl Environ Microbiol 60:19–25

Koskella J, Stotzky G (1997) Microbial utilization of free and clay-bound insecticidal toxins from *Bacillus thuringiensis* and their retention of insecticidal activity after incubation with microbes. Appl Environ Microbiol 63:3561–3568

Koskella J, Stotzky G (2002) Larvicidal toxins from *Bacillus thuringiensis* subspp. *kurstaki*, *morrisoni* (strain *tenebrionis*), and *israelensis* have no microbicidal or microbiostatic activity against selected bacteria, fungi, and algae in vitro. Can J Microbiol 48:262–267

Kowalchuk GA, Bruinsma M, van Veen JA (2003) Assessing responses of soil microorganisms to GM plants. Trends Ecol Evol 18:403–410

Koziel MG, Beland GL, Bowman C et al (1993) Field performance of elite transgenic maize plants expressing an insecticidal protein derived from *Bacillus thuringiensis*. Bio/Technology 11:194–200

Kramarz PE, De Vaufleury A, Carey M (2007) Studying the effect of exposure of the snail *Helix aspersa* to the purified Bt toxin, Cry1Ab. Appl Soil Ecol 37:169–172

Lacey LA, Frutos R, Kaya HK et al (2001) Insect pathogens as biological control agents: do they have a future? Biol Control 21:230–248

Lee L, Saxena D, Stotzky G (2003) Activity of free and clay-bound insecticidal proteins from *Bacillus thuringiensis* subsp. *israelensis* against the mosquito *Culex pipiens*. Appl Environ Microbiol 69:4111–4115

Li J, Derbyshire DJ, Promdonkoy B et al (2001) Structural implications for the transformation of the *Bacillus thuringiensis* delta-endotoxins from water-soluble to membrane-inserted forms. Biochem Soc Trans 29:571–577

Lilley AK, Bailey MJ, Cartwright C et al (2006) Life in earth: the impact of GM plants on soil ecology? Trends Biotechnol 24:9–14

Llewellyn D, Cousins Y, Mathews A et al (1994) Expression of *Bacillus thuringiensis* insecticidal protein genes in transgenic crop plants. Agric Ecosyst Environ 49:85–93

Losey JE, Rayor LS, Carter ME (1999) Transgenic pollen harms monarch larvae. Nature 399:214–214

Manachini B, Lozzia GC (2002) First investigations into the effects of Bt corn crop on Nematofauna. Bollettino di Zoologia Agraria e di Bachicoltura 34:85–96

Martin PAW, Travers RS (1989) Worldwide abundance and distribution of *Bacillus thuringiensis* isolates. Appl Environ Microbiol 55:2437–2442

Marvier M, McCreedy C, Regetz J et al (2007) A meta-analysis of effects of Bt cotton and maize on nontarget invertebrates. Science 316:1475–1477

Masson L, Mazza A, Sangadala S et al (2002) Polydispersity of *Bacillus thuringiensis* Cry1 toxins in solution and its effect on receptor binding kinetics. Biochim Biophys Acta 1594:266–275

McGaughey WH (1978) Response of *Plodia interpunctella* Lepidoptera-Pyralidae and *Ephestia cautella* Lepidoptera-Pyralidae larvae to spores and parasporal crystals of *Bacillus thuringiensis*. J Econ Entomol 71:687–688

McGaughey WH (1985) Insect resistance to the biological insecticide *Bacillus thuringiensis*. Science 229:193–194

McGaughey WH, Whalon ME (1992) Managing insect resistance to *Bacillus thuringiensis* toxins. Science 258:1451–1455

McGuire MR, Shasha BS, Lewis LC et al (1990) Field evaluation of antigranulocytes starch formulations of *Bacillus thuringiensis* against *Ostrinia nubilalis* (Lepidoptera, Pyralidae). J Econ Entomol 83:2207–2210

Meadows MP, Ellis DJ, Butt J et al (1992) Distribution, frequency, and diversity of *Bacillus thuringiensis* in an animal feed mill. Appl Environ Microbiol 58:1344–1350

Miyasono M, Inagaki S, Yamamoto M et al (1994) Enhancement of delta-endotoxin activity by toxin-free spore of *Bacillus thuringiensis* against the diamondback moth, *Plutella xylostella*. J Invertebr Pathol 63:111–112

Moar WJ (2003) Breathing new life into insect-resistant plants. Nat Biotechnol 21:1152–1154

Moar WJ, Trumble JT, Federici BA (1989) Comparative toxicity of spores and crystals from the NRD-12 and HD-1 strains of *Bacillus thuringiensis* subsp. *kurstaki* to neonate beet armyworm (Lepidoptera, Noctuidae). J Econ Entomol 82:1593–1603

Moar WJ, Pusztaicarey M, Vanfaassen H et al (1995) Development of *Bacillus thuringiensis* CryIC resistance by *Spodoptera exigua (*Hubner) (Lepidoptera, Noctuidae). Appl Environ Microbiol 61:2086–2092

Muchaonyerwa P, Chenu C, Pantani OL et al (2002) Adsorption of the insecticidal toxin from *Bacillus thuringiensis* subspecies *tenebrionis* to clay fractions of tropical soils. In: Violante A, Huang PM, Bollag J-M (eds) Soil mineral-organic matter-microorganism interactions and ecosystem health 28B. Elsevier, Amsterdam, pp 59–68

Muchaonyerwa P, Chevallier T, Pantani OL et al (2006) Adsorption of the pesticidal toxin from *Bacillus thuringiensis* subsp *tenebrionis* on tropical soils and their particle-size fractions. Geoderma 133:244–257

Naef A, Zesiger T, Defago G (2006) Impact of transgenic Bt maize residues on the mycotoxigenic plant pathogen *Fusarium graminearum* and the biocontrol agent *Trichodearma atroviride*. J Environ Qual 35:1001–1009

Nayak P, Basu D, Das S et al (1997) Transgenic elite indica rice plants expressing CryIAc delta-endotoxin of *Bacillus thuringiensis* are resistant against yellow stem borer (*Scirpophaga incertulas*). Proc Natl Acad Sci USA 94:2111–2116

Obrycki JJ, Losey JE, Taylor OR et al (2001) Transgenic insecticidal corn: beyond insecticidal toxicity to ecological complexity. Bioscience 51:353–361

O'Callaghan M, Glare TR, Burgess EPJ et al (2005) Effects of plants genetically modified for insect resistance on nontarget organisms. Annu Rev Entomol 50:271–292

Ostry V, Ovesna J, Skarkova J et al (2010) A review on comparative data concerning Fusarium mycotoxins in Bt maize and non-Bt isogenic maize. Mycotoxin Res 26:141–145

Pagel-Wieder S, Niemeyer J, Fischer WR et al (2007) Effects of physical and chemical properties of soils on adsorption of the insecticidal protein (Cry1Ab) from *Bacillus thuringiensis* at Cry1Ab protein concentrations relevant for experimental field sites. Soil Biol Biochem 39:3034–3042

Palm CJ, Donegan K, Harris D et al (1994) Quantification in soil of *Bacillus thuringiensis* var. *kurstaki* delta-endotoxin from transgenic plants. Mol Ecol 3:145–151

Palm CJ, Schaller DL, Donegan KK et al (1996) Persistence in soil of transgenic plant produced *Bacillus thuringiensis* var *kurstaki* delta-endotoxin. Can J Microbiol 42:1258–1262

Parker MW, Feil SC (2005) Pore-forming protein toxins: from structure to function. Prog Biophys Mol Biol 88:91–142

Peferoen M (1997) Progress and prospects for field use of Bt genes in crops. Trends Biotechnol 15:173–177

Perlak FJ, Deaton RW, Armstrong TA et al (1990) Insect resistant cotton plants. Bio/Technology 8:939–943

Perlak FJ, Stone TB, Muskopf YM et al (1993) Genetically improved potatoe – protection from damage by colorado potato beetles. Plant Mol Biol 22:313–321

Poncet S, Delecluse A, Klier A et al (1995) Evaluation of synergistic interactions among the CryIVA, CryIVB, and CryIVD toxic components of *Bacillus thuringiensis* subsp *israelensis* crystals. J Invertebr Pathol 66:131–135

Pont B, Nentwig W (2005) Quantification of Bt-protein digestion and excretion by the primary decomposer *Porcellio scaber*, fed with two Bt-corn varieties. Biocontrol Sci Technol 15:341–352

Quiquampoix H (2008) Enzymes and proteins, interactions with soil-constituent surfaces. In: Chesworth W (ed) Encyclopedia of soil science. Springer, Dordrecht, pp 210–216

Quiquampoix H, Ratcliffe RG (1992) A P-31 NMR-study of the adsorption of bovine serum albumin on montmorillonite using phosphate and the paramagnetic cation Mn^{2+} – modification of conformation with pH. J Colloid Interface Sci 148:343–352

Quiquampoix H, Staunton S, Baron MH et al (1993) Interpretation of the pH-dependence of protein adsorption on clay mineral surfaces and its relevance to the understanding of extracellular enzyme activity in soil. Colloids Surf, A Physicochem Eng Asp 75:85–93

Ramessar K, Capell T, Twyman RM et al (2009) Calling the tunes on transgenic crops: the case for regulatory harmony. Mol Breed 23:99–112

Ramessar K, Capell T, Twyman RM et al (2010) Going to ridiculous lengths-European coexistence regulations for GM crops. Nat Biotechnol 28:133–136

Ravoahangimalala O, Charles JF, Schoellerraccaud J (1993) Immunological localization of *Bacillus thuringiensis* serovar i*sraelensis* toxins in midgut cells of intoxicated *Anopheles gambiae* larvae (Diptera, Culicidae). Res Microbiol 144:271–278

Romeis J, Meissle M, Bigler F (2006a) Ecological risk assessment for Bt crops – response. Nat Biotechnol 24:751–753

Romeis J, Meissle M, Bigler F (2006b) Transgenic crops expressing *Bacillus thuringiensis* toxins and biological control. Nat Biotechnol 24:63–71

Roush RT (1994) Managing pests and their resistance to *Bacillus thuringiensis* – can transgenic crops be better than sprays. Biocontrol Sci Technol 4:501–516

Roush RT (1998) Two-toxin strategies for management of insecticidal transgenic crops: can pyramiding succeed where pesticide mixtures have not? Philos Trans R Soc Lond B Biol Sci 353:1777–1786

Rui YK, Yi GX, Zhao J et al (2005) Changes of Bt toxin in the rhizosphere of transgenic Bt cotton and its influence on soil functional bacteria. World J Microbiol Biotechnol 21:1279–1284

Sander M, Madliger M, Schwarzenbach RP (2010) Adsorption of transgenic insecticidal Cry1Ab protein to SiO2. 1. Forces driving adsorption. Environ Sci Technol 44:8870–8876

Saxena D, Stotzky G (2000) Insecticidal toxin from *Bacillus thuringiensis* is released from roots of transgenic Bt corn *in vitro* and *in situ*. FEMS Microbiol Ecol 33:35–39

Saxena D, Stotzky G (2001a) *Bacillus thuringiensis* (Bt) toxin released from root exudates and biomass of Bt corn has no apparent effect on earthworms, nematodes, protozoa, bacteria, and fungi in soil. Soil Biol Biochem 33:1225–1230

Saxena D, Stotzky G (2001b) Bt corn has a higher lignin content than non-Bt corn. Am J Bot 88:1704–1706

Saxena D, Flores S, Stotzky G (1999) Transgenic plants – insecticidal toxin in root exudates from Bt corn. Nature 402:480–480

Saxena D, Flores S, Stotzky G (2002) Bt toxin is released in root exudates from 12 transgenic corn hybrids representing three transformation events. Soil Biol Biochem 34:133–137

Saxena D, Stewart CN, Altosaar I et al (2004) Larvicidal Cry proteins from *Bacillus thuringiensis* are released in root exudates of transgenic *B. thuringiensis* corn, potato, and rice but not of *B. thuringiensis* canola, cotton, and tobacco. Plant Physiol Biochem 42:383–387

Schnepf HE (1995) *Bacillus thuringiensis* toxins – regulation, activities and structural diversity. Curr Opin Biotechnol 6:305–312

Schnepf E, Crickmore N, Van Rie J et al (1998) *Bacillus thuringiensis* and its pesticidal crystal proteins. Microbiol Mol Biol Rev 62:775–806

Schrader S, Munchenberg T, Baumgarte S et al (2008) Earthworms of different functional groups affect the fate of the Bt-toxin Cry1Ab from transgenic maize in soil. Eur J Soil Biol 44:283–289

Shan GM, Embrey SK, Herman RA et al (2005) Biomimetic extraction of *Bacillus thuringiensis* insecticidal crystal proteins from soil based on invertebrate gut fluid chemistry. J Agric Food Chem 53:6630–6634

Shelton AM, Tang JD, Roush RT et al (2000) Field tests on managing resistance to Bt-engineered plants. Nat Biotechnol 18:339–342

Shelton AM, Zhao JZ, Roush RT (2002) Economic, ecological, food safety, and social consequences of the deployment of Bt transgenic plants. Annu Rev Entomol 47:845–881

Shelton AM, Naranjo SE, Romeis J et al (2009) Setting the record straight: a rebuttal to an erroneous analysis on transgenic insecticidal crops and natural enemies. Transgenic Res 18:317–322

Shen RF, Cai H, Gong WH (2006) Transgenic Bt cotton has no apparent effect on enzymatic activities or functional diversity of microbial communities in rhizosphere soil. Plant Soil 285:149–159

Sims SR (1995) *Bacillus thuringiensis* var *kurstaki* CryIA(c) protein expressed in transgenic cotton: effects on beneficial and other non-target insects. Southwest Entomol 20:493–500

Sims SR, Berberich SA (1996) *Bacillus thuringiensis* CryIA protein levels in raw and processed seed of transgenic cotton: determination using insect bioassay and ELISA. J Econ Entomol 89:247–251

Sims SR, Holden LR (1996) Insect bioassay for determining soil degradation of *Bacillus thuringiensis* subsp. kurstaki CryIA(b) protein in corn tissue. Environ Entomol 25:659–664

Sims SR, Ream JE (1997) Soil inactivation of the *Bacillus thuringiensis* subsp. *kurstaki* CryIIA insecticidal protein within transgenic cotton tissue: laboratory microcosm and field studies. J Agric Food Chem 45:1502–1505

Sims SR, Berberich SA, Nida DL et al (1996) Analysis of expressed proteins in fiber fractions from insect-protected and glyphosate-tolerant cotton varieties. Crop Sci 36:1212–1216

Smith RA, Couche GA (1991) The phylloplane as a source of *Bacillus thuringiensis* variants. Appl Environ Microbiol 57:311–315

Stotzky G (2000) Persistence and biological activity in soil of insecticidal proteins from *Bacillus thuringiensis* and of bacterial DNA bound on clays and humic acids. J Environ Qual 29:691–705

Stotzky G (2004) Persistence and biological activity in soil of the insecticidal proteins from *Bacillus thuringiensis,* especially from transgenic plants. Plant Soil 266:77–89

Sun CX, Chen LJ, Wu ZJ et al (2007) Soil persistence of *Bacillus thuringiensis* (Bt) toxin from transgenic Bt cotton tissues and its effect on soil enzyme activities. Biol Fertil Soils 43:617–620

Sundaram KMS (1996) Sorptive interactions and binding of delta-endotoxin protein from *Bacillus thuringiensis* subsp *kurstaki* in forest soils. J Environ Sci Health B 31:1321–1340

Tabashnik BE (1994) Evolution of resistance to *Bacillus thuringiensis.* Annu Rev Entomol 39:47–79

Tabashnik BE (2001) Breaking the code of resistance. Nat Biotechnol 19:922–924

Tabashnik BE, Gould F, Carriere Y (2004) Delaying evolution of insect resistance to transgenic crops by decreasing dominance and heritability. J Evol Biol 17:904–912

Tabashnik BE, Gassmann AJ, Crowder DW et al (2008) Insect resistance to Bt crops: evidence versus theory. Nat Biotechnol 26:199–202

Tang JD, Shelton AM, VanRie J et al (1996) Toxicity of *Bacillus thuringiensis* spore and crystal protein to resistant diamondback moth (*Plutella xylostella*). Appl Environ Microbiol 62:564–569

Tapp H, Stotzky G (1995a) Dot-blot enzyme-linked-immunosorbent-assay for monitoring the fate of insecticidal toxins from *Bacillus thuringiensis* in soil. Appl Environ Microbiol 61:602–609

Tapp H, Stotzky G (1995b) Insecticidal activity of the toxins from *Bacillus thuringiensis* subspecies *kurstaki* and *tenebrionis* adsorbed and bound on pure and soil clays. Appl Environ Microbiol 61:1786–1790

Tapp H, Stotzky G (1997) Monitoring the insecticidal toxins from *Bacillus thuringiensis* in soil with flow cytometry. Can J Microbiol 43:1074–1078

Tapp H, Stotzky G (1998) Persistence of the insecticidal toxin from *Bacillus thuringiensis* subsp. *kurstaki* in soil. Soil Biol Biochem 30:471–476

Tapp H, Calamai L, Stotzky G (1994) Adsorption and binding of the insecticidal proteins from *Bacillus thuringiensis* subsp. *kurstaki* and subsp. *tenebrionis* on clay minerals. Soil Biol Biochem 26:663–679

Turrini A, Sbrana C, Nuti MP et al (2004) Development of a model system to assess the impact of genetically modified corn and aubergine plants on arbuscular mycorrhizal fungi. Plant Soil 266:69–75

Vacher C, Bourguet D, Rousset F et al (2003) Modelling the spatial configuration of refuges for a sustainable control of pests: a case study of Bt cotton. J Evol Biol 16:378–387

Vanfrankenhuyzen K, Gringorten JL, Milne RE et al (1991) Specificity of activated cryia proteins from bacillus-thuringiensis subsp kurstaki Hd-1 for defoliating forest Lepidoptera. Appl Environ Microbiol 57:1650–1655

Venkateswerlu G, Stotzky G (1992) Binding of the protoxin and toxin proteins of *Bacillus thuringiensis* subsp. *kurstaki* on clay minerals. Curr Microbiol 25:225–233

Vercesi ML, Krogh PH, Holmstrup M (2006) Can *Bacillus thuringiensis* (Bt) corn residues and Bt-corn plants affect life-history traits in the earthworm *Aporrectodea caliginosa*? Appl Soil Ecol 32:180–187

Vettori C, Paffetti D, Saxena D et al (2003) Persistence of toxins and cells of *Bacillus thuringiensis* subsp. *kurstaki* introduced in sprays to Sardinia soils. Soil Biol Biochem 35:1635–1642

Wang HY, Ye QF, Wang W et al (2006) Cry1Ab protein from Bt transgenic rice does not residue in rhizosphere soil. Environ Pollut 143:449–455

Watkinson I (1994) Global view of present and future markets for Bt products. Agric Ecosyst Environ 49:3–7

West AW, Burges HD, White RJ et al (1984) Persistence of *Bacillus thuringiensis* parasporal crystal insecticidal activity in soil. J Invertebr Pathol 44:128–133

Whalon ME, Wingerd BA (2003) Bt: mode of action and use. Arch Insect Biochem Physiol 54:200–211

Wu D, Johnson JJ, Federici BA (1994) Synergism of mosquitocidal toxicity between CytA and CryIVD proteins using inclusions produced from cloned genes of *Bacillus thuringiensis*. Mol Microbiol 13:965–972

Wu WX, Ye QF, Min H (2004a) Effect of straws from Bt-transgenic rice on selected biological activities in water-flooded soil. Eur J Soil Biol 40:15–22

Wu WX, Ye QF, Min H et al (2004b) Bt-transgenic rice straw affects the culturable microbiota and dehydrogenase and phosphatase activities in a flooded paddy soil. Soil Biol Biochem 36:289–295

Xue K, Luo HF, Qi HY et al (2005) Changes in soil microbial community structure associated with two types of genetically engineered plants analyzing by PLFA. J Environ Sci (China) 17:130–134

Yanni SF, Whalen JK, Ma JK (2010) Crop residue chemistry, decomposition rates, and CO_2 evolution in Bt and non-Bt corn agroecosystems in North America: a review. Nutr Cycl Agroecosyst 87:277–293

Yu L, Berry RE, Croft BA (1997) Effects of *Bacillus thuringiensis* toxins in transgenic cotton and potato on *Folsomia candida* (Collembola: Isotomidae) and *Oppia nitens* (Acari: Orbatidae). J Econ Entomol 90:113–118

Zambryski P, Joos H, Genetello C et al (1983) Ti-plasmid vector for the introduction of DNA into plant cells without alteration of their normal regeneration capacity. EMBO J 2:2143–2150

Zhao JZ, Fan YL, Fan XL et al (1999) Evaluation of transgenic tobacco expressing two insecticidal genes to delay resistance development of *Helicoverpa armigera*. Chin Sci Bull 44:1871–1874

Zhao J, Rui C, Lu M et al (2000) Monitoring and management of *Helicoverpa armigera* resistance to transgenic Bt cotton in Northern China. Resist Pest Manage 11:28–31

Zhou XY, Huang QY, Chen SW et al (2005) Adsorption of the insecticidal protein of *Bacillus thuringiensis* on montmorillonite, kaolinite, silica, goethite and Red soil. Appl Clay Sci 30:87–93

Zhou XY, Huang QY, Cai P et al (2007) Adsorption and insecticidal activity of toxin from *Bacillus thuringiensis* on rectorite. Pedosphere 17:513–521

Zwahlen C, Hilbeck A, Gugerli P et al (2003a) Degradation of the Cry1Ab protein within transgenic *Bacillus thuringiensis* corn tissue in the field. Mol Ecol 12:765–775

Zwahlen C, Hilbeck A, Howald R et al (2003b) Effects of transgenic Bt corn litter on the earthworm *Lumbricus terrestris*. Mol Ecol 12:1077–1086

Chapter 4
The Role of Bacteria and Archaea in Nitrification, Nitrate Leaching and Nitrous Oxide Emissions in Nitrogen-Rich Grassland Soils

Hong J. Di, Keith C. Cameron, Ju-Pei Shen, Chris S. Winefield, Maureen O'Callaghan, Saman Bowatte, and Ji-Zheng He

Abstract Nitrate (NO_3^-) leaching into fresh water and nitrous oxide (N_2O) greenhouse gas emissions are two serious environmental impacts that occur from intensively grazed grassland soils. The oxidation of ammonia (NH_3) to NO_3^- is a key process in the nitrogen (N) cycle which has implications both in influencing nitrous oxide emissions and NO_3^- leaching. We investigated the relationships between nitrification rate, NO_3^- leaching and N_2O emissions with ammonia oxidizing bacteria (AOB) and ammonia oxidizing archaea (AOA) in nitrogen rich grassland soils. Both AOA and AOB were detected in large numbers in these grassland soils. The AOB abundance grew by 3.2–10.4 fold and activity increased by 177 fold in response to the addition of urine-N, and the AOB growth was significantly inhibited by a nitrification inhibitor, dicyandiamide (DCD). However, neither the AOA abundance, nor activity, increased with the application of urine-N substrate. AOB prefer to grow under high nitrogen environments whereas AOA prefer to grow under low nitrogen environments. DCD decreased

H.J. Di (✉) • K.C. Cameron
Centre for Soil and Environmental Research, Lincoln University, PO Box 84, 7647 Lincoln, Christchurch, New Zealand
e-mail: hong.di@lincoln.ac.nz

J.-P. Shen • J.-Z. He
State Key Laboratory of Urban and Regional Ecology, Research Centre for Eco-Environmental Sciences, Chinese Academy of Sciences, 18 Shuangqing Road, Beijing 100085, China

C.S. Winefield
Faculty of Agriculture and Life Sciences, Lincoln University, PO Box 84, 7647 Lincoln, Christchurch, New Zealand

M. O'Callaghan
AgResearch, Research Centre, Private Bag 4749, 8140 Lincoln, Christchurch, New Zealand

S. Bowatte
AgResearch Ltd, Grasslands Research Centre, Tennent Drive, Private Bag 11008, 4442 Palmerston North, New Zealand

J. Xu and D.L. Sparks (eds.), *Molecular Environmental Soil Science*, Progress in Soil Science, DOI 10.1007/978-94-007-4177-5_4,

NO_3^- leaching by 59% and decreased N_2O emissions by 64% from animal urine patches. Significant quantitative relationships were found between the AOB abundance and the nitrification rate, NO_3^--N leaching losses, and N_2O emissions, whereas no such relationships were found with AOA. These findings suggest that nitrification, NO_3^- leaching and N_2O emissions are driven by bacteria rather than archaea in these nitrogen rich grassland soils.

Keywords Ammonia oxidizing bacteria • Ammonia oxidizing archaea • Nitrification • Nitrate leaching • Nitrous oxide emissions

4.1 Introduction

In New Zealand, the predominant land use is grazed pastures where animals graze outdoors all year round. There are two major environmental issues in intensively grazed pasture systems, (i) nitrate (NO_3^-) leaching and (ii) nitrous oxide (N_2O) emissions. Nitrate is a water contaminant and nitrous oxide (N_2O) is both a greenhouse gas and an ozone-depletion substance. The long-term global warming potential of N_2O is 298 times that of carbon dioxide (CO_2) (IPCC 1995). A recent study shows that N_2O has become and will remain the greatest threat to the ozone layer in the twenty-first century (Ravishankara et al. 2009).

In grazed grassland, most of the NO_3^- leaching and N_2O emissions come from the animal urine-N returned to the pasture by the animal during outdoor grazing (Di and Cameron 2002a). The N loading rate under a dairy cow urine patch in intensively grazed dairy grassland can be as high as 1,000 kg N ha^{-1} (Di and Cameron 2002a). Most of the N in the urine is urea which, when deposited to the soil, produces ammonium, which is subsequently subject to nitrification. A mitigation technology that has been shown to be effective in reducing N_2O emissions and NO_3^- leaching from animal urine patches in grazed grassland is the use of nitrification inhibitors (e.g. dicyandiamide or DCD) to treat the soil (Di and Cameron 2002b, 2003, 2004a, b, c, 2005, 2006, 2007, 2008; Di et al. 2007, 2009a, b, 2010a, b).

Nitrification, the microbial oxidation of ammonia (NH_3) to nitrite (NO_2^-), and then to nitrate (NO_3^-), is a key process in the nitrogen (N) cycle in soil with major environmental and ecological consequences. The first and rate-limiting step of nitrification, the conversion of ammonia to hydroxylamine, is performed by the key ammonia monooxygenase (AMO) enzyme which is encoded by the subunits of AMO genes. It has traditionally been assumed that this first step is carried out mainly by autotrophic ammonia oxidizing bacteria (AOB) of the β- and γ-subgroups of proteobacteria (Purkhold et al. 2000; Kowalchuk and Stephen 2001; Prosser and Nicol 2008). However, the *amoA* gene has also been discovered in archaea, thus raising the prospect of the presence of ammonia oxidizing archaea (AOA) in different ecosystems (Venter et al. 2004; Könneke et al. 2005; Francis et al. 2005; Wuchter et al. 2006; Prosser and Nicol 2008). In fact AOA populations have recently been found to be more abundant than AOB in a range of soils, suggesting a potentially greater role for AOA than AOB in nitrification (Leininger et al. 2006; He et al. 2007; Chen et al. 2008; Prosser and Nicol 2008).

However, our knowledge and understanding of the relative role of AOB and AOA in nitrification and in affecting NO_3^- leaching and N_2O emissions is still very limited, particularly in nitrogen-rich grassland soils. We therefore conducted a series of investigations into the abundance, activity, and dynamics of AOB and AOA in nitrogen-rich grassland soils as affected by the applications of animal urine and a nitrification inhibitor, DCD. We also studied the relationship between the abundance of AOB and AOA with NO_3^- leaching loss and N_2O emissions. This paper provides a summary of the results from these investigations.

4.2 Abundance, Activity and Dynamics of AOB and AOA

Soil samples (0–0.1 m depth) were taken from six different sites across New Zealand: Northland (NL), Waikato (WK), and Rotorua Lakes (RL) in the North Island; and Canterbury (CT), West Coast (WC) and Southland (SL) in the South Island. These soils range from sedimentary to volcanic in their parent materials and their physical and chemical properties vary widely (Table 4.1). All these soils were collected from intensively grazed dairy pasture land. The samples from each site were used in a laboratory incubation study to determine: (1) the variation of AOB and AOA populations in the six different soils; (2) the responses of AOB and AOA to animal urine-N substrate (to simulate animal urine deposition during grazing); and (3) the inhibitory effect of a nitrification inhibitor (NI) on the AOB and AOA populations. The abundance and activity of AOB and AOA were determined using real-time PCR and reverse-transcription PCR techniques (Di et al. 2009a). Nitrification rate was determined by monitoring changes in the NO_3^- concentration in the soils.

The results showed that both AOB and AOA *amoA* genes were present in large numbers in these intensively grazed grassland soils (Di et al. 2009a). Phylogenetic analysis showed that all the AOB clones recovered were closely related to the *Nitrosospira* species and no clones were closely aligned with the *Nitrosomonas* species. A majority of the AOA clones identified were closely aligned with the soil clade. Interestingly, a number of the clones were affiliated with cluster Marine, suggesting similarity with AOA of marine origin. The AOA to AOB ratio in this study varied from a high of 10.7 in the Waikato soil to a low of 0.2 in the Southland soil, showing that the abundance of the AOA gene is greater than that of the AOB in some soils, but smaller in other soils. This was in contrast to some other studies which showed the predominance of AOA over AOB in all the soils studied (Leininger et al. 2006; He et al. 2007; Shen et al. 2008).

Whilst the AOB populations in the Controls (no nitrogen substrate or nitrification inhibitor) of the six soils remained relatively stable over the 3 month incubation period, they increased by 3.2, 5.7, 9.4, 10.4, 4.7 and 5.4 fold following the

Table 4.1 Properties of the soils used for the study (From Di et al. 2009a)

Location	Northland (NL) 36°15′24″ S; 174°31′20″ E	Waikato (WK) 38°46′38″ S; 175°18′26″ E	Rotorua Lakes (RL) 38°00′14″ S; 176°10′45″ E	Canterbury (CT) 43°45′54″ S; 171°41′25″ E	West Coast (WC) 42°42′12″ S; 171°26′43″ E	Southland (SL) 46°19′21″ S; 168°16′02″ E
Soil type	Waikare silty clay loam	Horotiu silt loam	Oropi silt loam	Lismore stony silt loam	Harihari recent silt loam	Mataura recent sandy loam
Particle size (%)						
Sand	11.0	20.4	47.6	10.5	4.2	69.1
Silt	53.0	73.8	51.7	66.0	93.6	29.0
Clay	36.0	5.8	0.7	23.5	2.2	1.9
pH (H_2O)	5.7	5.9	6.1	5.9	5.4	5.6
Organic C (g kg^{-1})	84.7	40.4	36.5	36.5	68.8	62.6
Total N (g kg^{-1})	7.3	3.5	3.3	3.5	5.6	5.7
Olsen P (mg kg^{-1})	53.4	16.9	48.1	53.0	46.5	21.1
CEC ($cmol_c$ kg^{-1})	32.0	13.0	17.0	17.5	15.0	19.0
Exch. Ca^{2+} ($cmol_c$ kg^{-1})	20.8	7.1	12.0	10.3	5.0	6.7
Exch. Mg^{2+} ($cmol_c$ kg^{-1})	2.74	0.27	0.79	0.53	0.94	0.9
Exch. K^{+} ($cmol_c$ kg^{-1})	0.57	0.19	0.29	0.71	0.25	0.24
Base saturation (%)	75.3	60.6	76.8	66.0	40.4	42.5

application of the urine-N substrate in the NL, WK, RL, CT, WC and SL soils, respectively (Fig. 4.1a). When the nitrification inhibitor, DCD, was applied, the AOB population growth was significantly inhibited (Fig. 4.1a). In contrast, the AOA population abundance remained largely unchanged in all the soils irrespective of the urine-N and inhibitor treatments (Fig. 4.1b). Therefore, in all the soils the AOA populations did not grow with the supply of the large dose of urine-N substrate.

Following the application of the urine-N to the WK soil, the AOB RNA copy numbers increased by more than 177 times, from 1.1×10^2 in the Control soil to 2.4×10^4 in the urine-N treated soil (Di et al. 2009a). When DCD was applied, the AOB activity was significantly inhibited, with the RNA copy numbers remaining the same as that in the Control. In contrast, the AOA activity did not change with the urine-N application, with the RNA copy numbers remaining the same in the Control, Urine and Urine + DCD treatments. In the Urine treatment, the AOA RNA copy number was only 2% that of the AOB.

In all six soils, the addition of the urine-N substrate significantly increased the nitrification rate, as indicated by the rising NO_3^--concentrations, but the nitrification rates were reduced by the NI treatments (Fig. 4.1c).

In a subsequent study, Di et al. (2010a) determined the growth of AOB and AOA in the 0–0.2 m depth of the Canterbury soil with and without the application of urine (1,000 kg N ha^{-1}) and with and without the nitrification inhibitor DCD. It was shown that the AOB *amoA* gene copy numbers increased under the high urine-N environment, whereas the AOA *amoA* gene copy numbers did not increase under the high urine-N environment and only increased in the Control where no nitrogen was applied (Fig. 4.2).

Results from this study clearly show that AOB and AOA prefer different nitrogen conditions for their growth. Whilst AOB abundance and activity grew substantially when supplied with a high dose of urine substrate in the topsoils, the AOA community only grew in the Controls where no urine substrate was applied. In fact, the AOA community seemed to be inhibited by the high dose of urine substrate in two of the subsoils. This differential trend of AOB and AOA growth is supported by the transcription data of the *amoA* gene (RNA copy numbers) in the Canterbury topsoil (Di et al. 2010a). These findings suggest that AOB can grow under high nitrogen environments whereas AOA growth is favoured under low nitrogen environments. (Valentine 2007; Di et al. 2009a; Offre et al. 2009; Erguder et al. 2009).

4.3 Nitrate Leaching and Nitrous Oxide Emissions

Large undisturbed soil monolith lysimeters (0.5 m diameter and 0.7 m deep) were collected and used to determine NO_3^- leaching and N_2O emissions (Di et al. 2009b, 2010b). The lysimeters were installed in field lysimeter facilities where the tops of the lysimeters were exposed to the same climatic conditions as the soil and pasture

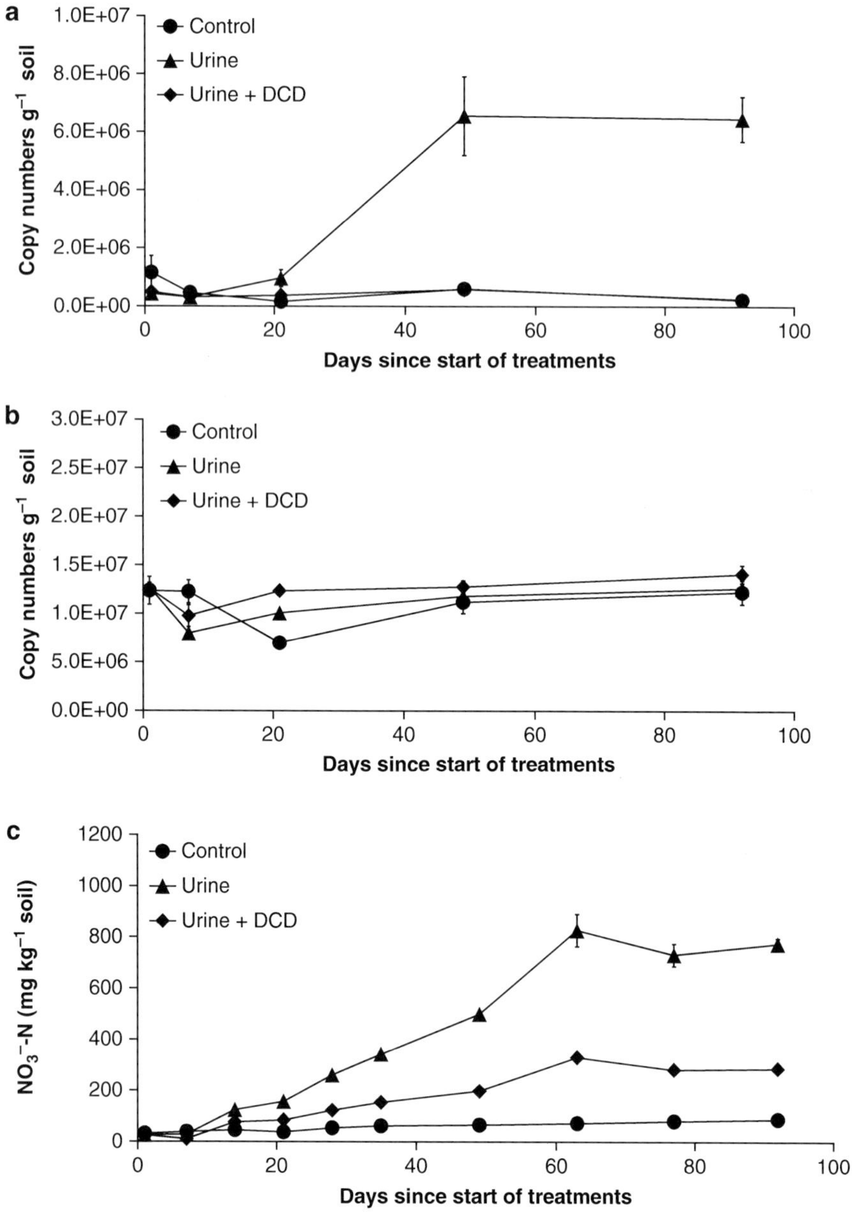

Fig. 4.1 (**a**) AOB *amoA* gene copy numbers; (**b**) AOA *amoA* gene copy numbers; and (**c**) Nitrate-N concentration in the Waikato Horotiu soil. Similar trends were found in the other soils (From Di et al. 2009a)

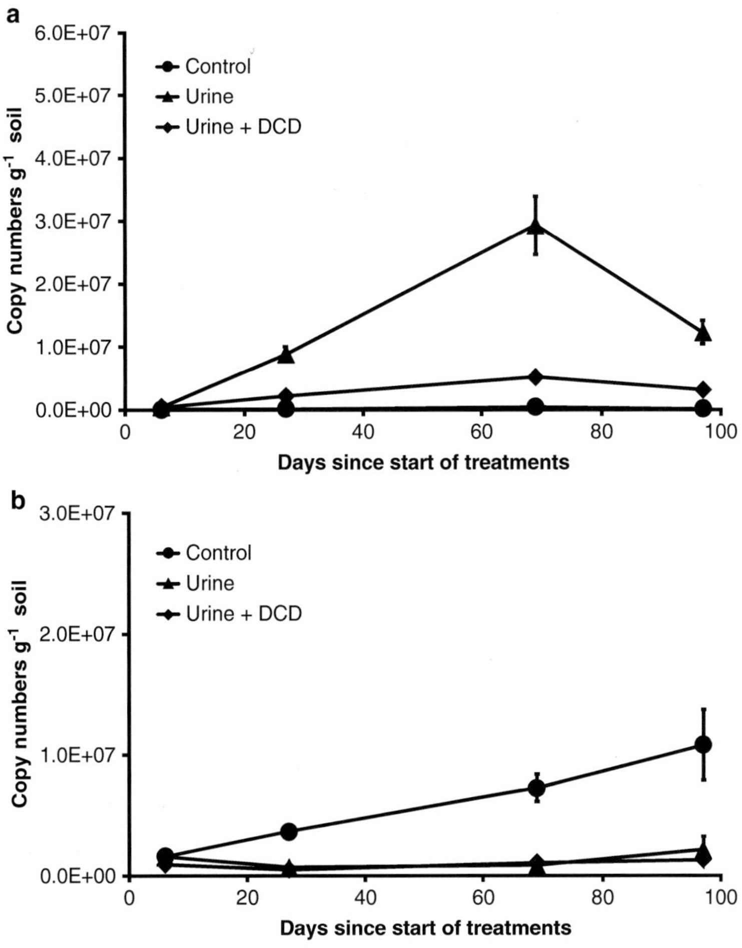

Fig. 4.2 (**a**) AOB *amoA* gene copy numbers; and (**b**) AOA *amoA* gene copy numbers, in the Canterbury topsoil (0–20 cm) as affected by urine and urine plus DCD (From Di et al. 2010a)

in the surrounding field. Two rainfall conditions (1,260 mm and 2,145 mm p.a.) were created using a rainfall simulation system in order to test the influence of rainfall inputs on NO_3^- leaching and N_2O emissions. Pasture was harvested at typical grazing heights and removed before the application of the treatments and was subsequently cut at typical grazing intervals thereafter. Leachates from the lysimeters were collected as required and analysed for NO_3^-, nitrite (NO_2^-), and NH_4^+ concentrations. A standard closed chamber method was used to determine N_2O emissions from the treated lysimeters (Di et al. 2010b).

Under the 1,260 mm rainfall treatment, total NO_3^--N leaching losses in the Urine treatment ranged from a low of 122.9 kg NO_3^--N ha^{-1} in the WC Harihari soil

Table 4.2 Total NO_3^--N leaching losses (From Di et al. 2009b)

Soil	Rainfall (mm p.a.)	Treatment	NO_3^--N leached (±SE) (kg ha^{-1})	% reduction of NO_3^--N leaching
CT	1,260	Urine	399.0 (27.3)	–[a]
CT	1,260	Urine + DCD	176.5 (15.5)	56
SL	1,260	Urine	435.8 (13.7)	–
SL	1,260	Urine + DCD	142.2 (0.9)	67
WC	1,260	Urine	122.9 (59.1)	–
WC	1,260	Urine + DCD	35.8 (17.4)	71
SL	2,145	Urine	457.0 (53.4)	–
SL	2,145	Urine + DCD	257.4 (22.3)	44
WC	2,145	Urine	67.7 (25.1)	–
WC	2,145	Urine + DCD	29.7 (11.5)	56
SED for treatment			20.0	
Average reduction of NO_3^--N leaching by DCD				59

[a]Not applicable

to a high of 435.8 kg NO_3^--N ha^{-1} in the SL Mataura soil (Table 4.2) (Di et al. 2009b). These losses were significantly decreased to between 35.8 and 176.5 kg NO_3^--N ha^{-1} when DCD was applied ($P < 0.05$). The application of DCD therefore reduced the NO_3^- leaching losses by between 56 and 71%. Under the 2,145 mm rainfall condition, the total NO_3^--N leaching losses in the Urine treatment varied from 67.7 kg NO_3^--N ha^{-1} in the WC Harihari soil to 457.0 kg NO_3^--N ha^{-1} in the SL Mataura soil. These losses were decreased to 29.7 and 257.4 kg NO_3^--N ha^{-1} with the application of DCD. The application of DCD therefore reduced these NO_3^- leaching losses by between 44 and 56%. The difference in the amount of NO_3^--N leached between the two rainfall conditions was not statistically significant ($P > 0.05$). The average reduction in NO_3^--N leaching loss by the DCD treatment under both rainfall conditions was 59%.

Total N_2O emissions varied significantly between the different soils, with those in the urine treatment ranging from a low of 13.9 kg N_2O-N ha^{-1} in the WC Harihari soil under the higher rainfall condition to a high of 39.8 kg N_2O-N ha^{-1} in the CT Lismore soil under the higher rainfall condition (Di et al. 2010b). The different water inputs did not result in significantly different total N_2O emissions ($P > 0.05$). However, the DCD treatment decreased the total N_2O emissions from all of the four soils. The emission factor from the urine (EF_3) varied from 1.4 to 3.0% (averaging 2.2%), and this was decreased to between 0.3 and 1.4% (averaging 0.8%) with the DCD treatment. Therefore, the DCD treatment resulted in a reduction of the average EF_3 by 64%.

4.4 Relationships Between Nitrogen Losses with AOB and AOA

Regression analysis showed a significant exponential relationship between the NO_3^--N concentration and the abundance of the AOB population (Di et al. 2009a). Therefore, as the AOB population abundance increased the nitrification rate increased exponentially. However, no quantitative relationship was found between the NO_3^--N concentration and the abundance of the AOA populations. A significant relationship was found between total NO_3^--N leaching loss from both rainfall treatments of the soils and the AOB *amoA* gene copy numbers (Di et al. 2009b), but no relationship was found with the AOA populations. A significant exponential relationship was also found between the total N_2O emissions and the AOB populations after 49 days of incubation (when the AOB population peaked in most of the urine treatments) (Di et al. 2010b) but no quantitative relationship was found with AOA populations.

4.5 Conclusions

AOB and AOA populations were present in large numbers in the New Zealand grassland soils. However, in the nitrogen rich urine patch areas of these soils, only the AOB population grew and activity increased when supplied with the high dose of ammonium substrate in the urine. The AOB population and activity were also significantly inhibited by the nitrification inhibitor DCD. The AOA population abundance and activity did not grow when supplied with the high dose of ammonia substrate (urine), rather the AOA grew in some of the Controls where no nitrogen was applied. The nitrification inhibitor DCD significantly decreased nitrate leaching by an average of 59% under two contrasting rainfall conditions (1,260 mm and 2,145 mm p.a.), and significantly decreased nitrous oxide emissions by an average of 64%. Significant quantitative relationships were found between the nitrification rate, nitrate leaching losses and nitrous oxide emissions with AOB abundance, but not with the AOA abundance. These results suggest that nitrification, and thus nitrate leaching and nitrous oxide emissions, are driven by AOB rather than AOA in these nitrogen-rich grassland soils that receive urine-N inputs from grazing animals.

Acknowledgments We thank the New Zealand Foundation for Research, Science and Technology (FRST), Ravensdown Fertiliser Co-operative Ltd, Ministry of Agriculture and Forestry (MAF), and the Pastoral21 Consortium for funding, Emily Gerard and Shona Brock of AgResearch and Jie Lei, Steve Moore, Carole Barlow, Trevor Hendry, Neil Smith, Nigel Beale, and Roger Atkinson of Lincoln University, for technical support.

References

Chen XP, Zhu YG, Xia Y, Shen JP, He JZ (2008) Ammonia-oxidizing archaea: important players in paddy rhyzosphere soil? Environ Microbiol 10:1978–1987

Di HJ, Cameron KC (2002a) Nitrate leaching in temperate agroecosystems: sources, factors and mitigating strategies. Nutr Cycl Agroecosyst 46:237–256

Di HJ, Cameron KC (2002b) The use of a nitrification inhibitor, dicyandiamide (DCD), to reduce nitrate leaching and nitrous oxide emissions in a simulated grazed and irrigated grassland. Soil Use Manag 18:395–403

Di HJ, Cameron KC (2003) Mitigation of nitrous oxide emissions in spray-irrigated grazed grassland by treating the soil with dicyandiamide, a nitrification inhibitor. Soil Use Manag 19:184–290

Di HJ, Cameron KC (2004a) Effects of temperature and application rate of a nitrification inhibitor, dicyandiamide (DCD) on nitrification rate and microbial biomass in a grazed pasture soil. Aust J Soil Res 42:927–932

Di HJ, Cameron KC (2004b) Effects of the nitrification inhibitor, diacyandiamide on potassium, magnesium and calcium leaching in grazed grassland. Soil Use Manag 20:2–7

Di HJ, Cameron KC (2004c) Treating grazed pasture soil with a nitrification inhibitor, eco-n^{TM}, to decrease nitrate leaching in a deep sandy soil under spray irrigation. N Z J Agric Res 47: 351–361

Di HJ, Cameron KC (2005) Reducing environmental impacts of agriculture by using a fine particle suspension nitrification inhibitor to decrease nitrate leaching from grazed pastures. Agric Ecosyst Environ 109:202–212

Di HJ, Cameron KC (2006) Nitrous oxide emissions from two dairy pasture soils as affected by different rates of a fine particle suspension nitrification inhibitor, dicyandiamide (DCD). Biol Fertil Soils 42:472–480

Di HJ, Cameron KC (2007) Nitrate leaching losses and pasture yields as affected by different rates of animal urine nitrogen returns and application of a nitrification inhibitor – a lysimeter study. Nutr Cycl Agroecosyst 79:281–290

Di HJ, Cameron KC (2008) Sources of nitrous oxide from ^{15}N-labelled animal urine and urea fertilizer with and without a nitrification inhibitor, dicyandiamide (DCD). Aust J Soil Res 46:76–82

Di HJ, Cameron KC, Sherlock RR (2007) Comparison of the effectiveness of a nitrification inhibitor, dicyandiamide (DCD), in reducing nitrous oxide emissions in four different soils under different climatic and management conditions. Soil Use Manag 23:1–9

Di HJ, Cameron KC, Shen JP, He JZ, Winefield CS (2009a) Nitrate leaching from grazed grassland as affected by a nitrification inhibitor, dicyandiamide, and relationships with ammonia oxidizing bacteria and archaea. Soil Use Manag 25:454–461

Di HJ, Cameron KC, Shen JP, Winefield CS, O'Callaghan M, Bowatte S, He JZ (2009b) Nitrification driven by bacteria and not archaea in nitrogen rich grassland soils. Nat Geosci 2:621–624

Di HJ, Cameron KC, Shen JP, Winefield CS, O'Callaghan M, Bowatte S, He JZ (2010a) Ammonia oxidizing bacteria and archaea grow under contrasting soil nitrogen conditions. FEMS Microbiol Ecol 72:386–394. doi:10.1111/j.1574-6941.2010.00861.x

Di HJ, Cameron KC, Sherlock RR, Shen JP, He JZ, Winefield CS (2010b) Nitrous oxide emissions from grazed grassland as affected by a nitrification inhibitor, dicyandiamide, and relationships with ammonia oxidizing bacteria and archaea. J Soil Sediment 10:943–954. doi:10.1007/s11368-009-0174-x

Erguder TH, Boon N, Wittebolle L, Marzorati M, Verstraete W (2009) Environmental factors shaping the ecological niches of ammonia oxidizing archaea. FEMS Microbiol Rev 33:855–869

Francis CA, Roberts KJ, Beman JM, Santoro AE, Oakley BB (2005) Ubiquity and diversity of ammonia-oxidizing archaea in water columns and sediments of the ocean. Proc Natl Acad Sci USA 102:14683–14688

He JZ, Shen JP, Zhang LM, Zhu YG, Zheng YM, Xu MG, Di HJ (2007) Quantitative analyses of the abundance and composition of ammonia-oxidizing bacteria and ammonia-oxidizing archaea of a Chinese upland red soil under long-term fertilization practices. Environ Microbiol 9:2364–2374

IPCC (1995) Climate change 1995: the science of climate change. Contribution of Working Group 1 to the Second Assessment of the Intergovernmental Panel on Climate Change. Cambridge University Press, Cambridge

Könneke M, Bernhard AE, de la Torre JR, Walker CB, Waterbury JB, Stahl DA (2005) Isolation of an autotrophic ammonia-oxidizing marine archaeon. Nature 437:543–546

Kowalchuk GA, Stephen JR (2001) Ammonia-oxidizing bacteria: a model for molecular microbial ecology. Annu Rev Microbiol 55:485–529

Leininger S, Urich T, Schloter M, Schwark L, Qi J, Nicol GW, Prosser JI, Schuster SC, Schleper C (2006) Archaea predominate among ammonia-oxidizing prokaryotes in soils. Nature 442:806–809

Offre P, Prosser JI, Nicol GW (2009) Growth of ammonia-oxidizing archaea in soil microcosms is inhibited by acetylene. FEMS Microbiol Ecol 70:99–108

Prosser JI, Nicol GW (2008) Relative contributions of archaea and bacteria to aerobic ammonia oxidation in the environment. Environ Microbiol 10:2931–2941

Purkhold U, Pommerening-Röser A, Juretschko S, Schmid MC, Koops HJ, Wagner M (2000) Phylogeny of all recognized species of ammonia oxidizers based on comparative 16 S rRNA and *amoA* sequence analysis: implications for molecular diversity surveys. Appl Environ Microbiol 66:5368–5382

Ravishankara AR, Daniel JS, Portmann RW (2009) Nitrous oxide (N2O): the dominant ozone-depleting substance emitted in the 21st century. Science 326:123–125. doi:10.1126/science.1176985

Shen JP, Zhang LM, Zhu YG, Zhang JB, He JZ (2008) Abundance and composition of ammonia-oxidizing bacteria and ammonia-oxidizing archaea communities of an alkaline sandy loam. Environ Microbiol 10:1601–1611

Valentine DL (2007) Adaptation to energy stress dictate the ecology and evolution of the Archaea. Nat Rev Microbiol 5:316–323

Venter JC, Remington K, Heidelberg JF, Halpern AL, Rusch D (2004) Environmental genome shotgun sequencing of the Sargasso Sea. Science 3–4:66–74

Wuchter C, Abbas B, Coolen MJL, Herfort L, van Bleijswijk J, Timmers P, Strous M, Teira E, Herndl GJ, Middleburge JJ, Schouten S, Sinninghe Damste JS (2006) Archaeal nitrification in the ocean. Proc Natl Acad Sci USA 103:12317–12322

Part II
Heavy Metals in Soil

Chapter 5
Advances in the Use of Synchrotron Radiation to Elucidate Environmental Interfacial Reaction Processes and Mechanisms in the Earth's Critical Zone

Donald L. Sparks

Abstract Reactions occurring at biogeochemical interfaces such as mineral/water, mineral/microbe, and plant/soil, greatly influence a number of important processes in soil and water environments. These include: sorption, desorption, oxidation-reduction, and precipitation, dissolution. It is useful to study these processes over a range of spatial and temporal scales. The use of molecular scale techniques, especially those that are in-situ and synchrotron-based, have provided a wealth of information on reactivity at biogeochemical interfaces. This review focuses on the use of these techniques, especially X-ray absorption spectroscopy, to speciate and elucidate reaction mechanisms of metal(loids) in soils and plants.

Keywords Critical zone • Multi-scale approach • Kinetics • Synchrotron-based techniques • Molecular environmental science

5.1 Introduction

The Earth's Critical Zone is that region of Planet Earth where we live. It extends from the tops of the trees to the water table (Fig. 5.1). There are an array of spatial and temporal scales in the Critical Zone. The spatial scales extend from the atomic to the global and the temporal scales range from milliseconds to eons. Additionally, there are a plethora of physical, chemical, and biological reactions and processes that are coupled and influence some of the major environmental challenges of our time including: water, soil, and air quality, human health, the health of oceans, climate change, and indeed economic vitality and development. There are also a number of significant biogeochemical interfaces such as the

D.L. Sparks (✉)
Department of Plant and Soil Sciences, Delaware Environmental Institute, University of Delaware, Newark, DE 19716, USA
e-mail: dlsparks@udel.edu

J. Xu and D.L. Sparks (eds.), *Molecular Environmental Soil Science*, Progress in Soil Science, DOI 10.1007/978-94-007-4177-5_5,

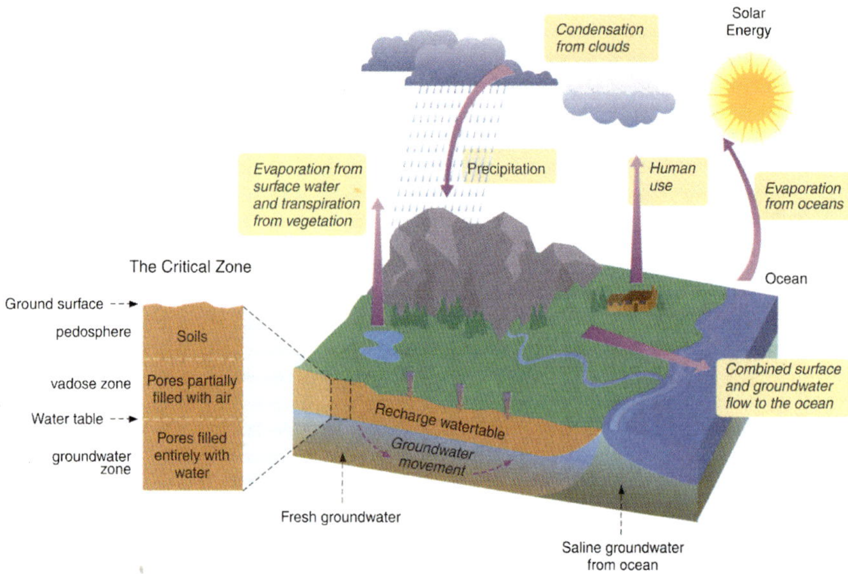

Fig. 5.1 Schematic of the Critical Zone. The Critical Zone includes the land surface and its canopy of vegetation, rivers, lakes, and shallow seas, and it extends through the pedosphere, unsaturated vadose zone, and saturated groundwater zone. Interactions at this interface between the solid Earth and its fluid envelopes determine the availability of nearly every life-sustaining resource (Reprinted with permission NRC 2001)

mineral/water, microbe/mineral, and plant/soil that play pivotal roles in the fate, transport, toxicity, speciation (form), and bioavailability of nutrients and inorganic and organic contaminants in the environment (Sparks 2005a). It is for all of these reasons that an enhanced understanding of biogeochemical processes and reactivity at Critical Zone interfaces, over a range of spatial and temporal scales, is fundamental to sustaining and enhancing Planet Earth.

This review will focus on the use of synchrotron radiation to further advance our understanding of the reactivity and mechanisms at important environmental biogeochemical interfaces. It will include past successes and future opportunities. Arguably, the use of synchrotron-based techniques such as X-ray absorption spectroscopy has revolutionized the soil and environmental sciences over the past two decades by greatly advancing discovery research and forging important inter- and multi-disciplinary collaborations that are integral to solving some of the grand, complex challenges we face globally. As a result of these collaborations the interdisciplinary field of molecular environmental science, which involves the study of the chemical and physical forms and distribution of contaminants in soils, sediments, waste materials, natural waters, and the atmosphere at the molecular level (Sparks 2002), is a vibrant interdisciplinary field of research.

There are a number of areas where the application of molecular environmental science is propelling major frontiers. These include speciation of contaminants, which is essential for understanding release mechanisms, spatial resolution, chemical transformations, toxicity, bioavailability, and ultimate impacts on human health; mechanisms of microbial transformations on mineral surfaces; phytoremediation; humic substance structure and chemistry; air and terrestrial emanated particulate reactivity and composition; soil structure; development of predictive models; effective remediation and waste management strategies; and risk assessment (Sparks 2002; Ginder-Vogel and Sparks 2010; Sparks and Ginder-Vogel 2011).

This review will focus on the use of X-ray absorption spectroscopy (XAS), X-ray fluorescence (XRF) spectroscopy, and X-ray computed microtomography to elucidate important information on speciation, binding mechanisms, and contaminant distribution and associations in mineral, soil, and plant systems with emphasis on biogeochemical interfacial processes.

5.2 Synchrotron Radiation

Intense light is produced at a synchrotron facility. Synchrotron radiation is produced over a wide range of energies from the infrared region with energies <1 eV to the hard X-ray region with energies of 100 keV or more. There are more than 60 synchrotron light sources in over 30 countries (www.lightsources.org). For more information on synchrotron user facilities around the globe and their capabilities in the geosciences, the reader should consult Sutton (2006). It is not the purpose of this review to delve into detail about synchrotron radiation and techniques. The reader can refer to several excellent reviews on the use of synchrotron techniques in the environmental sciences that have been published in the past decade. These reviews focus on the use of synchrotron techniques in low-temperature geochemistry and environmental science (Brown and Parks 2001; Brown and Sturchio 2002; Fenter et al. 2002), clay and soil science (Schulze et al. 1999; Kelly et al. 2008; Ginder-Vogel and Sparks 2010; Sparks and Ginder-Vogel 2011), and the study of heavy metals in the environment (Sparks 2005a). However, some detail is provided on XAS, one of the most widely used synchrotron-based spectroscopic techniques used in the environmental sciences.

5.2.1 X-Ray Absorption Spectroscopy

XAS was developed in the early 1970s (Sayers et al. 1971). It has a number of desirable features that makes it the technique of choice for studying environmental samples such as minerals, humic materials, soils, biosolids, plants, and particulates. XAS can be used to study most elements in crystalline or non-crystalline solid, liquid, or gaseous states over a concentration range of a few ppm to the pure element. It is also an *in-situ* technique, which means that one can study reactions in the presence of water. This is a major advantage over many molecular scale techniques,

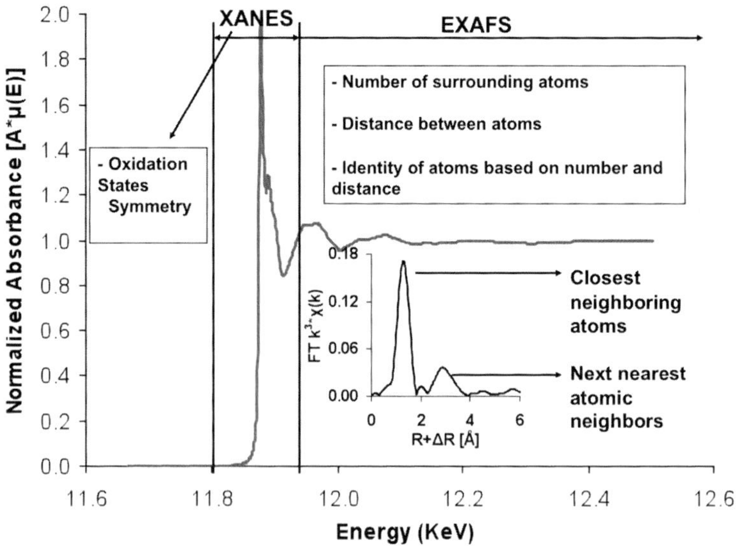

Fig. 5.2 Energy regions of a XAS spectrum

which are *ex-situ*, and often requiring drying of the sample material, placing it in an ultra-high vacuum (UHV), heating the sample, or employing particle bombardment. Such conditions can alter the sample, creating artifacts, and do not simulate most natural soil conditions (Ginder-Vogel and Sparks 2010; Sparks and Ginder-Vogel 2011).

An XAS experiment, which results in a spectrum (Fig. 5.2), consists of exposing a sample to an incident monochromatic beam of synchrotron X-rays, scanned over a range of energies below and above the absorption edge (K, L, M) of the element of interest (Sparks 2002). The energy region extending from just below to about 50 eV above the absorption edge is the XANES (X-ray absorption near edge structure) portion of the spectrum (Fig. 5.2). Fingerprint information (e.g., oxidation states) can be obtained from this region of the XAS spectrum. The XANES region of the spectrum, while not providing as much quantitative information as the extended X-ray absorption fine structure (EXAFS) region, can provide qualitative or semi-quantitative information on the oxidation state of the measured element (Brown et al. 1995). Such information can be obtained by comparing the features of the XANES spectrum of the sample with features of XANES spectra for well-characterized reference compounds. Some species, such as Cr, yield remarkably different, easily recognizable XANES spectra such as the pre-edge feature that is diagnostic for Cr(VI).

The energy region from 50 to 1,000 eV above the absorption edge is the EXAFS (extended X-ray absorption fine structure) portion of the spectrum (Fig. 5.2). Analysis of an EXAFS spectrum provides information on interatomic distances (R), coordination numbers (CN), and identity of first, second, and more distant shells of neighbors around an absorber (Brown et al. 1995; Sparks 2002; Ginder-Vogel and Sparks 2010; Sparks and Ginder-Vogel 2011).

Additional detail on XAS principles, methodology, sample preparation, and data analyses can be found in a number of excellent sources (Brown 1990; Manceau et al. 1992; Fendorf et al. 1994; Schulze and Bertsch 1995; O'Day 1999; Brown and Parks 2001; Bertsch and Hunter 2001; Fenter et al. 2002; Brown and Sturchio 2002; Kelly et al. 2008; Ginder-Vogel and Sparks 2010; Sparks and Ginder-Vogel 2011).

Other synchrotron-based techniques that allow for investigations of heterogeneous materials, which are characterized by multiple species of metal(loids) and nutrients, at micron and smaller scales include microfocused XAS and XRF and microtomography. Additionally, to study rapid processes in real-time at subsecond time scales and at the molecular level one can employ quick X-ray absorption spectroscopy (Q-XAS).

5.2.2 *Micro-focused XAS and XRF*

With micro-focused XAS and XRF one can investigate square micron areas and focus the beam to spatial resolutions of <5 um to determine speciation and elemental distribution and associations. Such ability is critical for understanding heterogeneous systems such as soils and plants. While bulk XAS can and should be used to provide what major species are found in such systems, one often needs to couple these studies with μ-XAS and μ-XRF investigations to more precisely quantify major species and determine associations and distributions of elements over small areas in heterogeneous systems where multiple species exist. Standard bulk XAS techniques explore an area of several square millimeters and provide information on the average local chemical environment. In heterogeneous natural systems where more than one type of surface species is present, bulk XAS will detect only the primary (or average) type of surface product/species in the bulk sample. Soils have particle sizes in the micrometer range, and metal speciation may vary over regions of a few 100 μm^2. Consequently, the presence of multiple species in soils results in overlapping atomic shells which makes precise speciation with bulk XAS challenging (McNear et al. 2005b; Ginder-Vogel and Sparks 2010; Sparks and Ginder-Vogel 2011).

Some excellent reviews have appeared on the experimental details and application of μ-XAS and μ-XRF to speciate contaminants in heterogeneous systems (Bertsch and Hunter 1998; Hunter and Bertsch 1998; Bertsch and Seaman 1999; Manceau et al. 2002). A number of investigations have appeared over the past 10 years on the speciation of metals and oxyanions in soils, plants, biosolids, and fly ash (see references in Brown and Sturchio 2002; Ginder-Vogel and Sparks 2010; Sparks and Ginder-Vogel 2011).

5.2.3 Quick XAS

Many important processes (e.g., adsorption, oxidation-reduction, precipitation) occurring at environmental surfaces are characterized by a rapid initial reaction on time-scales of milliseconds to minutes (Scheidegger and Sparks 1996). In many cases, a significant portion of the entire reaction process has occurred before the first measurements can be made using traditional batch and flow techniques. Thus, a knowledge of these initial reaction rates is critical to determining chemical kinetic rate constants and reaction mechanisms and understanding how the initial reactions impact succeeding processes. Chemical relaxation techniques such as pressure jump (p-jump) and concentration jump (c-jump such as stopped-flow) allow rapid data collection on time scales of milliseconds. However, rate "constants" are calculated from linearized rate equations that include parameters that were determined from equilibrium and modeling studies. Consequently, the rate "constants" are not directly determined (Ginder-Vogel et al. 2009; Ginder-Vogel and Sparks 2010; Sparks and Ginder-Vogel 2011). Direct, in situ, molecular- scale measurement of rapid reactions has, until recently been quite limited. Fendorf et al. (1993) used stop-flow electron paramagnetic resonance (SF-EPR) spectroscopy to measure Mn(II) sorption to birnessite (δ-MnO_2) on a time scale of milliseconds. More recently, Parikh et al. (2008) used in situ, Fourier Transform infrared (FTIR) spectroscopy to measure As(III) oxidation rates by hydrous manganese(IV) oxide (HMO) at a time scale of 2.5 s. However, both of these techniques suffer from limitations. EPR can only be used to measure EPR active nuclei, while FTIR requires both IR active functional groups and relatively high concentrations of the reactants being examined (Parikh et al. 2008).

Quick-scanning X-ray absorption spectroscopy overcomes both of these limitations. Depending on beamline instrumentation and flux, quick-scanning X-ray absorption spectroscopy (Q-XAS) can be used to probe most of the atoms on the periodic table and to relatively low concentrations (Khalid et al. 2010). However, using a unique, cam-operated, continuous-scanning setup at beamline X18B at the National Synchrotron Light Source (NSLS), it is possible to collect XANES and EXAFS spectra as the monochromator travels both up and down in energy (Khalid et al. 2010) on sub-second time scales. Landrot et al. (2010) employed Q-XAS at the NSLS to determine the initial kinetics of Cr(III) oxidation by HMO, and Ginder-Vogel et al. (2009) studied the rates and mechanisms of As(III) oxidation at the HMO mineral surface. In the study of Landrot et al. (2010), Cr(III) oxidation to Cr(VI) was exceedingly rapid as illustrated by the quick appearance of the pre-edge feature (Fig. 5.3), indicative of the appearance of Cr(VI). In both of the Q-XAS studies of Ginder-Vogel et al. (2009) and Landrot et al. (2010) there is strong indication that chemical kinetics are being measured. This is suggested in the study of Landrot et al. (2010), in which initial rate constants were similar as initial concentration of Cr(III) was increased (Sparks 1989).

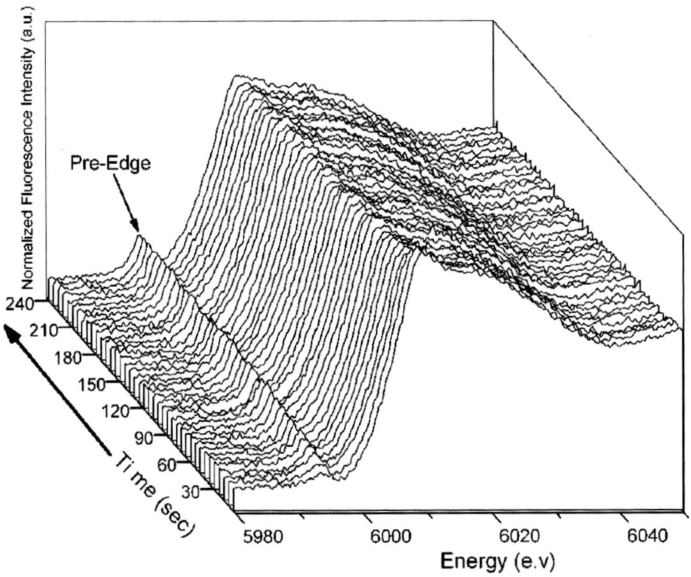

Fig. 5.3 Cr(III) oxidation kinetics using a Q-XAFS technique, at pH 2.5, [Cr(III)] = 100 mM, [HMO] = 20 g/L, and 0–240 s. Each XANES spectrum shown represents 3 s of the reaction (average of four 0.75 s spectra) (Reprinted with permission Landrot et al. 2010)

5.2.4 Applications of Synchrotron-Based Techniques to Elucidate Soil Chemical Processes and Reactions

5.2.4.1 Adsorption of Metals and Oxyanions on Soil Components

A multitude of studies, using bulk XAS to study mineral/water interfacial processes, have appeared in the scientific literature since the first published study by Hayes et al. (1987), who determined the surface complexation products for selenate and selenite on goethite (Sparks 1995, 2005b; Brown and Parks 2001; Brown and Sturchio 2002; Brown et al. 2006). These investigations have provided enlightening information on metal and oxyanion sorption on metal hydr(oxides), clay minerals, humic substances, and other natural materials, including structure, stiochiometry, attachment geometry (inner- vs. outer-sphere, monodentate vs. bidentate or tridentate), the presence of multinuclear complexes and precipitate phases, and the presence of ternary surface complexes when complexing ligands are present in solution (Brown and Parks 2001; Brown and Sturchio 2002; Sparks 2005b). The type of surface complexes on clay minerals and metal-(oxyhydr)oxides that occur with low atomic number elements, such as Al, B, Ca, Mg, S and Si, are not easy to ascertain using XAS under in-situ conditions. However, major advances are being made in the area of soft x-ray XAS spectroscopy that will enable one to directly determine the types of surface complexes that form with these metal(loids) (Ginder-Vogel and Sparks 2010; Sparks and Ginder-Vogel 2011).

Results from XAS studies, as well as from investigations using other in-situ molecular scale techniques such as Fourier transform infrared (FTIR) spectroscopy, allow one to make general conclusions on the predominate types of surface complexes that can occur on soil minerals. Although there are technical difficulties in analyzing alkaline earth cations, i.e., Mg^{2+}, Ca^{2+}, Sr^{2+}, and Ba^{2+}, they primarily form outer-sphere complexes. The divalent first-row transition metal cations Mn^{2+}, Fe^{2+}, Co^{2+}, Ni^{2+}, Cu^{2+}, and Zn^{2+}, and the divalent heavy metal cations such as Cd^{2+}, Hg^{2+}, and Pb^{2+} primarily form inner-sphere complexes. At higher metal loadings and pHs, sorption of metals such as Co, Cr, Ni, and Zn on phyllosilicates and metal-(oxyhydr) oxides can result in the formation of surface precipitates. The formation of these multinuclear and precipitate phases will be discussed in more detail later.

Experimental limitations are such that it is difficult to directly determine the type of surface complexes that NO_3^-, Cl^-, and ClO_4^- form on mineral surfaces. However, one can propose that they are sorbed as outer-sphere complexes and sorbed on surfaces that exhibit a positive charge. Some researchers have shown that SO_4^{2-} (Zhang and Sparks 1990) can be sorbed as an outer-sphere complex; however, there is other evidence that SO_4^{2-} can also be sorbed as an inner-sphere complex (Manceau and Charlet 1994). There is direct spectroscopic evidence to show that selenate can be sorbed as both an outer-sphere and an inner-sphere complex, depending on environmental factors (Hayes et al. 1987; Wijnja and Schulthess 2000).

Most other anions such as molybdate, arsenate, arsenite, selenite, phosphate, and silicate are strongly sorbed as inner-sphere complexes, and sorption occurs through a ligand exchange mechanism. The sorption maximum is often insensitive to ionic strength changes. Sorption of anions via ligand exchange results in a shift in the pzc of the sorbent to a more acidic value (Sparks 2002, 2005b).

However, bulk XAS probes an area of several square millimeters and only provides information on the average local chemical environment of a surface. Thus, if there is more than one type of surface species present, bulk XAS will detect only the primary (or average) type of surface product/species in the bulk sample. Therefore while the data may indicate that the primary surface product is an inner-sphere, this does not mean that outer-sphere complexes are not present. In fact, recent studies, such as those of Catalano et al. (2008), who employed X-ray scattering measurements to study metal(loid) binding on single crystal surfaces, showed that arsenate was sorbed simultaneously as inner- and outer-sphere complexes.

Environmental factors such as pH, surface loading, ionic strength, type of sorbent, and time all affect the type of sorption complex or product. Examples of this include the work of Strawn and Sparks (1999) who investigated Pb sorption on montmorillonite over an ionic strength (I) of 0.006–0.1 and a pH range of 4.48–6.77. At a pH of 4.48 and I of 0.006, outer-sphere complexation on basal planes in the interlayer regions of the montmorillonite predominated. At a pH of 6.77 and I of 0.1, inner-sphere complexation on edge sites of montmorillonite was most prominent, and at pH of 6.76, I of 0.006 and pH of 6.31, I of 0.1, both inner- and outer-sphere complexation occurred. These data are consistent with other findings that inner-sphere complexation is favored at higher pH and ionic strength. Recently, Lafferty et al. (2010), investigating As(III) oxidation kinetics on hydrous manganese oxide(HMO) showed that at shorter times (a few hours) As(V) was bound on HMO as bidentate binuclear and monodentate mononuclear complexes, but at longer times, a third arsenate complex, a bidentate-mononuclear complex appeared. Clearly, there is a continuum of adsorption complexes that can exist in soils (Sparks 2002, 2005b; Ginder-Vogel and Sparks 2010).

5.2.4.2 Metal(loid) Surface Precipitation/Dissolution

In addition to two-dimensional surface complexes, three dimensional Co^{2+}, Cr^{3+}, Cu^{2+}, Ni^{2+}, and Pb^{2+} metal hydroxide and mixed metal-Al surface precipitates, can form on metal oxides, phyllosilicates, soil clays, and soils (Charlet and Manceau 1992; Chisholm-Brause et al. 1994; Fendorf et al. 1994; Scheidegger et al. 1996a; Scheidegger et al. 1996b; Scheidegger et al. 1997; Scheidegger et al. 1998; Towle et al. 1997; Elzinga and Sparks 1999; Roberts et al. 1999; Thompson et al. 1999a; Thompson et al. 1999b; Ford and Sparks 2000; Scheckel and Sparks 2001b; Nachtegaal and Sparks 2004; McNear et al. 2006; Peltier et al. 2010). These metal hydroxide phases occur at metal loadings below theoretical monolayer coverage and in a pH range well below the pH where the formation of metal hydroxide precipitates would be expected, according to the thermodynamic solubility product (Scheidegger and Sparks 1996; Sparks 2002, 2005b; Borda and Sparks 2008; Ginder-Vogel and Sparks 2010).

Scheidegger et al. (1997) were the first to show that sorption of metals, such as Ni, on an array of phyllosilicates and Al-oxide, could result in the formation of mixed metal-Al hydroxide surface precipitates. The precipitate phase shares structural features common to the hydrotalcite group of minerals and the layered double hydroxides (LDH) observed in catalyst synthesis. The LDH structure is built of stacked sheets of edge-sharing metal octahedra, containing divalent and trivalent metal ions separated by anions between the interlayer spaces. The LDH structure exhibits a net positive charge x per formula unit which is balanced by an equal negative charge from interlayer anions A^{n-}, such as Cl^-, Br^-, I^-, NO_3^-, OH^-, ClO_4^-, and CO_3^{2-}; water molecules occupy the remaining interlayer space (Taylor 1984). Recently, Livi et al. (2009), using an array of microscopic techniques including analytical electron microscopy (AEM), high resolution transmission electron microscopy (HRTEM), and powder X-ray diffraction, conducted studies to elucidate the nature of Ni hydroxide precipitates, using the same environmental conditions employed by Scheidegger et al. (1996b, 1997) and reaction times ranging from 1 h to 5 years. While the precipitate phase had a bonding environment similar to Ni-Al LDH, the precipitate was amorphous.

The formation of metal hydroxide surface precipitates can significantly sequester metals (Scheckel and Sparks 2001a). As the surface precipitates age, metal release is greatly reduced (Fig. 5.4). Thus, the metals are less prone to leaching and being taken up by plants and microbes.

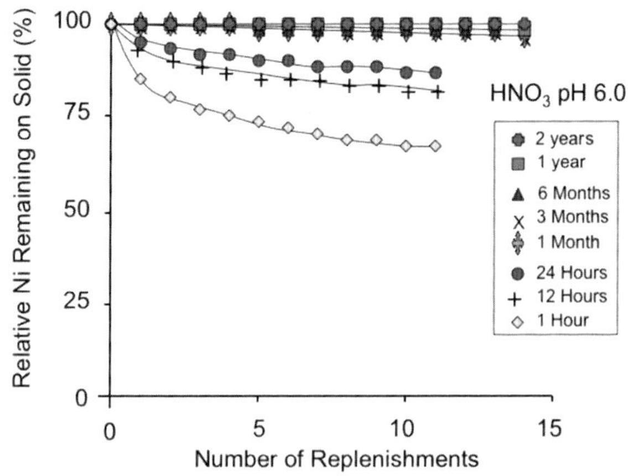

Fig. 5.4 Dissolution of Ni from surface precipitates formed on pyrophyllite at residence times of 1 h to 2 years. The figure shows the relative amount of Ni^{2+} remaining on the pyrophyllite surface following extraction for a 24 h period (each replenishment represents a 24 h extraction) with HNO_3 at pH 6.0 (Reprinted with permission Scheckel and Sparks 2001a)

The decrease in metal release and bioavailability is linked to the increasing silication of the interlayer of the LDH phases with increased residence time, resulting in a mineral transformation from a LDH phase to a precursor phyllosilicate surface precipitate (Ford et al. 1999; Ford and Sparks 2000). The mechanism for this transformation is attributed to diffusion of Si, originating from weathering of the sorbent, into the interlayer space of the LDH, replacing the anions such as NO_3^-. Peltier et al. (2006), using acid-solution calorimetry, and results from previous calorimetry studies, showed that the enthalpy of formation of LDH phases is more exothermic, indicating great stability, in the order of $Cl < NO_3 < SO_4 < CO_3 < Si$ of interlayer anionic composition, and that LDH phases were much more stable than a $Ni(OH)_2$ phase. The results of these and other studies show that with time, metal sorption on soil minerals involves a continuum of processes, from adsorption to precipitation to solid phase transformation, particularly in the case of certain metals, such as Co^{2+}, Ni^{2+}, and Zn^{2+}. The formation of metal surface precipitates is a way that metals can be naturally attenuated such that they are less mobile and bioavailable. Such products must be considered when modeling the fate and mobility of metals like Co^{2+}, Mn^{2+}, Ni^{2+}, and Zn^{2+} in soil and water environments (Sparks 2002; Sparks 2005b; Borda and Sparks 2008; Ginder-Vogel and Sparks 2010).

5.2.5 Speciation of Metal(loids) in Soils

Some examples of studies that have used μ-XAS and μ-XRD to speciate metal(loids) in soils include those by Hunter and Bertsch (1998), Manceau et al. (2000), Isaure et al. (2002), Strawn et al. (2002), Ginder-Vogel et al. (2005), Nachtegaal et al. (2005), Roberts et al. (2005), Arai et al. (2006), McNear et al. (2007), and Gräfe et al. (2008).

Nachtegaal et al. (2005) investigated the speciation of Zn in smelter contaminated soils, employing u-XAS and u-XRF, from a large site in Belgium in which part of the site had been remediated by adding beringite, an aluminosilicate material, compost, and planting metal tolerant plants. The other portion of the site was not treated. The objectives of the study were to determine how Zn speciation differed in the remediated (treated) and non-remediated (non-treated) soils, to investigate if Zn-LDH phases were present in the soils, and to establish the stability of the zinc under different environmental conditions.

A number of μ-XAS spectra for the treated (Fig. 5.5b) and non-treated (Fig. 5.5d) soils were collected at various regions of interest from the XRF spectra (Fig. 5.5a, treated and Fig. 5.5c, non-treated soils), as well as spectra for reference mineral, sorbed, and solution phases that were probable species in the soils. The μ-XAS spectra of the soils were analyzed via Principle Component Analysis, target transformation, and linear least squares fitting (LLSF) to determine the quantity (%) of each standard species within the individual sample spectra constituting the dataset (Manceau et al. 2002; Nachtegaal et al. 2005). Nachtegaal et al. (2005) found that both mineral (e.g., willemite, hemimorphite, spalerite) and sorbed (Zn-LDH) Zn species predominated in the treated and non-treated soils. The speciation differences in the treated and non-treated soils were minimal, with the major difference being the presence of kerolite, a Zn phyllosilicate phase, that was present in the treated soil (Fig. 5.5b). Significant quantities of Zn-LDH phases were formed in the non-treated soil (Fig. 5.5d). Desorption studies showed that the Zn in both remediated and non-remediated soils was quite stable, reflecting again the role that metal surface precipitates, i.e., Zn-LDH phases, play in sequestering metals such that mobility and bioavailability are diminished.

5.2.6 Speciation of Metals in Hyperaccumulator Plants

A very important biogeochemical interface in the Earth's Critical Zone is the plant/soil interface. Despite numerous studies, the biochemistry and dynamics of the rhizosphere are still poorly understood. As a way to advance our understanding of the plant/soil interface, scientists have used synchrotron-based XAS, XRF, XRD, and microtomography to investigate plant/soil interfacial processes and reactivity (Sarret et al. 2002, 2006; Küpper et al. 2004; Scheckel et al. 2004; McNear et al. 2005a, b; Tappero et al. 2007). Of particular interest is elucidating the speciation and uptake mechanisms of metals into hyperaccumulator plants. Such an enhanced comprehension could prove invaluable in increasing the effectiveness of phytoremediation in decontaminating metal polluted sites around the world. Phytoremediation is a "green" technology that uses plants to remove contaminants from the environment. Phytoextraction depends on unique plants, capable of accumulating higher than normal metal concentrations (e.g., >1,000 ppm for Ni and Co and >10,000 ppm for Zn) (Baker 1981; McNear et al. 2005a, b). To better understand the mechanisms involved in metal hyperaccumulation and tolerance,

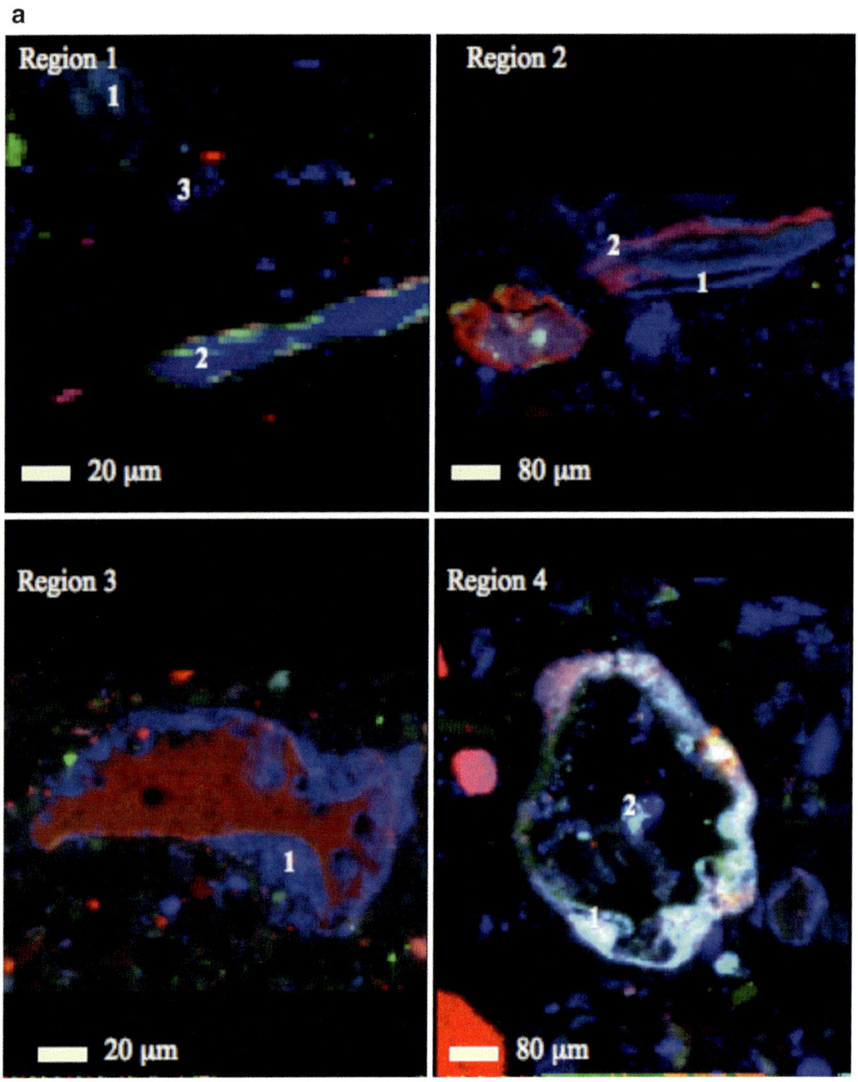

Fig. 5.5 (**a**) μ-SXRF tricolor maps for the treated soil samples. The numbers indicate the spots where μ-EXAFS spectra were collected. *Red* is indicative of the distribution of iron, *green* of copper and *blue* of zinc. (**b**) μ-EXAFS spectra from selected spots on thin sections from the treated soil. (**c**) μ-SXRF tricolor maps for the non-treated soil samples. (**d**) μ-EXAFS spectra from selected spots on thin sections from the non-treated soil. The *solid line* indicates the raw $k^3c(k)$ data and the *dotted line* indicates the best fits obtained with a linear fitting approach (Reprinted with permission Nachtegaal et al. 2005) (Color figure online)

it is critical to know whether accumulated metals are bound by strong (specific) ligands or loosely associated with common organic acids (i.e., speciation) as well as where these metals are stored (i.e., localization or compartmentalization). The use of synchrotron-based methods to explore metal speciation and uptake mechanisms

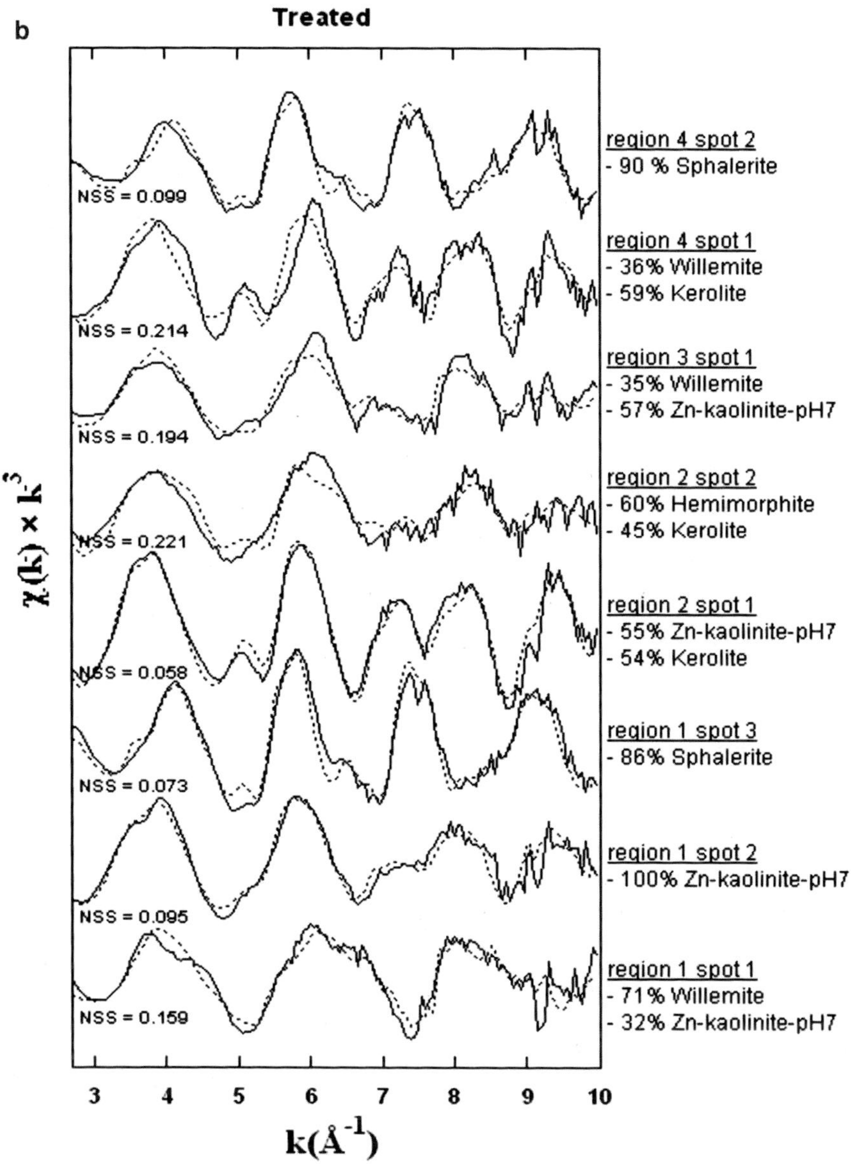

Fig. 5.5 (continued)

in plants is especially attractive because one can use "fresh" hyperaccumulator plant tissues (in vivo) with micrometer resolution and investigate distribution and association of the metals in the plant tissue on two-dimensional (μ-XRF) and three dimensional (microtomography) scales and determine direct metal speciation (μ-XAS). Several studies have employed these techniques to elucidate Ni and Co

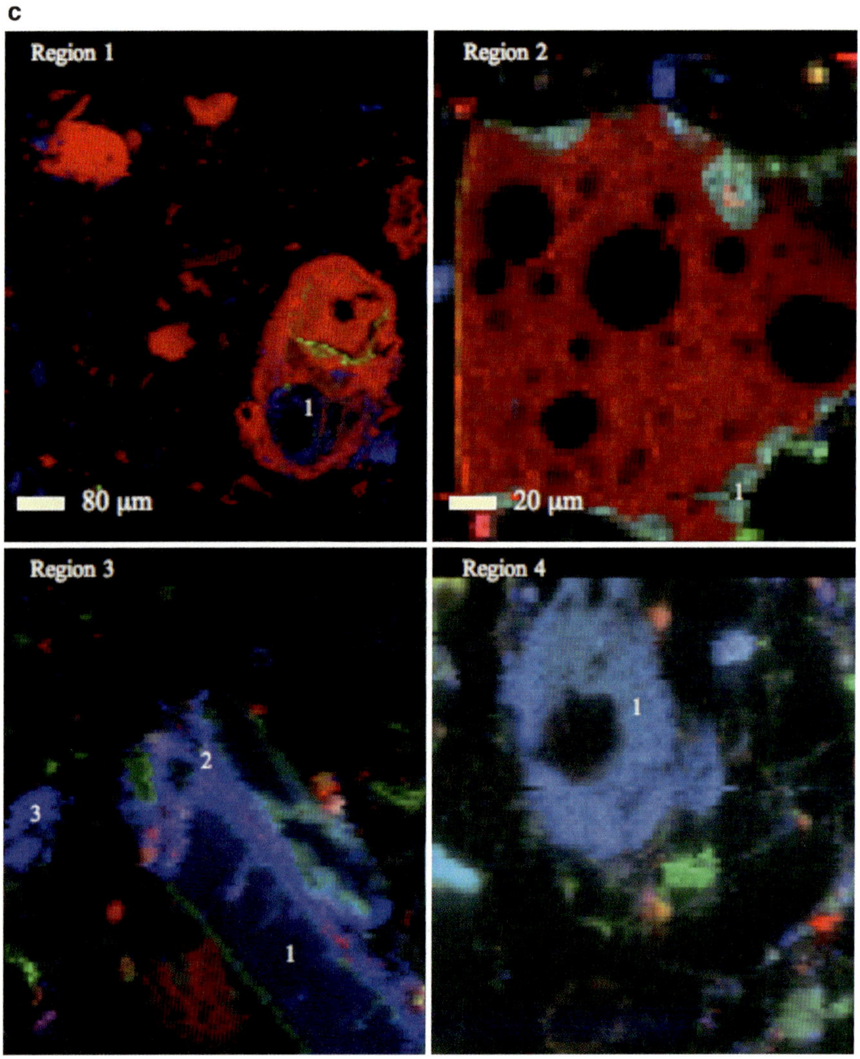

Fig. 5.5 (continued)

speciation in *Alyssum murale,* an important Ni hyperaccumulator plant from the Brassicaceae family that is native to serpentine soils throughout Mediterranean southern Europe (McNear et al. 2005a, b; Tappero et al. 2007; Ginder-Vogel and Sparks 2010; Sparks and Ginder-Vogel 2011).

Tappero et al. (2007) studied metal localization and elemental associations in *Alyssum murale* plants exposed to metal (Ni, Zn, and Co) co-contaminants using synchrotron-based μ-XRF, μ-XAS, and computed microtomography. Two dimensional XRF images of *Alyssum murale* leaves showed a marked localization pattern for Co compared to Ni and Zn. The Ni distribution was predominately

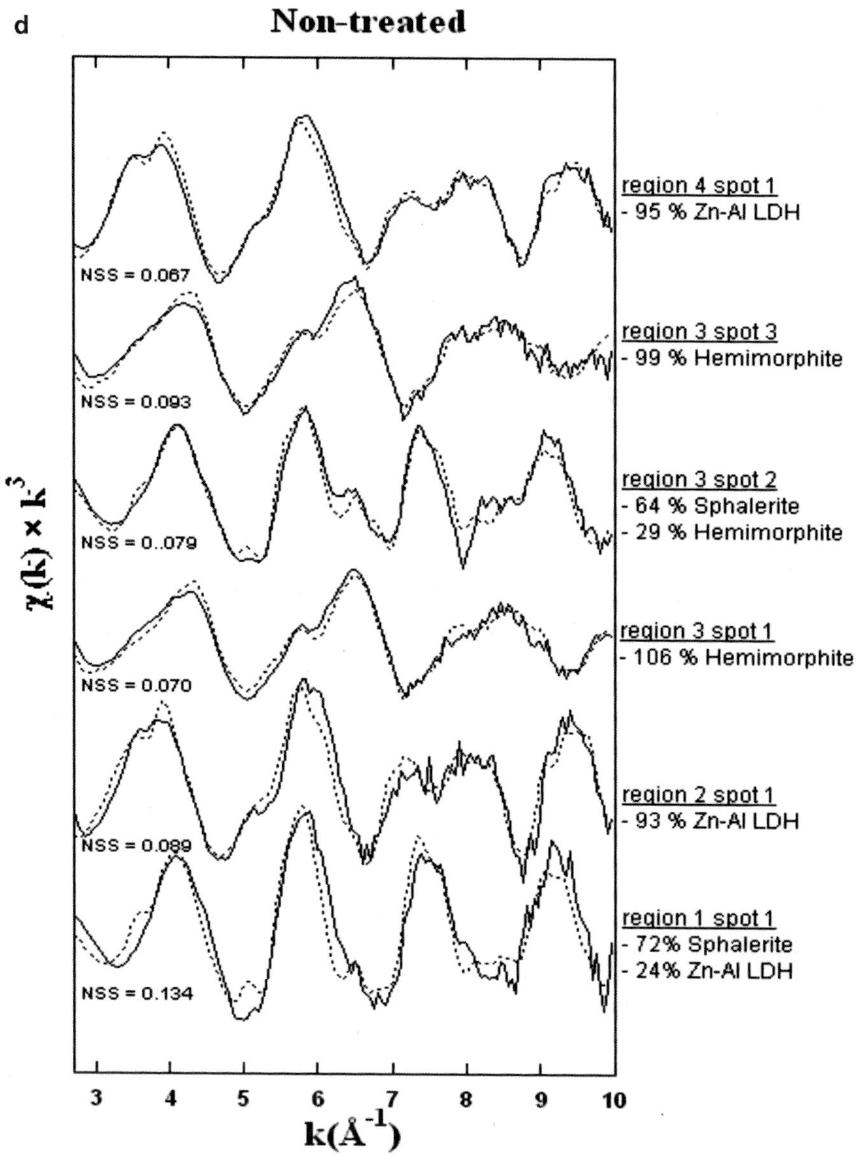

Fig. 5.5 (continued)

uniform (Fig. 5.6) which is consistent with previous findings that Ni is compartmentalized in epidermal tissues, i.e., vacuolar sequestration (Krämer et al. 1997, 2000; Broadhurst et al. 2004a, b). Zinc distribution was similar to Ni and was not hyperaccumulated. Cobalt however, was preferentially localized at the tips and margins of *A. murale* leaves as precipitate phases (Fig. 5.7). Differential absorption

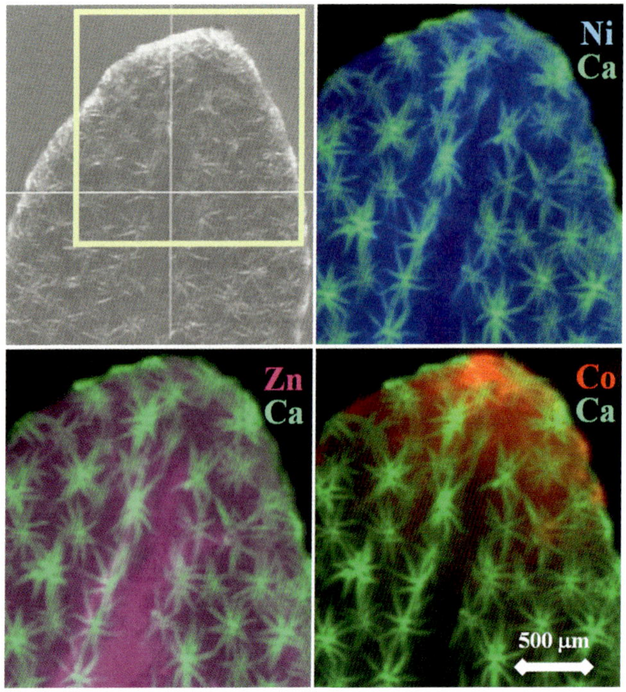

Fig. 5.6 Synchrotron X-ray microfluorescence (μ-SXRF) images of the nickel (Ni), cobalt (Co), and zinc (Zn) distributions in a hydrated *Alyssum murale* leaf from the Ni + Co + Zn treatment. Leaf trichomes are depicted in the Ca channel. The camera image shows the leaf region selected for SXRF imaging (Reprinted with permission Tappero et al. 2007)

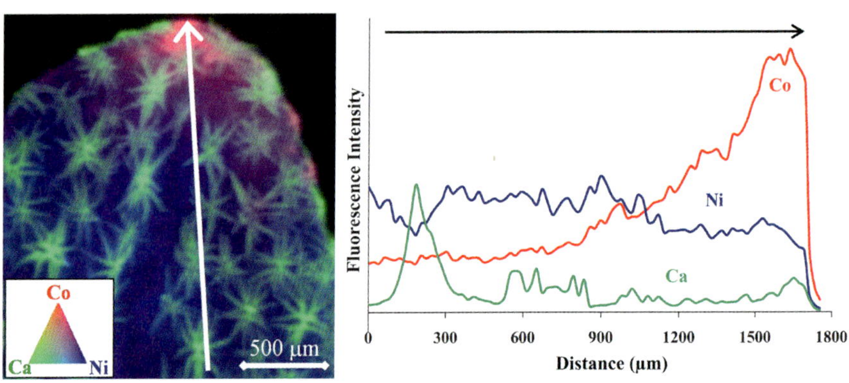

Fig. 5.7 Synchrotron X-ray microfluorescence (μ-SXRF) tricolor image (nickel (Ni), cobalt (Co), and calcium (Ca)) of a hydrated *Alyssum murale* leaf from the Ni + Co + Zn treatment, plus a line profile (fluorescence intensity vs. position) for a segment from the leaf center towards the leaf tip (indicated by a *white arrow*) (Reprinted with permission Tappero et al. 2007)

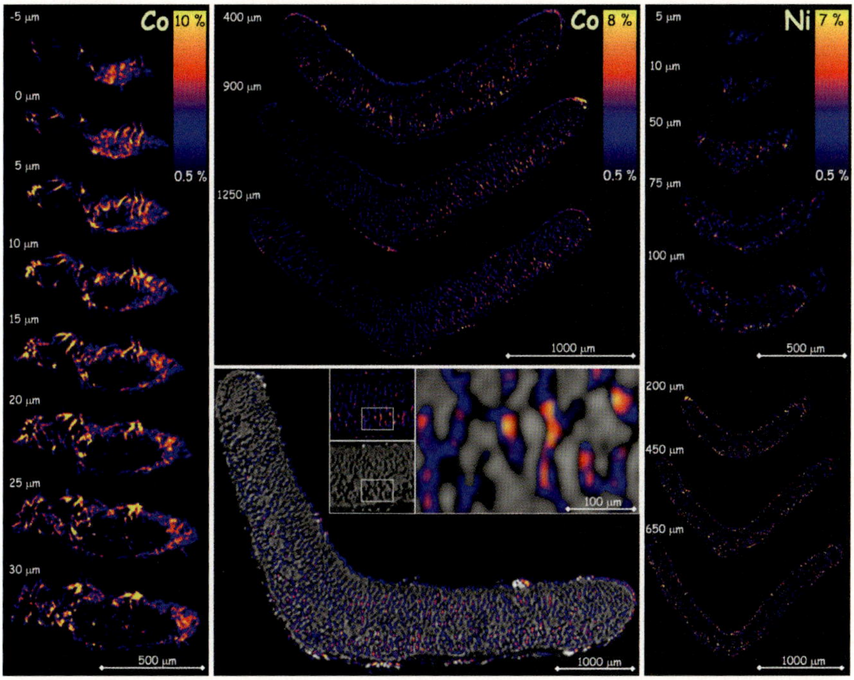

Fig. 5.8 Differential absorption (DA-CMT) tomographic projections (5.1 μm slices) of hydrated *Alyssum murale* leaves depicting (**a**) cobalt (Co) distribution in the leaf-tip region, (**b**) Co distribution in the bulk-leaf region, (**c**) Co distribution in relation to the leaf cell structure (*grey*), and (**d**) nickel (Ni) distribution in the leaf-tip and bulk-leaf regions. Leaves were collected from a Co-treated plant (**a**–**c**) and from a Ni-treated plant (**d**). Sinograms recorded above and below the Co or Ni K-edge energy (+30 and −100 eV, respectively) were computationally reconstructed and the resulting projections were subtracted (above – below) to reveal the metal distribution in leaves. Distances are relative to the leaf tissue at the tip as determined from leaf structure images (i.e. below-edge projections) (Reprinted with permission Tappero et al. 2007)

computed microtomography (DA-CMT) images (Fig. 5.8) of hydrated *A. murale* shows leaf Co enrichment in apoplasm tissue. Cobalt near leaf tips was localized primarily on the leaf exterior and the hyperaccumulation mechanism was via exocellular sequestration (Tappero et al. 2007).

5.2.7 Future Needs

As our knowledge of environmental systems continues to advance, it is critical that synchrotron-based techniques continue advancing with the field. Currently, a primary limitation of the application of these techniques to environmental systems is the availability of adequate amounts of beamtime. This problem can begin to be resolved by advances in both user support and instrumentation. Several improvements in user administration and services, including standardized data collection software, state-of-the-art data analysis software, increased beamline support staff, and improvements in laboratory support facilities would allow for the more efficient use of limited beamtime. The heterogeneity of environmental samples requires both the application of a wide array of techniques, including X-ray spectroscopic and traditional ones, and more time using each individual technique. The enhanced intensity and flux of third generation synchrotron light-sources allow for the analysis of both lower elemental concentrations and smaller samples; however, many environmental samples also require a broad energy range (5–50 keV), high energy resolution (0.1 eV) and a range of spatial resolutions (bulk, micro, and nano). Additionally, unique endstation capabilities, including flow-through reaction cells and anaerobic gloveboxes, coupled with fast-scanning, high energy resolution fluorescent detection and simultaneous collection of XRD and XAS data, will take advantage of the unique ability of X-ray absorption to analyze samples in situ, allowing a new generation of complex environmental problems to be solved (Ginder-Vogel and Sparks 2010; Sparks and Ginder-Vogel 2011).

References

Arai Y, Lanzirotti A, Sutton SR, Newville M, Dyer J, Sparks DL (2006) Spatial and temporal variability of arsenic solid-state speciation in historically lead arsenate contaminated soils. Environ Sci Technol 40(3):673–679

Baker AJM (1981) Accumulators and excluders – strategies in the response of plants to heavy metals. J Plant Nutr 3:643–654

Bertsch PM, Hunter DB (1998) Elucidating fundamental mechanisms in soil and environmental chemistry: the role of advanced analytical, spectroscopic, and microscopic methods. In: Huang PM (ed) Future prospects for soil chemistry. Soil Science Society of America, Madison, pp 103–122

Bertsch PM, Hunter DB (2001) Applications of synchrotron based X-ray microprobes. Chem Rev 101(6):1809–1842

Bertsch PM, Seaman JC (1999) Characterization of complex mineral assemblages: implications for contaminant transport and environmental remediation. Proc Natl Acad Sci USA 96(7):3350–3357

Borda MJ, Sparks DL (2008) Kinetics and mechanisms of sorption – desorption in soils: a multiscale assessment. In: Violante A, Huang PM, Gadd GM (eds) Biophysico-chemical processes of heavy metals and metalloids in soil environments. Wiley, New York, pp 94–124

Broadhurst CL, Chaney RL, Angle JS, Erbe EF, Maugel TK (2004a) Nickel localization and response to increasing Ni soil levels in leaves of the Ni hyperaccumulator Alyssum murale. Plant Soil 265(1):225–242

Broadhurst CL, Chaney RL, Angle JS, Maugel TK, Erbe EF, Murphy CA (2004b) Simultaneous hyperaccumulation of nickel, manganese, and calcium in Alyssum leaf trichomes. Environ Sci Technol 38(21):5797–5802

Brown GE Jr (1990) Spectroscopic studies of chemisorption reaction mechanisms at oxide-water interfaces. In: Hochella MF, White AF (eds) Mineral-water interface geochemistry. Mineralogical Society of America, Washington, DC, pp 309–353

Brown GE Jr, Parks GA (2001) Sorption of trace elements on mineral surfaces: modern perspectives from spectroscopic studies, and comments on sorption in the marine environment. Int Geogr Rev 43:963–1073
Brown GE Jr, Sturchio NC (2002) An overview of synchrotron radiation applications to low temperature geochemistry and environmental science. In: Fenter PA, Rivers ML, Sturchio NC, Sutton SR (eds) Applications of synchrotron radiation in low-temperature geochemistry and environmental sciences. Mineralogical Society of America, Washington, DC, pp 1–115
Brown GE Jr, Parks GA, O'Day PA (1995) Sorption at mineral-water interfaces: macroscopic and microscopic perspectives. In: Vaughan DJ, Pattrick RAD (eds) Mineral surfaces. Chapman & Hall, London, pp 129–183
Brown GE Jr, Calas G, Hemley RJ (2006) Scientific advances made possible by user facilities. Elements 2(1):23–30
Catalano JG, Park C, Fenter P, Zhang Z (2008) Simultaneous inner- and outer-sphere arsenate adsorption on corundum and hematite. Geochim Cosmochim Acta 72(8):1986–2004
Charlet L, Manceau A (1992) X-ray absorption spectroscopic study of the sorption of Cr(III) at the oxide-water interface. II: adsorption, coprecipitation and surface precipitation on ferric hydrous oxides. J Colloid Interface Sci 148:443–458
Chisholm-Brause CJ, Conradson SD, Buscher CT, Eller PG, Morris DE (1994) Speciation of uranyl sorbed at multiple binding sites on montmorillonite. Geochim Cosmochim Acta 58:3625–3631
Elzinga EJ, Sparks DL (1999) Nickel sorption mechanisms in a pyrophyllite-montmorillonite mixture. J Colloid Interface Sci 213(2):506–512
Fendorf SE, Sparks DL, Franz JA, Camaioni DM (1993) Electron paramagnetic resonance stopped-flow kinetic study of manganese (II) sorption–desorption on birnessite. Soil Sci Soc Am J 57(1):57–62
Fendorf SE, Lamble GM, Stapleton MG, Kelley MJ, Sparks DL (1994) Mechanisms of chromium (III) sorption on silica: 1. Cr(III) surface structure derived by extended x-ray absorption fine structure spectroscopy. Environ Sci Technol 28(2):284–289
Fenter PA, River ML, Sturchio NC, Sutton SR (eds) (2002) Applications of synchrotron radiation in low-temperature geochemistry and environmental science, vol 49, Reviews in mineralogy and geochemistry. The Mineralogical Society of America, Washington, DC
Ford RG, Sparks DL (2000) The nature of Zn precipitates formed in the presence of pyrophyllite. Environ Sci Technol 34(12):2479–2483
Ford RG, Scheinost AC, Scheckel KG, Sparks DL (1999) The link between clay mineral weathering and the stabilization of Ni surface precipitates. Environ Sci Technol 33(18):3140–3144
Ginder-Vogel M, Sparks DL (2010) The impacts of X-ray absorption spectroscopy on understanding soil processes and reaction mechanisms. In: Singh B, Grafe M (eds) Developments in soil science 34: synchrotron-based techniques in soils and sediments. Elsevier, Burlington, pp 1–26
Ginder-Vogel M, Borch T, Mayes MA, Jardine PM, Fendorf S (2005) Chromate reduction and retention processes within arid subsurface environments. Environ Sci Technol 39(20):7833–7839
Ginder-Vogel M, Landrot G, Fischel JS, Sparks DL (2009) Quantification of rapid environmental redox processes with quick-scanning x-ray absorption spectroscopy (Q-XAS). Proc Natl Acad Sci USA 106(38):16124–16128
Gräfe M, Tappero RV, Marcus MA, Sparks DL (2008) Arsenic speciation in multiple metal environments – II. Micro-spectroscopic investigation of a CCA contaminated soil. J Colloid Interface Sci 321(1):1–20
Hayes KF, Roe AL, Brown GE Jr, Hodgson KO, Leckie JO, Parks GA (1987) In situ x-ray absorption study of surface complexes: selenium oxyanions on α-FeOOH. Science 238:783–786
Hunter DB, Bertsch PM (1998) In situ examination of uranium contaminated soil particles by micro-X-ray absorption and micro-fluorescence spectroscopies. J Radioanal Nucl Chem 234(1–2):237–242
Isaure MP, Laboudigue A, Manceau M et al (2002) Quantitative Zn speciation in a contaminated dredged sediment by μ-PIXE, μ-SXRF, EXAFS spectroscopy and principal component analysis. Geochim Cosmochim Acta 66(9):1549–1567

Kelly SD, Hesterberg D, Ravel B (2008) Analysis of soils and minerals using X-ray absorption spectroscopy. In: Ulery AL, Drees LR (eds) Methods of soil analysis part 5 – mineralogical methods. Soil Science Society of America, Madison, pp 387–464
Khalid S, Caliebe WA, Siddons P et al (2010) Quick extended x-ray absorption fine structure instrument with millisecond time scale, optimized for in-situ applications. Rev Sci Instrum 81(1):015105
Krämer U, Smith RD, Wenzel WW, Raskin I, Salt DE (1997) The role of metal transport and tolerance in nickel hyperaccumulation by *Thlaspi goesingense Halacsy*. Plant Physiol 115(4):1641–1650
Krämer U, Pickering IJ, Prince RC, Raskin I, Salt DE (2000) Subcellular localization and speciation of nickel in hyperaccumulator and non-accumulator *Thlaspi* species. Plant Physiol 122(4):1343–1353
Küpper H, Mijovilovich A, Meyer-Klaucke W, Kroneck PMH (2004) Tissue- and age-dependent differences in the complexation of cadmium and zinc in the cadmium/zinc hyperaccumulator *Thlaspi caerulescens* (Ganges ecotype) revealed by X-ray absorption spectroscopy. Plant Physiol 134(2):748–757
Lafferty B, Ginder-Vogel M, Sparks DL (2010) Arsenite oxidation by a poorly crystalline manganese oxide. 1. Stirred-flow experiments. Environ Sci Technol 44:8460–8466
Landrot G, Ginder-Vogel M, Sparks DL (2010) Kinetics of chromium(III) oxidation by manganese(IV) oxides using quick scanning X-ray absorption fine structure spectroscopy (Q-XAFS). Environ Sci Technol 44:143–149
Livi KJT, Senesi GS, Scheinost AC, Sparks DL (2009) Microscopic examination of nano-sized mixed Ni-Al hydroxide surface precipitates on pyrophyllite. Environ Sci Technol 43(5):1299–1304
Manceau A, Charlet L (1994) The mechanism of selenate adsorption on goethite and hydrous ferric oxide. J Colloid Interface Sci 168:87–93
Manceau A, Charlet L, Boisset MC, Didier B, Spadini L (1992) Sorption and speciation of heavy metals on hydrous Fe and Mn oxides. From microscopic to macroscopic. Appl Clay Sci 7:201–223
Manceau A, Lanson B, Schlegel ML et al (2000) Quantitative Zn speciation in smelter-contaminated soils by EXAFS spectroscopy. Am J Sci 300(4):289–343
Manceau A, Marcus MA, Tamura N (2002) Quantitative speciation of heavy metals in soils and sediments by synchrotron X-ray techniques. In: Fenter P, Sturchio NC (eds) Applications of synchrotron radiation in low-temperature geochemistry and environmental science. Mineralogical Society of America, Washington, DC, pp 341–428
McNear DH, Peltier E, Everhart J et al (2005a) Application of quantitative fluorescence and absorption-edge computed microtomography to image metal compartmentalization in *Alyssum murale*. Environ Sci Technol 39(7):2210–2218
McNear DH Jr, Tappero R, Sparks DL (2005b) Shining light on metals in the environment. Elements 1:211–216
McNear DH, Tappero R, Sparks DL (2006) Elemental analysis of roots and fungi. In: Luster J, Finlay R (eds) Handbook of methods used in rhizosphere research. Swiss Federal Research Institute, Birmensdorf, pp 204–205
McNear DH, Chaney RL, Sparks DL (2007) The effects of soil type and chemical treatment on nickel speciation in refinery enriched soils: a multi-technique investigation. Geochim Cosmochim Acta 71(9):2190–2208
Nachtegaal M, Sparks DL (2004) Effect of iron oxide coatings on zinc sorption mechanisms at the clay-mineral/water interface. J Colloid Interface Sci 276:13–23
Nachtegaal M, Marcus MA, Sonke JE et al (2005) Effects of in situ remediation on the speciation and bioavailability of zinc in a smelter contaminated soil. Geochim Cosmochim Acta 69(19):4649–4664
NRC (2001) Basic research opportunities in earth science. National Academy Press, Washington, DC
O'Day PA (1999) Molecular environmental geochemistry. Rev Geophys 37(2):249–274

Parikh SJ, Lafferty BJ, Sparks DL (2008) An ATR-FTIR spectroscopic approach for measuring rapid kinetics at the mineral/water interface. J Colloid Interface Sci 320(1):177–185
Peltier E, Allada R, Navrotsky A, Sparks DL (2006) Nickel solubility and precipitation in soils: a thermodynamic study. Clays Clay Miner 54(2):153–164
Peltier E, van der Lelie D, Sparks DL (2010) Formation and stability of Ni-Al hydroxide phases in Delaware soils. Environ Sci Technol 44:302–308
Roberts DR, Scheidegger AM, Sparks DL (1999) Kinetics of mixed Ni-Al precipitate formation on a soil clay fraction. Environ Sci Technol 33(21):3749–3754
Roberts D, Nachtegaal M, Sparks DL (2005) Speciation of metals in soils. In: Tabatabai MA, Sparks DL (eds) Chemical processes in soils. Soil Science Society of America, Madison, pp 619–654
Sarret G, Saumitou-Laprade P, Bert V et al (2002) Forms of zinc accumulated in the hyperaccumulator *Arabidopsis halleri*. Plant Physiol 130(4):1815–1826
Sarret G, Harada E, Choi YE et al (2006) Trichomes of tobacco excrete zinc as zinc-substituted calcium carbonate and other zinc-containing compounds. Plant Physiol 141(3):1021–1034
Sayers DE, Stern EA, Lytle FW (1971) New technique for investigating noncrystalline structures: fourier analysis of extended X-ray – absorption fine structure. Phys Rev Lett 27(18):1204–1207
Scheckel KG, Sparks DL (2001a) Dissolution kinetics of nickel surface precipitates on clay mineral and oxide surfaces. Soil Sci Soc Am J 65(3):685–694
Scheckel KG, Sparks DL (2001b) Temperature effects on nickel sorption kinetics at the mineral-water interface. Soil Sci Soc Am J 65(3):719–728
Scheckel KG, Lombi E, Rock SA, McLaughlin NJ (2004) In vivo synchrotron study of thallium speciation and compartmentation in *Iberis intermedia*. Environ Sci Technol 38(19):5095–5100
Scheidegger AM, Sparks DL (1996) A critical assessment of sorption – desorption mechanisms at the soil mineral/water interface. Soil Sci 161(12):813–831
Scheidegger AM, Fendorf M, Sparks DL (1996a) Mechanisms of nickel sorption on pyrophyllite: macroscopic and microscopic approaches. Soil Sci Soc Am J 60(6):1763–1772
Scheidegger AM, Lamble GM, Sparks DL (1996b) Investigation of Ni sorption on pyrophyllite: an XAFS study. Environ Sci Technol 30(2):548–554
Scheidegger AM, Lamble GM, Sparks DL (1997) Spectroscopic evidence for the formation of mixed-cation hydroxide phases upon metal sorption on clays and aluminum oxides. J Colloid Interface Sci 186(1):118–128
Scheidegger AM, Strawn DG, Lamble GM, Sparks DL (1998) The kinetics of mixed Ni-Al hydroxide formation on clay and aluminum oxide minerals: a time-resolved XAFS study. Geochim Cosmochim Acta 62(13):2233–2245
Schulze DG, Bertsch PM (1995) Synchrotron x-ray techniques in soil, plant, and environmental research. Adv Agron 55:1–66
Schulze DG, Stucki JW, Bertsch PM (1999) Synchrotron x-ray methods in clay science, vol 98, CMS Workshop Lectures. Clay Minerals Society, Boulder
Sparks DL (1989) Kinetics of soil chemical processes. Academic, New York
Sparks DL (1995) Environmental soil chemistry. Academic, San Diego
Sparks DL (2002) Environmental soil chemistry, 2nd edn. Academic, San Diego
Sparks DL (2005a) Toxic metals in the environment: the role of surfaces. Elements 1:193–197
Sparks DL (2005b) Metal and oxyanion sorption on naturally occurring oxide and clay mineral surfaces. In: Grassian VH (ed) Environmental catalyis. Taylor & Francis, Boca Raton, pp 33–36
Sparks DL, Ginder-Vogel M (2011) The role of synchrotron radiation in elucidating the biogeochemistry of metal(loids) and nutrients at critical zone interfaces. In: Sumner ME, Li Y (eds) Handbook of soil sci. Taylor & Francis, Boca Raton
Strawn DG, Sparks DL (1999) The use of XAFS to distinguish between inner- and outer-sphere lead adsorption complexes on montmorillonite. J Colloid Interface Sci 216:257–269
Strawn D, Doner H, Zavarin M, McHugo S (2002) Microscale investigation into the geochemistry of arsenic, selenium, and iron in soil developed in pyritic shale materials. Geoderma 108:237–257

Sutton SR, Caffee MW, Dove MT (2006) Synchrotron radiation, neutron, and mass spectrometry techniques at user facilities. Elements 2:15–21

Tappero R, Peltier E, Grafe M et al (2007) Hyperaccumulator *Alyssum murale* relies on a different metal storage mechanism for cobalt than for nickel. New Phytol 175(4):641–654

Taylor RM (1984) The rapid formation of crystalline double hydroxy salts and other compounds by controlled hydrolysis. Clay Miner 19(4):591–603

Thompson HA, Parks GA, Brown GE Jr (1999a) Ambient-temperature synthesis, evolution, and characterization of cobalt-aluminum hydrotalcite-like solids. Clays Clay Miner 47(4):425–438

Thompson HA, Parks GA, Brown GE Jr (1999b) Dynamic interactions of dissolution, surface adsorption, and precipitation in an aging cobalt(II)-clay-water system. Geochim Cosmochim Acta 63(11–12):1767–1779

Towle SN, Bargar JR, Brown GE, Parks GA (1997) Surface precipitation of Co(II)(aq) on Al_2O_3. J Colloid Interface Sci 187(1):62–82

Wijnja H, Schulthess CP (2000) Vibrational spectroscopic study of selenate and sulfate adsorption mechanisms on Fe and A1 (hydr)oxides surfaces. J Colloid Interface Sci 229:286–297

Zhang PC, Sparks DL (1990) Kinetics and mechanisms of sulfate adsorption desorption on goethite using pressure-jump relaxation. Soil Sci Soc Am J 54(5):1266–1273

Chapter 6
Microbial Roles in Mineral Transformations and Metal Cycling in the Earth's Critical Zone

Geoffrey Michael Gadd

Abstract Microbes play key geoactive roles in the biosphere particularly in the areas of element biotransformations and biogeochemical cycling, metal and mineral transformations, decomposition, bioweathering, soil and sediment formation. All kinds of microbes, including prokaryotes and eukaryotes and their symbiotic associations with each other and "higher organisms", can contribute actively to geological phenomena, and central to many such geomicrobial processes are metal and mineral transformations. Microbial roles in mineral transformations and metal cycling are especially important in the Earth's Critical Zone (CZ) and microbes possess a variety of properties that can effect changes in metal speciation, toxicity and mobility, mineral formation or mineral dissolution or deterioration. Such mechanisms are important components of natural biogeochemical cycles for metals as well as associated elements in organic matter, soil, rocks and minerals, e.g. sulfur and phosphorus, and metalloids, actinides and metal radionuclides. Apart from being important in natural biosphere processes, metal and mineral transformations can have beneficial or detrimental consequences for human society. Bioremediation refers to the application of biological systems to the clean-up of organic and inorganic pollution with bacteria and fungi being the most important organisms for reclamation, immobilization or detoxification of metallic and radionuclide pollutants. Some biominerals or metallic elements deposited by microbes have catalytic and other properties in nanoparticle, crystalline or colloidal forms, and these are relevant to the development of novel biomaterials for structural, technological, environmental and antimicrobial purposes. In contrast, metal and mineral transformations by microbes may result in spoilage and destruction of natural and synthetic materials, rock and mineral-based building materials, e.g. concrete, acid mine drainage and associated metal pollution, biocorrosion of metals, alloys,

G.M. Gadd (✉)
Division of Molecular Microbiology, College of Life Sciences,
University of Dundee, Dundee DD15EH, UK
e-mail: g.m.gadd@dundee.ac.uk

J. Xu and D.L. Sparks (eds.), *Molecular Environmental Soil Science*,
Progress in Soil Science, DOI 10.1007/978-94-007-4177-5_6,

and related substances, and adverse effects on radionuclide speciation, mobility and containment, all with immense social and economic consequences. In view of the ubiquity and importance of microbes in such global processes, it can be proposed that an additional definition of the CZ could be "that portion of the terrestrial environment characterized by a significant microbial influence on metal and mineral transformations, organic matter decomposition, and the cycling of other elements".

Keywords Metals • Minerals • Radionuclides • Metalloids • Geomicrobiology • Geomycology • Bioremediation • Nanoparticles • Bioleaching • Biocorrosion • Biodeterioration

6.1 Microbes as Geoactive Agents

Microbes interact with metals and minerals altering their physical and chemical state, with metals and minerals also affecting microbial growth, activity and survival. In addition, many minerals are biogenic in origin, and biomineral formation is of global geological and industrial significance, as well as providing structural components for many organisms (Ehrlich 1996; Gadd and Raven 2010). Geomicrobiology can simply be defined as the roles of microbes in geological processes (Banfield and Nealson 1997; Banfield et al. 2005; Konhauser 2007; Ehrlich and Newman 2009; Gadd 2010). Metal-mineral-microbe interactions are of key importance within the framework of geomicrobiology and also fundamental to microbial biomineralization processes. The term biomineralization represents the collective processes by which organisms form minerals (Bazylinski 2001; Dove et al. 2003), and most biominerals are calcium carbonates, silicates, and iron oxides or sulfides (Baeuerlein 2000, 2001). Metal-mineral-microbe interactions are especially important in the so-called terrestrial "Earth's Critical Zone" (CZ) defined as "the heterogeneous, near-surface environment in which complex interactions involving rock, soil, water, air, and living organisms regulate the natural habitat and determine the availability of life sustaining resources" (Sparks 2005; Chorover et al. 2007; Amundson et al. 2007; Brantley et al. 2007). The CZ is therefore a complex mixture of air, water, biota, organic matter, and Earth materials, within which chemical and biochemical reactions occur which determine the provision of nutrients and energy for the biota and maintenance of terrestrial ecosystems, and characterized by pronounced physical, chemical and biological gradients (Brantley et al. 2007). It can also be defined as "the portion of the biosphere that lies at the interface of the lithosphere, atmosphere, and hydrosphere", and it encompasses soils and terrestrial ecosystems (Amundson et al. 2007). Clearly, processes occurring in the CZ can affect and influence processes in the hydrosphere and atmosphere, and *vice versa*, so to some extent, CZ is rather an artificial and loose term. Nevertheless, the importance of terrestrial system processes to global well-being is emphasized in the term, not least because of current challenges of population growth, climate change, ecosystem destruction and environmental pollution. Within this context, metal-mineral-microbe interactions play a very important role.

While the activities of microbes in transforming organic and inorganic substrates has long been known by microbiologists, there is growing awareness of the geochemical significance of microbes among researchers in geology, mineralogy, geochemistry and related disciplines (Banfield and Nealson 1997; Warren and Haack 2001; Macalady and Banfield 2003; Bottjer 2005; Gleeson et al. 2007; Konhauser 2007; Gadd 2008a, 2010; Uroz et al. 2009). Key topics within the geomicrobiology framework include biogeochemical cycling of the elements, mineral formation, mineral deterioration (which can include bioweathering and processes leading to soil and sediment formation), and chemical transformations of metals, metalloids and radionuclides (Ehrlich 1996). All kinds of microbes, including prokaryotes and eukaryotes and their symbiotic associations with each other and "higher organisms", can contribute to geological phenomena (Macalady and Banfield 2003; Bottjer 2005; Chorover et al. 2007; Konhauser 2007; Gleeson et al. 2007; Gadd 2008a, 2010), and most of these processes involve metal and mineral transformations (Table 6.1, Fig. 6.1). Examples of geomicrobially-important groups of microbes directly involved in geochemical transformations include iron-oxidizing and -reducing bacteria, manganese-oxidizing and -reducing bacteria, sulfate-reducing bacteria, sulfur-oxidizing and -reducing bacteria, and many other pro- and eukaryotes that can form or degrade silicates, carbonates, phosphates and other minerals (see Gadd 2007; Kim and Gadd 2008; Gadd and Raven 2010). Root-inhabiting rhizosphere microbes, including mycorrhizal fungi, have a major influence on plant nutrition by effects on phosphate availability and metal circulation (Amundson et al. 2007). In fact, during the early phases of soil formation the contribution of microbial activities (including the activities of lichens) to rock weathering, mineral dissolution and element cycling is also intimately related to metal movements and microbial strategies for metal transformations (Purvis and Pawlik-Skowronska 2008; Gilmour and Riedel 2009; Uroz et al. 2009). It should also be emphasized that general metabolic activities of all microbes affect metal distribution and bioavailability because of the metabolic essentiality of many metals and the existence of specific cellular uptake mechanisms, but also through decomposition or biodeterioration of organic and inorganic substrates (Warren and Haack 2001; Huang et al. 2004; Gadd 2007). Apart from being important in natural biosphere processes, metal- and mineral transformations can have beneficial or detrimental consequences in a human context. Bioremediation refers to the application of biological systems to the clean-up of organic and inorganic pollution with bacteria and fungi being the most important organisms in this context for reclamation, immobilization or detoxification of metallic pollutants. Some biominerals or metallic elements deposited by microbes have catalytic and other properties in nanoparticle, crystalline or colloidal forms (Lloyd et al. 2008; Theng and Yuan 2008; Petkov et al. 2009; Hennebel et al. 2009). In contrast, metal and mineral transformations by microbes may result in spoilage or destruction of natural and synthetic materials, rock and mineral-based building materials, as

Table 6.1 Microbial roles in the environmental cycling of metals and other elements

Element(s)	Microbial roles in elemental cycles
C, H, O	Uptake, assimilation, degradation and metabolism of organic and inorganic compounds; respiration (CO_2 production); photosynthesis; photorespiration; CO_2 fixation; biosynthesis of polymers, organic and inorganic metabolite excretion; humus formation; CN^- production; carbonate formation; oxalate formation; oxalate-carbonate cycle; dissolution of carbonates; methanotrophy; methanogenesis (archaea); hydrocarbon degraders; organometal(loid) degradation; metal(loid) biomethylation and demethylation; xenobiotic oxidation; CO utilization; water uptake; water transport, translocation and conduction (fungal mycelium); hydrogen oxidation and production
N	Decomposition of nitrogenous compounds; assimilation and transformations of organic and inorganic N compounds; N_2 fixation (prokaryotes only); nitrification and denitrification; ammonia and nitrite oxidation; anaerobic nitrification; biosynthesis of N-containing biopolymers, e.g. chitin; production of N-containing metabolites and gases, e.g. N_2O; ammonia fermentation under anaerobic conditions; mycorrhizal N transfer to plants (fungi); fixed N transfer to plants (symbiotic N_2 fixers)
P	Dissolution of inorganic phosphates and P-containing minerals in soils and rocks; decomposition of P-containing organic compounds; formation of insoluble P, e.g. polyphosphate, secondary phosphate minerals; release of organically-bound P by phosphatases; assimilation and transformation of inorganic P species; oxidation of reduced forms of phosphate, e.g. phosphate; transformations of soil organic P; production of diphosphates and phosphonates; P transfer to plants (mycorrhizas)
S	Degradation of organic S-containing compounds; organic-inorganic S transformations; uptake and assimilation of organic and inorganic S compounds; sulfidogenesis; S(0) accumulation; SO_4^{2-} reduction and assimilation; S(0) reduction; oxidation of reduced S compounds, e.g. S(0), thiosulfate, tetrathionate; oxidation of H_2S to S(0); reduction of S(0) to H_2S; dissolution of S-containing minerals in soils and rocks, e.g. sulfides, sulfates
Fe	Bioweathering of iron-containing minerals in rocks and soils; iron solubilization by siderophores, organic acids, metabolites etc.; Fe(III) reduction to Fe(II); Fe(II) oxidation to Fe(III); Fe biomineralization, e.g. oxides, hydroxides, carbonates, sulfides; metal sorption to Fe oxides
Mn	Mn(II) oxidation and immobilization as Mn(IV) oxides; Mn(IV) reduction; indirect Mn(IV)O_2 reduction by metabolites, e.g. oxalate; bioaccumulation of Mn oxides to surfaces and exopolymers; contribution to desert varnish formation; biosorption; accumulation; intracellular precipitation; Mn biomineralization, e.g. oxides, carbonates, sulfides, oxalates; metal sorption to Mn oxides
Cr	Cr(VI) reduction to Cr(III); Cr(III) oxidation; accumulation of Cr oxyanions
Mg, Ca, Co, Ni, Zn, Cd, Sr	Bioweathering of minerals in rocks and soil; biosorption; uptake and accumulation; bioprecipitation, e.g. oxalates, sulfides, phosphates, carbonate; Co(III) reduction
Ag	Reduction of Ag(I) to Ag(0); biosorption; accumulation
K, Na, Cs	Uptake and accumulation; translocation through mycelium (fungi); concentration in fruit bodies (fungi); mobilization from minerals in soil

(continued)

Table 6.1 (continued)

Element(s)	Microbial roles in elemental cycles
Cu	Mobilization from Cu-containing minerals in rocks and soils; CuS formation; biosorption; uptake and accumulation; bioprecipitation, e.g. oxalates
Se	Reductive transformation of Se oxyanions, e.g. Se(VI) to Se(IV) to Se(0); Se(0) oxidation; biomethylation and demethylation of Se compounds; assimilation of organic and inorganic Se compounds
Te	Reductive transformation of Te oxyanions, e.g. Te(VI) to Te(IV) to Te(0); Biomethylation; assimilation of organic and inorganic Te compounds
Pb	Biosorption; lead oxalate formation; biomethylation
Cl, Br, I	Dehalorespiration; biomethylation; accumulation in biomass
Sn	Organotin degradation; sorption and accumulation of soluble Sn species; biomethylation
Au	Reduction of soluble Au species to Au(0); Au mineral dispersion and solubilization
As	Biomethylation of arsenic species, e.g. arsenite to trimethylarsine; reduction of As oxyanions, e.g. arsenate to arsenite; oxidation of As oxyanions, e.g. arsenite to arsenate
Hg	Hg biomethylation; reduction of Hg(II) to Hg(0); oxidation of Hg(0) to Hg(II); Hg volatilization as Hg(0); degradation of organomercurials; biosorption; accumulation
Al	Al mobilization from Al-containing minerals in soils and rocks; aluminosilicate dissolution; Al precipitation as oxides (early stage of bauxitization); biosorption
Si	Uptake of soluble Si species; organic Si complex formation from inorganic silicates; organic siloxane formation; silica, silicate and aluminosilicate degradation; Si mobilization through production of chelators, acids, bases, exopolymers; silicification; structural biomineralization (some algae, protozoa)
U, Th	Biosorption; deposition of hydrolysis products; intracellular precipitation; U(VI) reduction to U(IV); U(IV) oxidation to U(VI); U biomineralization, e.g. phosphates; UO_2 formation
Tc	Pertechnetate accumulation; Tc(VII) reduction to Tc(IV); oxide formation
V	Vanadate accumulation; V(V) reduction to V(IV)

Some of the major or representative microbial rôles in metal and other elemental cycles are indicated without reference to their relative global significance. Only representative elements are shown: note virtually all elements in the Periodic Table (including actinides, lanthanides, radionuclides) can be accumulated by or be associated with microbial biomass depending on the context and environment. Microbes possess transport systems for essential metals; inessential metal species can also be taken up. Microbes are also capable of mediating metal- and mineral bioprecipitation by, e.g. metabolite production, changing the physico-chemical microenvironmental conditions around the biomass, and also by the indirect release of metal-precipitating substances from other activities, e.g. phosphate from organic decomposition or phosphate mineral solubilization. Many different metal-containing minerals formed as a direct/indirect result of microbial activity, e.g. various carbonates, phosphates etc., are omitted from the Table. Microbial cell walls, outer layers, and exoploymers can sorb, bind or entrap many soluble and insoluble metal species as well as, e.g. clay minerals, colloids, oxides etc. which also have significant metal sorption properties. Redox transformations are also widespread in microbial metabolism, some also mediated by the chemical activity of structural components

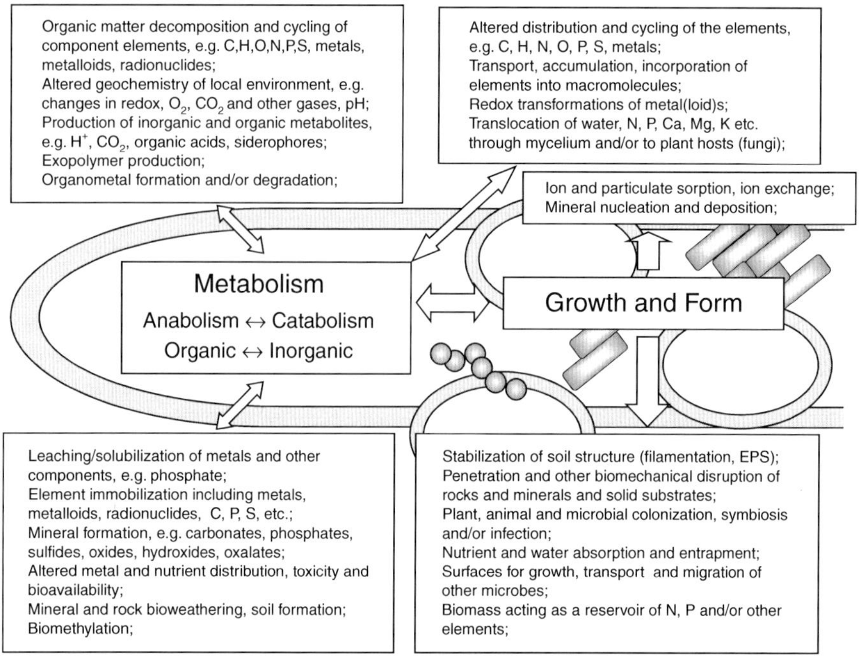

Fig. 6.1 Some of the important microbial roles and activities in biogeochemical processes leading to altered metal distributions and speciation in the CZ. Such activities take place in the terrestrial (and aquatic) environment, as well as in artificial and man-made systems, their relative importance depending on the populations present and physico-chemical factors. Most if not all of these roles are inter-linked, almost all directly or indirectly depending on the mode of growth, e.g. unicellular, filamentous, colonial, biofilm, and including symbiotic partnerships and mode of metabolism, in turn dependent on nutrient and energy source(s) for biosynthesis and energy, essential elements, such as C, N, O, P, S and many metals, for structural and cellular components (Adapted from Gadd 2008a, 2010)

well as acid mine drainage and associated metal pollution, biocorrosion of metals, alloys, and related substances, and adverse effects on radionuclide speciation, mobility and containment. In view of the ubiquity and importance of microbes in biosphere processes, it can easily be argued that geomicrobiology is one of the most important concepts within microbiology, and requiring an interdisciplinary approach involving physical, chemical and biological disciplines. This especially appears to apply to an understanding of the CZ (Chorover et al. 2007). This chapter seeks to outline some of the main ways that microbes (chiefly bacteria and fungi) interact with metals and minerals, their importance in geological and environmental processes, and their applied significance within the concept of CZ function. I hope to demonstrate that another definition of the CZ could be "that portion of the terrestrial environment characterized by a significant microbial influence on metal and mineral transformations, organic matter decomposition, and the cycling of other elements".

6.2 Metals and Minerals

Metals comprise about 75% of the known elements and are ubiquitous in the biosphere. Since the industrial revolution, metals are increasingly redistributed in the environment and are often associated with adverse effects on the biota and human health. Thirteen trace metals and metalloids (Ag, As, Be, Cd, Cr, Cu, Hg, Ni, Pb, Sb, Se, Tl, Zn) are considered to be priority pollutants (Sparks 2005), and originate from natural sources such as rocks and metalliferous minerals, and anthropogenic inputs (Landa 2005; Gilmour and Riedel 2009). Atmospheric deposition is a major mechanism for metal input to soils. In terrestrial ecosystems, soils are the major sink for metal contaminants, while sediments are the major sink for metals in aquatic systems. Metals in soil can impact aquatic systems through runoff, leaching and transport via mobile colloids (Adriano 2001; Adriano et al. 2004a, b). Metals are significant natural components of all soils in the mineral fraction and represent a pool of potentially-mobile metal species, many essential nutrients for plants and microbes, and important solid components that can have a fundamental effect on soil biogeochemical processes, e.g. clays, minerals, iron and manganese oxides (Huang et al. 2004). Metals are also present in organic soil fractions, frequently in bound forms, with metal recycling occurring as a result of organic matter degradation. The aqueous phase provides a mobile medium for metal transfer and circulation through the soil to the aquatic environment (Warren and Haack 2001). The aquatic environment also contains a vast pool of metals in various chemical forms, with many accumulating in sediments (Warren and Haack 2001; Gilmour and Riedel 2009). Other contaminants related to metals are metalloids, organometals, and organometalloids, while many radionuclides that enter the environment are metals.

Minerals are naturally-occurring inorganic solids of definite chemical composition with an ordered internal structure. Rocks can be considered to be any solid mass of mineral or mineral-like material. Silicates are the most common minerals with non-silicates constituting <10% of the Earth's crust, the most common being carbonates, oxides, sulfides and phosphates. Rocks and minerals represent a vast reservoir of elements, many essential for life, and these elements must be released in forms that may be assimilated by the biota. These include essential metals as well as nutrient elements like S and P (Gadd 2007; Gadd et al. 2005, 2007). Physical, chemical and biological mechanisms contribute to weathering of rocks and minerals, and for the latter, metal-microbe interactions are involved in the majority of cases. Activities of microbes (and other organisms) in rock and mineral deterioration can be termed bioweathering.

The vast majority of minerals contain metals, and there can be a range of mineral types for each metal, of varying distribution in the environment and physico-chemical properties. For example, manganese is found as a major or minor component in more than 100 naturally-occurring minerals of which major accumulations are as oxides, carbonates, and silicates (Ehrlich and Newman 2009). In addition, metals may be common components of many minerals as impurities from natural and industrial sources: potentially-toxic and other metals may be found

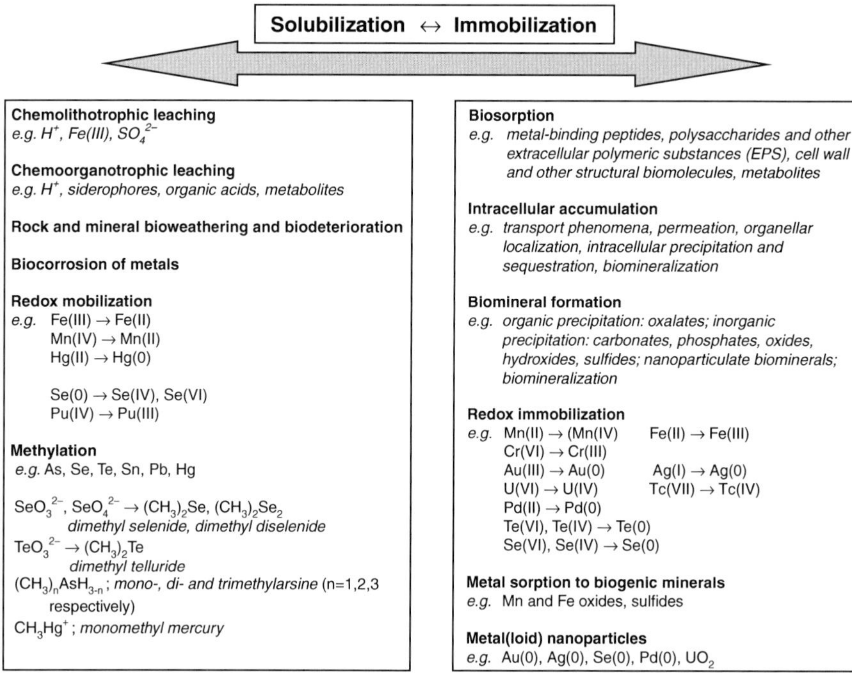

Fig. 6.2 Diagram depicting the major mechanisms of microbial metal transformations between soluble and insoluble metal species. The relative balance between such processes depends on the environment and associated physico-chemical conditions and the microbe(s) involved as well as relationships with plants, animals, and anthropogenic activities. Chemical equilibria between soluble and insoluble phases are influenced by abiotic components, including dead biota and their decomposition products, as well as other physico-chemical components of the environmental matrix, e.g. pH, water, inorganic and organic ions, molecules, compounds, colloids, and particulates. Solubilization can occur by chemolithotrophic (autotrophic) and chemoorganotrophic (heterotrophic) leaching; siderophores and other complexing agents; redox reactions; methylation and demethylation; biodegradation of organo-radionuclide complexes. Immobilization can occur by biosorption to cell walls, exopolymers, other structural components and derived/excreted products; precipitation can be a result of metabolite release, e.g. sulfide, oxalate, or reduction; transport, accumulation, intracellular deposition, localization and sequestration; adsorption and entrapment of colloids and particulates. The overall scheme is also affected by reciprocal interactions between biotic and abiotic components of the ecosystem such as abiotic influence on microbial diversity, numbers and metabolic activity; ingestion of particulates and colloids (including bacteria) by phagotrophs; biotic modification of physico-chemical parameters including redox potential, pH, O_2, CO_2, other gases and metabolites, temperature, and nutrient depletion (see Gadd 2010)

in, e.g. silicates, sulfides, oxides, carbonates and phosphates. Regardless of their geologic or biogenic origin, microbes can interact with minerals and affect their structure by mechanical and biochemical mechanisms, and affect the speciation of component metals (Ehrlich 1996; Edwards et al. 2005; Landa 2005; Gleeson et al. 2007; Lian et al. 2008a, b; Gadd 2008b; Brown et al. 2008; Uroz et al. 2009) (Table 6.1, Fig. 6.2).

In contrast to mineral deterioration, dissolution or fragmentation, all groups of microbes can mediate mineral formation by direct and indirect mechanisms. Such minerals can be termed biominerals, to emphasise the involvement of living organisms although the chemical structure of the minerals so produced may be identical to minerals produced by geochemical mechanisms. Biomineralization may particularly refer to processes where biomineral formation is a highly-directed process and fundamental to the formation of a cellular structure, e.g. magnetosomes composed of magnetite in magnetotactic bacteria (Bazylinski and Moskowitz 1997; Bazylinski 2001; Posfai and Dunin-Borkowski 2009), silicaceous cell walls of diatoms (algae) and radiolarians (protozoa), and carbonate tests of various amoebae and algae (e.g. coccolithophores) (Banfield and Nealson 1997; Dove et al. 2003; Banfield et al. 2005; Ehrlich and Newman 2009; Gadd and Raven 2010). Other biominerals may arise from redox transformations of metals, sorption phenomena, and metabolic activities where organic and inorganic metabolites e.g. oxalate, respiratory CO_2, and sulfide, may precipitate metals in the cellular microenvironment, or effect chemical changes in the substrate which also lead to secondary mineral formation (Ehrlich 1996; Hamilton 2003; Glasauer et al. 2004; Konhauser 2007; Ehrlich and Newman 2009).

6.3 Microbes, Metals and Minerals

Metals are directly and/or indirectly involved in all aspects of microbial growth, metabolism and differentiation (Gadd 1992a). Metals and their compounds interact with microbes in various ways depending on the metal species, organism and environment, while structural components and metabolic activity also influence metal speciation and therefore solubility, mobility, bioavailability, and toxicity (Gadd and Griffiths 1978; Gadd 1992a, 1993a, 2004, 2005, 2007) (Table 6.1, Fig. 6.2). Many metals are essential for life, e.g. Na, K, Cu, Zn, Co, Ca, Mg, Mn, and Fe, but all exert toxicity when present above certain threshold concentrations. Other metals, e.g. Cs, Al, Cd, Hg and Pb, have no known essential metabolic functions but can still be accumulated. Microbes are intimately associated with the biogeochemical cycling of metals, and associated elements, where their activities can result in mobilization and/or immobilization depending on the mechanism(s) involved and the microenvironment where the organism(s) are located (Gadd 2004, 2007; Violante et al. 2008; Ehrlich and Newman 2009).

Metals exhibit a range of microbial toxicities, and while toxic effects can arise from natural geochemical events, toxic effects on microbial communities are more commonly associated with anthropogenic contamination of ecosystems from aerial and aquatic sources, as well as agricultural and industrial activities, and domestic and industrial wastes (Gadd 2009a; Violante et al. 2008). It is commonly accepted that toxic metals, their chemical derivatives, metalloids, and organometals, can have significant effects on microbial populations and, under toxic conditions, almost every index of microbial activity can be affected (Giller et al. 2009). However,

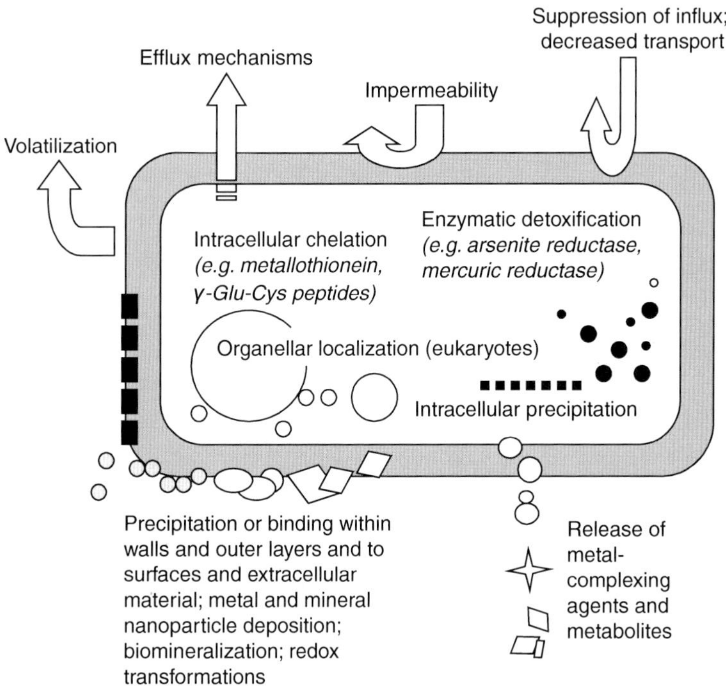

Fig. 6.3 Mechanisms involved in detoxification and transformation of metals including mechanisms that restrict entry into the cell and intracellular detoxification or organellar compartmentation, the latter occurring in some eukaryotes, e.g. fungi. Operation of a number of mechanisms is possible depending on the organism and the cellular environment, some dependent and/or independent of metabolism. A variety of mechanisms may be involved in transport phenomena contributing to decreased uptake and/or efflux. A variety of specific or non-specific mechanisms may also effect redox transformations, intracellular chelation, and intracellular precipitation. Biomineral formation (biomineralization) may be biologically-induced, i.e. caused by physico-chemical environmental changes mediated by the microbes, or biologically-controlled (*solid rectangles*) (Adapted from Gadd 2009a, 2010)

metal toxicity is greatly affected by the physico-chemical nature of the environment and the chemical behaviour of the metal species (Gadd and Griffiths 1978, 1980). Despite potential toxicity, many microbes grow and even flourish in apparently metal-polluted locations and a variety of mechanisms, both active and incidental, contribute to resistance (Gadd and Griffiths 1978; Mowll and Gadd 1984; Gadd et al. 1984; Avery 2001; Holden and Adams 2003; Fomina et al. 2005c; Silver and Phung 2009). (Fig. 6.3).

Most microbial metal resistance mechanisms depend on some change in metal speciation leading to decreased or increased mobility (Osman and Cavet 2008; Silver and Phung 2009). These include redox transformations, the production of metal-binding peptides and proteins (e.g. metallothioneins, phytochelatins), organic and inorganic precipitation, active transport, efflux and intracellular compartmentalization, while cell walls and other structural components have significant metal binding abilities (Mowll and Gadd 1984; White and Gadd 1998; Gadd 2004, 2005, 2006) (Figs. 6.2 and 6.3). Other microbial properties lead to metal solubilization from organic and inorganic sources (Gadd 2007). Such metal transformations are central to metal biogeochemistry, and emphasize the link between microbial responses and geochemical cycles for metals (Ehrlich 1997; Gilmour and Riedel 2009). Such metal-mineral-microbe interactions are especially important in the CZ (Sparks 2005; Chorover et al. 2007; Amundson et al. 2007).

Many important microbial processes can be influenced by minerals including energy generation, nutrient acquisition, cell adhesion and biofilm formation (Hochella 2002; Brown et al. 2008). Essential nutrients may be acquired from mineral surfaces and this concentrates these substances above surrounding environmental levels, e.g. C, N, P, Fe, essential metals, and various organic compounds (Vaughan et al. 2002). Environmental contaminants, including metals, may also be sorbed to mineral surfaces and these can be displaced by microbial activity (Kraemer et al. 1999; Huang et al. 2004; Chorover et al. 2007; Theng and Yuan 2008). Potentially toxic metals released from minerals as a result of physico-chemical and biological processes may also affect microbial communities (Fomina et al. 2005c; Gadd 2005). Such properties of mineral surfaces as microtopography, surface composition, surface charge and hydrophobicity play an important role in thigmotropism, microbial attachment and detachment, and are therefore critical for colonization and biofilm formation, and the ecology of microbial populations associated with mineral substrates (Vaughan et al. 2002; Gleeson et al. 2005, 2006, 2010; Bowen et al. 2007, 2008). Interactions of soil minerals with organic substances and microbes also have an impact on the formation and transformation of metal oxides. Al and Fe oxides, especially in their nanoparticulate forms, are among the most reactive component of acidic and neutral soils. Such metal oxides are ubiquitous and play a significant role in influencing soil behaviour, e.g. mineral catalysis of humic substance formation, and influence on enzymatic stability and microbial activity, and, together with microbial activities in metal and mineral transformations have a great impact on soil physical, chemical, and biological processes (Huang et al. 2005).

6.4 Metal Mobilization

Metal mobilization from rocks, minerals, soil and other substrates can be achieved by protonolysis, complexation by excreted metabolites and Fe(III)-binding siderophores, chemical oxidation or reduction, indirect Fe(III) attack, and methylation which can result in volatilization to the atmosphere (Fig. 6.2). Other excreted metabolites with metal-complexing properties, e.g. amino acids, phenolic compounds, and organic acids may also play a role. Low molecular weight carboxylic acids can play an important role in chemical attack of minerals providing protons as well as a metal-chelating anion (Burgstaller and Schinner 1993; Jacobs et al. 2002a, b; Huang et al. 2004; Lian et al. 2008a, b). Oxalic acid can leach metals that form soluble oxalate complexes, e.g. Al and Fe (Strasser et al. 1994) and such activity may be involved in the weathering of silicate minerals such as feldspars (Drever and Stillings 1997). Solubilization mechanisms may have adverse consequences when metals are mobilized from toxic metal-containing minerals, e.g. pyromorphite ($Pb_5(PO_4)_3Cl$), contaminated soil and other solid wastes (Sayer et al. 1999; Fomina et al. 2004, 2005a, b).

Microbes can also mobilize metals and attack mineral surfaces by redox processes (Ehrlich 1996; Lloyd and Lovley 2001; Holden and Adams 2003; Schröder et al. 2003; Lloyd et al. 2003): Fe(III) and Mn(IV) solubility is increased by reduction to Fe(II) and Mn(II) respectively (Table 6.1, Fig. 6.2). Microbial reduction of Fe(III) and Mn(IV) may also be a means for releasing contaminant metals absorbed to Fe(III) and Mn(IV) oxides and this process may be enhanced by humic materials, or related compounds (Lovley and Coates 1997; Lloyd et al. 2003). Bacterial Fe(III) reduction resulted in release of, e.g. Mn and Co, from goethite (Bousserrhine et al. 1999), Pu from contaminated soils (Rusin et al. 1993) and Ra from uranium mine tailings (Landa and Gray 1995). Mercuric ion (Hg^{2+}) can be enzymatically reduced to metallic mercury by bacteria and fungi which serves as a resistance mechanism as Hg^0 is volatile (Gadd 1993b; Lloyd et al. 2003; Barkay and Wagner-Dobler 2005). Metallic mercury may also be oxidized to Hg^{2+} interaction with metabolic by-products (Barkay and Wagner-Dobler 2005; Ehrlich and Newman 2009). Enzymatic reduction of plutonium(IV) to more soluble plutonium(III) under anaerobic conditions was demonstrated for *Geobacter metallireducens* GS-15 and *Shewanella oneidensis* MR-1 (Boukhalfa et al. 2007). Ferric iron, Fe(III), can be enzymatically reduced to ferrous iron with a suitable electron donor (Schröder et al. 2003). Many Fe(III) reducers are heterotrophs and such Fe(III) respiration may be a more important mechanism of carbon source decomposition in some anaerobic environments than sulfate reduction (Ehrlich and Newman 2009). Some ferric iron reduction can be the result of metabolic products such as H_2S or formate, or other secondary metabolites. Naturally-occurring microbially-produced metal chelators that may solubilize Fe(III) include oxalate, citrate, humic acids, and tannins.

Methylated derivatives of several elements naturally arise in the environment as a result of chemical and biological methylation, microbes playing significant roles in the latter process (Thayer 1989; Gadd 1993b). Methylation of Hg, Sn, and Pb, and the metalloids As, Se, and Te (see later), can be mediated by a range of microbes, including clostridia, methanogens, and sulfate-reducing bacteria under anaerobic conditions, and principally fungi under aerobic conditions, such as *Penicillium* and *Alternaria* spp., as well as a variety of bacteria, including pseudomonads. Methyl groups are enzymatically transferred to the metal, and a given species may transform a number of different metal(loid)s. Methylated metal compounds formed by these processes differ in their solubility, volatility, and toxicity. Volatile methylated species are often lost from the soil. There are examples of methyl-accumulating reactions for Tl, Pd, Pt, Au and Cr but the involvement of microbial/abiotic components is unclear. Mercuric ion (Hg^{2+}) may be methylated by bacteria and fungi to methylmercury [$(CH_3)Hg^+$] which is more toxic than the mercuric ion

(Barkay and Wagner-Dobler 2005). Some bacteria can methylate methylmercury, forming volatile dimethylmercury. Methylmercury as well as phenylmercury can be enzymatically reduced to volatile metallic mercury (Hg^0) by some bacteria. Phenylmercury can also be microbially converted to diphenylmercury (Barkay and Wagner-Dobler 2005).

6.4.1 Siderophores

Siderophores are highly specific Fe(III) ligands (formation constants often greater than 10^{30}) that can bind and transport, or shuttle, Fe. These low-molecular weight coordination molecules are excreted by a wide variety of fungi and bacteria to aid Fe assimilation. Siderophores can complex other metals apart from iron, e.g. actinides. Because of such metal-binding abilities, there are potential applications for siderophores in medicine, reprocessing of nuclear fuel, bioremediation and treatment of industrial wastes (Renshaw et al. 2002).

6.5 Metal Immobilization

Microbial biomass provides a metal sink, either by biosorption to cell walls, pigments and extracellular polysaccharides, intracellular accumulation, or precipitation of metal compounds in and/or around cells, hyphae or other structures (Gadd and Mowll 1985; Gadd 1986, 1993a, 2000a, 2001a, b, 2007, 2010; Baldrian 2003; Fomina et al. 2007a, b) (Table 6.1, Figs. 6.2 and 6.4). All microbial material can be effective metal biosorbents except for mobile alkali metal cations like Na^+ and K^+, and this can be an important passive process in living and dead organisms (Gadd 1993a, 2009b; Sterflinger 2000; Wang and Chen 2009).

In natural systems, metal bioavailability is determined by interactions with environmental components, such components including clay and other minerals, humic substances, soil colloidal materials, biogenic debris and exudates, and living organisms. Sorption is one of the most important reactions that influences bioavailability, and metal sorption to cells is likely to play an important role in all microbe-metal-mineral interactions (Burford et al. 2003a), taking place over a massive range of timescales from msec to years (Borda and Sparks 2008; Theng and Yuan 2008). Further, metal interactions with specific cell-surface groups may also enhance or inhibit metal transport, metal transformations and biomineralization processes (Barkay and Schaefer 2001). Soil contains a vast amount of metal-sorbing material with high surface area to volume ratios: bacteria have the highest surface area:volume ratios of any living organisms and microbes are major biotic components of the soil (Stumm and Morgan 1996). Microbial cells, on a specific unit area basis, can exhibit higher sorption values for metals than even, e.g. clay minerals (Garnham et al. 1993; Morley and Gadd 1995). It is possible

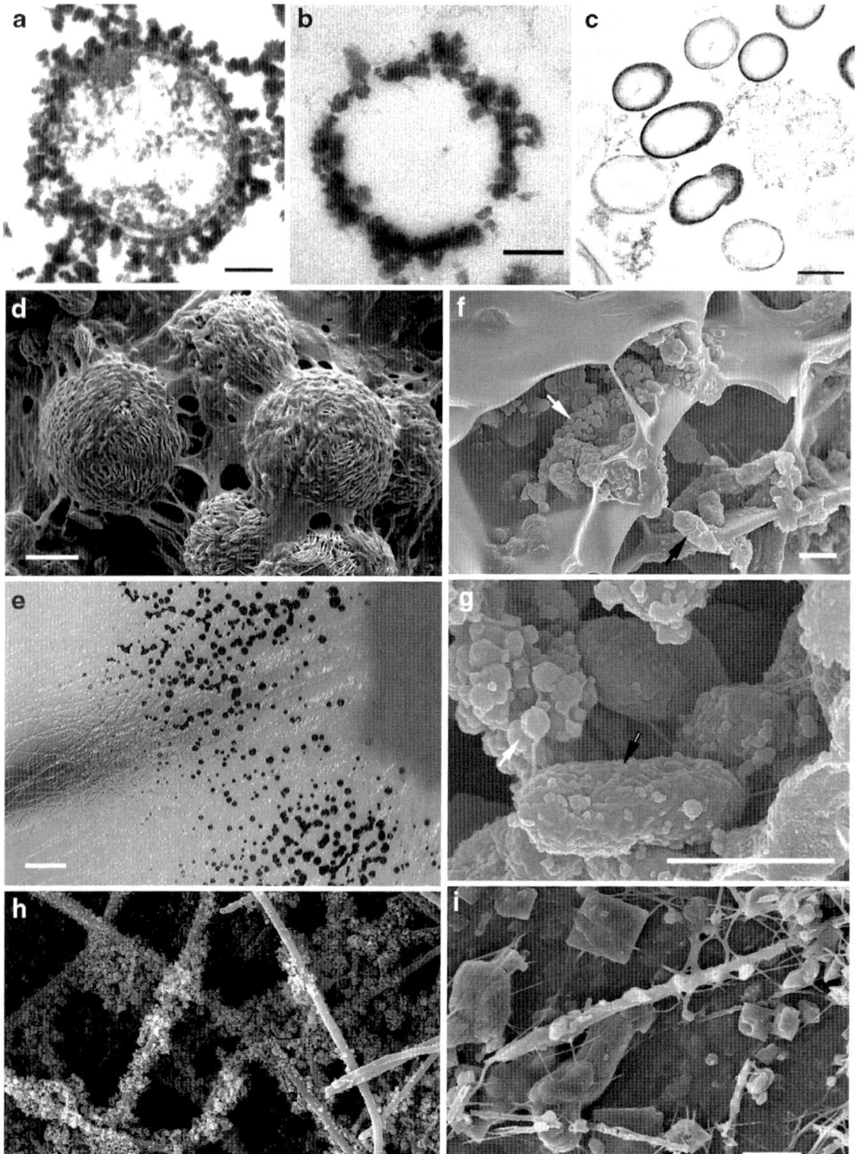
a
b
c
d
f
e
g
h
i

that biosorption phenomena have a more significant role in metal/radionuclide speciation, bioavailability and mobility in the biosphere than has previously been supposed (Krantz-Rulcker et al. 1993, 1996; Ledin et al. 1996; McLean et al. 2002) and it should be emphasised that this may also accompany or precede nucleation, precipitation and biomineral formation (Burford et al. 2003a; Gadd 2007, 2009a, b). It has been stated that sorption is among the most important processes controlling the fate, mobility, and bioavailability of molecules in the CZ (Chorover et al. 2007), and metal biosorption has also perhaps been unappreciated in this context (Gadd 2009a, b, 2010).

Where microbial reduction of a metal to a lower redox state occurs, mobility and toxicity may be reduced for several elements (Lovley 2001; Lloyd and Lovley 2001; Finneran et al. 2002a, b; Lloyd et al. 2003; Holden and Adams 2003; Wall and Krumholz 2006), e.g. U(VI) to U(IV) and Cr(VI) to Cr(III) (Phillips et al. 1995; Smith and Gadd 2000). U(VI) reduction to U(IV) can be the basis of U removal from contaminated waters and leachates as well as the formation of uranium ores such as uraninite (UO_2) (Lovley and Coates 1997; Lovley 2001; Finneran et al. 2002a, b; Lloyd 2003; Lloyd and Renshaw 2005; Landa 2005). Anaerobically, hexavalent uranium(VI) can be reduced to tetravalent uranium(IV) by a number of bacteria using either H_2 or one of various organic electron donors (Lovley and Coates 1997; Landa 2005; Wall and Krumholz 2006). Aerobic or anaerobic microbial reduction of Cr(VI) to Cr(III) is widespread (Smith and Gadd 2000; McLean and Beveridge 2001). Sulfur and sulfate-reducing bacteria are particularly important in reductive precipitation of, e.g. U(VI), Cr(VI), Tc(VII), and Pd(II) (Aubert et al. 1998; Lloyd et al. 1999a, b; Lloyd and Macaskie 1998; Lloyd 2003; Lloyd and Renshaw 2005). Some sulfate-reducing bacteria like *Desulfotomaculum reducens* share

---

Fig. 6.4 Some examples of metal(loid) and metal mineral deposition by microbes. (**a**) Transmission electron microscopy (*TEM*) *thin section* showing abiotically precipitated goethite nanocrystals sorbed to the cell wall of *Shewanella putrefaciens*. Bar = 150 nm. (**b**) TEM *thin section* of an unknown bacterial cell, taken from a Yellowstone hot spring, embedded in metal-rich precipitates sample. Bar = 150 nm. (**c**) TEM *thin section* showing uranium precipitates formed in the periplasmic space of *Geobacter metallireducens* cells during dissimilatory reduction of U(VI). Bar = 150 nm (**a**, **b**, **c** adapted from Glasauer et al. 2004). (**d**) moolooite (copper oxalate) deposition within fungal biofilm consisting of aggregated hyphae and exoplymeric matrix. Bar = 20 μm (Fomina, M. and Gadd, G.M., unpublished) (**e**) MnO_x deposition in agar media around a colony of *Acremonium strictum* growing in the presence of $MnSO_4$. Bar = 0.5 cm (Wei, Z. and Gadd, G.M., unpublished) (**f**, **g**) precipitation of selenium-sulphur granules within a sulfate-reducing biofilm matrix. Individual cells of *Desulfomicrobium norvegicum* (*black arrows*) form colonies within the matrix, with abundant Se-S nanoparticles (*white arrows*) precipitated beneath the biofilm canopy. Bars = 1 μm (Adapted from Hockin and Gadd 2003) (**h**) U biomineralization of fungal hyphae with uranium phosphates after growth in the presence of U_3O_8. Bar = 10 μm (Adapted from Fomina et al. 2007a) (**i**) a crust of calcium oxalate (weddelite and whewellite) crystals and tubular crystalline sheath around fungal hyphae formed on medium containing $CaCO_3$ and $Cu_3(PO_4)_2$. Bar = 20 μm (Adapted from Gadd et al. 2007; Gadd 2010; Fomina, M. and Gadd, G.M., unpublished)

physiological properties of both sulfate- and metal-reducing groups of bacteria, and can use Cr(VI), Mn(IV), Fe(III) and U(IV) as sole electron acceptors (Tebo and Obraztsova 1998). Such direct processes may accompany indirect mechanisms of reductive metal precipitation, e.g. in sulfate-reducing bacterial systems where reduction of Cr(VI) can be a result of indirect reduction by Fe^{2+} and the produced sulfide. Elemental silver (Ag^0) and gold (Au^0) nanoparticles result during microbial reduction of ionic silver and gold species (Kierans et al. 1991; Holden and Adams 2003; Southam et al. 2009). Other redox transformations of metals such as Mo, V, Sb are also known which must play a role in their speciation although rather less is known about rarer elements. Microbes can also mediate formation of several inorganic and organic biominerals, e.g. oxalates, phosphates, sulfides, oxides and carbonates, which lead to metal immobilization (Gadd 2007) (Table 6.1, Figs. 6.2 and 6.4).

Weathering of iron-containing minerals in rocks, soils, and sediments is promoted partly by bacterial and fungal action and partly by chemical activity (Lovley 2000). Mobilized ferrous (Fe(II)) may be biologically or abiotically oxidized to ferric iron at pH >5 under anaerobic or partially or fully aerobic conditions. Some bacteria can oxidize ferrous iron enzymatically with the generation of energy, e.g. acidophiles like *Acidithiobacillus ferrooxidans*, *Leptospirillum ferrooxidans*, *Sulfolobus* spp., *Acidianus brierleyi*, and *Sulfobacillus thermosulfidooxidans*. Fe(II) is least susceptible to autoxidation below pH 5. Some bacteria growing at circumneutral pH can also oxidize ferrous iron enzymatically under partially reduced conditions, e.g. the stalked bacterium *Gallionella ferruginea* and sheathed bacteria like *Leptothrix* spp. (Ehrlich and Newman 2009). Ferrous iron can also be oxidized non-enzymatically by microbes when their metabolic activities alter the miocroenvironment to favour autoxidation. Some Fe(III) precipitation may also arise from the destruction of ferric iron chelates. Ferric iron may also be locally concentrated by adsorption to microbial surfaces and metal oxides. Microbial formation of hydrous iron oxides in aqueous environments may cause accumulation of other metal ions by coprecipitation or adsorption: such adsorbed metals may be remobilized by reduction of the iron oxides or acidification (Ehrlich and Newman 2009).

6.6 Organic Matter Decomposition and Metal Cycling

Organic matter decomposition is one of the most important microbial activities in the biosphere and the ability of microbes, mainly bacteria and fungi, to utilize a wide spectrum of organic compounds is well-known. These range from simple compounds such as sugars, organic acids, and amino acids to more complex molecules which may be broken down by extracellular enzymes before cellular uptake and metabolism. These latter compounds include cellulose, pectin, lignin, lignocellulose, chitin and starch, and also hydrocarbons, pesticides, and other xenobiotics that may be anthropogenically produced. Degradation of such substances results in redistribution of component elements between organisms and environmental compartments. The vast majority of elements in plant, animal and microbial biomass (>95%) comprise carbon, hydrogen, oxygen, nitrogen, phosphorus and sulfur, and, as well as these, several other elements are typically found in living organisms most with essential biochemical and structural functions, e.g. K, Ca, Mg, B, Cl, Fe, Mn, Zn, Cu, Mo, Ni, Co, Se, Na, and Si. However, all 90 or so naturally-occurring elements may be found in plants, animals and microbes, including Au, As, Hg, Pb, Cd and U. Some of these elements will be taken up as contaminants in food and from the environment. Therefore, it should be stressed that all decomposition, degradative and pathogenic microbial activities are linked to cycling of these constituent elements, most of which are metals and some of which may be radionuclides accumulated from anthropogenic sources.

Biodegradation of organometallic (and organometalloid) compounds, still widely used in agriculture and industry, can result from direct enzymatic action, or by microbial facilitation of abiotic degradation, e.g. by alteration of pH and excretion of metabolites (Gadd 1993b, 2000b). Organotins, such as tributyltin oxide and tributyltin naphthenate, may be degraded to mono- and dibutyltins, inorganic Sn(II) being the ultimate product (Gadd 2000b). Organomercury compounds may be detoxified by organomercury lyase, the resultant Hg^{2+} being subsequently reduced to less toxic, diffusible and volatile Hg^0 by mercuric reductase (Gadd 1993b).

6.7 Metalloid Transformations

Microbes can transform certain metalloids (elements with properties intermediate between those of metals and non-metals). Elements included in the chemical classification of metalloids are boron, silicon, germanium, arsenic, antimony, tellurium, and polonium. Selenium is also often considered under this heading. Microbial oxidation, reduction and methylation influence the bioavailability and toxicity of many metalloids. For example, methylated selenium derivatives are volatile and less toxic than inorganic forms while reduction of metalloid oxyanions, such as selenite or tellurite to amorphous elemental selenium or tellurium respectively, results in immobilization and detoxification. Arsenic and selenium are readily metabolized by prokaryotes, participating in a full range of metabolic functions including assimilation, methylation, detoxification, and anaerobic respiration. Arsenic speciation and mobility is affected by microbes through oxidation/reduction reactions as part of resistance and respiratory processes (Stolz et al. 2006).

For selenium, some bacteria can use $SeO_4{}^{2-}$ as a terminal e^- acceptor in dissimilatory reduction and also reduce and incorporate Se into organic components, e.g. selenoproteins. Selenate ($SeO_4{}^{2-}$) and selenite ($SeO_3{}^{2-}$) can be reduced to Se^0, with $SeO_3{}^{2-}$ reduction appearing more ubiquitous than $SeO_4{}^{2-}$ reduction. However, only $SeO_4{}^{2-}$ can support bacterial growth under anaerobic conditions: $SeO_4{}^{2-}$ reduction to Se^0 is a major sink for Se oxyanions in anoxic sediments (Oremland et al. 1989; Stolz and Oremland 1999; Oremland and Stolz 2000). Reduction to Se^0 can be considered a detoxification mechanism (Dungan and Frankenberger 1999; Yee and Kobayashi 2008). The opposite process of Se^0 oxidation can also occur in soils and sediments (Dowdle and Oremland 1998; Losi and Frankenberger 1998).

Methylation of Se is a ubiquitous microbial property and can occur in soils, sediments and water (Gadd 1993b). Bacteria and fungi are the most important Se-methylaters in soil (Karlson and Frankenberger 1988) with the most ubiquitous volatile being dimethyl selenide ($(CH_3)_2Se$; DMSe) (Karlson and Frankenberger 1988, 1989; Thompson-Eagle et al. 1989). Other volatiles produced in smaller amounts include dimethyl diselenide ($(CH_3)_2Se_2$; DMDSe) (Dungan and Frankenberger 1999). The opposite process of demethylation can also occur in soil and water systems (Oremland et al. 1989). Tellurium may also be transformed by reduction and methylation (Chasteen and Bentley 2003). Reduction of tellurite (TeO_3^{2-}) to Te^0 results in grey to black colouration of microbial colonies and extracellular and intracellular precipitation (Gharieb et al. 1999). Dimethyl telluride (DMTe) is the main product of Te methylation (Chasteen and Bentley 2003). Arsenic methylation can be mediated by many organisms with compounds having the general structure $(CH_3)_nAsH_{3-n}$ and mono-, di- and trimethylarsine ($n = 1$, 2, 3 respectively) being major volatile compounds (Bentley and Chasteen 2002). Methylation pathway(s) involve a series of steps in which the reduction of the pentavalent form of As is followed by the oxidative addition of a methyl group (Dombrowski et al. 2005) generating a growing methylated series of methyl arsenite (MMA), dimethyl arsenate (DMA-V), dimethyl arsenite (DMA-III) and trimethyl arsine oxide (TMAO). Two different arsenic reduction activities have been found in bacteria. One is the dissimilatory reduction for anaerobic respiration, which involves arsenate as the terminal electron acceptor. The second detoxifies arsenic by converting arsenate to arsenite, the substrate of the arsenic efflux pumps and a determinant of arsenic resistance (Bentley and Chasteen 2002; Paez-Espino et al. 2009). Microbial oxidation of arsenite to arsenate by arsenite oxidase enzymes could be a detoxification mechanism (Tamaki and Frankenberger 1992; Lebrun et al. 2003) with arsenite oxidation serving as an electron donor.

6.8 Mineral Transformations

Microbes are involved in the formation and deterioration of minerals, including those produced biogenically. The majority of such interactions are likely to be accompanied by changes in metal speciation and mobility.

6.8.1 Mineral Formation

Biomineralization refers to the processes by which organisms form minerals. Mineral synthesis can be categorized into biologically-induced mineralization (BIM) and biologically-controlled mineralization (BCM). Biologically-induced mineralization is where an organism modifies its local microenvironment creating conditions such that there is extracellular chemical precipitation of mineral phases

(Bazylinski 2001; Hamilton 2003; Dupraz et al. 2009) (Fig. 6.4). The organism does not appear to control the biomineralization process in BIM while a great degree of control is exerted by the organism in BCM, e.g. magnetosome formation in magnetotactic bacteria, and complex cellular biomineral structures in certain eukaryotes (Bazylinski and Schubbe 2007; Posfai and Dunin-Borkowski 2009; Gadd and Raven 2010). Organomineralization has also been used as a term encompassing biomineralization processes (Dupraz et al. 2009). The term 'organomineral' has been used to refer to minerals that are affected by organic substances, mostly life-related, but not directly produced by living cells. Biominerals are defined as being directly produced by an organism while organominerals are formed under the influence of adjacent organic matter (Perry et al. 2007). However, such definitions are not universally accepted because of the imprecision of almost all such terms (Altermann et al. 2009).

Most biomineralization examples referred to here relate to biologically-induced mineralization. This can result from microbial oxidation or reduction of a metal species, metabolite excretion, e.g. sulfide, oxalate, and other metabolism-dependent transformations of metal species, with microbial surfaces providing chemically-reactive sites for sorption (≡biosorption) which can also lead to the nucleation and formation of mineral precipitates around biomass (Beveridge 1989; Fortin et al. 1997; McLean et al. 2002; Lloyd et al. 2008; Gadd 2009b, 2010). Highly reactive bacterial cell surfaces can bind metal ions (Beveridge 1989), inorganic anions (e.g. silicates; Urrutia and Beveridge 1994) and even preformed nanominerals (Glasauer et al. 2001) leading to fine-grained mineral phases on cell surfaces (Lloyd et al. 2008). This reactivity arises from the array of ionizable groups, such as carboxylates and phosphates, present in the lipopolysaccharides (LPS) of Gram-negative cell walls and the peptidoglycan, teichuronic acids and teichoic acids of Gram-positive bacteria (Beveridge 1989). Sometimes nanomineral phases can form directly in cytoplasmic components of bacteria, such as magnetite and greigite (Bazylinski and Moskowitz 1997) or amorphous iron hydroxide (Glasauer et al. 2002). External metal precipitation and biomineralization may also result from indirect effects of bacterial metabolism on microenvironmental geochemistry (Lee and Beveridge 2001; McLean et al. 2002; Lloyd et al. 2008). Precipitation can also be due to the production of reactive components which either bind metals or catalyze precipitation (Macaskie et al. 1994; Mossman et al. 1999; Wang et al. 2001; Hockin and Gadd 2003, 2006, 2007; Gadd 2007, 2010) (Fig. 6.4). The nature of the resultant mineral(s) may depend on the nature of the cell surface, the cellular microenvironment, and the presence of reactive anions, e.g. sulfide, carbonate, and phosphate. However, such biomineralization can also occur independent of microbial activity and on dead biomass. Mineral phases may undergo further changes in crystallinity with time and some minerals may incorporate other metals into their structure (Watson et al. 1995, 2000; Brown et al. 1999; McLean et al. 2002; Lloyd et al. 2008). The formation of solid mineral phases by bacterial processes may result in mineral deposition over geological time scales (Beveridge et al. 1983). Oxidative and reductive reactions mediated by microbes play important roles in the iron cycle and the formation of some sedimentary iron deposits has been attributed directly to microbial iron oxidation. Bacterial iron reduction has been associated with the formation of magnetite and siderite (Ehrlich and Newman 2009).

While much work on microbial mineral formation has been carried out using bacteria, it should be stressed that all major microbial groups have roles in metal immobilization and mineral formation, e.g. cyanobacteria, microalgae, protozoa and fungi. While cyanobacterial and microalgal processes are generally of greater significance in aquatic environments, free-living and symbiotic fungi have especially important roles in the terrestrial environment regarding mineral dissolution, metal and anion cycling, and also mineral formation in free-living and symbiotic forms (see later) (Fig. 6.4). As in bacteria, formation of secondary minerals by fungi can occur through metabolism-independent and -dependent processes (Gadd 2007). Precipitation, nucleation and deposition of crystalline material on and within cell walls are influenced by factors such as pH and wall composition. Cationic species of Si(IV), Fe(III), Al(III) and Ca(II), that may be released through dissolution mechanisms, stimulate precipitation of compounds that may bond soil particles: precipitation of carbonates, phosphates and hydroxides increases soil aggregation. Hyphae and any associated exoplymeric material can also entrap soil particles, and also release organic metabolites that enhance aggregate stability (Bronick and Lal 2005). Apart from the biomineral examples that follow, a variety of other secondary minerals have been recorded in association with fungal biomass, e.g. birnessite, ferrihydrite, iron gluconate, calcium formate, forsterite, goethite, halloysite, hydrocerussite, todorokite, moolooite, montmorillonite, and uranium phosphates (Burford et al. 2003a, b; Gadd 2007; Fomina et al. 2007a, b, 2008) (Fig. 6.4).

6.8.2 *Mineral Biodeterioration*

Direct and indirect physical, chemical and biochemical mechanisms are involved in mineral biodeterioration by microbes (Sand 1997; Edwards et al. 2005; Amundson et al. 2007; Fomina et al. 2007c; Lian et al. 2008a, b; Uroz et al. 2009) (Figs. 6.1 and 6.2). Biomechanical deterioration of rocks and minerals can occur through penetration, boring and burrowing into decaying material and along crystal planes in, e.g. calcitic and dolomitic rocks (Sterflinger 2000; Golubic et al. 2005; Smits 2006; Gadd 2007; Cockell and Herrera 2008). Cleavage penetration can also occur with lichens (Banfield et al. 1999; Chen et al. 2000; Adamo and Violante 2000). It should also be appreciated that plants and animals that grow through or burrow in terrestrial substrates also mediate some geomorphic changes (Amundson et al. 2007) while plant-and animal-associated and symbiotic microorganisms will be further involved in biophysical and biochemical transformations. Biochemical actions are believed to be more significant processes than mechanical degradation. Biochemical weathering of rocks and minerals can occur through excretion of, e.g., H^+, organic acids and other metabolites. This can result in changes in the mineral microtopography through pitting and etching of surfaces to complete dissolution (Drever and Stillings 1997; Ehrlich 1998; Gharieb et al. 1998; Kumar and Kumar 1999; Adamo and Violante 2000; Adeyemi and Gadd 2005; Edwards et al. 2005; Uroz et al. 2009; Wei et al. 2012a). Mineral dissolution may result in release of toxic (Sayer et al. 1999) or essential metals like K (Lian et al. 2008b). Fungi generally acidify their micro-environment via a number of mechanisms, which include the excretion of protons and organic acids, while respiratory CO_2 can result in carbonic acid formation. In addition, fungi excrete a variety of other metal-complexing metabolites (e.g. siderophores, carboxylic acids, amino acids and phenolic compounds) (Burgstaller and Schinner 1993). The origin of fungal tunnels within soil minerals has been explained as a result of dissolution and "burrowing" within the mineral matrix – the "rock-eating fungi" hypothesis (Jongmans et al. 1997; Landeweert et al. 2001; Golubic et al. 2005; Cockell and Herrera 2008). An additional "rock-building fungi" hypothesis has also been presented which explains how fungal tunnels may occur inside minerals (Fomina et al. 2010). The "rock-building fungi" hypothesis particularly highlights fungal involvement in mineral formation. It is likely that fungi explore and burrow within pre-existing cracks, fissures, pores and weak points in weatherable minerals and, during colonization and mineral dissolution, build a matrix of secondary minerals often of the same chemical composition as the host-rock, e.g. secondary $CaCO_3$ precipitation in calcareous soil and rock (Verrecchia 2000). As a result, fissures and cracks become secondarily cemented with mycogenic minerals. In less easily weathered silicate minerals such as feldspars, fungal-mediated precipitation of amorphous aluminosilicates (Van Breemen et al. 2000), may modify the appearance of pre-existing cracks and fissures in an analogous manner. After death and degradation of fungal hyphae, tunnels may be left within surface patinas and crusts providing distinct signatures of previous biogeochemical fungal activity.

6.9 Common Mineral and Biomineral Transformations by Microbes

6.9.1 Bauxite

Aluminum is the third most abundant element in the Earth's crust after silicon and oxygen. Several kinds of microbes are involved in the formation of some aluminum-containing minerals through bioweathering. Formation of bauxite (bauxitization) involves two stages where microbes are involved. The major constituents of bauxite are Al_2O_3, Fe_2O_3 and SiO_2 or aluminosilicate in various forms, and the source material for bauxitization may be volcanic and other aluminosilicate rocks, limestone and alluvium. Weathering of source rock (formation of protobauxite) is promoted by those activities of bacteria and fungi that mobilize Al, Fe, and Si that are then subsequently precipitated as oxides, silica, and silicate minerals. Maturation of protobauxite to bauxite is promoted by iron-reducing and fermentative bacteria under anaerobic conditions, which selectively mobilize iron oxides and silica or silicate, and enrich the bauxite in aluminum (Ehrlich and Newman 2009).

6.9.2 Carbonates

A significant portion of insoluble carbonate at the Earth's surface is of biogenic origin. Certain bacteria, cyanobacteria, and fungi can deposit calcium carbonate extracellularly (Verrecchia et al. 1990; Chafetz and Buczynski 1992). Mineralized carbonate precipitates are also found in association with bacterial biofilms (Glasauer et al. 2002). Calcium, as well as some magnesium ions, can combine with carbonate of biogenic origin sometimes to massive scales, e.g. the White Cliffs of Dover, UK, is a biogenic carbonate deposit in the form of chalk. Carbonate precipitation is possibly the most important process that impacts global carbon cycling (Dupraz et al. 2009).

Calcium carbonate associated with stromatolites, originating from types of cyanobacterial mats, may be a result of calcium carbonate entrapment or deposition as well as cyanobacterial photosynthesis and bacterial activities. Calcium carbonate associated with travertine (a porous limestone) and lacustrine carbonate crusts and nodules can result from cyanobacterial photosynthesis in freshwater environments. Calcareous nodules are formed around rounded rocks, stones, pebbles, shells etc. to which calcium carbonate–depositing cyanobacteria are attached. Most bacteria, including cyanobacteria, and some algae cause precipitation of $CaCO_3$ close to or at their cell surface. In contrast, some algae and protozoa form $CaCO_3$ intracellularly and then export it to the cell surface to become support structures. These include the coccolithophores (green algae) and foraminifera (Protozoa, Sarcodinae), the mineral form of calcium carbonate deposited being calcite or aragonite. Apart from Ca and Mg carbonates, others that may have a microbial involvement in their formation include hydromagnesite ($Mg_5(CO_3)_4(OH)_2.4H_2O$), $SrCO_3$, siderite ($FeCO_3$), rhodochrosite ($MnCO_3$), sodium carbonate (natron, $Na_2CO_3.10H_2O$) (Ehrlich and Newman 2009). The range of composition and crystallography of carbonate precipitates produced by microbial communities is influenced by environmental conditions and community species composition (Dupraz and Visscher 2005). Key components of carbonate biomineralization in microbial mat systems are the "alkalinity" engine (microbial metabolism and environmental conditions which impact on the calcium carbonate saturation index) and microbially-produced extracellular polymeric substances (EPS) which provide a template for carbonate nucleation (Dupraz et al. 2009). Alkalinity produced by SRB has a prominent role in such carbonate deposition (Dupraz et al. 2009).

Insoluble carbonates may be broken down by microbial attack. This is usually the result of organic and inorganic acid formation but may also involve physical processes (Lian et al. 2008a, b). Various bacteria, fungi, lichens, cyanobacteria and algae have been implicated (Schneider and Le Campion-Alsumard 1999;

Adamo and Violante 2000; Hoppert et al. 2004; Cockell and Herrera 2008; Lian et al. 2008a). Such activity is evident on limestones and marble used in building construction, but it is also evident in natural limestone formations such as coral reefs, where limestone-boring cyanobacteria, algae, and fungi are active in the breakdown process (Golubic et al. 2005; Cockell and Herrera 2008). Bacteria and fungi contribute to the discolouration and destruction of structural limestone and marble and are also involved in patina formation. Fungal attack on carbonate substrates (dolomites and limestones) can result in significant microbial diagenesis of these substrates to neo-dolomite, glushinskite, weddellite, whewellite and possibly struvite, as well as intense substrate "de-micritization" and "micritization" with oxalates, grain bridging and cementation, open space filling, formation of intergranular and intragranular porosity, and permeability enhancement. Advanced stages of diagenesis were characterized by dissolution and replacement of the original minerals by the new substrates produced by fungal biomineralization processes (Kolo et al. 2007).

6.9.3 Oxalates

Calcium oxalate is the most common form of oxalate in the environment, mostly occurring as the dihydrate (weddellite) or the more stable monohydrate (whewellite) (Gadd 1999). Calcium oxalate crystals are commonly associated with free-living, pathogenic and plant symbiotic fungi, and lichens, and are formed by the precipitation of solubilized calcium as the oxalate (Gadd 1999; Gharieb et al. 1998; Adamo and Violante 2000; Adamo et al. 2002). Biotic fungal calcium oxalate can exhibit a variety of crystalline forms (tetragonal, bipyramidal, plate-like, rhombohedral or needles) (Arnott 1995). Calcium oxalate precipitation has an important influence on biogeochemical processes in soils, acting as a reservoir for calcium, and also influencing phosphate availability. Fungi can also produce other metal oxalates on interacting with a variety of different metals and metal-bearing minerals, e.g. Ca, Cd, Co, Cu, Mg, Mn, Sr, Zn, Ni and Pb (White et al. 1997; Gadd 1999, 2007 Sayer et al. 1999; Sayer and Gadd 1997; Wei et al. 2012b) (Fig. 6.4). The formation of toxic metal oxalates may also provide a mechanism enabling fungi to tolerate high concentrations of toxic metals (Gadd 1993a; Jarosz-Wilkołazka and Gadd 2003).

In many arid and semi-arid regions, calcareous soils and near surface limestones (calcretes) are secondarily cemented with calcite ($CaCO_3$) and whewellite (calcium oxalate monohydrate, $CaC_2O_4.H_2O$). The presence of fungal filaments mineralized with calcite ($CaCO_3$), together with whewellite (calcium oxalate monohydrate, $CaC_2O_4.H_2O$), has been reported in limestone and calcareous soils (Verrecchia 2000). Calcium oxalate can also be degraded to calcium carbonate, and this may cement pre-existing limestones (Verrecchia et al. 2006). During decomposition of fungal hyphae, calcite crystals can act as sites of further secondary calcite precipitation. Calcite will also readily nucleate on chitin, an important component of fungal cell walls. Other experimental work has demonstrated fungal precipitation of secondary calcite, whewellite, and glushkinskite ($MgC_2O_4.2H_2O$) (Burford et al. 2003a, b, 2006; Gadd 2007). Fungal attack on dolomitic and seawater substrates resulted in the formation of Ca-oxalates (weddellite, $CaC_2O_4.2H_2O$; whewellite, $CaC_2O_4.H_2O$) and glushinskite ($MgC_2O_4.2H_2O$) (Kolo and Claeys 2005).

6.9.4 Oxides

A good example of microbial oxide formation is provided by manganese. Many bacterial species can oxidize manganese which is then deposited on cells, sheaths, or spores as oxides (Tebo et al. 2005). Some promote oxidation non-enzymatically, others enzymatically with possible involvement of a multicopper oxidase system, and include spore-forming and non-spore-forming rods, sheathed and appendaged bacteria as well as the usual morphological forms of Gram-positive and -negative bacteria from a diverse range of freshwater, marine and terrestrial ecosystems (Tebo et al. 2005). Several fungi can also promote Mn(II) oxidation to Mn(IV)O_2 including *Acremonium* spp. (Miyata et al. 2004, 2006, 2007; Saratovsky et al. 2009) (Fig. 6.4). In many cases, fungal oxidation is probably non-enzymatic and due to interaction with a metabolic product (e.g. a hydroxy acid) or a cellular component (Ehrlich and Newman 2009) although the involvement of laccase and/or multicopper oxidases have been shown in ascomycetes which are ubiquitous in natural environments (Miyata et al. 2004, 2007; Tebo et al. 2005). For some basidiomycete white-rot fungi like *Phanerochaete chrysosporium*, redox transformations of Mn occur during lignin degradation but this oxidation is thought to be of minor significance regarding manganese redistribution in the environment (Ehrlich and Newman 2009). The MnO_x material produced by *Acremonium* KR21-2 is manifest as small crystalline particles which adopt a todorokite-like tunnel structure, which is in striking contrast to previously reported microbial MnO_x materials which adopt layered birnessite-type structures (Saratovsky et al. 2009). Non-enzymatic microbial Mn^{2+} oxidation may be effected through production of metabolites, e.g. hydroxycarboxylic acids such as citrate, lactate, malate, gluconate, or tartrate. Some microbes can oxidize Mn(II) and Fe(II) in metal-bearing minerals such as siderite ($FeCO_3$) and rhodochrosite ($MnCO_3$) and precipitate them as oxides (Grote and Krumbein 1992).

Manganese and iron oxides are major components (20–30%) along with clay (~60%) and various trace elements in the brown-to-black veneers known as desert varnish or rock varnish (Grote and Krumbein 1992; Gorbushina 2007). The prevalence of manganese oxidizers in desert varnish implicates a role for these bacteria in its formation (Ehrlich and Newman 2009). Mn oxide phases have high sorption capacities for numerous metal cations (e.g., Ni, Zn, Cu, Co, Mn, Pb, and Cd), and also serve as strong oxidants for inorganic [e.g., As(III) to As(V); Cr(III) to Cr(IV)] and organic compounds such as humic substances (Tebo et al. 2004; Miyata et al. 2007). Furthermore, in anoxic environments, Mn oxides as well as Fe(III) (hydr)oxides become terminal electron acceptors for microbial metal respiration, potentially controlling the fates of a wide variety of organic compounds such as organic acids, fatty acids, and aromatics in the environment (Lovley 2000).

Conversely, manganese-reducing microbes may mobilize oxidized or fixed manganese, releasing it into the aqueous phase. A number of different, taxonomically-unrelated bacteria can reduce manganese enzymatically or non-enzymatically. The bacteria that reduce manganese enzymatically often do so as a form of respiration where oxidized manganese serves as a terminal electron acceptor and is reduced to Mn(II) (Lovley 2000). Some bacteria can reduce the oxidized manganese aerobically or anaerobically, whereas others can reduce it only anaerobically. Microbial reduction of oxidized manganese can also be enzymatic or non-enzymatic. Some bacteria and most of those fungi that reduce Mn(IV) oxides such as MnO_2 reduce them indirectly (non-enzymatically) with the likely mechanism being the production of metabolic products that can act as reductants for Mn(IV) oxides such as formic acid, pyruvate, H_2S, sulfite, Fe^{2+} (bacteria) and oxalate (fungi) (Ehrlich and Newman 2009; Wei et al. 2012b). Microbial reduction of manganese oxides may also lead to the formation of manganous carbonate. Ferromanganese nodules on parts of the ocean floor are inhabited by Mn-oxidizing and -reducing bacteria and these are likely to contribute to nodule formation (Ehrlich and Newman 2009).

Many bacteria can precipitate and deposit Fe(III) oxides and hydroxides, e.g. FeOOH, Fe_3O_4, around their cells by enzymatic, e.g. *Gallionella* sp., and non-enzymatic processes, e.g. *Leptothrix* sp. (Ehrlich and Newman 2009).

6.9.5 Phosphates

Phosphorus occurs primarily as organic phosphate esters and as inorganic forms, e.g. calcium, aluminum, and iron phosphates. Organic phosphates are hydrolyzed by phosphatases which liberate orthophosphate during microbial decomposition of organic material. Microbes also liberate free orthophosphate from insoluble inorganic phosphates by producing organic or mineral acids or chelators, e.g. such as gluconate and 2-ketogluconate, citrate, oxalate, and lactate, which complex the metal resulting in dissociation, or, for iron phosphates by producing H_2S. Phosphate-solubilizing activity is very important in the plant rhizosphere (Whitelaw et al. 1999).

Microbes can also play a role in the formation of phosphate minerals such as vivianite ($Fe_3(PO_4)_2.8H_2O$), strengite ($FePO_4.2H_2O$), and variscite ($AlPO_4.2H_2O$). Here, the orthophosphate may arise from organic phosphate degradation while Fe or Al may arise from microbial solubilization of other minerals. Such formation of phosphate minerals is probably most common in soil (Ehrlich and Newman 2009). Secondary mycogenic uranium mineral precipitates on fungal mycelia growing in the presence of uranium oxides or depleted uranium were found to be uranyl phosphate minerals of the meta-autunite group, uramphite and/or chernikovite (Fomina et al. 2007a, 2008) (Fig. 6.4). Several free-living fungi were found to mediate the formation of pyromorphite on interaction with metallic lead (Rhee et al. 2012).

6.9.6 Silicates

Silica, silicates, and aluminosilicates in rocks are weathered by biological, chemical, and physical processes (Brehm et al. 2005). Silicon liberated as soluble silicate can be assimilated by several microbial groups in order to fabricate cell-support structures. They include diatoms, some chrysophytes, silicoflagellates, some xanthophytes, radiolarians, and actinopods. Biosilicification of natural microbial mats has been shown to be a microbially-mediated geochemical process, with a requirement for colloidal silica, an acidic pH and exposed organic surfaces both of which favour colloidal silica sorption (Amores and Warren 2007).

Silicon dioxide, when combined with oxides of magnesium, aluminium, calcium and iron, forms the silicate minerals in rocks and soil (Bergna 1994). Silicates are the largest class of minerals comprising 30% of all minerals and making up 90% of the Earth's crust (Ehrlich 1998; Ehrlich and Newman 2009). Silicate minerals are unstable in the biosphere and break down readily to form clays (Adamo et al. 2002; Tazaki 2006). Many kinds of bacteria, fungi and lichens play an important role in the dissolution of silicates, and therefore in the genesis of clay minerals, and in soil and sediment formation (Barker and Banfield 1996, 1998; Rodriguez Navarro et al. 1997; Banfield et al. 1999; Adamo and Violante 2000; Arocena et al. 1999, 2003; Tazaki 2006; Theng and Yuan 2008; Ehrlich and Newman 2009; Cockell et al. 2009a, b). Even silicates of great physical and chemical resistance can be attacked, e.g. quartz sand, crystalline quartz and commercial glass (Brehm et al. 2005). Microbial action is mainly indirect, either through the production of chelates or the production of acids (mineral or organic), or other metabolites, together with biomechanical effects (Cromack et al. 1979; De la Torre et al. 1993; Mandal et al. 2002; Wei et al. 2012a). In bioweathering of rock silicates and aluminosilicates, cleavage of Si–O–Si (siloxane) or Al–O bonds or removal of cations from the silicate crystal lattice may cause collapse of the silicate lattice structure. The mechanisms of attack may include microbially-produced (1) ligands of cations; (2) organic or inorganic acids (a source of protons); (3) alkali (ammonia or amines); or (4) extracellular polysaccharides that act at acidic pH. Such agents may be excreted into the bulk phase but may also involve attached biofilm microbes on surfaces of silica or silicates resulting in etching (Bennett et al. 1996, 2001; Wei et al. 2012a). Such mechanisms of silicate dissolution may release limiting nutrients like bound P and Fe (Rogers and Bennett 2004). For several bacteria, dissolution of silicates results from complexation of cationic components by 2-ketogluconate. Quartz (SiO_2) can be subject to slow dissolution by organic acids such as citric and oxalic acid (Bennett et al. 1988), the mechanism of action being chelation rather than protonation. Hydration of respiratory or fermentative CO_2 to give the weak carbonic acid H_2CO_3 can also result in solubilization of silicates. Alkaline conditions can mobilize silicon from silicates, ammonia production from urea hydrolysis being one bacterial mechanism (Ehrlich and Newman 2009).

In lichen weathering of silicate minerals, calcium, potassium, iron clay minerals and nanocrystalline aluminous iron oxyhydroxides become mixed with fungal organic polymers (Barker and Banfield 1998), while biotite was interpenetrated by fungal hyphae along cleavages, partially converting it to vermiculite (Barker and Banfield 1996). The fungal partner has also been reported to be involved in the formation of secondary silicates, such as opal and forsterite, in lichen thalli (Gorbushina et al. 2001). The transformation rate of mica and chlorite to 2:1 expandable clays was pronounced in ectomycorrhizosphere soil and was probably a result of the high production of organic acids and direct extraction of K^+ and Mg^{2+} by fungal hyphae (Arocena et al. 1999).

Silicon compounds in the form of clays (aluminosilicates) can exert many effects on microbes in soil and stimulate or inhibit metabolism (Marshall 1971; Marshman and Marshall 1981a, b; Weaver and Dugan 1972; Theng and Yuan 2008). Effects of clays are mostly indirect and arise from physicochemical effects of clays on the microenvironment, e.g. action as buffers, and as sorptive agents for cells, metabolites, ions and enzymes (Tazaki 2006; Ehrlich and Newman 2009). Clay minerals (bentonite, palygorskite and kaolinite) can also influence the size, shape and structure of fungal mycelial pellets in liquid culture (Fomina and Gadd 2002a).

6.9.7 Sulfides

Most non-ferrous sulfides are formed abiotically but some sedimentary deposits are of biogenic origin. Sulfate-reducing bacteria play an important role in some sedimentary environments in mediating the formation of certain sulfides, especially iron pyrite (FeS_2). Microbial roles in the biogenesis of sulfide deposits arises from the generation of H_2S, usually from bacterial reduction of sulfate (Ehrlich and Newman 2009), and reaction of metal ions with the biogenic sulfide, although some sulfide may also result from decomposition of sulfur-containing organic compounds.

Metal sulfides are subject to oxidation by bacteria such as *Acidithiobacillus ferrooxidans*, *Leptospirillum ferrooxidans*, *Sulfolobus* spp., and *Acidianus brierleyi*. Bacterial action may involve direct oxidative attack of the crystal lattice of a metal sulfide or indirect oxidative attack by generation of acid ferric sulfate which oxidizes the metal sulfide chemically. The indirect mechanism is of primary importance in the solubilization of uraninite (UO_2). Microbial oxidation of metal sulfides is industrially exploited in extracting metals from low-grade metal sulfide ores and uraninite (see later). In bituminous coal seams that are exposed as a result of mining activity, pyrite oxidation by these bacteria is an environmentally deleterious process and the source of acid mine drainage (AMD) (Rawlings et al. 2003; Ehrlich and Newman 2009; Jerez 2009).

6.10 Fungal Symbioses in Mineral Transformations

A remarkable adaptation of fungi for exploitation of the terrestrial environment is their formation of mutualistic partnerships with plants (mycorrhizas) and algae or cyanobacteria (lichens). The symbiotic fungi are provided with carbon by the photosynthetic partners (photobionts), while the fungi may protect the symbiosis from harsh environmental conditions (e.g., desiccation, metal toxicity), increase the absorptive area, and provide increased access to inorganic nutrients.

6.10.1 Lichens

Lichens are fungi that exist in facultative or obligate symbioses with one or more photosynthesizing partners, and play an important role in many biogeochemical processes. The lichen symbiosis formed between the fungal partner (mycobiont) and the photosynthesizing partner (algal or cyanobacterial photobiont) enables lichens to grow in practically all surface terrestrial environments: an estimated 6% of the Earth's land surface is covered by lichen-dominated vegetation (Haas and Purvis 2006). Lichens are pioneer colonizers of fresh rock outcrops. Globally, lichens play an important role in the retention and distribution of nutrient (e.g. C, N) and trace elements, in soil formation, and in rock weathering. Alteration of bedrock minerals and synthesis of biominerals in the proximity of lichens gives rise to different chemical microenvironments and emphasises their participation in mineral nutrient cycling (Banfield et al. 1999; Chen et al. 2000; Adamo and Violante 2000). Lichens can accumulate metals such as lead (Pb), copper (Cu), and many other elements of environmental concern, including radionuclides, to high levels (Purvis and Pawlik-Skowronska 2008). They can also form a variety of metal-organic biominerals, e.g. oxalates, especially during growth on metal-rich substrates (Chen et al. 2000; Adamo et al. 2002). For example, on copper-sulfide bearing rocks, precipitation of copper oxalate (moolooite) can occur within the lichen thallus (Purvis 1996; Purvis and Halls 1996).

6.10.2 Mycorrhizas

Nearly all land plants depend on symbiotic mycorrhizal fungi (Smith and Read 1997). Two main types of mycorrhizas include endomycorrhizas where the fungus colonizes the interior of host plant root cells (e.g. ericoid and arbuscular mycorrhizas) and ectomycorrhizas where the fungus is located outside plant root cells. Mycorrhizal fungi are involved in proton-promoted and ligand-promoted metal mobilization from mineral sources, metal immobilization within biomass, and extracellular precipitation of mycogenic metal oxalates (Fomina et al. 2004, 2005b).

Biogeochemical activities of mycorrhizal fungi lead to changes in the physico-chemical characteristics of the root environment and enhanced weathering of soil minerals resulting in metal release. Ectomycorrhizal mycelia may respond to the presence of different soil silicate and phosphate minerals (apatite, quartz, potassium feldspar) by regulating their growth and activity, e.g. colonization, carbon allocation and substrate acidification (Rosling et al. 2004a, b).

During growth, mycorrhizal fungi often excrete low molecular weight carboxylic acids (Martino et al. 2003; Fomina et al. 2004). The weathering of hornblendes, feldspars and granitic bedrock in certain soils has been attributed to oxalic, citric, succinic, formic and malic acid excretion by ectomycorrhizal hyphae which can produce micro- to millimolar concentrations of these organic acids in their microenvironments. Weathering and the supply of nutrients derived from minerals to plants is known to be stimulated by plant symbiotic mycorrhizal fungi. Ectomycorrhizal fungi can form narrow cylindrical pores in weatherable minerals in podzol E horizons, probably by exuding low-molecular weight organic acids at their hyphal tips, causing local dissolution of Al silicates (Jongmans et al. 1997; van Breemen et al. 2000).

Ectomycorrhizal fungi (*Suillus granulatus* and *Paxillus involutus*) can release elements from apatite and wood ash (K, Ca, Ti, Mn, Pb) and accumulate them in the mycelia (Wallander et al. 2003). Ericoid mycorrhizal and ectomycorrhizal fungi can dissolve a variety of cadmium, copper, zinc and lead-bearing minerals including metal phosphates (Leyval and Joner 2001; Martino et al. 2003; Fomina et al. 2004, 2005b). Mobilization of phosphorus from inorganic phosphate is generally regarded as one of the most important functions of mycorrhizal fungi, and this will also result in movement and redistribution of associated metals, and also the formation of other secondary minerals including phosphates.

6.11 Environmental and Applied Significance of Mineral-Metal-Microbe Interactions

6.11.1 Biocorrosion of Metals

The deterioration of metal due to microbial activity is termed biocorrosion or microbially-influenced corrosion (MIC) (Beech and Sunner 2004). The main types of microbes associated with metals in terrestrial and aquatic habitats are sulfate-reducing bacteria (SRB), sulfur-oxidizing bacteria, iron oxidizing/reducing bacteria, manganese-oxidizing bacteria, bacteria secreting organic acids and slime, and various algae and fungi, and many develop complex biofilm communities on surfaces of materials (Beech and Sunner 2004; Gu 2009). Mechanisms of corrosion are complicated and include depolarization of metals, biomineral formation, complexation by exopolymeric materials, H_2 embrittlement, acidic attack, and electron shuttling (Gu 2009).

6.11.2 Bioleaching of Metals from Ores

Solubilization mechanisms provide a route for removal of metals from industrial wastes and by-products, low-grade ores, and metal-bearing minerals, which is relevant to bioremediation of soil matrices and solid wastes, metal recovery and recycling (Burgstaller and Schinner 1993; Gadd 2000a; Gadd and Sayer 2000; Brandl 2001; Suzuki 2001; Rawlings 2002; Kartal et al. 2006). Industrial scale bioleaching of metals is used to extract metals like copper, gold, uranium, and others from their ores (Suzuki 2001; Rawlings 2002; Rawlings et al. 2003; Jerez 2009). This process, also termed biomining, employs chemolithoautotrophic microbes with the most common leaching bacteria belonging to the genus *Acidithiobacillus*, e.g. *A. ferrooxidans* and *A. thiooxidans*. Most bioleaching operations for gold or copper use some kind of irrigation process involving percolation of leaching solutions through the crushed ore contained in a column, heap or dump (Rawlings 2002; Rawlings et al. 2003). In an environmental context, acidophilic microbes can mobilize metals from waste ores, abandoned mines and dumps, and generate acid mine drainage (AMD), itself a serious environmental problem (Jerez 2009).

6.11.3 Bioweathering of Rocks and Minerals: Soil Formation

Weathering is a process in which rock is eroded or broken down into smaller particles and finally to constituent minerals, and this ultimately leads to mineral soil formation (Tazaki 2006; Ehrlich and Newman 2009). Surface soil is in fact a CZ region of intense metabolic and geochemical activity and contains the highest microbial biomass (including plant-root symbiotic mycorrhizas), the highest number of plant roots and the largest pool of natural organic matter (Chorover et al. 2007). Physical, chemical and biological processes are involved: bioweathering can be defined as the erosion and decay of rocks and minerals mediated by living organisms, and in microbes is mediated by the kind of direct and indirect physical and chemical mechanisms detailed previously. Some researchers prefer use of the term biodeterioration to bioweathering (Gorbushina and Krumbein 2005). Microbes are effective biological weathering agents and many can tolerate extremes of light, salinity, pH, and water potential as well as adopting a variety of growth, metabolic and morphological strategies (Gorbushina et al. 1993; Sterflinger 2000; Verrecchia 2000; Burford et al. 2003a, b; Gadd 2007; Gorbushina and Broughton 2009). Many organisms can scavenge nutrients from the atmosphere and rainwater, and also use organic and inorganic residues on surfaces or within cracks and fissures, waste products of other microbes, decaying plants and insects, dust particles, aerosols and animal faeces as nutrient sources (Sterflinger 2000; Gorbushina and Broughton 2009). Many organisms receive protection from the presence of pigments in their cell walls, e.g. melanins in fungi, and by the production of extracellular polymeric materials (EPS) during adhesion with biofilm formation providing extra protection (Gadd 1993a; Gorbushina 2007). Microbes are probably associated with all rocks and minerals, building stone and concrete (Burford et al. 2003a, b; Walker and Pace 2007; Gleeson et al. 2005, 2006, 2010; Gorbushina 2007; Fomina et al. 2007c; Gorbushina and Broughton 2009). Lichens, a fungal growth form, are highly significant bioweathering agents (Adamo and Violante 2000; Adamo et al. 2002). Lithobiotic biofilm communities can interact with mineral substrates both geophysically and geochemically: as well as deterioration this can also result in the formation of patinas, films, varnishes, crusts and stromatolites (Gadd 2007; Gorbushina 2007; Fomina et al. 2010). Elements found in soil therefore largely reflect the composition of the Earth's crust, though some modification occurs by further weathering, biogenic and anthropogenic activities: chemical changes include mineral dissolution while biological activity causes enrichment of C, N and S. Elements and minerals that remain can reform as secondary minerals. Mycorrhizal fungi are also very important in mineral weathering and dissolution of insoluble metal compounds in the soil, along with free-living bacteria and fungi. Bioweathering is a primary source of essential elements for organisms, except nitrogen and carbon (Hoffland et al. 2004). Acidification is an important fungal bioweathering mechanism with low molecular weight organic anions (LMWOAs) being especially significant (Gadd 1999, 2007; Hoffland et al. 2004). Biotic roles in soil weathering also impact on the composition and reactivity of streams and oceans (Amundson et al. 2007).

Many microbes can attack silicates, therefore playing a role in the genesis of clay minerals, and in soil and sediment formation. Microbe–clay mineral interactions therefore play an important role in soil evolution, as well as aggregation and stabilization (Burford et al. 2003a; Tazaki 2006; Theng and Yuan 2008). The production of EPS by cyanobacteria, bacteria and fungi is significant while filamentous fungi entangle soil particles in their hyphae forming stable microaggregates and also take part in polysaccharide aggregation (Lunsdorf et al. 2000; Ritz and Young 2004). Interactions between clay minerals and microbes alter the adsorptive properties of both clays and biomass (Fomina and Gadd 2002b; Tazaki 2006). As well as being important in stages of mineral soil and sediment formation and development, bioweathering mechanisms also affect structural decay of rock and mineral-based building materials, monuments and other structures (Scheerer et al. 2009). The most common stone types affected are marble, limestone, sandstone, and granite, with many overlaps between calcareous and siliceous rocks. Concrete is also subject to microbial deterioration (Fomina et al. 2007c). External stone surfaces are a complex ecosystem, including not only cyanobacteria, algae, bacteria, fungi, protozoa and but also "higher" organisms such as small animals, and plants (Scheerer et al. 2009). Stone-inhabiting microbes may grow on the surface (epilithic), in more protected habitats such as crevices and fissures (chasmolithic), or may penetrate some millimetres or even centimetres into the rock pore system (endolithic). Microbial colonization generally initiates with phototrophic cyanobacteria and algae, usually in a biofilm, probably followed by lichens, and then general heterotrophs (Hoppert et al. 2004) although establishment of heterotrophic rock communities is possible without initial phototroph involvement (Roeselers et al. 2007). Highly deteriorated stone surfaces provide appropriate conditions (a "proto-soil") for colonization by mosses, ferns and higher plants. Ceramic materials, concrete and cement can also be deteriorated by microbes (Gu et al. 1998; Nica et al. 2000; Zhdanova et al. 2000; Gu 2009; Scheerer et al. 2009). It should be stressed that the activities of all groups of microbes and interactions between them should be considered in bioweathering, and indeed other geomicrobiological, studies. Fungal–bacterial interactions in the root environment are likely to be significant in mineral weathering and plant nutrition for example (Balogh-Brunstad et al. 2008; Koele et al. 2009).

6.12 Bioremediation, Biotechnology and Bioprocessing

Many microbial metal and mineral transformations have potential for the treatment of environmental pollution, and some processes are in commercial operation (Gadd 1992b, 2004, 2005; Lloyd and Lovley 2001; Pumpel and Paknikar 2001; Lloyd and Renshaw 2005). (e.g. Gadd 2001a, b; Pumpel and Paknikar 2001; Barkay and Wagner-Dobler 2005; Hockin and Gadd 2007; Gadd 2009b, 2010). In addition to bioremediation, microbe-metal-mineral transformations have applications in other areas of biotechnology and bioprocessing, including biosensors, biocatalysis, electricity generation and nanotechnology.

6.12.1 Bioleaching

Microbiological solubilization of metals from solid minerals (bioleaching) is a well established process in the mining industry (see previously). In addition, other metal and mineral wastes, including contaminated soil, may be subject to microbial bioleaching for metal recovery, recycling and bioremediation purposes (White et al. 1998; Brandl 2001; Brandl and Faramarzi 2006).

6.12.2 Bioprecipitation

Bioprecipitation of metals by both sulfides and phosphates has been investigated because of the low solubility of many of their metal compounds (Macaskie 1991; White and Gadd 1998; Boswell et al. 1999, 2001; Renninger et al. 2001, 2004; Hockin and Gadd 2007). Sulfide produced from sulfate reduction plays a major role in metal sulfide immobilization in sediments but has also been applied to bioremediation of metals in waters and soil leachates (White and Gadd 1998, 2000; White et al. 2003; Hockin and Gadd 2007).

6.12.3 *Biosorption and Bioaccumulation*

Biosorption is a physico-chemical process, simply defined as "the removal of substances from solution by biological material", is a property of both living and dead organisms (and their components), and has been heralded as a promising biotechnology for removal (and/or recovery) of metals, radionuclides, organic pollutants for many years because of its simplicity, analogous operation to conventional ion exchange technology, apparent efficiency and availability of biomass and waste bio-products (Macaskie 1991; Gadd 1986, 2001a, b, 2009b; Volesky 1990; Garnham et al. 1992; Gadd and White 1990, 1993; Wang and Chen 2009). Most biosorption studies are carried out on microbial systems, chiefly bacteria, microalgae and fungi, and with toxic metals and radionuclides, particularly actinides and lanthanides. The term is now applied to all manner of organic and inorganic pollutants and biosorption can describe "any system where a sorbate (e.g. an atom, molecule, a molecular ion) interacts with a biosorbent (i.e. a solid surface of a biological matrix) resulting in an accumulation at the sorbate-biosorbent interface, and therefore a reduction in the solution sorbate concentration" (Gadd 2009b). Biosorption also has application for recovery and use of precious metals, e.g. gold (Volesky 1990; Gadd 2009b; Wang and Chen 2009). However, despite dramatic increases in published research on biosorption there has been little or no exploitation in an industrial context.

Biosorptive processes are a component of varying significance in any form of primary or secondary biological treatment process for process streams including domestic, municipal and industrial wastes, and solid wastes. Sewage treatment, activated sludge plants, biofilters, biofilm reactors, fixed and suspended film systems, lagoon treatments, stream meanders, nitrification and denitrification treatments, biological phosphate removal processes, wetlands and reed-bed technologies, composting, *in situ* and *ex situ* bioremediation processes all rely on the activities of microbes to break down organic substances. Many wastes contain metals in addition to organics and therefore biosorption of metals must play a part in the overall process although the significance of such a role is difficult to establish. In biological treatment processes, a clear advantage is that metabolic activities contribute to the degradation of organic substances (Gadd and White 1993).

6.12.4 *Metalloid Bioremediation*

The ability of fungi, along with bacteria, to transform metalloids has been utilized successfully in the bioremediation of contaminated land and water. Selenium methylation results in volatilization, a process which has been used to remove selenium from the San Joaquin Valley and Kesterson Reservoir, California, using evaporation pond management and primary pond operation (Thomson-Eagle and Frankenberger 1992). Microbial responses to arsenic of potential in bioremediation include chelation, compartmentalization, exclusion, and immobilization (Dhankher et al. 2002; Song et al. 2003; Tsai et al. 2009).

6.12.5 Mycoremediation and the Mycorrhizosphere

Mycorrhizal associations may also be used for metal clean-up in the general area of phytoremediation (van der Lelie et al. 2001; Rosen et al. 2005; Gohre and Paszkowski 2006). However, many complicatory factors affect successful exploitation (Meharg 2003). The potential impact of mycorrhizal fungi on bioremediation may be conditional and dependent on the metal tolerance of fungal strains, their mycorrhizal status, and the nutritional status of contaminated soils (Meharg 2003). It is generally concluded that local conditions in metal-contaminated sites may determine the cost-benefit relationship between the plant and arbuscular-mycorrhizal (AM) fungi since detrimental, neutral or beneficial interactions have all been documented (Meharg and Cairney 2000). A protective metal-binding effect of ectomycorrhizal fungi (EcM) has been postulated (e.g. Leyval et al. 1997). Some ectomycorrhizal fungi persistently fixed Cd(II) and Pb(II), and formed an efficient biological barrier that reduced movement of these metals in birch tissues (Krupa and Kozdroj 2004). Such mycorrhizal metal immobilization around plant roots, including biomineral formation, may also assist soil remediation and revegetation. The insoluble glycoprotein, glomalin, produced in copious amounts on hyphae of arbuscular mycorrhizal fungi can sequester such metals, and could be considered a useful stabilization phenomenon in remediation of polluted soils (Gonzalez-Chavez et al. 2004). Phytostabilization strategies may be suitable to reduce the dispersion of uranium (U) and the environmental risks of U-contaminated soils. AM fungi and root hairs improved not only P acquisition but also root uptake of U, and the mycorrhiza generally decreased U translocation from plant root to shoot (Rufyikiri et al. 2004; Chen et al. 2005a, b).

For ericaceous mycorrhizas, clear host protection has been observed in, e.g. *Calluna*, *Erica*, and *Vaccinium* spp. growing on Cu- and Zn-polluted and/or naturally metalliferous soils, the fungus preventing metal translocation to plant shoots (Bradley et al. 1981, 1982). The development of stress-tolerant plant-mycorrhizal associations may therefore be a promising strategy for phytoremediation and soil amelioration (Schutzendubel and Polle 2002). Because of the symbiosis with ericoid mycorrhizal fungi, ericaceous plants are able to grow in highly polluted environments, where metal ions can reach toxic levels in the soil substrate (Perotto et al. 2002; Martino et al. 2003). Ericoid mycorrhizal fungal endophytes, and sometimes their plant hosts, can evolve toxic metal resistance which enables ericoid mycorrhizal plants to colonize polluted soil (Cairney and Meharg 2003).

Free-living and symbiotic microbial populations associated with plant roots significantly alter the physico-chemical characteristics of the rhizosphere which may have consequences for the biogeochemical mobility of metals and associated elements (Wenzel et al. 1994; Olsson and Wallander 1998; Whitelaw et al. 1999). The importance of mycorrhizas in plant phosphorus nutrition has been appreciated for a long time and their ability to dissolve and transform calcium-containing insoluble compounds and minerals (calcium phosphates, carbonate and sulfate) has been widely studied (Callot et al. 1985a, b; Lapeyrie et al. 1990, 1991; Gharieb and Gadd 1999). However, toxic metal mineral solubilization has received little attention, though this should be considered in any revegetation, natural attenuation, or phytoremediation strategies. The ectomycorrhizal fungi *Suillus granulatus* and *Pisolithus tinctorius* can promote the release of cadmium and phosphorus from rock phosphate (Leyval and Joner 2001) while the ericoid mycorrhizal fungus *Oidiodendron maius* can solubilize zinc oxide and phosphate (Martino et al. 2003). Many ericoid mycorrhizal and ectomycorrhizal fungi can solubilize zinc, cadmium, copper phosphates and lead chlorophosphate (pyromorphite) releasing phosphate and metals (Fomina et al. 2004, 2006).

6.12.6 Phytoremediation

Phytoremediation is the use of plants to remove or detoxify environmental pollutants (Baker and Brooks 1989; Salt et al. 1998). Although free-living and symbiotic microbes influence plant productivity, metal bioavailability and interactions, there are few integrated studies, and many phytoremediation studies are carried out without reference to contributory microbial processes. Phytoremediation can be divided into phytoextraction (pollutant removal from soil into shoots and leaves), phytodegradation (pollutant degradation by plant-microbe systems), rhizofiltration (absorption of pollutants by plant roots), phytostabilization (plant-mediated reduction of pollutant bioavailability), phytovolatilization (plant-mediated volatilization of pollutants) and phytoscrubbing (plant removal of atmospheric pollutants). Most attention has focussed on metals with two remediation strategies being chelate-assisted and continuous phytoextraction. Application of various synthetic chelates can enhance plant metal accumulation (Huang et al. 1997; Salt et al. 1998). Continuous phytoextraction of metals relies on properties of plants that lead to accumulation in aerial plant tissues. However, many natural 'hyperaccumulators' often exhibit low biomass, slow growth rates and none are known for important elements like Pb, Cd, As and U (Salt et al. 1998). Ni, Zn and Se appear to be the elements accumulated to the highest levels (Salt et al. 1998).

6.12.7 Reductive Transformations, Nanoparticle Formation and Nano-biotechnology

Microbial activities in anaerobic, subsurface environments also offer possibilities for metal and radionuclide bioremediation (Lloyd 2003; Lloyd et al. 2003; Lloyd and Renshaw 2005). Metal(loid)s that form insoluble precipitates when reduced include Se(0), Cr(III), Tc(IV) and U(IV) (Lovley and Coates 1997; Stolz and Oremland 1999; Thompson-Eagle and Frankenberger 1992; Hockin and Gadd 2003, 2006; Wall and Krumholz 2006; Yee and Kobayashi 2008). Microbial reduction of U(VI) to U(IV) has been proposed as a bioremediation strategy for uranium-contaminated groundwaters (Lovley et al. 1991; Lovley 1995; Lloyd et al. 2003) as reduction of U(VI) under anaerobic conditions produces U(IV), which precipitates as the insoluble mineral, uraninite (Wall and Krumholz 2006). Biogenic uraninite is an important nanoscale biogeological material and crucial to the viability of microbial bioremediation strategies for subsurface uranium contamination by stimulated uranium reduction because it is orders of magnitude less soluble than most other U species. Successful uranium bioremediation requires continued strictly anaerobic conditions, with the biomass-associated nanoparticles tending to aggregate making it unlikely that they will be transported in groundwater (Bargar et al. 2008).

Metal micro-/nano-particles, with appropriate chemical modification, have applications as new ceramic–metal (cermet) or organic–metal (orgmet) composites or structured materials for a variety of applications. The use of metal-accumulating microbess for the production of nanoparticles, and their assembly, may allow control over size, morphology, composition, and crystallographic orientation of the particles. The potential of such biomimetic materials appears great and is relevant to production of new advanced materials, with applications in metal and radionuclide bioremediation, antimicrobial treatments (e.g. nano-silver), solar energy and electrical battery applications, and microelectronics (Dameron et al. 1989; Klaus-Joerger et al. 2001). Because of their high specific surface area and high catalytic properties, biogenic metal products also offer potential for sorption and degradation of organic contaminants, as well as a variety of other applications, e.g. electricity generation in fuel cells, novel catalysts and sensors. Biogenic Mn oxides can sequester metals like Pb, Zn, Co, Ni, As and Cr and also oxidize certain organic pollutants (Hennebel et al. 2009). Biogenic Fe oxides have larger specific surface areas than chemically synthesized Fe oxides and may be more efficient adsorbents for metals (Hennebel et al. 2009). Biogenic Fe oxides, such as nanoscale magnetite (Fe_3O_4), are also reactive in dehalogenation reactions, which may be several orders of magnitude faster than enzymic degradation (Hennebel et al. 2009).

6.12.8 Soil Treatment Processes and Microbial Influence

Application to soils of certain amendments that immobilize metals, e.g. lime or phosphate treatment, have demonstrated, under field conditions, enhanced natural remediation resulting in improved vegetation growth, increased microbial activity and diversity, and reduced offsite metal transport. However, while long-term stability of certain metal complexes and compounds has been shown in model systems (Adriano et al. 2004a, b), the influence of plant roots and its microbial and mycorrhizal associations on such stability has often been neglected. For example, pyromorphite ($Pb_5(PO_4)_3Cl$), which can form in urban and industrially-contaminated soils, can be solubilized by phosphate-solubilizing fungi, with concomitant production of lead oxalate (Sayer et al. 1999; Fomina et al. 2004). The ability of free-living and mycorrhizal fungi to transform pyromorphite (and other toxic metal-containing minerals) should clearly be taken into account in risk assessments of the long-term environmental consequences of *in situ* chemical remediation techniques, revegetation strategies or natural attenuation of contaminated sites. The bioweathering potential of fungi has been envisaged as a possible means for the bioremediation of asbestos rich soils. Several fungi could extract iron from asbestos mineral fibres thereby removing the reactive iron ions responsible for DNA damage (Daghino et al. 2006).

6.13 Conclusions

Microbes play key geoactive roles in the biosphere in element biotransformations and biogeochemical cycling, metal and mineral transformations, including mineral formation and mineral deterioration, organic matter decomposition, bioweathering, soil and sediment formation. These processes are all highly important in the CZ and fundamental to optimal function. All kinds of microbes, including prokaryotes and eukaryotes and their symbiotic associations with each other and "higher organisms", contribute actively to such geological phenomena, and central to these geomicrobial processes are transformations of metals and minerals. An additional definition of the CZ therefore could be that portion of the terrestrial environment "characterized by a significant microbial influence on metal and mineral transformations, as well as on the cycling of other elements." Apart from being important in natural biosphere processes, metal- and mineral transformations can have beneficial or detrimental consequences in a human context. Bioremediation is the application of biological systems to the clean-up of organic and inorganic pollution with bacteria and fungi being the most important organisms in this context for reclamation, immobilization or detoxification of metallic and radionuclide pollutants. Some biominerals or metallic elements deposited by microbes have catalytic and other properties in nanoparticle, crystalline or colloidal forms. In contrast, several metal and mineral transformations by microbes result in spoilage and destruction of natural and synthetic materials, rock and mineral-based building materials, e.g. concrete, acid mine drainage and associated metal pollution, biocorrosion of metals, alloys, and related substances, and adverse effects on radionuclide speciation, mobility and containment. In view of the fundamental importance of microbes in biosphere processes, it is clear that an interdisciplinary approach involving physical, chemical and biological disciplines is necessary to fully understand processes occurring within the CZ, their environmental and applied significance, and to underpin exploitation in agriculture, biotechnology and ecosystem management.

Acknowledgements The author gratefully acknowledges research support from the Biotechnology and Biological Sciences Research Council, the Natural Environment Research Council, the Royal Societies of London and Edinburgh, CCLRC Daresbury SRS, and British Nuclear Fuels plc. I would also like to thank key members of my research group who contributed immensely to several aspects of the research reviewed here, especially the late Dr Chris White, and Drs Jacqueline Sayer (now Sugden), Euan Burford, Mohammed Gharieb, Simon Hockin, Helen Jacobs, and Marina Fomina. Thanks are also due to Martin Kierans for invaluable assistance with electron microscopy, and Euan Starke for technical support over many years.

References

Adamo P, Violante P (2000) Weathering of rocks and neogenesis of minerals associated with lichen activity. Appl Clay Sci 16:229–256

Adamo P, Vingiani S, Violante P (2002) Lichen-rock interactions and bioformation of minerals. Dev Soil Sci 28B:377–391

Adeyemi AO, Gadd GM (2005) Fungal degradation of calcium-, lead- and silicon-bearing minerals. Biometals 18:269–281

Adriano DC (2001) Trace elements in the terrestrial environment: biogeochemistry, bioavailability and risks of metals, 2nd edn. Springer, New York

Adriano DC, Bolan NS, Vangronsveld J, Wenzel WW (2004a) Heavy metals. In: Hillel D (ed) Encyclopedia of soils in the environment. Elsevier, Amsterdam, pp 175–182

Adriano DC, Wenzel WW, Vangronsveld J, Bolan NS (2004b) Role of assisted natural remediation in environmental cleanup. Geoderma 122:121–142

Altermann W, Böhmer C, Gitter F, Heimann F, Heller I, Läuchli B, Putz C (2009) Defining biominerals and organominerals: direct and indirect indicators of life. Perry et al., Sedimentary geology, 201, 157–179. Sediment Geol 213:150–151

Amores DR, Warren LA (2007) Identifying when microbes biosilicify: the interconnected requirements of acidic pH, colloidal SiO_2 and exposed microbial surface. Chem Geol 240:298–312

Amundson R, Richter DD, Humphreys GS, Jobbagy EG, Gaillardet J (2007) Coupling between biota and earth materials in the critical zone. Elements 3:327–332

Arnott HJ (1995) Calcium oxalate in fungi. In: Khan SR (ed) Calcium oxalate in biological systems. CRC Press, Boca Raton, pp 73–111

Arocena JM, Glowa KR, Massicotte HB, Lavkulich L (1999) Chemical and mineral composition of ectomycorrhizosphere soils of subalpine fir (*Abies lasiocarpa* (Hook.) Nutt.) in the AE horizon of a Luvisol. Can J Soil Sci 79:25–35

Arocena JM, Zhu LP, Hall K (2003) Mineral accumulations induced by biological activity on granitic rocks in Qinghai Plateau, China. Earth Surf Proc Landf 28:1429–1437

Aubert C, Lojou E, Bianco P, Rousset M, Durand M-C, Bruschi M, Dolla A (1998) The *Desulfuromonas acetoxidans* triheme cytochrome c_7 produced in *Desulfovibrio desulfuricans* retains its metal reductase activity. Appl Environ Microbiol 64:1308–1312

Avery SV (2001) Metal toxicity in yeast and the role of oxidative stress. Adv Appl Microbiol 49:111–142

Baeuerlein E (2000) Biomineralization. Wiley-VCH, Weinheim

Baker AJM, Brooks RR (1989) Terrestrial higher plants which accumulate metallic elements – a review of their distribution, ecology and phytochemistry. Biorecovery 1:81–126

Baldrian P (2003) Interaction of heavy metals with white-rot fungi. Enzyme Microb Technol 32:78–91

Balogh-Brunstad Z, Keller CK, Gill RA, Bormann BT, Li CY (2008) The effect of bacteria and fungi on chemical weathering and chemical denudation fluxes in pine growth experiments. Biogeochemistry 88:153–167

Banfield JF, Nealson KH (eds) (1997) Geomicrobiology: interactions between microbes and minerals, reviews in mineralogy and geochemistry, vol 35. Mineralogical Society of America, Washington, DC

Banfield JP, Barker WW, Welch SA, Taunton A (1999) Biological impact on mineral dissolution: application of the lichen model to understanding mineral weathering in the rhizosphere. Proc Nat Acad Sci USA 96:3404–3411

Banfield JF, Cervini-Silva J, Nealson KH (eds) (2005) Molecular geomicrobiology, reviews in mineralogy and geochemistry, vol 59. Mineralogical Society of America, Washington, DC
Bargar JR, Bernier-Latmani R, Glammar DE, Tebo BM (2008) Biogenic uraninite nanoparticles and their importance for uranium remediation. Elements 4:407–412
Barkay T, Schaefer J (2001) Metal and radionuclide bioremediation: issues, considerations and potentials. Curr Opin Microbiol 4:318–323
Barkay T, Wagner-Dobler I (2005) Microbial transformations of mercury: potentials, challenges, and achievements in controlling mercury toxicity in the environment. Adv Appl Microbiol 57:1–52
Barker WW, Banfield JF (1996) Biologically versus inorganically mediated weathering reactions: relationships between minerals and extracellular microbial polymers in lithobiotic communities. Chem Geol 132:55–69
Barker WW, Banfield JF (1998) Zones of chemical and physical interaction at interfaces between microbial communities and minerals: a model. Geomicrobiol J 15:223–244
Bazylinski DA (2001) Bacterial mineralization. In: Buschow KHJ, Chan RW, Flemings MC, Ilschner B, Kramer EJ, Mahajan S (eds) Encyclopedia of materials: science and technology. Elsevier, Amsterdam, pp 441–448
Bazylinski DA, Moskowitz BM (1997) Microbial biomineralization of magnetic iron minerals: microbiology, magnetism, and environmental significance. Rev Miner 35:181–223
Bazylinski DA, Schubbe S (2007) Controlled biomineralization by and applications of magnetotactic bacteria. Adv Appl Microbiol 62:21–62
Beech IB, Sunner J (2004) Biocorrosion: towards understanding interactions between biofilms and metals. Curr Opin Biotechnol 15:181–186
Bennett PC, Siegel DI, Melcer ME, Hassett JP (1988) The dissolution of quartz in dilute aqueous solutions of organic acids at 25 C. Geochim Cosmochim Acta 52:1521–1530
Bennett PC, Hiebert FK, Choi WJ (1996) Microbial colonization and weathering of silicates in petroleum-contaminated groundwater. Chem Geol 132:45–53
Bennett PC, Rogers JA, Hiebert FK, Choi WJ (2001) Silicates, silicate weathering, and microbial ecology. Geomicrobiol J 18:3–19
Bentley R, Chasteen TG (2002) Microbial methylation of metalloids: arsenic, antimony and bismuth. Micro Mol Biol Rev 66:250–271
Bergna HE (1994) Colloid chemistry of silica – an overview. Colloid Chem Silica 234:1–47
Beveridge TJ (1989) Role of cellular design in bacterial metal accumulation and mineralization. Ann Rev Microbiol 43:147–171
Beveridge TJ, Meloche JD, Fyfe WS, Murray RGE (1983) Diagenesis of metals chemically complexed to bacteria: laboratory formation of metal phosphates, sulfides and organic condensates in artificial sediments. Appl Environ Microbiol 45:1094–1108
Borda MJ, Sparks DL (2008) Kinetics and mechanisms of sorption-desorption in soils: a multiscale assessment. In: Violante A, Huang PM, Gadd GM (eds) Biophysico-chemical processes of heavy metals and metalloids in soil environments. Wiley, Hoboken, pp 97–124
Boswell CD, Dick RE, Macaskie LE (1999) The effect of heavy metals and other environmental conditions on the anaerobic phosphate metabolism of *Acinetobacter johnsonii*. Microbiology 145:1711–1720
Boswell CD, Dick RE, Eccles H, Macaskie LE (2001) Phosphate uptake and release by *Acinetobacter johnsonii* in continuous culture and coupling of phosphate release to heavy metal accumulation. J Indust Microbiol Biotechnol 26:333–340
Bottjer DJ (2005) Geobiology and the fossil record: eukaryotes, microbes, and their interactions. Palaeogeo Palaeoclim Palaeoecol 219:5–21
Boukhalfa H, Icopini GA, Reilly SD, Neu MP (2007) Plutonium(IV) reduction by the metal-reducing bacteria *Geobacter metallireducens* GS15 and *Shewanella oneidensis* MR1. Appl Environ Microbiol 73:5897–5903
Bousserrhine N, Gasser UG, Jeanroy E, Berthelin J (1999) Bacterial and chemical reductive dissolution of Mn-, Co- Cr-, and Al-substituted geothites. Geomicrobiol J 16:245–258

Bowen AD, Davidson FA, Keatch R, Gadd GM (2007) Induction of contour sensing in *Aspergillus niger* by stress and its relevance to fungal growth mechanics and hyphal tip structure. Fungal Gen Biol 44:484–491
Bradley R, Burt AJ, Read DJ (1981) Mycorrhizal infection and resistance to heavy metals. Nature 292:335–337
Bradley B, Burt AJ, Read DJ (1982) The biology of mycorrhiza in the Ericaceae. VIII. The role of mycorrhizal infection in heavy metal resistance. New Phytol 91:197–209
Brandl H (2001) Heterotrophic leaching. In: Gadd GM (ed) Fungi in bioremediation. Cambridge University Press, Cambridge, pp 383–423
Brandl H, Faramarzi MA (2006) Microbe-metal-interactions for the biotechnological treatment of metal-containing solid waste. China Partic 4:93–97
Brantley SL, Goldhaber MB, Ragnarsdottir KV (2007) Crossing disciplines and scales to understand the critical zone. Elements 3:307–314
Brehm U, Gorbushina A, Mottershead D (2005) The role of microorganisms and biofilms in the breakdown and dissolution of quartz and glass. Palaeogeo Palaeoclim Palaeoecol 219:117–129
Bronick CJ, Lal R (2005) Soil structure and management: a review. Geoderma 124:3–22
Brown GE, Foster AL, Ostergren JD (1999) Mineral surfaces and bioavailability of heavy metals: a molecular-scale perspective. Proc Natl Acad Sci USA 96:3388–3395
Brown GE, Trainor TP, Chaka AM (2008) Geochemistry of mineral surfaces and factors affecting their chemical reactivity. In: Nilsson A, Pettersson LGM, Norskov JK (eds) Chemical bonding at surfaces and interfaces. Elsevier, Amsterdam, pp 457–509
Burford EP, Fomina M, Gadd GM (2003a) Fungal involvement in bioweathering and biotransformation of rocks and minerals. Mineral Mag 67:1127–1155
Burford EP, Kierans M, Gadd GM (2003b) Geomycology: fungal growth in mineral substrata. Mycologist 17:98–107
Burford EP, Hillier S, Gadd GM (2006) Biomineralization of fungal hyphae with calcite ($CaCO_3$) and calcium oxalate mono- and dihydrate in carboniferous limestone microcosms. Geomicrobiol J 23:599–611
Burgstaller W, Schinner F (1993) Leaching of metals with fungi. J Biotechnol 27:91–116
Cairney JWG, Meharg AA (2003) Ericoid mycorrhiza: a partnership that exploits harsh edaphic conditions. Eur J Soil Sci 54:735–740
Callot G, Guyon A, Mousain D (1985a) Inter-relation entre aiguilles de calcite et hyphes mycéliens. Agronomie 5:209–216
Callot G, Mousain D, Plassard C (1985b) Concentrations de carbonate de calcium sur les parois des hyphes mycéliens. Agronomie 5:143–150
Chafetz HS, Buczynski C (1992) Bacterially induced lithification of microbial mats. Palaios 7:277–293
Chasteen TG, Bentley R (2003) Biomethylation of selenium and tellurium: microorganisms and plants. Chem Rev 103:1–26
Chen J, Blume H-P, Beyer L (2000) Weathering of rocks induced by lichen colonization – a review. Catena 39:121–146
Chen BD, Jakobsen I, Roos P, Zhu YG (2005a) Effects of the mycorrhizal fungus *Glomus intraradices* on uranium uptake and accumulation by *Medicago truncatula* L. from uranium-contaminated soil. Plant Soil 275:349–359
Chen BD, Zhu YG, Zhang XH, Jakobsen I (2005b) The influence of mycorrhiza on uranium and phosphorus uptake by barley plants from a field-contaminated soil. Environ Sci Poll Res 12:325–331
Chorover J, Kretzschmar R, Garcia-Pichel F, Sparks DL (2007) Soil biogeochemical processes within the critical zone. Elements 3:321–326
Cockell CS, Herrera A (2008) Why are some microorganisms boring? Trends Microbiol 16:101–106
Cockell CS, Olsson K, Herrera A, Meunier A (2009a) Alteration textures in terrestrial volcanic glass and the associated bacterial community. Geobiology 7:50–65

Cockell CS, Olsson K, Herrera A, Kelly L, Thorsteinsson T, Marteinsson V (2009b) Bacteria in weathered basaltic glass, Iceland. Geomicrobiol J 26:491–507

Cromack K Jr, Solkins P, Grausten WC, Speidel K, Todd AW, Spycher G, Li CY, Todd RL (1979) Calcium oxalate accumulation and soil weathering in mats of the hypogeous fungus *Hysterangium crassum*. Soil Biol Biochem 11:463–468

Daghino S, Turci F, Tomatis M, Favier A, Perotto S, Douki T, Fubini B (2006) Soil fungi reduce the iron content and the DNA damaging effects of asbestos fibers. Environ Sci Technol 40:5793–5798

Dameron CT, Reese RN, Mehra RK, Kortan AR, Carroll PJ, Steigerwald ML, Brus LE, Winge DR (1989) Biosynthesis of cadmium sulphide quantum semiconductor crystallites. Nature 338:596–597

De La Torre MA, Gomez-Alarcon G, Vizcaino C, Garcia MT (1993) Biochemical mechanisms of stone alteration carried out by filamentous fungi living on monuments. Biogeochem 19:129–147

Dhankher OP, Li YJ, Rosen BP, Shi J, Salt D, Senecoff JF, Sashti NA, Meagher RB (2002) Engineering tolerance and hyperaccumulation of arsenic in plants by combining arsenate reductase and gamma-glutamylcysteine synthetase expression. Nat Biotechnol 20:1140–1145

Dombrowski PM, Long W, Parley KJ, Mahony JD, Capitani JF, Di Toro DM (2005) Thermodynamic analysis of arsenic methylation. Environ Sci Technol 39:2169–2176

Dove PM, De Yoreo JJ, Weiner S (eds) (2003) Biomineralization. Reviews in mineralogy and geochemistry, vol 54. Mineralogical Society of America, Washington, DC

Dowdle PR, Oremland RS (1998) Microbial oxidation of elemental selenium in soil slurries and bacterial cultures. Environ Sci Technol 32:3749–3755

Drever JI, Stillings LL (1997) The role of organic acids in mineral weathering. Coll Surf 120:167–181

Dungan RS, Frankenberger WT (1999) Microbial transformations of selenium and the bioremediation of seleniferous environments. Bioremed 3:171–188

Dupraz C, Visscher PT (2005) Microbial lithification in marine stromatolites and hypersaline mats. Trends Microbiol 13:429–438

Dupraz C, Reid RP, Braissant O, Decho AW, Norman RS, Visscher PT (2009) Processes of carbonate precipitation in modern microbial mats. Earth Sci Rev 96:141–162

Edwards KJ, Bach W, McCollom TM (2005) Geomicrobiology in oceanography: microbe-mineral interactions at and below the seafloor. Trends Microbiol 13:449–456

Ehrlich HL (1996) How microbes influence mineral growth and dissolution. Chem Geol 132:5–9

Ehrlich HL (1997) Microbes and metals. Appl Microbiol Biotechnol 48:687–692

Ehrlich HL (1998) Geomicrobiology: its significance for geology. Earth Sci Rev 45:45–60

Ehrlich HL, Newman DK (2009) Geomicrobiology, 5th edn. CRC Press/Taylor & Francis Group, Boca Raton

Finneran KT, Housewright ME, Lovley DR (2002a) Multiple influences of nitrate on uranium solubility during bioremediation of uranium-contaminated subsurface sediments. Environ Microbiol 4:510–516

Finneran KT, Anderson RT, Nevin KP, Lovley DR (2002b) Bioremediation of uranium-contaminated aquifers with microbial U(VI) reduction. Soil Sediment Conta 11:339–357

Fomina M, Gadd GM (2002a) Influence of clay minerals on the morphology of fungal pellets. Mycol Res 106:107–117

Fomina M, Gadd GM (2002b) Metal sorption by biomass of melanin-producing fungi grown in clay-containing medium. J Chem Technol Biotechnol 78:23–34

Fomina MA, Alexander IJ, Hillier S, Gadd GM (2004) Zinc phosphate and pyromorphite solubilization by soil plant-symbiotic fungi. Geomicrobiol J 21:351–366

Fomina M, Hillier S, Charnock JM, Melville K, Alexander IJ, Gadd GM (2005a) Role of oxalic acid over-excretion in toxic metal mineral transformations by *Beauveria caledonica*. Appl Environ Microbiol 71:371–381

Fomina MA, Alexander IJ, Colpaert JV, Gadd GM (2005b) Solubilization of toxic metal minerals and metal tolerance of mycorrhizal fungi. Soil Biol Biochem 37:851–866

Fomina M, Burford EP, Gadd GM (2005c) Toxic metals and fungal communities. In: Dighton J, White JF, Oudemans P (eds) The fungal community. Its organization and role in the ecosystem. CRC Press, Boca Raton, p 733

Fomina M, Charnock JM, Hillier S, Alexander IJ, Gadd GM (2006) Zinc phosphate transformations by the *Paxillus involutus*/pine ectomycorrhizal association. Microb Ecol 52:322–333

Fomina M, Charnock JM, Hillier S, Alvarez R, Gadd GM (2007a) Fungal transformations of uranium oxides. Environ Microbiol 9:1696–1710

Fomina M, Charnock J, Bowen AD, Gadd GM (2007b) X-ray absorption spectroscopy (XAS) of toxic metal mineral transformations by fungi. Environ Microbiol 9:308–321

Fomina M, Podgorsky VS, Olishevska SV, Kadoshnikov VM, Pisanska IR, Hillier S, Gadd GM (2007c) Fungal deterioration of barrier concrete used in nuclear waste disposal. Geomicrobiol J 24:643–653

Fomina M, Charnock JM, Hillier S, Alvarez R, Francis Livens F, Gadd GM (2008) Role of fungi in the biogeochemical fate of depleted uranium. Curr Biol 18:375–377

Fomina M, Burford EP, Hillier S, Kierans M, Gadd GM (2010) Rock-building fungi. Geomicrobiol J 27:624–629

Fortin D, Ferris FG, Beveridge TJ (1997) Surface-mediated mineral development by bacteria. In: Banfield J, Nealson KH (eds) Reviews in mineralogy, vol 35. Mineralogical Society of America, Washington, DC, pp 161–180

Gadd GM (1986) The uptake of heavy metals by fungi and yeasts: the chemistry and physiology of the process and applications for biotechnology. In: Eccles H, Hunt S (eds) Immobilisation of ions by bio-sorption. Ellis Horwood Ltd., Chichester, pp 135–147

Gadd GM (1992a) Metals and microorganisms: a problem of definition. FEMS Microbiol Lett 100:197–204

Gadd GM (1992b) Microbial control of heavy metal pollution. In: Fry JC, Gadd GM, Herbert RA, Jones CW, Watson-Craik I (eds) Microbial control of pollution. Cambridge University Press, Cambridge, pp 59–88

Gadd GM (1993a) Interactions of fungi with toxic metals. New Phytol 124:25–60

Gadd GM (1993b) Microbial formation and transformation of organometallic and organometalloid compounds. FEMS Microbiol Rev 11:297–316

Gadd GM (1999) Fungal production of citric and oxalic acid: importance in metal speciation, physiology and biogeochemical processes. Adv Microb Physiol 41:47–92

Gadd GM (2000a) Bioremedial potential of microbial mechanisms of metal mobilization and immobilization. Curr Opin Biotechnol 11:271–279

Gadd GM (2000b) Microbial interactions with tributyltin compounds: detoxification, accumulation, and environmental fate. Sci Total Environ 258:119–127

Gadd GM (2001a) Accumulation and transformation of metals by microorganisms. In: Rehm H-J, Reed G, Puhler A, Stadler P (eds) Biotechnology, a multi-volume comprehensive treatise, vol 10, Special processes. Wiley-VCH/Verlag GmbH, Weinheim, pp 225–264

Gadd GM (ed) (2001b) Fungi in bioremediation. Cambridge University Press, Cambridge

Gadd GM (2004) Microbial influence on metal mobility and application for bioremediation. Geoderma 122:109–119

Gadd GM (2005) Microorganisms in toxic metal polluted soils. In: Buscot F, Varma A (eds) Microorganisms in soils: roles in genesis and functions. Springer, Berlin, pp 325–356

Gadd GM (ed) (2006) Fungi in biogeochemical cycles. Cambridge University Press, Cambridge

Gadd GM (2007) Geomycology: biogeochemical transformations of rocks, minerals, metals and radionuclides by fungi, bioweathering and bioremediation. Mycol Res 111:3–49

Gadd GM (2008a) Bacterial and fungal geomicrobiology: a problem with communities? Geobiology 6:278–284

Gadd GM (2008b) Fungi and their role in the biosphere. In: Jorgensen SE, Fath B (eds) Encyclopedia of ecology. Elsevier, Amsterdam, p 1709

Gadd GM (2009a) Heavy metal pollutants: environmental and biotechnological aspects. In: Schaechter M (ed) Encyclopedia of microbiology. Elsevier, Oxford, pp 321–334

Gadd GM (2009b) Biosorption: critical review of scientific rationale, environmental importance and significance for pollution treatment. J Chem Technol Biotechnol 84:13–28
Gadd GM (2010) Metals, minerals and microbes: geomicrobiology and bioremediation. Microbiology 156:609–643
Gadd GM, Griffiths AJ (1978) Microorganisms and heavy metal toxicity. Microb Ecol 4:303–317
Gadd GM, Griffiths AJ (1980) Influence of pH on copper uptake and toxicity in *Aureobasidium pullulans*. Trans Brit Mycol Soc 75:91–96
Gadd GM, Mowll JL (1985) Copper uptake by yeast-like cells, hyphae and chlamydospores of *Aureobasidium pullulans*. Exp Mycol 9:230–240
Gadd GM, Raven JA (2010) Geomicrobiology of eukaryotic microorganisms. Geomicrobiol J 27:491–519
Gadd GM, Sayer JA (2000) Fungal transformations of metals and metalloids. In: Lovley DR (ed) Environmental microbe-metal interactions. American Society for Microbiology, Washington, DC, pp 237–256
Gadd GM, White C (1990) Biosorption of radionuclides by yeast and fungal biomass. J Chem Technol Biotechnol 49:331–343
Gadd GM, White C (1993) Microbial treatment of metal pollution – a working biotechnology? Trends Biotechnol 11:353–359
Gadd GM, Chudek JA, Foster R, Reed RH (1984) The osmotic responses of *Penicillium ochrochloron*: changes in internal solute levels in response to copper and salt stress. J Gen Microbiol 130:1969–1975
Gadd GM, Fomina M, Burford EP (2005) Fungal roles and function in rock, mineral and soil transformations. In: Gadd GM, Semple KT, Lappin-Scott HM (eds) Microorganisms in earth systems – advances in geomicrobiology. Cambridge University Press, Cambridge, pp 201–231
Gadd GM, Burford EP, Fomina M, Melville K (2007) Mineral transformation and biogeochemical cycles: a geomycological perspective. In: Gadd GM, Dyer P, Watkinson S (eds) Fungi in the environment. Cambridge University Press, Cambridge, pp 78–111
Garnham GW, Codd GA, Gadd GM (1992) Accumulation of cobalt, zinc and manganese by the estuarine green microalga *Chlorella salina* immobilized in alginate microbeads. Environ Sci Technol 26:1764–1770
Garnham GW, Codd GA, Gadd GM (1993) Uptake of cobalt and caesium by microalgal- and cyanobacterial-clay mixtures. Microb Ecol 25:71–82
Gharieb MM, Gadd GM (1999) Influence of nitrogen source on the solubilization of natural gypsum ($CaSO_4.2H_2O$) and the formation of calcium oxalate by different oxalic and citric acid-producing fungi. Mycol Res 103:473–481
Gharieb MM, Sayer JA, Gadd GM (1998) Solubilization of natural gypsum ($CaSO_4.2H_2O$) and the formation of calcium oxalate by *Aspergillus niger* and *Serpula himantioides*. Mycol Res 102:825–830
Gharieb MM, Kierans M, Gadd GM (1999) Transformation and tolerance of tellurite by filamentous fungi: accumulation, reduction and volatilization. Mycol Res 103:299–305
Giller KE, Witter E, McGrath SP (2009) Heavy metals and soil microbes. Soil Biol Biochem 41:2031–2037
Gilmour C, Riedel G (2009) Biogeochemistry of trace metals and metalloids. In: Likens GE (ed) Encyclopedia of inland waters. Elsevier, Amsterdam, pp 7–15
Glasauer S, Langley S, Beveridge TJ (2001) Sorption of Fe (hydr)oxides to the surface of *Shewanella putrefaciens*: cell-bound fine-grained minerals are not always formed de novo. Appl Environ Microbiol 67:5544–5550
Glasauer S, Langley S, Beveridge TJ (2002) Intracellular iron minerals in a dissimilatory iron-reducing bacterium. Science 295:117–119
Glasauer S, Beveridge TJ, Burford EP, Harper FA, Gadd GM (2004) Metals and metalloids, transformations by microorganisms. In: Hillel D, Rosenzweig C, Powlson DS, Scow KM, Singer MJ, Sparks DL, Hatfield J (eds) Encyclopedia of soils in the environment. Elsevier, Amsterdam, pp 438–447

Gleeson DB, Clipson NJW, Melville K, Gadd GM, McDermott FP (2005) Mineralogical control of fungal community structure in a weathered pegmatitic granite. Microb Ecol 50:360–368
Gleeson DB, Kennedy NM, Clipson NJW, Melville K, Gadd GM, McDermott FP (2006) Mineralogical influences on bacterial community structure on a weathered pegmatitic granite. Microb Ecol 51:526–534
Gleeson D, McDermott F, Clipson N (2007) Understanding microbially active biogeochemical environments. Adv Appl Microbiol 62:81–104
Gleeson DB, Melville K, McDermott FP, Clipson NJW, Gadd GM (2010) Molecular characterization of fungal communities in sandstone. Geomicrobiol J 27:559–571
Gohre V, Paszkowski U (2006) Contribution of the arbuscular mycorrhizal symbiosis to heavy metal phytoremediation. Planta 223:1115–1122
Golubic S, Radtke G, Le Campion-Alsumard T (2005) Endolithic fungi in marine ecosystems. Trends Microbiol 13:229–235
Gonzalez-Chavez MC, Carrillo-Gonzalez R, Wright SF, Nichols KA (2004) The role of glomalin, a protein produced by arbuscular mycorrhizal fungi, in sequestering potentially toxic elements. Environ Poll 130:317–323
Gorbushina AA (2007) Life on the rocks. Environ Microbiol 9:1613–1631
Gorbushina AA, Broughton WJ (2009) Microbiology of the atmosphere-rock interface: how biological interactions and physical stresses modulate a sophisticated microbial ecosystem. Ann Rev Microbiol 63:431–450
Gorbushina AA, Krumbein WE (2005) Role of organisms in wear down of rocks and minerals. In: Buscot F, Varma A (eds) Microorganisms in soils: roles in genesis and functions. Springer, Berlin, pp 59–84
Gorbushina AA, Krumbein WE, Hamann R, Panina L, Soucharjevsky S, Wollenzien U (1993) On the role of black fungi in colour change and biodeterioration of antique marbles. Geomicrobiol J 11:205–221
Gorbushina AA, Boettcher M, Brumsack HJ, Krumbein WE, Vendrell-Saz M (2001) Biogenic forsterite and opal as a product of biodeterioration and lichen stromatolite formation in table mountain systems (tepuis) of Venezuela. Geomicrobiol J 18:117–132
Grote G, Krumbein WE (1992) Microbial precipitation of manganese by bacteria and fungi from desert rock and rock varnish. Geomicrobiol J 10:49–57
Gu JD (2009) Corrosion, microbial. In: Schaechter M (ed) Encycolpedia of Microbiology, 3rd Edition. Elsevier, Amsterdam, pp 259–269
Gu JD, Ford TE, Berke NS, Mitchell R (1998) Biodeterioration of concrete by the fungus *Fusarium*. Int Biodet Biodegrad 41:101–109
Haas JR, Purvis OW (2006) Lichen biogeochemistry. In: Gadd GM (ed) Fungi in biogeochemical cycles. Cambridge University Press, Cambridge, pp 344–376
Hamilton WA (2003) Microbially influenced corrosion as a model system for the study of metal microbe interactions: a unifying electron transfer hypothesis. Biofouling 19:65–76
Hennebel T, Gusseme BD, Verstraete W (2009) Biogenic metals in advanced water treatment. Trends Biotechnol 27:90–98
Hochella MF (2002) Sustaining earth: thoughts on the present and future roles in mineralogy in environmental science. Miner Mag 66:627–652
Hockin S, Gadd GM (2003) Linked redox-precipitation of sulfur and selenium under anaerobic conditions by sulfate-reducing bacterial biofilms. Appl Environ Microbiol 69:7063–7072
Hockin S, Gadd GM (2006) Removal of selenate from sulphate-containing media by sulphate-reducing bacterial biofilms. Environ Microbiol 8:816–826
Hockin S, Gadd GM (2007) Bioremediation of metals by precipitation and cellular binding. In: Barton LL, Hamilton WA (eds) Sulphate-reducing bacteria: environmental and engineered systems. Cambridge University Press, Cambridge, pp 405–434
Hoffland E, Kuyper TW, Wallander H, Plassard C, Gorbushina AA, Haselwandter K, Holmstrom S, Landeweert R, Lundstrom US, Rosling A, Sen R, Smits MM, van Hees PAW, van Breemen N (2004) The role of fungi in weathering. Front Ecol Environ 2:258–264

Holden JF, Adams MWW (2003) Microbe-metal interactions in marine hydrothermal vents. Curr Opin Chem Biol 7:160–165

Hoppert M, Flies C, Pohl W, Gunzl B, Schneider J (2004) Colonization strategies of lithobiontic microorganisms on carbonate rocks. Environ Geol 46:421–428

Huang JWW, Chen JJ, Berti WR, Cunningham SD (1997) Phytoremediation of lead-contaminated soils: role of synthetic chelates in lead phytoextraction. Environ Sci Technol 31:800–805

Huang PM, Wang MC, Wang MK (2004) Mineral-organic-microbial interactions. In: Hillel D, Rosenzweig C, Powlson DS, Scow KM, Singer MJ, Sparks DL, Hatfield J (eds) Encyclopedia of soils in the environment. Elsevier, Amsterdam, pp 486–499

Huang P-M, Wang M-K, Chiu C-Y (2005) Soil mineral-organic matter-microbe interactions: impacts on biogeochemical processes and biodiversity in soils. Pedobiologia 49:609–635

Jacobs H, Boswell GP, Ritz K, Davidson FA, Gadd GM (2002a) Solubilization of metal phosphates by *Rhizoctonia solani*. Mycol Res 106:1468–1479

Jacobs H, Boswell GP, Ritz K, Davidson FA, Gadd GM (2002b) Solubilization of calcium phosphate as a consequence of carbon translocation by *Rhizoctonia solani*. FEMS Microbiol Ecol 40:65–71

Jarosz-Wilkołazka A, Gadd GM (2003) Oxalate production by wood-rotting fungi growing in toxic metal-amended medium. Chemosphere 52:541–547

Jerez CA (2009) Metal extraction and biomining. In: Schaechter M (ed) Encyclopedia of microbiology, 3rd edn. Elsevier, Amsterdam, pp 407–420

Jongmans AG, van Breemen N, Lundstrom US, van Hees PAW, Finlay RD, Srinivasan M, Unestam T, Giesler R, Melkerud P-A, Olsson M (1997) Rock-eating fungi. Nature 389:682–683

Karlson U, Frankenberger WT (1988) Effects of carbon and trace element addition on alkylselenide production by soil. Soil Sci Soc Am J 52:1640–1644

Karlson U, Frankenberger WT (1989) Accelerated rates of selenium volatilization from California soils. Soil Sci Soc Am J 53:749–753

Kartal SN, Katsumata N, Imamura Y (2006) Removal of copper, chromium, and arsenic from CCA-treated wood by organic acids released by mold and staining fungi. For Prod J 56:33–37

Kierans M, Staines AM, Bennett H, Gadd GM (1991) Silver tolerance and accumulation in yeasts. Biol Metals 4:100–106

Kim BH, Gadd GM (2008) Bacterial physiology and metabolism. Cambridge University Press, Cambridge

Klaus-Joerger T, Joerger R, Olsson E, Granquist C-G (2001) Bacteria as workers in the living factory: metal-accumulating bacteria and their potential for materials sciences. Trends Biotechnol 19:15–20

Koele N, Turpault M-P, Hildebrand EE, Uroz S, Frey-Klett P (2009) Interactions between mycorrhizal fungi and mycorrhizosphere bacteria during mineral weathering: budget analysis and bacterial quantification. Soil Biol Biochem 41:1935–1942

Kolo K, Claeys P (2005) *In vitro* formation of Ca-oxalates and the mineral glushinskite by fungal interaction with carbonate substrates and seawater. Biogeosciences 2:277–293

Kolo K, Keppens E, Preat A, Claeys P (2007) Experimental observations on fungal diagenesis of carbonate substrates. J Geophys Res 112:1–20

Konhauser K (2007) Introduction to geomicrobiology. Blackwell, Oxford

Kraemer SM, Cheah SF, Zapf R, Xu JD, Raymond KN, Sposito G (1999) Effect of hydroxamate siderophores on Fe release and Pb(II) adsorption by goethite. Geochim Cosmochim Acta 63:3003–3008

Krantz-Rulcker C, Allard B, Schnurer J (1993) Interactions between a soil fungus, *Trichoderma harzianum* and IIB metals – adsorption to mycelium and production of complexing metabolites. Biometals 6:223–230

Krantz-Rulcker C, Allard B, Schnurer J (1996) Adsorption of IIB metals by 3 common soil fungi – comparison and assessment of importance for metal distribution in natural soil systems. Soil Biol Biochem 28:967–975

Krupa P, Kozdroj J (2004) Accumulation of heavy metals by ectomycorrhizal fungi colonizing birch trees growing in an industrial desert soil. World J Microbiol Biotechnol 20:427–430

Kumar R, Kumar AV (1999) Biodeterioration of stone in tropical environments: an overview. The J. Paul Getty Trust, Madison
Landa ER (2005) Microbial biogeochemistry of uranium mill tailings. Adv Appl Microbiol 57:113–130
Landa ER, Gray JR (1995) US Geological Survey – results on the environmental fate of uranium mining and milling wastes. J Ind Microbiol 26:19–31
Landeweert R, Hoffland E, Finlay RD, Kuyper TW, Van Breemen N (2001) Linking plants to rocks: ectomycorrhizal fungi mobilize nutrients from minerals. Trends Ecol Evol 16:248–254
Lapeyrie F, Picatto C, Gerard J, Dexheimer J (1990) TEM Study of intracellular and extracellular calcium oxalate accumulation by ectomycorrhizal fungi in pure culture or in association with *Eucalyptus* seedlings. Symbiosis 9:163–166
Lapeyrie F, Ranger J, Vairelles D (1991) Phosphate-solubilizing activity of ectomycorrhizal fungi *in vitro*. Can J Bot 69:342–346
Lebrun E, Brungna M, Baymann F, Muller D, Lievremont D, Lett MC, Nitschke W (2003) Arsenite oxidase, an ancient bioenergetic enzyme. Mol Biol Evol 20:686–693
Ledin M, Krantz-Rulcker C, Allard B (1996) Zn, Cd and Hg accumulation by microorganisms, organic and inorganic soil components in multicompartment systems. Soil Biol Biochem 28:791–799
Lee J-U, Beveridge TJ (2001) Interaction between iron and *Pseudomonas aeruginosa* biofilms attached to sepharose surfaces. Chem Geol 180:67–80
Leyval C, Joner EJ (2001) Bioavailability of heavy metals in the mycorrhizosphere. In: Gobran GR, Wenzel WW, Lombi E (eds) Trace elements in the rhizosphere. CRC Press, Boca Raton, pp 165–185
Leyval C, Turnau K, Haselwandter K (1997) Effect of heavy metal pollution on mycorrhizal colonization and function: physiological, ecological and applied aspects. Mycorrhiza 7:139–153
Lian B, Chen Y, Zhu L, Yang R (2008a) Effect of microbial weathering on carbonate rocks. Earth Sci Front 15:90–99
Lian B, Wang B, Pan M, Liu C, Teng HH (2008b) Microbial release of potassium from K-bearing minerals by thermophilic fungus *Aspergillus fumigatus*. Geochim Cosmochim Acta 72:87–98
Lloyd JL (2003) Microbial reduction of metals and radionuclides. FEMS Microbiol Rev 27:411–425
Lloyd J, Lovley DR (2001) Microbial detoxification of metals and radionuclides. Curr Opin Biotechnol 12:248–253
Lloyd JR, Macaskie LE (1998) Enzymatic recovery of elemental palladium using sulfate-reducing bacteria. Appl Environ Microbiol 64:4607–4609
Lloyd JR, Renshaw JC (2005) Bioremediation of radioactive waste: radionuclide-microbe interactions in laboratory and field-scale studies. Curr Opin Biotechnol 16:254–260
Lloyd JR, Ridley J, Khizniak T, Lyalikova NN, Macaskie LE (1999a) Reduction of technetium by *Desulfovibrio desulfuricans*: biocatalyst characterization and use in a flow-through bioreactor. Appl Environ Microbiol 65:2691–2696
Lloyd JR, Thomas GH, Finlay JA, Cole JA, Macaskie LE (1999b) Microbial reduction of technetium by *Escherichia coli* and *Desulfovibrio desulfuricans*: enhancement via the use of high activity strains and effect of process parameters. Biotechnol Bioeng 66:122–130
Lloyd JR, Lovley DR, Macaskie LE (2003) Biotechnological application of metal-reducing microorganisms. Adv Appl Microbiol 53:85–128
Lloyd JR, Pearce CI, Coker VS, Pattrick RADP, van der Laan G, Cutting R, Vaughan DV, Paterson-Beedle M, Mikheenko IP, Yong P, Macaskie LE (2008) Biomineralization: linking the fossil record to the production of high value functional materials. Geobiology 6:285–297
Losi ME, Frankenberger WT (1998) Microbial oxidation and solubilization of precipitated elemental selenium in soil. J Environ Qual 27:836–843
Lovley DR (1995) Bioremediation of organic and metal contaminants with dissimilatory metal reduction. J Ind Microbiol 14:85–93
Lovley DR (2000) Fe(III) and Mn(IV) reduction. In: Lovley DR (ed) Environmental microbe-metal interactions. ASM Press, Washington, DC, pp 3–30

Lovley DR (2001) Anaerobes to the rescue. Science 293:1444–1446
Lovley DR, Coates JD (1997) Bioremediation of metal contamination. Curr Opin Biotechnol 8:285–289
Lovley DR, Phillips EJP, Gorby YA, Landa ER (1991) Microbial reduction of uranium. Nature 350:413–416
Lunsdorf H, Erb RW, Abraham WR, Timmis KN (2000) 'Clay hutches': a novel interaction between bacteria and clay minerals. Environ Microbiol 2:161–168
Macalady J, Banfield JF (2003) Molecular geomicrobiology: genes and geochemical cycling. Earth Planet Sci Lett 209:1–17
Macaskie LE (1991) The application of biotechnology to the treatment of wastes produced by the nuclear fuel cycle – biodegradation and bioaccumulation as a means of treating radionuclide-containing streams. Crit Rev Biotechnol 11:41–112
Macaskie LE, Jeong BC, Tolley MR (1994) Enzymically accelerated biomineralization of heavy metals: application to the removal of americium and plutonium from aqueous flows. FEMS Microbiol Rev 14:351–367
Mandal SK, Roy A, Banerjee PC (2002) Iron leaching from china clay by fungal strains. Trans Indian Inst Metals 55:1–7
Marshall KC (1971) Sorption interactions between soil particles and microorganisms. In: McLaren AD, Skujins J (eds) Soil biochemistry, vol 2. Marcel Dekker, New York, pp 409–445
Marshman NA, Marshall KC (1981a) Bacterial growth on proteins in the presence of clay minerals. Soil Biol Biochem 13:127–134
Marshman NA, Marshall KC (1981b) Some effects of montmorillonite on the growth of mixed microbial cultures. Soil Biol Biochem 13:135–141
Martino E, Perotto S, Parsons R, Gadd GM (2003) Solubilization of insoluble inorganic zinc compounds by ericoid mycorrhizal fungi derived from heavy metal polluted sites. Soil Biol Biochem 35:133–141
McLean J, Beveridge TJ (2001) Chromate reduction by a pseudomonad isolated from a site contaminated with chromated copper arsenate. Appl Environ Microbiol 67:1076–1084
McLean JS, Lee J-U, Beveridge TJ (2002) Interactions of bacteria and environmental metals, fine-grained mineral development, and bioremediation strategies. In: Huang PM, Bollag J-M, Senesi N (eds) Interactions between soil particles and microorganisms. Wiley, New York, pp 228–261
Meharg AA (2003) The mechanistic basis of interactions between mycorrhizal associations and toxic metal cations. Mycol Res 107:1253–1265
Meharg AA, Cairney JWG (2000) Co-evolution of mycorrhizal symbionts and their hosts to metal-contaminated environments. Adv Ecol Res 30:69–112
Miyata N, Tani Y, Iwahori K, Soma M (2004) Enzymatic formation of manganese oxides by an *Acremonium*-like hyphomycete fungus, strain KR21–2. FEMS Microbiol Ecol 47:101–109
Miyata N, Tani Y, Maruo K, Tsuno H, Sakata M, Iwahori K (2006) Manganese(IV) oxide production by *Acremonium* sp. strain KR21–2 and extracellular Mn(II) oxidase activity. Appl Environ Microbiol 72:6467–6473
Miyata M, Tani Y, Sakata M, Iwahori K (2007) Microbial manganese oxide formation and interaction with toxic metal ions. J Biosci Bioeng 104:1–8
Morley GF, Gadd GM (1995) Sorption of toxic metals by fungi and clay minerals. Mycol Res 99:1429–1438
Mossman DJ, Reimer T, Durstling H (1999) Microbial processes in gold migration and deposition: modern analogues to ancient deposits. Geosci Can 26:131–140
Mowll JL, Gadd GM (1984) Cadmium uptake by *Aureobasidium pullulans*. J Gen Microbiol 130:279–284
Nica D, Davis JL, Kirby L, Zuo G, Roberts DJ (2000) Isolation and characterization of microorganisms involved in the biodeterioration of concrete in sewers. Int Biodet Biodegrad 46:61–68
Olsson PA, Wallander H (1998) Interactions between ectomycorrhizal fungi and the bacterial community in soils amended with various primary minerals. FEMS Microbiol Ecol 27:195–205

Oremland R, Stolz J (2000) Dissimilatory reduction of selenate and arsenate in nature. In: Lovley DR (ed) Environmental microbe-metal interactions. ASM Press, Washington, DC, pp 199–224

Oremland RS, Hollibaugh JT, Maest AS, Presser TS, Miller LG, Culbertson CW (1989) Selenate reduction to elemental selenium by anaerobic bacteria in sediments and culture: biogeochemical significance of a novel sulfate-independent respiration. Appl Environ Microbiol 55:2333–2343

Osman D, Cavet JS (2008) Copper homeostasis in bacteria. Adv Appl Microbiol 65:217–247

Paez-Espino D, Tamames J, de Lorenzo V, Canovas D (2009) Microbial responses to environmental arsenic. Biometals 22:117–130

Perotto S, Girlanda M, Martino E (2002) Ericoid mycorrhizal fungi: some new perspectives on old acquaintances. Plant Soil 244:41–53

Perry RS, Mcloughlin N, Lynne BY, Sephton MA, Oliver JD, Perry CC, Campbell K, Engel MH, Farmer JD, Brasier MD, Staley JT (2007) Defining biominerals and organominerals: direct and indirect indicators of life. Sediment Geol 201:157–179

Petkov V, Ren Y, Saratovsky I, Pasten P, Gurr SJ, Hayward MA, Poeppelmeier KR, Gaillard JF (2009) Atomic-scale structure of biogenic materials by total X-ray diffraction: a study of bacterial and fungal MnO_x. ACS Nano 3:441–445

Phillips EJP, Landa ER, Lovley DR (1995) Remediation of uranium contaminated soils with bicarbonate extraction and microbial U(VI) reduction. J Ind Microbiol 14:203–207

Posfai M, Dunin-Borkowski RE (2009) Magnetic nanocrystals in organisms. Elements 5:235–240

Pumpel T, Paknikar KM (2001) Bioremediation technologies for metal-containing wastewaters using metabolically active microorganisms. Adv Appl Microbiol 48:135–169

Purvis OW (1996) Interactions of lichens with metals. Sci Prog 79:283–309

Purvis OW, Halls C (1996) A review of lichens in metal-enriched environments. Lichenologist 28:571–601

Purvis OW, Pawlik-Skowronska B (2008) Lichens and metals. In: Avery SV, Stratford M, van West P (eds) Stress in yeasts and filamentous fungi. Elsevier, Amsterdam, pp 175–200

Rawlings DE (2002) Heavy metal mining using microbes. Ann Rev Microbiol 56:65–91

Rawlings DE, Dew D, du Plessis C (2003) Biomineralization of metal-containing ores and concentrates. Trends Biotechnol 21:38–44

Renninger N, McMahon KD, Knopp R, Nitsche H, Clark DS, Keasling JD (2001) Uranyl precipitation by biomass from an enhanced biological phosphorus removal reactor. Biodegradation 12:401–410

Renninger N, Knopp R, Nitsche H, Clark DS, Jay D, Keasling JD (2004) Uranyl precipitation by *Pseudomonas aeruginosa* via controlled polyphosphate metabolism. Appl Environ Microbiol 70:7404–7412

Renshaw JC, Robson GD, Trinci APJ, Wiebe MG, Livens FR, Collison D, Taylor RJ (2002) Fungal siderophores: structures, functions and applications. Mycol Res 106:1123–1142

Rhee YJ, Hillier S, Gadd GM (2012) Lead transformation to pyromorphite by fungi. Curr Biol 22:237–241

Ritz K, Young IM (2004) Interaction between soil structure and fungi. Mycologist 18:52–59

Rodriguez Navarro C, Sebastian E, RodriguezGallego M (1997) An urban model for dolomite precipitation: authigenic dolomite on weathered building stones. Sediment Geol 109:1–11

Roeselers G, van Loosdrecht MCM, Muyzer G (2007) Heterotrophic pioneers facilitate phototrophic biofilm development. Microb Ecol 54:578–585

Rogers JR, Bennett PC (2004) Mineral stimulation of subsurface microorganisms: release of limiting nutrients from silicates. Chem Geol 203:91–108

Rosen K, Zhong WL, Martensson A (2005) Arbuscular mycorrhizal fungi mediated uptake of Cs-137 in leek and ryegrass. Sci Total Environ 338:283–290

Rosling A, Lindahl BD, Taylor AFS, Finlay RD (2004a) Mycelial growth and substrate acidification of ectomycorrhizal fungi in response to different minerals. FEMS Microbiol Ecol 47:31–37

Rosling A, Lindahl BD, Finlay RD (2004b) Carbon allocation to ectomycorrhizal roots and mycelium colonising different mineral substrates. New Phytol 162:795–802

Rufyikiri G, Huysmans L, Wannijn J, Van Hees M, Leyval C, Jakobsen I (2004) Arbuscular mycorrhizal fungi can decrease the uptake of uranium by subterranean clover grown at high levels of uranium in soil. Environ Poll 130:427–436

Rusin PA, Sharp JE, Oden KL, Arnold RG, Sinclair NA (1993) Isolation and physiology of a manganese-reducing *Bacillus polymyxa* from an Oligocene silver-bearing ore and sediment with reference to Precambrian biogeochemistry. Precamb Res 61:231–240

Salt DE, Smith RD, Raskin I (1998) Phytoremediation. Ann Rev Plant Physiol Plant Mol Biol 49:643–668

Sand W (1997) Microbial mechanisms of deterioration of inorganic substrates: a general mechanistic overview. Int Biodeter Biodegrad 40:183–190

Saratovsky I, Gurr SJ, Hayward MA (2009) The structure of manganese oxide formed by the fungus *Acremonium* sp. strain KR21-2. Geochim Cosmochim Acta 73:3291–3300

Sayer JA, Gadd GM (1997) Solubilization and transformation of insoluble metal compounds to insoluble metal oxalates by *Aspergillus niger*. Mycol Res 101:653–661

Sayer JA, Cotter-Howells JD, Watson C, Hillier S, Gadd GM (1999) Lead mineral transformation by fungi. Curr Biol 9:691–694

Scheerer S, Ortega-Morales O, Gaylarde C (2009) Microbial deterioration of stone monuments: an updated overview. Adv Appl Microbiol 66:97–139

Schneider J, Le Campion-Alsumard T (1999) Construction and destruction of carbonates by marine and freshwater cyanobacteria. Eur Phycol 34:417–426

Schröder I, Johnson E, de Vries S (2003) Microbial ferric iron reductases. FEMS Microbiol Rev 27:427–447

Schutzendubel A, Polle A (2002) Plant responses to abiotic stresses: heavy metal-induced oxidative stress and protection by mycorrhization. J Exp Bot 53:1351–1365

Silver S, Phung LT (2009) Heavy metals, bacterial resistance. In: Schaechter M (ed) Encyclopedia of microbiology. Elsevier, Oxford, pp 220–227

Smith WL, Gadd GM (2000) Reduction and precipitation of chromate by mixed culture sulphate-reducing bacterial biofilms. J Appl Microbiol 88:983–991

Smith SE, Read DJ (1997) Mycorrhizal symbiosis, 2nd edn. Academic, San Diego

Smits M (2006) Mineral tunnelling by fungi. In: Gadd GM (ed) Fungi in biogeochemical cycles. Cambridge University Press, Cambridge, pp 311–327

Song WY, Sohn EJ, Martinoia E, Lee YJ, Yang YY, Jasinski M, Forestier C, Hwang I, Lee Y (2003) Engineering tolerance and accumulation of lead and cadmium in transgenic plants. Nat Biotechnol 21:914–919

Southam G, Lengke MF, Fairbrother L, Reith F (2009) The biogeochemistry of gold. Elements 5:303–307

Sparks DL (2005) Toxic metals in the environment: the role of surfaces. Elements 1:193–197

Sterflinger K (2000) Fungi as geologic agents. Geomicrobiol J 17:97–124

Stolz JF, Oremland RS (1999) Bacterial respiration of arsenic and selenium. FEMS Microbiol Rev 23:615–627

Stolz JF, Basu P, Santini JM, Oremland RS (2006) Arsenic and selenium in microbial metabolism. Ann Rev Microbiol 60:107–130

Strasser H, Burgstaller W, Schinner F (1994) High yield production of oxalic acid for metal leaching purposes by *Aspergillus niger*. FEMS Microbiol Lett 119:365–370

Stumm W, Morgan JJ (1996) Aquatic chemistry. Chemical equilibria and rates in natural waters. Wiley, New York

Suzuki I (2001) Microbial leaching of metals from sulfide minerals. Biotechnol Adv 19:119–132

Tamaki S, Frankenberger WT (1992) Environmental biochemistry of arsenic. Rev Environ Contam Toxicol 124:79–110

Tazaki K (2006) Clays, microorganisms, and biomineralization. In: Bergaya F, Theng BKG, Lagaly G (eds) Handbook of clay science, developments in clay science, vol 1. Elsevier, Amsterdam, pp 477–497

Tebo BM, Obraztsova AY (1998) Sulfate-reducing bacterium grows with Cr(VI), U(VI), Mn(IV), and Fe(III) as electron acceptors. FEMS Microbiol Lett 162:193–198

Tebo BM, Bargar JR, Clement BG, Dick GJ, Murray KJ, Parker D, Verity R, Webb SM (2004) Biogenic manganese oxides: properties and mechanisms of formation. Ann Rev Earth Planet Sci 32:287–328
Tebo BM, Johnson HA, McCarthy JK, Templeton AS (2005) Geomicrobiology of manganese(II) oxidation. Trends Microbiol 13:421–438
Thayer JS (1989) Methylation: its role in the environmental mobility of heavy elements. Appl Organomet Chem 3:123–128
Theng BKG, Yuan G (2008) Nanoparticles in the soil environment. Elements 4:395–399
Thompson-Eagle ET, Frankenberger WT (1992) Bioremediation of soils contaminated with selenium. In: Lal R, Stewart BA (eds) Advances in soil science. Springer, New York, pp 261–309
Thompson-Eagle ET, Frankenberger WT, Karlson U (1989) Volatilization of selenium by *Alternaria alternata*. Appl Environ Microbiol 55:1406–1413
Tsai S-L, Singh S, Chen W (2009) Arsenic metabolism by microbes in nature and the impact on arsenic remediation. Curr Opin Biotechnol 20:1–9
Uroz S, Calvaruso C, Turpault M-P, Frey-Klett P (2009) Mineral weathering by bacteria: ecology, actors and mechanisms. Trends Microbiol 17:378–387
Urrutia MM, Beveridge TJ (1994) Formation of fine-grained silicate minerals and metal precipitates by a bacterial cell surface (*Bacillus subtilis*) and implications on global cycling of silicon. Chem Geol 116:261–280
Van Breemen N, Lundstrom US, Jongmans AG (2000) Do plants drive podzolization via rock-eating mycorrhizal fungi? Geoderma 94:163–171
Van der Lelie D, Schwitzguebel JP, Glass DJ, Vangronsveld J, Baker A (2001) Assessing phytoremediation's progress in the United States and Europe. Environ Sci Technol 35:446A–452A
Vaughan DJ, Pattrick RAD, Wogelius RA (2002) Minerals, metals and molecules: ore and environmental mineralogy in the new millenium. Miner Mag 66:653–676
Verrecchia EP (2000) Fungi and sediments. In: Riding RE, Awramik SM (eds) Microbial sediments. Springer, Berlin, pp 69–75
Verrecchia EP, Dumont JL, Rolko KE (1990) Do fungi building limestones exist in semi-arid regions? Naturwissenschaften 77:584–586
Verrecchia EP, Braissant O, Cailleau G (2006) The oxalate-carbonate pathway in soil carbon storage: the role of fungi and oxalotrophic bacteria. In: Gadd GM (ed) Fungi in biogeochemical cycles. Cambridge University Press, Cambridge, pp 289–310
Violante A, Huang PM, Gadd GM (eds) (2008) Biophysico-chemical processes of heavy metals and metalloids in soil environments. Wiley, Chichester
Volesky B (1990) Biosorption of heavy metals. CRC Press, Boca Raton
Walker JJ, Pace NR (2007) Endolithic microbial ecosystems. Ann Rev Microbiol 61:331–347
Wall JD, Krumholz LR (2006) Uranium reduction. Ann Rev Microbiol 60:149–166
Wallander H, Mahmood S, Hagerberg D, Johansson L, Pallon J (2003) Elemental composition of ectomycorrhizal mycelia identified by PCR-RFLP analysis and grown in contact with apatite or wood ash in forest soil. FEMS Microbiol Ecol 44:57–65
Wang J, Chen C (2009) Biosorbents for heavy metals removal and their future. Biotechnol Adv 27:195–226
Wang CL, Lum AM, Ozuna SC, Clark DS, Keasling JD (2001) Aerobic sulfide production and cadmium precipitation by *Escherichia coli* expressing the *Treponema denticola* cysteine desulfhydrase gene. Appl Microbiol Biotechnol 56:425–430
Warren LA, Haack EA (2001) Biogeochemical controls on metal behaviour in freshwater environments. Earth Sci Rev 54:261–320
Watson JHP, Ellwood DC, Deng QX, Mikhalovsky S, Hayter CE, Evans J (1995) Heavy metal adsorption on bacterially-produced FeS. Min Eng 8:1097–1108
Watson JHP, Cressey BA, Roberts AP, Ellwood DC, Charnock JM, Soper AK (2000) Structural and magnetic studies on heavy-metal-adsorbing iron sulphide nanoparticles produced by sulphate-reducing bacteria. J Magnetism Magn Mater 214:13–30

Weaver TL, Dugan PR (1972) Enhancement of bacteria methane oxidation by clay minerals. Nature 237:518

Wei Z, Kierans M, Gadd GM (2012a) A model sheet mineral system to study fungal bioweathering of mica. Geomicrobiol J 29:323–331

Wei Z, Hillier S, Gadd GM (2012b) Biotransformation of manganese oxides by fungi: solubilization and production of manganese oxalate biominerals. Environ Microbiol 14:1744–1753

Wenzel CL, Ashford AE, Summerell BA (1994) Phosphate-solubilizing bacteria associated with proteoid roots of seedlings of warratah [*Telopea speciosissima* (Sm.) R. Br.]. New Phytol 128:487–496

White C, Gadd GM (1998) Accumulation and effects of cadmium on sulphate-reducing bacterial biofilms. Microbiology 144:1407–1415

White C, Gadd GM (2000) Copper accumulation by sulphate-reducing bacterial biofilms and effects on growth. FEMS Microbiol Lett 183:313–318

White C, Sayer JA, Gadd GM (1997) Microbial solubilization and immobilization of toxic metals: key biogeochemical processes for treatment of contamination. FEMS Microbiol Rev 20:503–516

White C, Sharman AK, Gadd GM (1998) An integrated microbial process for the bioremediation of soil contaminated with toxic metals. Nat Biotechnol 16:572–575

White C, Dennis JS, Gadd GM (2003) A mathematical process model for cadmium precipitation by sulphate-reducing bacterial biofilms. Biodegradation 14:139–151

Whitelaw MA, Harden TJ, Helyar KR (1999) Phosphate solubilization in solution culture by the soil fungus *Penicillium radicum*. Soil Biol Biochem 31:655–665

Yee N, Kobayashi DY (2008) Molecular genetics of selenate reduction by *Enterobacter cloacae* SLD1a-1. Adv Appl Microbiol 64:107–123

Zhdanova NN, Zakharchenko VA, Vember VV, Nakonechnaya LT (2000) Fungi from Chernobyl: mycobiota of the inner regions of the containment structures of the damaged nuclear reactor. Mycol Res 104:1421–1426

Chapter 7
Role of Biomolecules in Influencing Transformation Mechanisms of Metals and Metalloids in Soil Environments

Antonio Violante, J. Zhu, M. Pigna, A. Jara, V. Cozzolino, and M.L. Mora

Abstract The aim of this work is to provide information on the factors which affect the mobility of trace elements in soils and related environments The most important processes in which trace elements are involved include adsorption-desorption, solution complexation, precipitation-dissolution and redox reactions. Adsorption on and desorption from organic and inorganic sorbents, such as humic substances, phyllosilicates, variable charge minerals, and microorganisms and chemical complexation reactions in the presence of high and low molecular mass organic compounds play a key role on the bioavailability and toxicity of trace elements.. Special attention is also devoted to the influence of biomolecules on the transformation of some metals (e.g. Al and Fe) and the subsequent formation of short-range ordered or non-crystalline precipitates, which play a very important role on the mobility of toxic elements both in cationic and anionic form.

Keywords Metals • Metalloids • Sorption/desorption processes • Biomolecules • Al and Fe precipitates

A. Violante (✉) • J. Zhu • M. Pigna • V. Cozzolino
Department of Soil, Plant, Environment and Animal Production Science, Faculty of Agriculture, University of Naples "Federico II", Via University 100, 80055 Portici, Naples, Italy
e-mail: violante@unina.it

A. Jara • M.L. Mora
Departamento de Ciencias Quimicas y Recursos Naturales, Scientific and Technological Bioresource Nucleous, Universidad de La Frontera, Temuco, Chile

J. Xu and D.L. Sparks (eds.), *Molecular Environmental Soil Science*, Progress in Soil Science, DOI 10.1007/978-94-007-4177-5_7,

7.1 Introduction

Trace elements may be present in soil solution with positive or negative charges and in different redox states. They occur predominantly in cationic form (Pb, Cu, Zn, Ni, Cd, Hg, Cr(III), Co) but some trace elements are present in anionic form (As, Se, Cr(VI), Mo and B).

In soil environments, redox (both biotic and abiotic) and complex reactions, sorption/desorption reactions on/from inorganic and organic sorbents and microorganisms control the concentration of metals and metalloids in solution and affect their bioavailability, leaching and toxicity. Sorption-desorption processes of elements in cationic form differ greatly from those in anionic form. These reactions are affected by many factors, such as pH, nature of the sorbents, presence and concentration of organic and inorganic ligands, including humic and fulvic acid, root exudates and nutrients.

The aim of this paper is to provide information on the factors which affect the mobility of trace elements. Special attention is devoted to the influence of biomolecules on the sorption/desorption processes of trace elements in cationic and anionic forms on/from soil components, soils and selected synthetic minerals and on the transformation of some metals (e.g. Al and Fe) and the subsequent formation of short-range ordered or non-crystalline precipitates, which play a very important role on the mobility and bioavailability of toxic elements.

7.2 Biomolecules in Soil Environments

Soil organic compounds are of two major types: (1) a series of acidic, yellow-to black-colored substances formed by secondary synthesis reactions, which are referred to as humic and fulvic acids; and (2) well-defined biochemical substances produced by microorganisms and plants. In the rhizosphere, a wide variety and considerable amounts of organic compounds are released by plants or microorganisms, which are present in significantly greater amounts than in the bulk soil. High-and low-molecular-mass substances, such as mucilages, polysaccharides, proteins, carbohydrates, phenolics, phytosiderophores, amino acids and organic acids, are released at the soil–root interface (Marschner 1995; Huang and Germida 2002; Hinsinger and Courchesne 2008). Low-molecular-mass organic acids (LMMOAs) are important constituents of soil solution and natural waters and include volatile and non-volatile aliphatic acids and, less commonly, aromatic acids. The most common LMMOAs in soil environments are those participating in the tricarboxylic acid cycle, including citric, α-ketoglutaric, fumaric and oxalacetic acid. Other aliphatic acids of common occurrence on plants and soils are formic, acetic, butyric, oxalic, and lactic acids.

The amount and composition of exudates released into the rhizosphere are highly variable and dependent on plant species, variety and age, and physicochemical environment. The concentrations of LMMOAs in the soil solution reported in the literature are usually low, in the range of 0.1–100 μmol/L. However, higher concentrations are found in the rhizosphere (Marschner 1995; Hinsinger and Courchesne 2008).

Studies have shown that rhizosphere soil exhibits differences in weathering, physical characteristics, and mineralogy compared with bulk soil. Larger amounts of Al and Fe were extracted from rhizosphere than from non-rhizosphere clay samples (Violante et al. 2002). Organic ligands interact with Fe and Al released from primary and secondary minerals, promoting the formation of short-range-ordered precipitates and organomineral complexes (as discussed below). Biomineralization of non-crystalline Al-oxides common in the cells of mature root bodies has been described earlier by April and Keller (1990). Organics retained on the surfaces or present in the network of short-range-ordered precipitates strongly inhibit the crystallization of metal hydroxides or oxyhydroxides (Huang and Violante 1986; Violante et al. 2002). Crystalline, non-crystalline, or short-range-ordered Fe, Al and Mn hydroxides and oxyhydroxides as well as some short-range-ordered aluminosilicates (allophane and imogolite) react readily with heavy metals and metalloids.

7.3 Influence of Biomolecules on the Formation and Transformation of Metal Oxides

Metal oxides are ubiquitous in soil. They play an important role in influencing soil behaviour and thus have great impact on the ecosystem. They may exist as crystalline minerals, as short-range ordered minerals, or non-crystalline precipitates, which are partly present as coating on clay minerals and humic substances. The short-range ordered Al, Fe and Mn oxides are among the most reactive inorganic components. These precipitates exist in soils as nanoparticles (size lower than 100 nm). The high surface area of nanoparticles means that they are particularly influential in controlling the fate and transport of pollutants as well as the bioavailability of key nutrients such as phosphate.

As reported before root exudates and biomolecules produced by microorganisms as well as humic and fulvic acids are involved in the weathering of minerals and in the subsequent release and transformations of Fe and Al at the root–soil interface. Aluminum and Fe are released to soil solution and natural waters through chemical and biochemical processes, undergoes hydrolysis and may give rise to precipitation products (Fig. 7.1). Clay minerals, pH, and organic and inorganic ligands are the most important factors which influence the phase transformation of Al and Fe and the mineralogy, order, particle size, specific surface and reactivity of the metal precipitation products towards nutrients and pollutants (Huang and Violante 1986; Hsu 1989; Violante et al. 2002; Huang et al. 2002a, b).

The influence of organic (as well as inorganic ligands) on Al nucleation reactions, Al transformations and properties of Al precipitation products has been widely studied. Fulvic and humic acids, and dissolved organic ligands that form stable

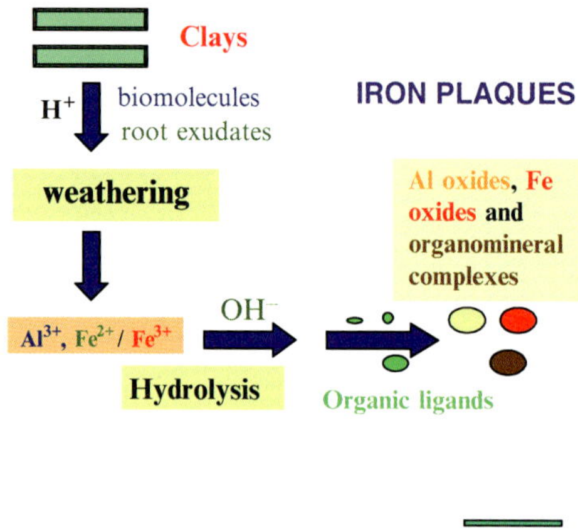

Fig. 7.1 Interaction between clay minerals and biomolecules (e.g. root exudates). Biomolecules are involved in the weathering of clay minerals, in the release of Al and Fe ions and in the formation of poorly crystalline Al and/or Fe precipitation products and organo-mineral complexes. On plant roots short-range ordered Fe precipitates are formed, usually called iron plaques

complexes with Al have been found to compete with OH^- ions, strongly perturbing Al hydrolytic reactions, and retarding and inhibiting the crystallization of Al-hydroxides through occupation of aluminum-hydroxide coordination sites (Huang and Violante 1986; Violante et al. 2002; Colombo et al. 2004).

The rate of the Al transformations in neutral or lightly acidic or alkaline environments

$$\text{Short-range noncrystalline materials} \rightarrow \text{boehmites } (AlOOH) \rightarrow Al(OH)_3 \text{ polymorphs}$$

is controlled mainly by pH, foreign ligands, clay minerals and temperature. In acid environments, the formation of Al precipitates occurs through hydrolytic reactions of Al monomers and polymers (Sposito 1996; Huang et al. 2002a, b; Violante et al. 2002).

Organic complexing anions delay the crystallization of $Al(OH)_3$ to varying degrees. The chemical composition, the molecular structure and the nature of functional groups of each anion control the kinetics of crystallization. The effectiveness of each ligand in inhibiting Al-hydroxides formation increases when its concentration rises and the pH decreases.

The relative effectiveness of some organic ligands in retarding or inhibiting $Al(OH)_3$ crystallization is approximately as follows (Violante et al. 2002): chetones < alcohols < acetate < glutarate < succinate = phthalate < glycine < tricarballilate < malonate < acetylacetone < glutamate < aspartate < oxalate < salicylate = malate < tannate < citrate < tartrate.

Fulvic and humic acids seem to have a behaviour almost similar to tannate. Many pieces of evidence indicate that ligands with a strong affinity for Al inhibit and/or perturb the crystallization of Al oxides more than ligands with a poor or moderate affinity for Al. The inhibition of crystallization by organic acids is often related to the stability of the soluble ligand-Al complex. However, it is not the higher the chelating power or affinity of a ligand for Al, the higher its influence in retarding

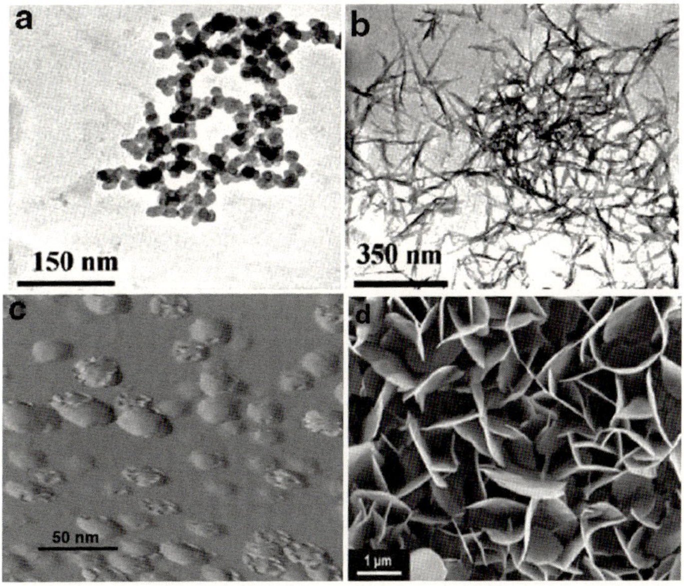

Fig. 7.2 Transmission electron micrographs of (**a**) a non-crystalline Al precipitation product formed in the presence of tannate at pH 5.0 and tannate/Al molar ratio of 0.1 after 9 years of aging at 50°C; (**b**) poorly crystalline boehmite formed in the presence of tannate at pH 8.0 and tannate/Al molar ratio of 0.05 after 9 years of aging at 50°C; (**c**) an atomic force micrograph of a ferrihydrite formed at pH 7.0 and citrate/Fe molar ratio of 0.1 and (**d**) a scanning electron micrograph of a double layered hydroxide

the kinetics of $Al(OH)_3$ crystallization. Clearly, polydentate and large ligands such as tannate, fulvate and humate often have a stronger influence on stabilizing non-crystalline materials than those with fewer functional groups or smaller size even if the latter have a stronger affinity for Al than the former. In fact, polydentate and large ligands may promote not only structural distortion but also aggregation between the particles of the non-crystalline materials, preventing solubilization and recrystallization. Within certain ranges of optimal concentrations perturbing ligands promote and stabilize the formation of short-range ordered boehmite or non-crystalline precipitates over $Al(OH)_3$ polymorphs (Fig. 7.2).

Colombo et al. (2004) studied the influence of tannate (tannic acid/Al molar ratio [R] of 0.01–0.1), pH (5.0 and 8.0), and aging for many years at different temperatures on the nature, size of the particles and crystallinity of synthetic Al precipitation products. Electron microscope studies showed that some precipitates were noncrystalline materials even after a long aging period (9 year at 50°C) and usually appeared to be agglomerates of spherical particles (Fig. 7.2a, b) whose size ranged from 10 to 40 nm. Other studies demonstrated that the nature and concentration of organic ligands also affected the morphology of short-range ordered precipitates (Violante et al. 2002, and references there in).

Short-range ordered Al-oxyhydroxides (poorly crystalline boehmite) also formed and showed fibrous morphology, but there were differences in the degree of fiber development depending on the ligand/Al molar ratio, pH, time of aging and temperature. These materials have a large surface area and a great reactivity.

The effect of biomolecules on the mechanisms of the formation and nature of the transformation products of Al(oxy)hydroxides at the atomic and molecular levels and the impact on their nanoscale surface chemistry remain to be uncovered. Only recently, some authors (Hu et al. 2008) studied the structure of Al in Al (oxy)hydroxides formed under the influence of varying concentrations of tannic acid and LMMOAs such as citric, malic, salicylic and acetic acids and a humic acid (HA) with X-ray absorption near edge structure (XANES) spectroscopy, Fourier transform infrared (FTIR) spectroscopy and X-ray diffraction analysis. The Al K- and L-edge XANES spectra showed that with increasing tannate or LMMOAs concentration the coordination number of Al changed from 6-fold to a mixed six, five and/or fourfold coordination, except for acetate because acetate was unable to perturb the formation of Al (oxy)hydroxides even at the acetate/Al molar ratio (MR) = 0.1. The proportion of 4-fold to 6-fold coordinated Al in the Al precipitation products depended on the structure and functionality of the LMMOAs. The incorporation of the LMMOA into the network structure of Al (oxy)hydroxides prevented the formation of sheets/inter-layer H-bonding that was required for the formation of crystalline Al (oxy)hydroxides. The HA only slightly perturbed the crystallization of the Al (oxy)hydroxides. The Al K-edge data showed that Al coordination number had not been altered in the presence of HA.

A number of iron oxides (hematite, goethite, lepidocrocite, maghemite, and magnetite) or short-range ordered precipitates (ferrihydrite) may be found in soils. In soil environments, mainly in the rhizosphere, iron undergoes a series of reactions that can increase or decrease its mobility and availability. Many studies have been carried out on the influence of organics, usually released in the rhizosphere by plants and microorganisms, on the transformation of Fe-oxides and on their surface properties.

Organic ligands interact with Fe released from primary and secondary minerals (as discussed before), promoting the formation of short-range ordered precipitates (mainly ferrihydrite) and organo-mineral complexes. This behaviour is similar to that ascertained in the weathering interface between lichens and rocks, leading to the first stages in the formation of soil. Rock weathering by lichens results in poorly ordered secondary products and although these may become better crystallized with time, this has far rarely been observed. The biotic weathering is usually considered to be combined with an absence of well ordered weathering products. Poorly ordered and crystalline Fe species have been revealed. Iron plaques are commonly formed on the surfaces of rice roots as a result of the release of oxygen and oxidants into the rhizosphere (Fig. 7.1). Numerous studies suggest that the predominant Fe root plaque phase is Fe(III)-oxihydroxides, including ferrihydrite (mainly), goethite and lepidocrocite. Despite a long history of observing Fe plaque on wetland plant roots and understanding the basics of plaque formation (i.e. reaction of plant-transported O_2 with Fe(II) in soils/sediments), it was largely assumed that plaque formation is predominately an abiotic (i.e. chemical) process because the kinetics of chemical oxidation can be extremely rapid. However, recent evidence has shown that populations of lithotrophic Fe(II)-oxidizing bacteria are associated with Fe plaques and may play a role in plaque deposition.

Ferrihydrite covers a range of poorly ordered compounds whose degree of ordering depends on the rate of hydrolysis and the time of aging. Neither the formula nor the structure of ferrihydrite has been fully established. The degree of ordering of ferrihydrite is variable and a range of x-ray diffraction (XRD) patterns may be obtained. The two extremes of ordering are referred as 2-line (0.252 and 0.148 nm) and 6-line ferrihydrite (additional lines at 0.225, 0.197, 0.172, and 0.151 nm), because the XRD patterns range from 2 to 6 reflections as structural order increases. Ferrihydrite is wide-spread in soils, especially Podsols (B horizons), Andisols and placic horizons. Its formation is favored by rapid oxidation of Fe, particularly in the presence of high concentrations of silicates, phosphate, organic matter and some cations including Al and Mn, which inhibit the formation of crystalline Fe oxides. Ferrihydrite forms very small particles (4–6 nm in size) more or less spherical in shape. Single particles are visible in 6-line ferrihydrite (Fig. 7.2c) and aluminous ferrihydrite (Fig. 7.2d), whereas they are difficult to identify in the 2-line ferrihydrite, being strongly aggregated each other (probably 2–3 nm in size).

Crystalline Fe and Al oxides usually have specific surface of 15–70 $m^2\ g^{-1}$, whereas short-range ordered precipitates may have a specific surface area even greater than 300 $m^2\ g^{-1}$. On these materials, hydroxylated or hydrated surfaces, positive or negative charges are developed by adsorption or desorption of H^+ or OH^- ions. The pH at which the net variable charge on the surfaces of these components is zero is called the point of zero charge (PZC). The reported PZC of Fe-oxides range from pH 7.0 to 9.5, whereas that of Al-oxides ranges from pH 8.0 to 9.2. (Hsu 1989; Cornell and Schwertmann 1996; Huang et al. 2002a, b; Violante et al. 2005).

Little is known of the effect of LMMOAs on the formation of short-range ordered aluminosilicates (allophanes, imogolite). Inoue and Huang (1984) reported that citric acid greatly perturbs the interactions of hydroxyl Al ions with orthosilicic acid and thus hinders the formation of imogolite and allophanes.

The manganese oxides and hydroxides are also important constituents in soils. They have, as Fe- and Al-oxides, a high capacity to sorb heavy metals and metalloids. Manganese is an essential element for plants and animals, and, to a large extent, the oxidation and reduction of Mn controls the amount available for uptake. Oxidizing conditions and high pH decrease the availability of Mn promoting the formation of precipitates, while reduction under acidic conditions may result in toxic levels of Mn in the soil (McKenzie 1989). The mineralogy of Mn is particularly complicated by the formation of many oxides and hydroxides. The influence of microorganisms and organic compounds additionally increases the complexity of Mn oxide formation in soil environments (Kampf et al. 2000). In soil environments these minerals are present in low concentrations and usually show a poor crystallinity. Manganese and Fe oxides are usually associated in nodules.

Double layered hydroxides (LDHs) are present in natural environments. In soils green rusts (Fe(II)-Fe(III) hydroxides that have a pyroaurite-stype structure) are formed by a number of abiotic and biotic processes under circumneutral to alkaline conditions in suboxic environments. Biomolecules may be sorbed on these minerals and may affect the sorption of toxic elements in anionic form (As and Se) (Violante, unpublished data).

In soil environments, non-crystalline or short-range metal precipitates (mainly Al and Fe oxides) are usually held on the surfaces of phyllosilicates. Organic matter, phyllosilicates, OH-Fe and OH-Al species, and short-range ordered metal oxides are constantly in close association with each other, forming organo-mineral complexes, which have different shape, size, charge and reactivity towards nutrients and pollutants, including metals and metalloids (Violante and Gianfreda 2000; Yuan and Theng 2011). Organo-mineral complexes have peculiar chemical and physico-chemical properties and reactivity, which differ from those of each soil component present in these aggregates (Violante and Gianfreda 2000; Zhu et al. 2010).

7.4 Redox and Complexes Reactions of Metals and Metalloids

Metals and metalloids such as Cr, Co, Ni, Cu, Pb, As, Se and Sb exist in more than one oxidation state. Redox reactions are important in controlling their transformation, mobility and toxicity (Huang and Germida 2002; Sparks 2002). They also control the transformation of Mn and Fe oxides, which strongly sorb metals and metalloids.

Clays and oxides often display the ability to catalyze electron transfer reactions. Abiotic redox processes occur on the surfaces of humic substances, Fe(III)- and Mn-oxides as depicted in Fig. 7.3. Iron and manganese oxides and layer silicates with structural Fe(III) are the most active in this regard. Redox reactions also control the transformation and Fe- reactivity and Mn-oxides in soils, which are the major sinks for heavy metals and metalloids. The efficiency of Mn oxides as an "electron pump"

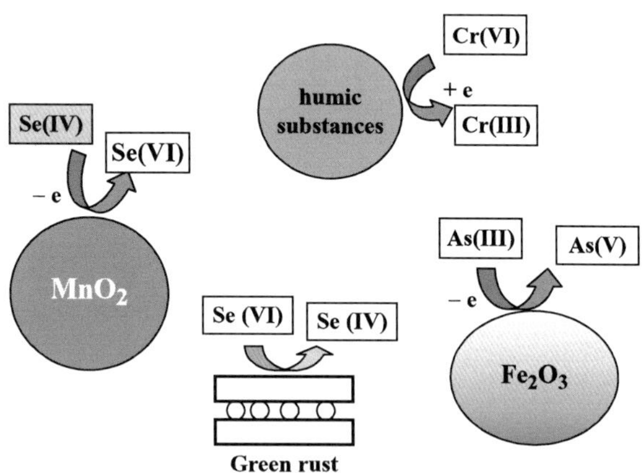

Fig. 7.3 Possible abiotic redox processes occurring on the surfaces of humic substances, Fe- and Mn-oxides and green rusts

for a wide range of redox reactions is unique among common soil minerals (Borch et al. 2010; and references therein).

Redox reactive metals often do have different degrees of toxicity depending on the specific metal oxidation state. Arsenic exists in two oxidation states, namely As(III) and As(V), with As(III) being more toxic and bioavailable than As(V). Manganese oxides are capable of oxidizing As(III) to As(V), so this oxidation process has important implications for the mobility and toxicity of arsenic. Oscarson et al. (1981) evidenced using colorimetry and x-ray photoelectron spectroscopy that a redox reaction between Fe(III)-oxide and As(III) did not occur within 72 h, indicating that the kinetics of the redox reaction between As(III) and Fe(III) is relatively slow. More recently, De Vitre et al. (1991) reported that As(III) oxidation by Fe oxyhydroxides was observed at high Fe/As ratios. Sun and Doner (1996) demonstrated that oxidation of sorbed As(III) to As(V) on Fe-oxides may occur to a limited extent (about 20%).

Manganese oxides catalize the oxidation of other heavy metals such as Co, Pb, Ni, and Cu. Reducing conditions may release As into solution through reduction of Fe and Mn oxides and reduction of As(V) to As(III). The simultaneous presence of high concentrations of dissolved As and Fe(II) in anoxic groundwaters has led to the conclusion that reductive dissolution of As-rich Fe(III) (hydr)oxides mobilizes geogenic As. The As mobility, bioavailability, toxicity and environmental fate are controlled by biogeochemical transformations that either form or destroy As-bearing carrier phases, or modify the redox state and chemical As speciation (see Figure 2 and Table 1 in Borch et al. 2010).

Chromate is toxic to plants, animals and humans, and is a suspected carcinogen, whereas Cr(III) is not toxic to plants and is necessary in animal nutrition, so that reactions that reduce Cr(VI) to Cr(III) are of great importance. Furthermore, Cr(VI) is mobile in soils and readily available. Organic materials, sulfides and ferrous species appear to be the dominant reductants. Very stable Cr(III)-organic complexes form when Cr(VI) is reduced by soil organic matter (Fendorf 1995). Microorganisms may directly reduce many highly toxic metals (e.g., Cr, Hg and U) via detoxification pathways. Microbial reduction of certain metals to a lower redox state may reduce mobility and toxicity of these metals. Such processes may accompany other metal precipitation mechanisms. Aerobic and anaerobic reduction of Cr(VI) to Cr(III) is widespread in microorganisms (Gadd 2008; Borch et al. 2010). The Cr(III) redox transformation on pyrolusite has been also demonstrated (Sparks 2002).

Selenium can exist in a wide range of oxidation states. Selenium is an essential element in a wide range of organisms from bacteria to human at low concentration, but it is extremely toxic at high concentrations. Birnessite is able to oxidize Se(IV) to Se(VI). The geochemical cycling of Se parallels that of sulphur so that selenium is predominantly cycled through biological pathways (McLean et al. 2002). However, many suboxic soil environments contain green rust, which is a mixed Fe(II)-Fe(III) oxide, and is able to reduce Se from an oxidation state of VI to 0. Similar green-rust-mediated abiotic redox reactions are probably involved in the mobility of several other trace metals in the environment, such as Cr(VI), Se(VI) and V(VI) (Fig. 7.3).

Precipitation of Fe-, Mn-, and Al-(hydr)oxides may efficiently capture dissolved trace metals (Violante et al. 2006, 2007, 2009a, b). If the precipitates formed during oxidative precipitation are nanoparticulate and colloidal (Fig. 7.2a, b), they may greatly enhance the mobility of associated trace metals in aquatic and subsurface environments. Because, the wetland rhizosphere is an interface between oxic and anoxic environments, it is a site where microbial reactions can cycle iron between oxidized and reduced forms. Furthermore, other important microbial and chemical processes can interact with microbial iron transformations near plant roots. Iron plaque, a rusty colored coating found on many wetland plant roots, often contains high concentrations of metals such as Cr, Pb, Cu and Zn, and the metalloids As and Si (Si is largely inert and will not be considered further). The sequestration of metals and metalloids can reduce the availability of these potentially toxic elements. However, iron plaque may not be a permanent sink for these elements. As plant roots senesce and iron reduction begins to dominate over iron oxidation, iron plaque can be dissolved and elements associated with the plaque released in a more bio-available form. Microbial processes can accelerate iron cycle between oxidized and reduced forms but the role of microbes in the exchanges of elements between iron plaque and soil has not received sufficient attentions (Neubauer et al. 2008).

Sulfate reduction in contaminated soils may mobilize Cu, Pb and Cd through the formation of Cu-rich sulphide colloids. Sulphate-reducing bacteria (SRB) oxidise organic compounds and produce sulphide. The solubility product of most heavy metal sulphides, is very low so that even a moderate output of sulphide can remove metals from solution. As reviewed by Borch et al. (2010) biogeochemical oxidation processes are driven by O_2 entering anoxic systems. Certain oxidation processes such as Fe(II) oxidation by O_2 at neutral to alkaline pH proceed rapidly abiotically, but many slower processes are facilitated by chemotrophic microorganisms.

The soil solution is the medium through which metals and metalloids are taken up, either actively or passively by plant roots and microbial cells. The soil processes (adsorption-desorption, complexation, precipitation-dissolution or oxidation-reduction) control solution composition. Microorganisms and organic matter influence the composition of the soil solution. Microorganisms breakdown organic matter and produce new biochemical reaction products. Soluble organics originate from plant roots and soil microorganisms. Organic matter, especially LMMOAs, greatly enhance the solubility and mobility of metals in soils. In particular, metal complexation reactions with ligands affect the chemical reactivity, bioavailability and toxicity of metals in soil solution. The most important complexing functional groups of soil organic matter can be classified based on their affinity for hard, transition and soft metals (Sparks 2002; Senesi and Loffredo 2008). A substantial fraction of the soluble metal ions in the soil solution may be complexed with a series of biomolecules commonly present in microbial metabolites and root exudates.

The toxicity or bioavailability of a metal has been related to the activity of the free aquo ion. However, some evidence is now emerging that free metal ion hypothesis may not be valid in all situations. Plant uptake of metals seems to vary with the kind of chelators present in solution at the same free metal activity (Huang and Germida 2002 and references there in).

7.5 Sorption of Heavy Metals and Metalloids in the Absence or Presence of Biomolecules

The soil components responsible for trace element sorption include soil organic matter, phyllosilicates, carbonates, microorganisms and variable charge minerals whose charge varies with the pH of the soil solution. Soil components differ greatly in their sorption capacities, their cation and anion exchange capacities, and the binding energies of their sorption sites. Two general surface complexes exist and are described by the configuration geometry of the adsorbate at the adsorbent surface. These include *inner-* and *outer-sphere surface complexes* and are defined by the presence, or absence, of the hydration sphere of the adsorbate molecule upon interaction (Sposito 1986; Sparks 2002; Borda and Sparks 2008). As summarized by Sparks (2002) outer-sphere complexes involve electrostatic coulombic interactions, and are thus weak compared to inner-sphere and are reversible. Inner-sphere complexation is usually slower than outer-sphere complexation and is often not reversible. Anyway, inner- and outer-sphere complexations can occur simultaneously.

The sorption/desorption processes of trace elements in cationic form differ greatly from those of elements in anionic form.

Heavy metal cations show typical ion exchange behaviour on phyllosilicates. Surface bonding is prevalently electrostatic, and the high selectivity of some phyllosilicates for heavy metals (relative to metals as Ca or Mg) may indicate the presence of a few sites (–SiOH or –AlOH groups) capable of chemisorbing these metals. Sorbed heavy metals on the edge sites and planar sites of montmorillonite have been also identified by EXAFS as inner-sphere complexes and outer-sphere complexes, respectively (Morton et al. 2001; Sparks 2002, 2005). The sites at crystal edges of phyllosilicates may easily sorb toxic elements in anionic form as arsenic, selenium, molybdate. Kaolinite, halloysite and chlorite have much greater capacity to sorb anions than illite or montmorillonite. Lin and Puls (2000) demonstrated that halloysite and chlorite sorbed As(V) 35 fold more than illite or montmorillonite. Angove et al. (1997) revealed that Cd adsorption on kaolinite is due to two distinct processes: ion-exchange at the permanent charge sites on silanol faces and complexation with aluminol and silanol groups at crystal edges. The modelling of adsorption and potentiometric titration data also suggested that the adsorption of Cu, Pb, Zn, Cd, Ni, Co and Mn on kaolinite involved both electrostatic attraction with permanent negatively charged sites on the silanol faces and inner-sphere complexes with hydroxyl groups on the variable charged crystal edges (Gu and Evans 2008).

Organic ligands (e.g. organic acids) which form strong complexes with heavy metals usually prevent or reverse their association with phyllosilicates, by forming stable negatively charged complexes (Fig. 7.5). In contrast, the presence of certain (phyto)siderophores produced by microorganisms and exuded by plants may promote the formation of positive complexes and, consequently, the sorption of trace elements onto phyllosilicates (Violante et al. 2008; see their Figures 5.3 and 5.4).

Many studies have demonstrated that organic matter and metal oxides are much more effective scavengers of heavy metals in cationic form, because specific sorption and other complexation processes are the dominant binding mechanisms (Huang and Germida 2002; Violante et al. 2008).

Humic substances compose of humic, fulvic and humin fractions. Humic substances contain a large number of complexing sites and thus behave as a natural "multiligand" complexing system (Senesi and Loffredo 2008). They contain a variety of functional groups such as carboxyls (–COOH), phenols (–OH), thiols (–SH) and amines (–NH_2) and show strong binding ability for trace metals in cationic form (Senesi and Loffredo 2008). The high degree of selectivity of soil organic matter for most of trace metals in cationic form indicates that they form *inner-sphere* complexes with the functional groups, often forming internal five- or six-member ring on structures (Senesi 1992; Huang and Germida 2002; Sparks 2002). In a Cu-contaminated agricultural soil in Hubbardsville, New York, most of Cu was found to be adsorbed on soil organic matter via bidentate inner-sphere coordination with carboxyl or amine ligands rather than on metal oxides, silicates, phosphates or carbonates (Strawn and Baker 2008), suggesting the importance of humic substances in scavenging cationic trace elements. Alvarez-Puebla et al. (2004) found that electrostatic retention rather than specific retention was the most important mechanism of Cu, Ni and Co retention on brown humic acids at very low cation concentrations, but this mechanism decreased progressively with the increase of cation concentrations. During this process, the humic acids modified their conformation by decreasing the surface area to coordinate with cations. In the last decade, many studies confirmed that the formation of inner-sphere complexes was the most accurate mechanism of trace elements sorption on humic substances (Karlsson et al. 2006; Prado et al. 2006; Manceau and Matynia 2010).

By considering the coordination of Cu in first, second and third coordination shell in EXAFS spectra, Karlsson et al. (2006) suggested that Cu was complexed by either one or two five-membered chelate rings involving possible combinations of amino, carboxyl or carbonyl functional groups in soil organic matter samples, collected from four different locations. Recently, Manceau and Matynia (2010) confirmed that Cu formed a five-membered Cu$(\text{malate})_2$-like ring chelate at 100–300 ppm Cu concentration and a six-membered Cu $(\text{malonate})_{1-2}$-like ring chelate at higher concentration at pH 4.5 and 5.5 with four representative natural organic matter. Moreover, the malate-type chelate showed stronger binding strength owed to an –OH for –H substitution on the α carbon. Nitrogen-containing functional groups similar or dissimilar to amino acids were also likely to be engaged in the complexation of Cu at low concentration in aquatic humic substances (Croué et al. 2003). Karlsson et al. (2005) found that Cd complexed in soil organic matter was a mixture of a 4-coordination with S (thiols) and 4- and 6-coordinations with O/N ligands, but Cd-S associations on average were stronger than Cd-O/N associations, suggesting that reduced S ligands were involved in the complexation of Cd by natural organic matter.

Trace elements which exist in anionic form are not sorbed on soil organic matter, but certain elements (e.g. borate, arsenate, arsenite) are found to be bound

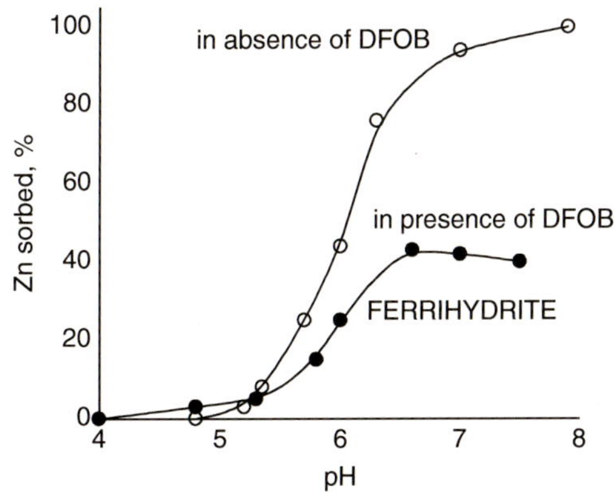

Fig. 7.4 Sorption of Zn on ferrihydrite as a function of pH in the absence and presence of hydroxamate siderophore desferrioxamine-B (DFOB). Dissolved concentration of Zn in ferrihydrite suspension was 10 mM without and with 300 mM DFOB (Modified from Neubauer et al. 2002)

to humic acids (McBride 2000; Violante et al. 2008). The content of organic matter in soils has been usually correlated with the B adsorption. Yermiyahu et al. (1995) showed that the organic matter is a better adsorbent for B than the mineral constituents, including variable charge minerals. Furthermore, Yermiyaho et al. (1988), Yermiyahu et al. (2001) and Sharma et al. (2006) demonstrated that addition of increasing amounts of organic manures to the soils promoted boron fixation. The mechanism of boron adsorption on organic matter is by ligand exchanged (Yermiyahu et al. 1988) forming B-diol, monodentate and bidentate complexes.

Unlike the phyllosilicates the ability of variable charge minerals (crystalline and short range ordered Fe-, Al-, Mn-oxides, allophanes, imogolite) to retain trace metals is pH dependent. Sorption, which increases from 0 to 100% of the amount added over a narrow region of 1–2 pH units, is termed as "sorption edge" (Fig. 7.4). Most heavy metals (Pb, Cu, Cr, Ni, Co, Zn, Al, Fe and Mn) are often sorbed more strongly than alkaline earth cations. Spectroscopic techniques such as electron spin resonance (ESR) and Extended X-ray Absorption Fine Structure Spectroscopy (EXAFS) have been used for the identification of metal complexes at the surfaces of Al, Fe or Mn oxides, silicate clays and soil organic matter. The mechanism of metal ion association with hydrous oxide surfaces involves an ion-exchange process in which the sorbed cations replace bound protons. Usually, specifically sorbed cations raise the value of PZC of variable charge minerals. pH affects sorption of trace elements, either by changing the number of sites available for sorption (sorption increases by increasing pH) or by changing the concentration of cation species (Me^{2+}, $MeOH^+$, $Me(OH)_2$).

Experiments with various synthetic Fe, Al and Mn oxides showed that the affinity of trace elements for Mn oxide was usually much greater than that for Fe or Al oxides. However, the nature, crystallinity, size of the crystals, surface charge of metal oxides and mixed metal oxides (e.g., Fe-Al oxides) also play an important role in the sorption selectivity of trace elements in cationic form (McKenzie 1980; McBride 1982; Violante et al. 2003, 2008; Sparks 2002).

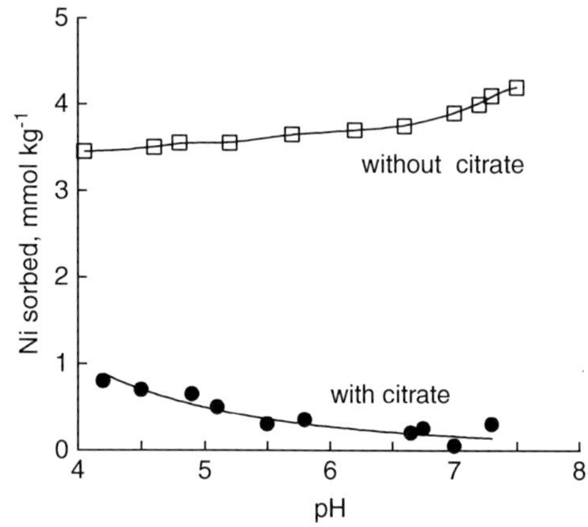

Fig. 7.5 Adsorption of Ni as a function of pH in the absence or presence of 100 μM citrate (Modified from Marcussen et al. 2009)

Sorption of anions onto variable charge minerals and soils varies with pH. With increasing pH, within a certain range, sorption decreases (due to a decrease of positive charge of minerals) or else increases to a maximum close to the pKa for anions of monoprotic conjugate acids and then decreases. Arsenite and selenite may be sorbed more easily at high pH values because they form weak acids at low pHs and consequently may be dissociated only in alkaline environments (Violante et al. 2008). Anions may be specifically or nonspecifically sorbed, forming *inner-sphere* or *outer-sphere* complexes. Trace elements, which form *inner-sphere* complexes, are arsenate, arsenite, molybdate, and selenite, which may form different surface complexes on inorganic soil components: monodentate, bidentate-binuclear and bidentate-mononuclear complex in different proportions depending on pH and surface coverage (Manning et al. 1998; Ona-Nguema et al. 2005). The EXAFS observation by Ona-Nguema et al. (2005) revealed that arsenite mainly formed bidentate mononuclear edge-sharing and bidentate binuclear corner-sharing complex on two-line ferrihydrite and hematite at high surface coverage, while a dominant amount of bidentate binuclear corner-sharing complex and a minor amount of monodentate mononuclear corner-sharing complex were formed on both goethite and lepidocrocite.

The EXAFS observation also revealed that both arsenate and arsenite formed inner-sphere complexes with a bidentate binuclear configuration on γ-Al_2O_3. However, additional outer-sphere arsenite complexes were detected by XANES spectra, particularly by increasing pH (5.5–8.0) and decreasing the ionic strength (Arai et al. 2001).

The processes which affect the sorption of trace element cations onto variable charge minerals in the presence of complexing agents, are particularly complex and are different from those onto phyllosilicates (Violante et al. 2008). Marcussen et al. (2009) observed that without citrate, montmorillonite shows higher Ni sorption than goethite. Citrate, strongly decreases Ni sorption to montmorillonite (Fig. 7.5); in presence of 100 um citrate, goethite become a stronger Ni sorbent

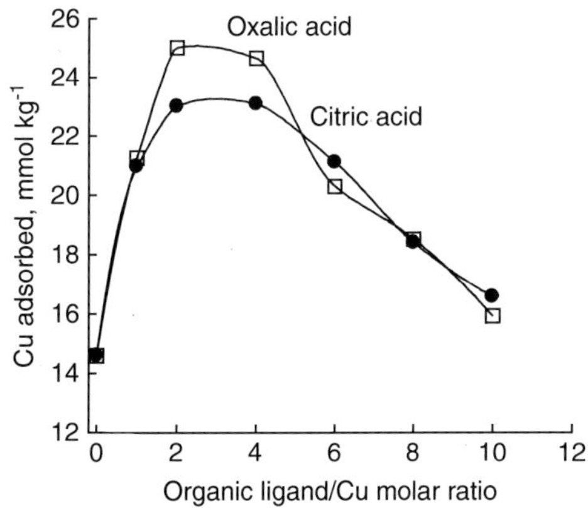

Fig. 7.6 Adsorption of Cu on goethite at pH 5 in the presence of increasing concentrations of oxalic or citric acid (initial ligand/Cu or Pb molar ratio from 0 to 10; 25 mmol of Cu added per kg of goethite)

than montmorillonite. The influence of different organic ligands on the sorption of heavy metal cations onto phyllosilicates and variable charge minerals have received attention as reviewed by Violante et al. (2008). Ha et al. (2009) prepared two hematite with different particle sizes [hematite nanoparticles (HN) (average diameter = 10.5 nm) and hematite microparticles (HM) (average diameter = 550 nm)] to investigate the interaction of Zn, oxalate and hematite in ternary system. By keeping constant the initial concentration of oxalate (8 mmol L^{-1}), an increase in Zn uptake was found on HN when the concentration of Zn was higher than 2 mmol L^{-1}, while an inhibition in the sorption of Zn was observed on HM when the concentration of Zn was lower than 6 mmol L^{-1}. These differences must be attributed to the large surface area of HN.

Recently, Perelomov et al. (2011) studied the influence of oxalate and citrate on Cu sorption onto a goethite (Fig. 7.6). By increasing the initial organic ligand/Cu molar ratio (r) from 0 to 10, the sorption of the metals initially increased and then decreased.

The sorption of heavy metals in cationic form in the presence of LMMOAs onto the surfaces of variable charge minerals may be affected by several processes, which may occur simultaneously: (i) increased retention of heavy metals for the decrease of the positive charge of the mineral when an organic ligand is adsorbed, (ii) increased retention due to the formation of negatively charged complexes between a metal and an organic ligand, which have a higher affinity for the sorbent, (iii) decreased retention due to the block of adsorption sites by the organic ligands on the minerals, (iv) decreased retention due to the strong complexation of metal in solution in nonsorbing form, with lowered free metal activity leading to desorption of surface-bound metal, (v) dissolution of the mineral in the presence of relatively high concentration of chelating organic ligands, leading to lesser metal retention (Cornell and Schwertmann 1996; Sparks 2002; Violante et al. 2008 and references there in).

Certain siderophores (e.g., DFOB) strongly affect the sorption of heavy metals onto variable minerals, but their behaviour is often different from that of LMMOAs.

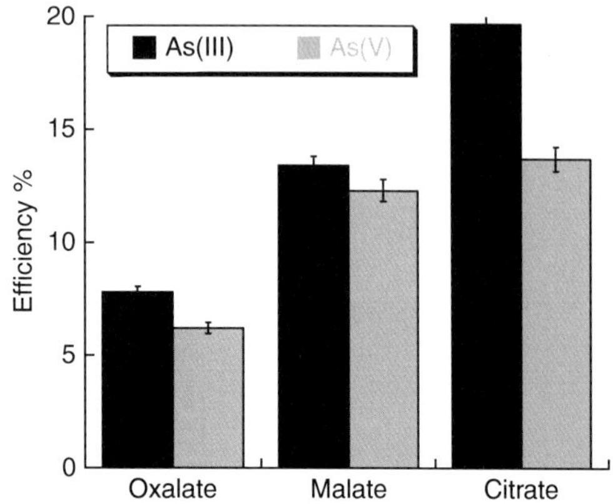

Fig. 7.7 Influence of selected organic ligands on the sorption of As(III) and As(V) onto ferrihydrite. Arsenite or As(V) and ligands were added together and at an initial molar ratio of As/ligands = 1 at pH 6.0

Neubauer et al. (2002) found that DFOB strongly prevented Zn sorption on ferrihydrite, because positively charged DFOB complexes were dominant up to pH 8.0. Because the complexes are not attracted by the positively charged ferrihydrite, DFOB decreased the sorption of Zn (Fig. 7.4), in contrast to montmorillonite suspensions (Violante et al. 2008; see their Fig. 5.A).

Organic ligands interact with variable charge minerals and soils and then affect the sorption of trace elements in anionic form by competing for available sorption sites and/or reducing the surface charge of the sorbents. The competition depends on the affinity of the anions for the surfaces of the sorbents as well as the nature and surface properties of the minerals and soils.

Zhu et al. (2011) studied the influence of different organic ligands (malate, tartrate and citrate) in preventing the sorption of As(III) and As(V) onto ferrihydrite at pH 5.0 and ligand/As(III) or As(V) molar ratio of 1 (Fig. 7.7). All the ligands showed a stronger capacity in preventing As(III) than As(V) (Fig. 7.7), in spite of more As(III) than As(V) was sorbed on the Fe-oxide, evidencing that As(V) form more stable complexes on the surfaces of the oxides

Dynes and Huang (1997) showed that the ability of 12 LMMOAs to inhibit selenite sorption on poorly crystalline Al-hydroxides was oxalate > malate > citrate > succinate > glycolate > aspartate > salycilate > *p*-hydroxybenzoate > glycine = formiate = acetate.

Trace elements may be sorbed on other abiotic and biotic sorbents. Evidence on the sorption of trace elements on microorganisms has been reported. Free-living bacteria and their extra-cellular macromolecular products (e.g., fibrils) can accumulate trace elements and may have mineral coatings with bound metals on their surfaces (Beveridge 1989; Jackson and Leppard 2002 and references therein). All microorganisms contain biopolymers such as proteins, nucleic acids, and polysaccharides which provide reactive sites for binding metal ions. The cell surfaces of all bacteria are largely negatively charged containing different types of negatively charged functional groups, such as carboxyl, hydroxyl and phosphoryl that can adsorb metal cations, and retain them by mineral nucleation. Biosorption comprises a variety of processes including ion exchange, chelation, adsorption and diffusion through cell walls and membranes all of which are dependent on the species used, the biomass origin and solution chemistry (Gavrilescu 2004). Differently from humic substances, besides the formation of outer-sphere and inner-sphere complexes on cell wall, considered a passive uptake in biosorption, active uptake which transports the metal ions into the cell is another and a peculiar mechanism for scavenging trace elements by living bacteria (Vijayaraghavan and Yun 2008). In passive uptake, some studies found or inferred that nonspecific sorption was the primary mechanism (Shuttleworth and Unz 1993; Özdemir et al. 2009), whereas other researches demonstrated that specific sorption was the main mechanism of heavy metal retention (Boyanov et al. 2003; Guiné et al. 2006). By monitoring the release of Mg^{2+}, Ca^{2+} and H^{+} during the adsorption of Cu^{2+}, Cd^{2+}, Zn^{2+} and Mn^{2+} on a fungal cell (*Saccharomyces cerevisiae*), Avery and Tobin (1993) confirmed that covalent bonding (H^{+} displacement) of the metal was greater at low metal concentrations, while weaker electrostatic interactions (Mg^{2+} plus Ca^{2+} displacement) became increasingly important at higher metal concentrations.

Layered double hydroxides (LDHs), which consist of positively charged layers and negatively charged interlayer anions, are known as "anionic clays" and postulated to be intermediate phases in the formation or dissolution of iron- or aluminum-(oxyhydro) oxides in natural environments (Schwertmann and Cornell 2000; Livi et al. 2009). Trace elements in anionic form may be easily sorbed by "anionic clays" (layered double hydroxides; LDHs) (Goh et al. 2008; Violante et al. 2009b). There are many published literature reporting the effect of competitive inorganic and organic anions on oxyanion sorption by LDHs (Goh et al. 2008, and references therein). Recently, Caporale et al. (2011) showed that oxalate, tartrate and citrate were effective in preventing As(V) sorption on synthetic LDHs. Citrate prevented As(V) sorption more than tartrate and oxalate in the order listed.

7.6 Kinetics of Sorption

Sorption (and desorption) of heavy metals and metalloids in soils can occur over wide time scales. Sorption is initially rapid (from milliseconds to hours) and can be ascribed to chemical reaction (adsorption) (Sparks 2002). During this rapid reaction process, a large portion of the sorption may occur. Many studies have shown that sorption/desorption of metals, oxyanions, radionuclides, and organic chemicals on soils can be slow, and may demonstrate a residence time effect (contact time between trace element and sorbent). The mechanisms for these lower reaction rates have been ascribed to diffusion into micropores of inorganic minerals such as some metal oxides and into humic substances, retention on sites of varying reactivity, and to surface nucleation/precipitation (Borda and Sparks 2008; and references there in). Reviews on kinetics and mechanisms of sorption-desorption in soils include that of Borda and Sparks (2008).

Soil and its minerals are porous and contain both macropores (>2 nm) and micropores (<2 nm), which can be between aggregates (inter-particle) or within an individual particle (intra-particle) (Sparks et al. 1999). Pore diffusion and solid diffusion thus were often proposed to account for the slow sorption of trace elements on soil components (Bruemmer et al. 1988; Papelis 1995; Axe and Trivedi 2002). By observing the sorption kinetics of Ni, Zn and Cd on goethite during a time period of 2 h to 42 day at pH 6, Bruemmer et al. (1988) proposed that the adsorption of trace elements on goethite involves three steps: (i) adsorption of trace elements on external surfaces, (ii) solid-state diffusion of trace elements from external to internal binding sites, (iii) trace elements binding and fixation at positions inside the goethite particles. Moreover, they suggested that the second step was the rate-limiting step. On the other hand, the minerals may have binding sites with different affinities.

As the amount of metal cations or anions sorbed on a surface increases, sorption can proceed from mononuclear adsorption to surface precipitation (a three dimensional phase). There is often a continuum between surface complexation (adsorption) and surface precipitation. At low surface coverages surface complexation (e.g., *outer-* and *inner-sphere* adsorption) tends to dominate. As surface loadings increase, nucleation occurs, resulting in the formation of distinct entities or aggregates on the surface. As surface loadings increase further, surface precipitation becomes the dominant mechanism (Sparks 2002, 2005).

Surface precipitates can start to form on time scale of minutes then growing in amount for hours or days. The initial precipitates are stabilized upon aging. This stabilization (residence time effect) has been attributed to the transformation of a mixed LDH phase into a more stable phyllosilicate phase. In other words, the metal sorption on soil minerals often results in a continuum of processes from adsorption to precipitation to solid phase transformation. As a consequence, heavy metals become less mobile and bioavailable with time. Spectroscopic evidence indicates that the formation of LDH phases can be suppressed by the presence of humic acids or LMMOAs (Yamaguchi et al. 2002).

Several models such as zero-order, first-order, second-order, parabolic diffusion, power function and Elovich equation have been applied to describe the sorption kinetics of trace elements on various soil components, but the suitability depends on the concrete sorbate-sorbent system condition. For example, Raven et al. (1998) reported that the adsorption of arsenite and arsenate on ferrihydrite was described best by parabolic diffusion equation, while Pigna et al. (2006) found that the sorption of arsenate on four typical metal oxides [noncrystalline $Al(OH)_x$, gibbsite, ferrihydrite and goethite] followed Elovich equation best. First order equation was found to be the best to describe the adsorption of Pb on iron oxides formed under the influence of citrate (Liu and Huang 2003), while the second-order equation was found to be more appropriate in fitting the adsorption of selenite on hydroxyaluminum- and hydroxyaluminosilicate-montmorillonite complexes (Saha et al. 2004), the adsorption of Cd on aluminum precipitation products formed under the influence of tannate (Yu et al. 2006), the adsorption of Cd, Pb, Cu and Cd on peat (Qin et al. 2006) and the adsorption of Cu on goethite and hematite nano-photocatalysts (Chen and Li 2010).

7.7 Desorption of Metals and Metalloids by Biomolecules

In contrast to sorption studies, relatively little information is available on the desorption of metals and metalloids from soils or soil components as affected by organic ligands, in spite of the wide occurrence of these reactions in soil environments particularly in the rhizosphere (Hinsinger and Courchesne 2008; as discussed below).

The presence of LMMOAs secreted by plant roots and rhizosphere microbes (acetic, citric, fumaric, malic, oxalic and succinic acids) was shown to enhance the rate of Cd release from contaminated tropical soils (Onyatta and Huang 2003). The amounts of Cd released from the soils increased with the log stability constant values of the Cd-LMMOAs complexes, indicating that Cd was brought into solution as Cd-LMMOA complexes.

Qin et al. (2004) studied the effects of LMMOAs on the desorption of Cu, Cd and Pb from two soils, and found that the desorption of all the heavy metals studied followed the decreasing order: citric acid > malic acid > acetic acid > $CaCl_2$ > $NaNO_3$, which was consistent with the order of stability of Cu-, Cd-, and Pb-LMMOAs complexes. The desorption of metals by organic ligands showed different trends in response to pH change due to their different complexing abilities. Desorption of metals decreased with increasing residence time of metals in soils. The effect of acetate, citrate and desferrioxamine B (DFOB) on the kinetics of Cd desorption from two fibrous silicate clay minerals (palykorskite and sepiolite) was studied (Shirvani et al. 2007). The presence of organic ligands to the systems induced higher amounts of Cd desorbed. Formation of soluble complexes between ligand anions and Cd or direct competition of ligands with Cd ions for surface sites was considered as a mechanism responsible for this findings. However, Yuan et al. (2007) studied the desorption of Cu and Cd from soils in the presence of selected organic ligands such as EDTA, citric acid, oxalic acid and tartaric acid as affected by pH and explained the desorption mechanism as the competition of complexation, adsorption and precipitation. The effect of complex contributed to the desorption, while the effect of adsorption and precipitation retarded the desorption. These effects were highly affected by pH. The net effect determined the desorption efficiency.

Glover et al. (2002) demonstrated that Pb and Cd removal from goethite by salicylate and oxalate at pH 4.5 was slow compared with the sorption reaction, and was not reversible, except for short-term Cd in the presence of salycilate. These authors hypothesized that the greater effectiveness of salicylate compared to oxalate was due to the capacity of oxalate to form ternary complexes between the metal and the surfaces of goethite.

Many desorption studies of trace elements in anionic form from soils and soil components were carried out using phosphate anions as the replacing ligands, because phosphate is added as fertilizer to soils and has a very high affinity for soil sorbents (Violante et al. 2008). In contrast, scant attention has been devoted to the influence of biomolecules in replacing xenobiotics.

Liu and Huang (2005) studied the desorption kinetics of Cd following its sorption on iron oxides formed under the influence of citric acid and ascertained that rate of desorption of Cd by nitrate, chloride, acetate or citrate was greatly influenced by the increase of their degree of crystal disorder, specific surface, and microporosity. The desorption kinetics can be divided into the fast and slow reactions. For the fast reaction, the rate of Cd desorption is generally in accord with the stability constants of Cd-extractant ligand complexes. For the slow reaction, the rate of Cd desorption is apparently influenced by the size of the extractant molecules: citrate > acetate ≥ chloride > nitrate. According to these authors a longer induction period for an extractant with a larger molecule is required for Cd desorption from the micropore surface.

Xu et al. (2008) studied the desorption kinetics by phosphate of arsenate adsorbed on Al (oxy)hydroxides formed under the influence of tannic acid. The kinetic data indicated multiple rate characteristics; a fast reaction period (0.083–3 h) and a slow reaction period (3–24 h). Aluminum precipitation products formed in the presence of small amounts of tannic acid (tannate/Al molar ratio 0.001 and 0.01) resulted in slower desorption rates of the adsorbed arsenate compared with the precipitates obtained in the absence of the organic ligands. These findings were ascribed to the tannate-induced structural perturbation, development of microporosity, enhanced surface area and reactivity. However, when the tannate/Al molar ratio was increased to 0.1 the desorption of As steadily increased, due to the decrease in PZC and the electrostatic repulsion from coprecipitated tannate which would weaken the binding of arsenate on the Al precipitates.

Acknowledgments This research was supported by the Italian Research Program of National Interest (PRIN), year 2008. DiSSPAPA Number 006.

References

Alvarez-Puebla RA, Valenzuela-Calahorro C, Carrido JJ (2004) Retention of Co(II), Ni(II), and Cu(II) on a purified brown humic acid. Modeling and characterization of the sorption process. Langmuir 20:3657–3664

Angove MJ, Johnson BB, Wells JD (1997) Adsorption of cadmium(II) on kaolinite. Colloids Surf A Physicochem Eng Asp 126:137–147

April R, Keller D (1990) Mineralogy of the rhizosphere in forest soils of the eastern United States. Biogeochem 9:1–12

Arai Y, Elzinga EJ, Sparks DL (2001) X-ray absorption spectroscopic investigation of arsenite and arsenate adsorption on the aluminum oxide-water interface. J Coll Interf Sci 235:80–88

Avery SV, Tobin JM (1993) Mechanism of adsorption of hard and soft metal ions to *Saccharomyces cerevisiae* and influence of hard and soft anions. Appl Environ Microbiol 59:2851–2856

Axe L, Trivedi P (2002) Intraparticle surface diffusion of metal contaminants and their attenuation in microporous amorphous Al, Fe, and Mn oxides. J Colloid Interface Sci 247:259–265

Beveridge TJ (1989) Role of cellular design in bacterial metal accumulation and mineralization. Annu Rev Microbiol 43:147–171

Borch T, Kretzschmar R, Kappler A, Van Cappelen P, Ginder-Vogel M, Voegelin A, Campbell K (2010) Biogeochemical redox processes and their impact on contaminant dynamics. Environ Sci Technol 44:15–23

Borda MJ, Sparks DL (2008) Mobility of trace elements in soil environments. In: Violante A, Huang PM, Gadd GM (eds) Biophysico-chemical processes of metals and metalloids in soil environments. Wiley, Hoboken

Boyanov MI, Kelly SD, Kemner KM, Bunker BA, Fein JB, Fowle DA (2003) Adsorption of cadmium to *Bacillus subtilis* bacterial cell walls: a pH-dependent X-ray absorption fine structure spectroscopy study. Geochim Cosmochim Acta 67:3299–3311

Bruemmer GW, Gerth J, Tiller KG (1988) Reaction kinetics of the adsorption and desorption of nickel, zinc and cadmium by goethite. I. Adsorption and diffusion of metals. Eur J Soil Sci 39:37–52

Caporale AG, Pigna M, Dynes JJ, Cozzolino V, Zhu J, Violante A (2011) Effect of inorganic and organic ligands on the sorption/desorption of arsenate on/from Al-Mg and Fe-Mg layered double hydroxides. J Hazard Mater 198: 291–298

Chen YH, Li FA (2010) Kinetic study on removal of copper(II) using goethite and hematite nano-photocatalysts. J Colloid Interface Sci 347:277–281

Colombo C, Ricciardella M, Di Cerce A, Maiuro L, Violante A (2004) Effect of tannate, pH, sample preparation, ageing and temperature on the formation and nature of Al oxyhydroxides. Clays Clay Miner 52:721–733

Cornell RM, Schwertmann U (1996) The iron oxides. Structure, properties, reactions and uses. VCH Publ, New York

Croué JP, Benedetti MF, Violleau D, Leenheer JA (2003) Characterization and copper binding of humic and nonhumic organic matter isolated from the South Platte River: evidence for the presence of nitrogenous binding site. Environ Sci Technol 37:328–336

De Vitre R, Belize N, Tessier A (1991) Speciation and adsorption of arsenic on diagenetic iron hydroxides. Limnol Oceanogr 36:1480–1485

Dynes JJ, Huang PM (1997) Influence of organic acids on selenite sorption by poorly ordered aluminum hydroxides. Soil Sci Soc Am J 61:772–783

Fendorf SE (1995) Surface reactions of chromium in soils and waters. Geoderma 67:55–71

Gadd JM (2008) Transformation and mobilization of metals, metalloids, and radionuclides by microorganisms. In: Violante A, Huang PM, Gadd GM (eds) Biophysico-chemical processes of metals and metalloids in soil environments, vol 1. Wiley, Hoboken

Gavrilescu M (2004) Removal of heavy metals from the environment by biosorption. Eng Life Sci 4:219–232

Glover LJ, Eick MJ, Brady PV (2002) Desorption kinetics of cadmium^{2+} and lead^{2+} from goethite: influence of time and organic acids. Soil Sci Soc Am J 66:797–804

Goh KH, Lim TT, Dong Z (2008) Application of layered double hydroxides for removal of oxyanions: a review. Water Res 42:1343–1368

Gu X, Evans LJ (2008) Surface complexation modelling of Cd(II), Cu(II), Ni(II), Pb(II) and Zn(II) adsorption onto kaolinite. Geochim Cosmochim Acta 72:267–276

Guiné V, Spadini L, Sarret G, Muris M, Delolme C, Gaudet JP, Martins JMF (2006) Zinc sorption to three gram-negative bacteria: combined titration, modeling, and EXAFS study. Environ Sci Technol 40:1806–1813

Ha J, Trainor TP, Farges F, Brown GE Jr (2009) Interaction of Zn(II) with hematite nanoparticles and microparticles: part 2. ATR-FTIR and EXAFS study of the aqueous Zn(II)/oxalate/hematite ternary system. Langmuir 25:5586–5593

Hinsinger P, Courchesne F (2008) Biogeochemistry of metals and metalloids at the soil-root interface. In: Violante A, Huang PM, Gadd GM (eds) Biophysico-chemical processes of heavy metals and metalloids in soil environments. Wiley, Hoboken

Hsu PH (1989) Aluminum hydroxides and oxyhydroxides. In: Dixon JB, Weed SB (eds) Minerals in soil environments, 2nd edn. Soil Science Society of America, Madison

Hu YF, Xu RK, Dynes JJ, Blyth RIR, Yu G, Kozak LM, Huang PM (2008) Coordination nature of aluminum (oxy)hydroxides formed under the influence of tannic acid studied by X-ray absorption spectroscopy. Geochim Cosmochim Acta 72:1959–1969

Huang PM, Germida JJ (2002) Chemical and biochemical processes in the rhizosphere: metal pollutants. In: Huang PM, Bollag JM, Senesi N (eds) Interactions between soil particles and microorganisms: impact on the terrestrial ecosystem. Wiley, New York

Huang PM, Violante A (1986) Influence of organic acids on crystallization and surface properties of precipitation products of aluminium. In: Huang PM, Schnitzer M (eds) Interactions of soil minerals with natural organics and microbes, Special publication, no. 17. Soil Science Society of America, Madison

Huang PM, Bollag JM, Senesi N (2002a) Interactions between soil particles and microorganisms: impact on the terrestrial ecosystem. Wiley, New York

Huang PM, Wang MK, Kampf N, Schulze DG (2002b) Aluminum hydroxides. In: Dixon JB, Schulze DG (eds) Soil mineralogy with environmental application. Soil Science Society of America, Madison

Inoue K, Huang PM (1984) Influence of citric acid on the natural formation of imogolite. Nature 308:58–60

Jackson TA, Leppard GG (2002) Energy dispersive x-ray microanalysis and its applications in biological research. In: Violante A, Huang PM, Bollag JM, Gianfreda L (eds) Soil mineral-organic-matter-microorganism interactions and ecosystem health: dynamics, mobility and transformations of pollutants and nutrients, developments in Soil Sci, vol 28. Elsevier, New York

Kampf N, Scheinost AC, Schultze DG (2000) Oxides minerals. In: Sumner ME (ed) Handbook of soil science. CRC Press, Boca Raton

Karlsson T, Persson P, Skyllberg U (2005) Extended X-ray absorption fine structure spectroscopy evidence for the complexation of cadmium by reduced sulfur groups in natural organic matter. Environ Sci Technol 39:3048–3055

Karlsson T, Persson P, Skyllberg U (2006) Complexation of copper(II) in organic soils and in dissolved organic matter-EXAFS evidence for chelate ring structures. Environ Sci Technol 40:2623–2628

Lin Z, Puls RW (2000) Adsorption, desorption and oxidation of arsenic affected by clay minerals and aging process. Environ Geol 39:753–759

Liu C, Huang PM (2003) Kinetics of lead adsorption by iron oxides formed under the influence of citrate. Geochim Cosmochim Acta 67:1045–1054

Liu C, Huang PM (2005) Kinetics of cadmium desorption from iron oxides formed under the influence of citrate. In: Huang PM, Gobran GR (eds) Biogeochemistry of trace elements in the rhizosphere. Elsevier, Amsterdam

Livi KJT, Senesi GS, Scheinost AC, Sparks DL (2009) Microscopic examination of nanosized mixed Ni-Al hydroxide surface precipitates on pyrophyllite. Environ Sci Technol 43:1299–1304

Manceau A, Matynia A (2010) The nature of Cu bonding to natural organic matter. Geochim Cosmochim Acta 74:2556–2580

Manning BA, Fendorf SE, Goldberg S (1998) Surface structures and stability of arsenic (III) on goethite. Spectroscopic evidence for inner-sphere complexes. Environ Sci Technol 32:2383–2388

Marcussen H, Holm PE, Strobel BW, Hansen HCB (2009) Nickel sorption to goethite and montmorillonite in presence of citrate. Environ Sci Technol 43:1122–1127

Marschner H (1995) Mineral nutrition of higher plants. Academic, New York

McBride MB (1982) Cu^{2+}-adsorption characteristics of aluminum hydroxide and oxyhydroxides. Clays Clay Miner 30:21–28

McBride MB (2000) Chemisorption and precipitation reactions. In: Summer ME (ed) Handbook of soil science. CRC Press, Boca Raton

McKenzie RM (1980) The adsorption of lead and other heavy metals on oxides of manganese and iron. Aust J Soil Res 21:505–513

McKenzie RM (1989) Manganese oxides and hydroxides. In: Dixon JB, Weed SB (eds) Minerals in soil environments, 2nd edn, Soil Science Society of America, Madison

McLean JS, Lee JU, Beveridge TJ (2002) Interactions of bacteria and environmental metals, fine-grained mineral development, and bioremediation strategies. In: Huang PM, Bollag JM, Senesi N (eds) Interactions between soil particles and microorganisms. Wiley, New York

Morton JD, Semrau JD, Hayes KF (2001) An X-ray absorption spectroscopy study of the structure and reversibility of copper adsorbed to montmorillonite clay. Geochim Cosmochim Acta 65:2709–2722

Neubauer U, Furrer G, Schulin R (2002) Heavy metal sorption on soil minerals affected by the siderophore desferrioxamine B: the role of Fe(III) (hydr)oxides and dissolved Fe(III). Eur J Soil Sci 53:45–55

Neubauer SC, Emerson D, Megonigal JP (2008) Microbial oxidation and reduction of iron in the root zone and influences on metal mobility. In: Violante A, Huang PM, Gadd G (eds) Biophysico-chemical processes of metals and metalloids in soil environments. Wiley, Hoboken

Ona-Nguema G, Morin G, Juillot F, Calas G, Brown GE Jr (2005) EXAFS analysis of arsenite adsorption onto two-line ferrihydrite, hematite, goethite, and lepidocrocite. Environ Sci Technol 39:9147–9155

Onyatta JO, Huang PM (2003) Kinetics of cadmium release from selected tropical soils from Kenya by low-molecular-weight organic acids. Soil Sci 168:234–252

Oscarson DW, Huang PM, Defose C, Herbillon A (1981) Oxidative power of Mn(IV) and Fe(III) oxides with respect to As(III) in terrestrial and aquatic environments. Nature 291:50–51

Özdemir S, Kilinc E, Poli A, Nicolaus B, Güven K (2009) Biosorption of Cd, Cu, Ni, Mn and Zn from aqueous solutions by thermophilic bacteria, *Geobacillus toebii* sub.sp. *decanicus* and *Geobacillus thermoleovorans* sub.sp. *stromboliensis*: equilibrium, kinetic and thermodynamic studies. Chem Eng J 152:195–206

Papelis C (1995) X-ray photoelectron spectroscopic studies of cadmium and selenite adsorption on aluminum oxides. Environ Sci Technol 29:1526–1533

Perelomov L, Cozzolino V, Pigna M, Violante A (2011) Adsorption of Cu and Pb on goethite in the presence of low-molecular mass aliphatic acids. Geomicrobiol J 28:582–589

Pigna M, Krishnamurti GSR, Violante A (2006) Kinetics of arsenate sorption-desorption from metal oxides: effect of residence time. Soil Sci Soc Am J 70:2017–2027

Prado AGS, Torres JD, Martins PC, Pertusatti J, Bolzon LB, Faria EA (2006) Studies on copper(II)- and Zinc(II)- mixed ligand complexes of humic acid. J Hazard Mater B136:585–588

Qin F, Shan XQ, Wei B (2004) Effects of low-molecular-weight organic acids and residence time on desorption of Cu, Cd, and Pb from soils. Chemosphere 57:253–263

Qin F, Wen B, Shan XQ, Xie YN, Liu T (2006) Mechanisms of competitive adsorption of Pb, Cu, and Cd on peat. Environ Poll 144:669–680

Raven KP, Jain A, Loeppert RH (1998) Arsenite and arsenate adsorption on ferrihydrite: kinetics, equilibrium, and adsorption envelopes. Environ Sci Technol 32:344–349

Saha UK, Liu C, Kozak LM, Huang PM (2004) Kinetics of selenite adsorption on hydroxyaluminum- and hydroxyaluminosilicate-montmorillonite complexes. Soil Sci Soc Am J 68:1197–1209

Schwertmann U, Cornell RM (2000) Iron oxides in the laboratory: preparation and characterization. Wiley-VCH, Weinheim

Senesi N (1992) Metal-humic substance complexes in the environment. Molecular and mechanistic aspects by multiple spectroscopic approach. In: Adriano DC (ed) Biogeochemistry of trace elements. Lewis, Boca Raton

Senesi N, Loffredo E (2008) Spectroscopic techniques for studying metal-humic substance complexes in soil: an overview. In: Violante A, Huang PM, Gadd GM (eds) Biophysico-chemical processes of heavy metals and metalloids in soil environments. Wiley, Hoboken

Sharma KR, Srivastava PC, Srivastava P, Singh VP (2006) Effect of farmyard application on boron adsorption-desorption characteristics of some soils. Chemosphere 65:769–777

Shirvani M, Shariatmadari H, Kalbasi M (2007) Kinetics of cadmium desorption from fibrous silicate clay minerals: influence of organic ligands and aging. Appl Clay Sci 37:175–184

Shuttleworth KL, Unz RF (1993) Sorption of heavy metals to the filamentous bacterium thiothrix strain A1. Appl Environ Microbiol 59:1274–1282

Sparks DL (2002) Environmental soil chemistry. Academic, New York
Sparks DL (2005) Metal and oxyanion sorption on naturally occurring oxide and clay mineral surfaces. In: Grassian VH (ed) Environmental catalysis. Taylor & Francis, Boca Raton
Sparks DL, Scheidegger AM, Strawn DG, Scheckel KG (1999) Kinetics and mechanisms of metal sorption at the mineral-water interface. In: Sparks DL, Grundl TJ (eds) Kinetics and Mechanisms of Reactions at the Mineral/Water Interface. American Chemical Society Symposium Series 715
Sposito G (1986) Distribution of potentially hazardous trace metals. In: Siegel H (ed) Metal ions in biological systems, vol 20. Marcel Dekker, New York
Sposito G (1996) The environmental chemistry of aluminum. CRC Press, Boca Raton
Strawn DG, Baker LL (2008) Speciation of Cu in a contaminated agricultural soil measured by XAFS, μ-XAFS, and μ-XRF. Environ Sci Technol 42:37–42
Sun X, Doner HE (1996) An investigation of arsenate and arsenite bonding structures on goethite by FTIR. Soil Sci 161:865–872
Vijayaraghavan K, Yun YS (2008) Bacterial biosorbents and biosorption. Biotechnol Adv 26:266–291
Violante A, Gianfreda L (2000) Role of biomolecules in the formation of variable-charge minerals and organo-mineral complexes and their reactivity with plant nutrients and organics in soil. In: Bollag JM, Stotzky G (eds) Soil biochemistry, vol 10. Marcel Dekker, New York
Violante A, Krishnamurti GSR, Huang PM (2002) Impact of organic substances on the formation of metal oxides in soil environments. In: Huang PM, Bollag JM, Senesi N (eds) Interactions between soil particles and microorganism: impact on the terrestrial ecosystem. Wiley, New York
Violante A, Ricciardella M, Pigna M (2003) Adsorption of heavy metals on mixed Fe-Al oxides in the absence or presence of organic ligands. Water Air Soil Pollut 145:289–306
Violante A, Ricciardella M, Pigna M, Capasso R (2005) Effects of organic ligands on the sorption of trace elements onto metal oxides and organo-mineral complexes. In: Gobran GR, Huang PM (eds) Biogeochemistry of trace elements in the rhizosphere. Elsevier B.V, Amsterdam
Violante A, Ricciardella M, Del Gaudio S, Pigna M (2006) Coprecipitation of arsenate with metal oxides: nature, mineralogy and reactivity of aluminum precipitates. Environ Sci Technol 40:4961–4967
Violante A, Pigna M, Del Gaudio S, Ricciardella M, Banerjee D (2007) Coprecipitation of arsenate with metal oxides: nature, mineralogy and reactivity of iron(III) precipitates. Environ Sci Technol 41:8275–8280
Violante A, Krishnamurti GSR, Pigna M (2008) Mobility of trace elements in soil environments. In: Violante A, Huang PM, Gadd G (eds) Biophysico-chemical processes of metals and metalloids in soil environments. Wiley, Hoboken
Violante A, Cozzolino V, Del Gaudio S, Benerjee D, Pigna M (2009a) Coprecipitation of arsenate within metal oxides: 3 nature, mineralogy and reactivity of iron(III)/Al precipitates. Environ Sci Technol 43:1515–1521
Violante A, Pucci M, Cozzolino V, Zhu J, Pigna M (2009b) Sorption/desorption of arsenate on/from Mg-Al layered double hydroxides: influence of phosphate. J Colloid Interface Sci 333:63–70
Xu RX, Yu G, Kozak LM, Huang PM (2008) Desorption kinetics of arsenate adsorbed on Al (oxy)hydroxides formed under the influence of tannic acid. Geoderma 148:55–62
Yamaguchi NU, Scheinost AS, Sparks DL (2002) Influence of gibbsite surface area and citrate on Ni sorption mechanisms at pH 7.5. Clays Clay Miner 50:784–790
Yermiyahu U, Keren R, Chen Y (1988) Boron sorption on composted organic matter. Soil Sci Soc Am J 52:1309–13313
Yermiyahu U, Keren R, Chen Y (1995) Boron sorption by soil in presence of composted organic matter. Soil Sci Soc Am J 59:405–409
Yermiyahu U, Keren R, Chen Y (2001) Effect of composted organic matter on boron uptake by plants. Soil Sci Soc Am J 65:1436–1441

Yu G, Saha UK, Kozak LM, Huang PM (2006) Kinetics of cadmium adsorption on aluminum precipitation products formed under the influence of tannate. Geochim Cosmochim Acta 70:5134–5145

Yuan G, Theng BKG (2011) Clay-organic interactions in soil environments, Chapter 2. In: Summer ME, Li Y (eds) Handbook of soil science. CRC Press, Boca Raton

Yuan S, Xi Z, Jiang Y, Wan J, Wu C, Zheng Z, Lu X (2007) Desorption of copper and cadmium from soils enhanced by organic acids. Chemosphere 68:1289–1297

Zhu J, Pigna M, Cozzolino V, Caporale GA, Violante A (2010) Competitive sorption of copper(II), chromium(III) and lead(II) on ferrihydrite and two organomineral complexes. Geoderma 159:409–416

Zhu J, Pigna M, Cozzolino V, Caporale AG, Violante A (2011) Sorption of arsenite and arsenate on ferrihydrite: effect of organic and inorganic ligands. J Hazard Mater 15;189:564–571

Chapter 8
Interaction between Heavy Metals and Variable Charge Surfaces

Ren-kou Xu

Abstract Large areas of variable charge soils are distributed in tropical and subtropical regions of the world. Adsorption reactions of heavy metals are important factors controlling their mobility in soils. Low cation exchange capacity and pH create conditions conducive to increase mobility of heavy metal cations, but decrease the mobility of anions such as arsenate and chromate in these soils. In this Chapter, the adsorption mechanisms of heavy metals and their competition for adsorption sites on variable charge surfaces as well as the effects of ionic strength, pH, surface chemical properties and accompanying anions on the adsorption in this type of soil are summarized. The results presented here can help to understand the adsorption processes of heavy metals in variable charge soils and provide useful references for the control of soil heavy metals pollutions and remediation of heavy metal-contaminated soils in tropical and subtropical regions.

Keywords Variable charge soil • Heavy metals • Electrostatic adsorption • Specific adsorption • Surface charge

The concentration of heavy metal in soil solution is of great importance for all ecological consideration because plants are likely to take up the available metals from soil solution. The transport of metals within the soil or even to groundwater also depends on the metal concentration of the solution phase (Brümmer et al. 1986). It was suggested that sorption and desorption reactions on the surface of soil and oxides are two important processes controlling the concentration of heavy metals in solution (Brümmer et al. 1988; Ainsworth et al. 1994; Backes et al. 1995; McLaren et al. 1998). It is well known, tropical and subtropical regions are distributed with large areas of variable charge soils. These soils contain relatively higher amount of

R. Xu (✉)
Institute of Soil Science, Chinese Academy of Sciences, P.O. Box 821, Nanjing, China
e-mail: rkxu@issas.ac.cn

J. Xu and D.L. Sparks (eds.), *Molecular Environmental Soil Science*,
Progress in Soil Science, DOI 10.1007/978-94-007-4177-5_8,

iron and aluminum oxides and usually carry both positive and negative charges on their surfaces, therefore can adsorb both anions and cations (Yu 1997). The surface chemical characteristics of variable charge soils are different from these of constant charge soils from temperate regions and the adsorption behaviors of heavy metals on variable charge soils are also different from these of constant charge soils. For example, the constant charge soils carry larger amount of negative surface charge and have higher soil pH than variable charge soils, thus can adsorb more heavy metal cations but less amounts of anions such as chromate and arsenate compared with variable charge soils. However, variable charge soils have high adsorptive capacity for these two anions.

Soil organic matter can either enhance or inhibit the adsorption of heavy metals, depending on the other soil properties or experimental conditions, such as pH, cation exchange capacity, and accompanying ions (Harter and Naidu 1995). The effect of organic matter on adsorption of heavy metals through the addition of foreign organic compounds into soil systems is widely studied. In this regards the low-molecular weight organic acids and humic acid were studied extensively. For variable charge soils, below the zero point of charge (ZPC), organic compounds generally increased the adsorption of heavy metal cations through complexation between metals and adsorbed organic molecules (Harter and Naidu 1995). However, with pH above the ZPC values, these organic compounds could not facilitate the retention of heavy metals by soils through the formation of soluble complexes between metals and organic compounds (Harter and Naidu 1995). For natural soil organic matter (OM), investigations indicated that removal of OM decreased the adsorption of Cu(II), Cd(II) and Zn(II) by acid soils (Silveira et al. 2002; Agbenin and Olojo 2004; Zhong et al. 2010). The content of organic matter of paddy soils derived from upland Ultisol increased with extension of cultivation years and the adsorption of Cu(II) and Pb(II) by the paddy soils increased consequently (Ma et al. 2010). Moreover, the OM in acid soils caused relatively more effect on Cu(II) adsorption than on Pb(II), Cd(II) and Zn(II) (Silveira et al. 2002; Agbenin and Olojo 2004; Ma et al. 2010; Zhong et al. 2010). This is because Cu(II) has a stronger affinity for organics than the other heavy metals.

Therefore, in soils dominated by permanent charge surfaces, heavy metal cations are not mobile but in variable charge soils, the low surface charge density creates conditions conducive to increased mobility of heavy metal cations (Naidu et al. 1997). However, the mobility of heavy metals can be decreased through adjusting the adsorption behaviors of the metal cations in variable charge soils. In this Chapter, the adsorption mechanisms of heavy metals and their competition for adsorption sites on variable charge surfaces as well as the effects of ionic strength, pH, surface chemical properties and accompanying anions on the adsorption in this type of soil are summarized. The results presented here can help to understand the adsorption processes of heavy metals in variable charge soils and provide useful references for the control of soil pollutions by heavy metals and remediation of heavy metal-contaminated soils in tropical and subtropical regions, especially these lands around metal mines and metal smelting factories.

8.1 Adsorption Mechanisms of Heavy Metals on Variable Charge Surfaces

8.1.1 Adsorption of Heavy Metal Cations

8.1.1.1 Effect of Surface Charge

The surface charge density is a significant factor influencing the adsorption of heavy metal cations by tropical soils. A strong linear relationship between Cd(II) distribution coefficient and particle surface-charge density has been observed in strongly weathered soils (Naidu et al. 1998). Three tropical soils derived from basalt with different development degree were used to compare the adsorption of Cu(II) and Cd(II). The adsorption of both Cu(II) and Cd(II) by these soils increased with soil cation exchange capacity (CEC) (Zhong et al. 2010). This suggested that the adsorption of Cu(II) and Cd(II) increased with the increase in the value of soil negative surface charge. Similar trends were also found for the adsorption of Cu(II) and Pb(II) by paddy soils derived from an Ultisol (Ma et al. 2010). Heavy metal cations can be adsorbed by variable charge soils and minerals electrostatically and specifically. In soil chemistry, the adsorbed ions which can be desorbed by un-buffered salt such as KNO_3 are termed electrostatically adsorbed ions. Thus the electrostatic adsorption and non-electrostatic adsorption of heavy metals can be distinguished by desorption experiments with un-buffered salt. The desorption rate of heavy metals pre-adsorbed (the percentage of heavy metals desorbed from the pre-adsorbed heavy metals) can be estimated from adsorption and desorption of a heavy metal and it represents the relative contribution of electrostatic adsorption to the total amount of the heavy metals adsorbed. The adsorption, desorption and desorption rate for Cu(II) in three tropical soils with different development degrees are presented in Table 8.1. These soils were derived from basalts. The development degree of the soils was increased with basalt age and soil CEC decreased with basalt age (Zhong et al. 2010). The desorption rate of pre-adsorbed Cu(II) decreased with the increase in basalt age (Table 8.1). Thus the higher development degree and lower CEC of the soils led to the lower desorption rate of Cu(II) and smaller contribution of electrostatic adsorption to Cu(II) adsorption by the soils (Zhong et al. 2010). The adsorption and desorption of Cd(II) in these soils were also investigated (Table 8.2). Although both adsorption and desorption of Cd(II) decreased with the increase in basalt age and soil development degree due to the reduction of soil CEC, the desorption rate of Cd(II) changed slightly with soil development degree. The desorption rate of pre-adsorbed Cd(II) was over 87.5% which is much higher than that for Cu(II) in these soils. The results of the desorption rate of Cd(II) suggested that it was adsorbed by these tropical soils mainly through electrostatic mechanisms. When the desorption rate of Cu(II) was <50% (Table 8.1), the Cu(II) was mainly adsorbed by these soils through non-electrostatic adsorption pathway. Therefore, the relative contribution of electrostatic and non-electrostatic mechanisms to heavy metal adsorption was different for different heavy metal cations.

Table 8.1 Adsorption, desorption and desorption rate for Cu(II) in the tropical soils derived from the basalts with different ages (Zhong et al. 2010)

	10-Kiloyear basalt			1330-Kiloyear basalt			2290-Kiloyear basalt		
Cu(II) in equilibrium solution (mM)	Adsorption (mmol/kg)	Desorption (mmol/kg)	Desorption rate(%)	Adsorption (mmol/kg)	Desorption (mmol/kg)	Desorption rate(%)	Adsorption (mmol/kg)	Desorption (mmol/kg)	Desorption rate(%)
0.1	29.0	11.0	37.9	–	–	–	–	–	–
0.2	38.0	15.8	41.6	17.9	5.5	30.7	11.7	2.1	17.9
0.4	–	–	–	23.0	8.0	34.8	15.8	4.2	26.6
0.8	–	–	–	29.5	11.0	37.3	20.7	6.2	30.0

Table 8.2 Adsorption, desorption and desorption rate for Cd(II) in the tropical soils derived from the basalts with different ages (Zhong et al. 2010)

	10-Kiloyear basalt			1330-Kiloyear basalt			2290-Kiloyear basalt		
Cu(II) in equilibrium solution (mM)	Adsorption (mmol/kg)	Desorption (mmol/kg)	Desorption rate (%)	Adsorption (mmol/kg)	Desorption (mmol/kg)	Desorption rate (%)	Adsorption (mmol/kg)	Desorption (mmol/kg)	Desorption rate (%)
0.2	5.7	5.1	89.5	2.7	2.5	92.6	1.6	1.45	90.6
0.4	6.7	6.1	91.0	4.0	3.5	87.5	2.4	2.2	91.7
0.8	7.8	7.0	89.7	4.7	4.35	92.6	3.22	3.0	93.2

8.1.1.2 Effect of pH

Under field conditions, soils are subjected to pH changes either by farm management practices (e.g. liming, leguminous pastures) or by natural processes (e.g. acid rain) (Naidu et al. 1994). The change of pH has marked influences on heavy metal adsorption (Naidu et al. 1998). The effect of pH on heavy metal adsorption has been investigated extensively using both metal oxides (McKenzie 1980; Padmanabham 1983; Brümmer et al. 1988; Fischer et al. 2007) and soils (Harter 1983; Naidu et al. 1994). It has often been observed that sorption of heavy metal cations increases with an increase in pH. Hydrolysis models of metals have been used to describe the sorption mechanisms of heavy metal ions with the change of pH. Hydrolysis of metal ions result in the formation of metal-hydroxyl species which are sorbed preferentially over the free metal ion (Hodgson et al. 1964), as shown in Eqs. (8.1) and (8.2):

$$M^{2+} + H_2O = MOH^+ + H^+ \tag{8.1}$$

$$MOH^+ + Soil^- = Soil - MOH \tag{8.2}$$

The values of metal hydrolysis constants (pKa) characterize their adsorption behavior. With decreasing pKa and at equal pH values with increasing ionic size, heavy metal sorption decreases in the sequence: Hg (3.4) > Cu (7.7) ≫ Zn (9.0) > Co (9.7) > Ni (9.9) > Cd (10.1) where the values in parenthesis indicate the pKa of metals (Naidu et al. 1997). Based on pH_{50} values (the pH where the 50% of heavy metal added was adsorbed) obtained from detailed sorption reactions involving Cd, Zn and Ni with natural soil clays, Tiller et al. (1979) found that the adsorption of these metals is controlled by metal ion-hydrolysis reactions. In recent time, Fischer et al. (2007) studied the reactions of ten metals of Hg, Cr, Al, Cu, Pb, Zn, Co, Ni, Cd and Mn with goethite and also found that the sequence of pH_{50} for these metals is the same with their pKa. These results suggested that the greater the affinity of the metals for hydroxyl ions, the greater their affinity for the goethite surface (Fischer et al. 2007).

The effects of pH on adsorption of Cu(II) and Cd(II) by an Oxisol and an Ultisol are shown in Fig. 8.1. The adsorption of both metal cations increased with increasing pH. The pH had a bigger effect on Cu(II) adsorption than Cd(II) adsorption by these two variable charge soils, because the difference of curves between Cu(II) and Cd(II) increased with increasing pH (Fig. 8.1). This order is opposite to the sequence of pKa of the metals: Cu(II) ≪ Cd(II). The greater hydrolysis of Cu(II) creates more $Cu\text{-}OH^+$ than Cd(II) at equal pH and thus leads to larger enhancement of Cu(II) adsorption with increasing pH compared with Cd(II).

The increase in negative surface charge of variable charge soils and minerals is another important mechanism for the effect of pH on heavy metal adsorption by these variable charge surfaces (Naidu et al. 1998). Simple correlation analysis shows that most of the variation in Cd(II) sorption with pH can be explained by changes in

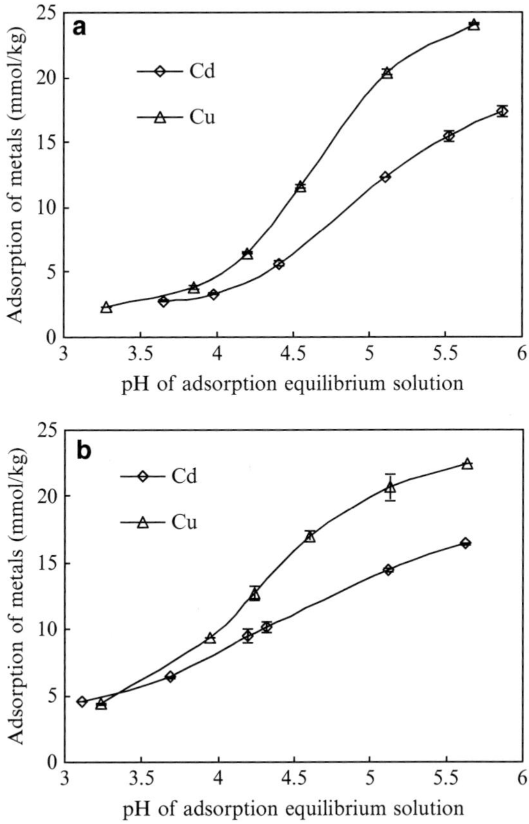

Fig. 8.1 Effect of pH on adsorption of Cu(II) and Cd(II) by an Oxisol from Kunming, China (**a**) and an Ultisol from Guiyang, China (**b**)

surface negative charge with pH (Naidu et al. 1994). A similar strong relationship between surface charge and sorption of heavy metals, Cd(II), Zn(II), Pb(II) and Co(II), was observed (Forbes et al. 1976). The desorption of pre-adsorbed Cu(II) and Cd(II) at different pH (Fig. 8.1) by 0.1 M KNO_3 was presented in Fig. 8.2. The desorption of pre-adsorbed Cd(II) by both soils increased with increasing pH of the adsorption equilibrium solution, and the desorption rate was in the range of 62–80% within the pH range from 4.0 to 5.5 for the Oxisol and in the range from 80 to 86% for the Ultisol in the same pH range (Table 8.3). The high desorption rate of adsorbed Cd(II) suggested that the increased Cd(II) adsorption with increasing pH was attributed to the increase in electrostatic adsorption of the metal cation due to the increase of negative surface charge of the soils with the increasing pH. On the other hand, the desorption of pre-adsorbed Cu(II) increased with the rise in pH, reaching a maximum value at a certain pH, and then decreased. The desorption rate of adsorbed Cu(II) decreased with the increasing pH of adsorption equilibrium solution, and the rate decreased markedly after the pH of the maximum desorption of Cu(II) and was much lower than the desorption rate of Cd(II) at the

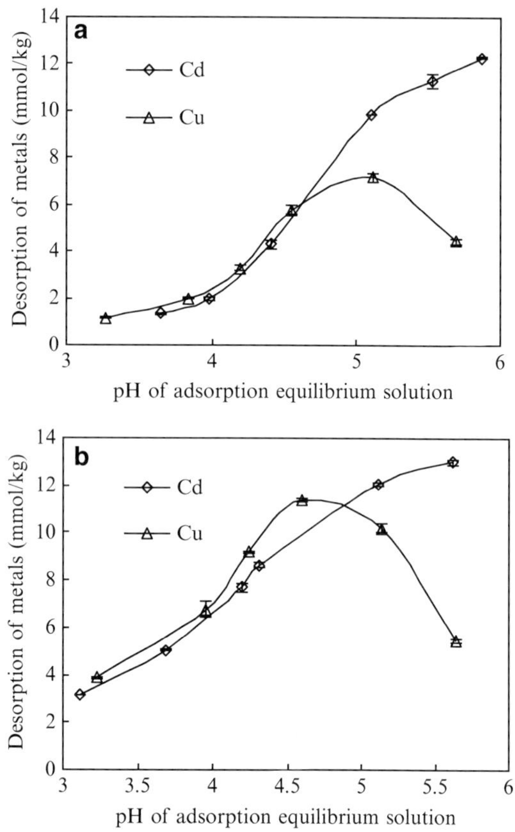

Fig. 8.2 Effect of pH on desorption of the pre-adsorbed Cu(II) and Cd(II) by an Oxisol from Kunming, China (**a**) and an Ultisol from Guiyang, China (**b**) using 0.1 M KNO_3 as desorbing solution

Table 8.3 Desorption rate of pre-adsorbed Cu(II) and Cd(II) at different pH in an Oxisol from Kunming, China and Ultisol from Guiyang, China

Soil	Heavy metal	pH 4.0	pH 4.5	pH 5.0	pH 5.5
Oxisol	Cu	53	50	37	24
	Cd	62	75	80	74
Ultisol	Cu	70	69	54	32
	Cd	80	86	83	81

same pH (Table 8.3). These results suggested that the enhanced Cu(II) adsorption with increasing pH was mainly due to the increase in negative surface charge below the pH of the maximum Cu(II) desorption, and related to the hydrolysis reaction of Cu(II) above the pH values. Therefore, both the increase of negative surface charge and hydrolysis mechanism of metal ions can contribute to the adsorption of heavy metals enhanced by increasing pH, the relative contribution of these two mechanisms depends on the pKa of metal cations and the range of pH.

Thus, in tropical and subtropical regions, increasing soil pH through addition of alkaline materials can enhance adsorption of heavy metals and decrease the mobility of heavy metals in polluted soils.

It is often observed that the adsorption of heavy metal cations by Fe/Al oxides and variable charge soils results in the release of H^+ (Forbes et al. 1976; Bleam and McBride 1986; Yu 1997). The mechanism of coordination adsorption or complexation of heavy metal cations was proposed to interpret the release of H^+. The proton of the hydroxyl group or the water molecule on the surface of the oxides and soils can undergo ligand exchange with heavy metal cations, forming monodentate, bidentate, or even tridentate surface complexes in which the metal ions combine with the oxygen atoms through chemical bonds. Because heavy metal cations normally react with these surface groups to form either a monodentate complex or a bidentate chelate, resulting in the release of one or two protons, the exchange ratio of H^+/M^{2+} varies within the range of 1–2 (Yu 1997). The second mechanism for the release of H^+ during adsorption of heavy metal cations is hydrolysis of these cations. Metal oxides and variable charge soils can preferentially adsorb hydroxyl metal ions and thus promote the hydrolysis of heavy metal cations, which also leads to the release of H^+. The ligand exchange of heavy metal cations with protons of the hydroxyl or the water molecular on surfaces of oxides and soils is a predominant mechanism for the release of H^+ at relative low pH; while at relatively high pH, the release of H^+ during the adsorption of heavy metals is mainly from hydrolytic reaction of the heavy metal cations.

8.1.1.3 Effect of Soil Organic Matter

Soil organic matter can influence the adsorption of heavy metals by variable charge soils. For natural soil organic matter (OM), investigations indicated that removal of OM decreased the adsorption of Cu(II) and Zn(II) by acid soils. Moreover, the OM in acid soils caused relatively more effect on Cu(II) adsorption than Zn(II) (Silveira et al. 2002; Agbenin and Olojo 2004). This is because Cu(II) has a stronger affinity for organics than Zn(II) (McBride et al. 1997). To study the effect of organic matter on adsorption and desorption of Cu(II) by a strongly weathered Oxisol from Hainan, China, the organic matter was removed from the soils of surface and subsurface layers. Further, the adsorption and desorption of Cu(II) in the original soils and the soils with organic matter removed were determined and compared. Table 8.4 revealed that the removal of organic matter decreased the adsorption of Cu(II) by the soils. After organic matter was removed, the amount of Cu(II) adsorbed by the surface soil was decreased by 34.1%. However, the removal of organic matter led to an increase in the desorption of Cu(II) from the soils, which was opposite to the change of Cu(II) adsorption. Therefore, the removal of soil organic matter significantly increased Cu(II) desorption rate compared to the original soils (Table 8.4). These results demonstrated that soil organic matter increased not only the amount of Cu(II) adsorption, but also the binding strength of Cu(II) to soil surface. Surface active groups of soil organic matter can form stable complexes with Cu(II), and increase the amount of Cu(II) specifically adsorbed by the soils (Guo et al. 2006). The Cu(II) adsorbed with this pathway could not be

Table 8.4 Effect of organic matter on adsorption and desorption of Cu(II) in an Oxisol from Hainan, China (Zhong et al. 2010)

Depth (cm)	Organic matter (g/kg)	Original soil		Organic matter removed		Desorption rate	
		Adsorption (mmol/kg)	Desorption (mmol/kg)	Adsorption (mmol/kg)	Desorption (mmol/kg)	Original soil (%)	Organic matter removed
0–20	34.0	17.3	3.2	11.4	5.4	18.4	47.4
20–50	19.1	16.0	3.3	9.7	5.0	20.2	51.0

desorbed by un-buffered salts. Therefore, the increasing soil organic matter through the incorporation of plant residues and green manure could eventually enhance the fixing ability of tropical soils for Cu(II) and reduce its mobility in the soils.

8.1.2 Adsorption of Chromate and Arsenate

8.1.2.1 Adsorption of Chromate

Chromium is a potential contaminant in soil, surface water, ground water and air. The most important oxidation states of chromium in the environment are Cr(III) and Cr(VI) (dichromate at low pH). Chromate is a relatively reactive form, more mobile and toxic than Cr(III) (Ross et al. 1981). The adsorption of chromate has been investigated extensively in systems of pure iron and aluminum oxides (Hsia et al. 1993; Aide and Cummings 1997; Fendorf et al. 1997; Abdel-Samad and Watson 1997; Garman et al. 2004). Microscopically, spectroscopic evidence has shown that chromate is able to form inner-sphere complexes (the specific adsorption) on goethite (Aide and Cummings 1997; Fendorf et al. 1997). Macroscopically, a number of studies have demonstrated that chromate exhibits fairly high mobility in the environment (Zachara et al. 1989; Selim et al. 1989).

The isotherms of the chromate adsorption and desorption curves of the pre-adsorbed chromate in an Ultisol and an Oxisol at pH 4.6 are examined. The adsorption of chromate and desorption of the pre-adsorbed chromate in the Oxisol from Guangdong Province, China is greater than that in the Ultisol from Jiangxi Province, China, which is consistent with the content of free iron oxides in these soils. Iron oxides are an important carrier of positive surface charges and consequently they were the absorbents for chromate in the soils (Xu et al. 2005). The free iron oxides of 108.3 g/kg in the Oxisol were much greater than that of 51.1 g/kg in the Ultisol. Thus, the higher content of free iron oxides in the soil led to the larger amount of chromate adsorbed by the soil.

The results of desorption of the pre-adsorbed chromate indicated that there could be at least two mechanisms responsible for the adsorption process: the non-electrostatic adsorption and the electrostatic adsorption. As mentioned above, electrostatically adsorbed ions are those that can be desorbed by an un-buffered neutral salt such as $NaNO_3$ (Xu et al. 2005). Therefore, the amount of chromate desorbed by $NaNO_3$ represents the electrostatic adsorption of chromate by the soils. The relative quantity of the non-electrostatic adsorption and the electrostatic adsorption could be calculated from the data of the adsorption and desorption of chromate. The percentage of the non-electrostatic adsorption was found to be 54.3–60.3%, 43.9–46.2% for the Oxisol and Ultisol at different equilibrium concentrations, respectively, and the percentage of the electrostatic adsorption was 39.7–45.7%, 53.8–56.1% for the two soils, respectively. It follows that both the non-electrostatic adsorption and the electrostatic adsorption contributed to the chromate adsorption in the variable charge soils, with the former being larger than the latter for the Oxisol, while a reverse trend seemed to hold for the Ultisol.

The amount of chromate adsorbed by variable charge soils and pure Fe/Al oxides decreased with increasing of pH (Aide and Cummings 1997; Grossl et al. 1997; Jiang et al. 2008). The desorption experimental data indicated that the relative fraction of the non-electrostatic adsorption to the electrostatic adsorption in variable charge soils changed to a lesser extent with changing of pH (Jiang et al. 2008). This suggests that similar mechanisms were probably responsible for the chromate adsorption by variable charge soils under acidic conditions.

During specific adsorption, exchange of anions with hydroxyl (–OH) or water ($-OH_2$) ligands on the surfaces of iron and aluminum oxides or soil colloids occurs and the inner-sphere complexes form between the anions and the solid surface; as a result, hydroxyl ions are released into the soil solution. Therefore, the release of hydroxyl is an important consequence of the specific adsorption of anions by variable charge soils and minerals. Jiang et al. (2008) reported that at pH 4.27, 4.80 and 5.40, the amount of hydroxyl released was 8.30, 7.40 and 4.58 mmol/kg for the Oxisol from Guangdong province, China and 2.27, 2.08 and 1.10 mmol/kg for the Ultisol from Jiangxi Province, China. The hydroxyl release during the chromate adsorption increased with decreasing of pH in variable charge soils, which is consistent with the change of chromate adsorbed by the soils with pH. At the same pH, the hydroxyl release increased with the increase in the content of free iron oxides in soils. These findings suggest that the higher content of iron oxides in the soils and larger amount of the chromate adsorption by the soils resulted in much more hydroxyl release from the soil surfaces. This also provides evidence for the occurrence of the specific adsorption of chromate by the variable charge soils. Therefore, the non-electrostatic adsorption of chromate by the variable charge soils as mentioned above includes the specific adsorption to some extent.

Based on the model for sulfate adsorption on iron and aluminum oxides (Rajan 1978; Parfitt and Smart 1978), a model for chromate adsorption on the variable charge soils is proposed as follow:

$$\left.\begin{matrix} M-OH_2 \\ M-OH \end{matrix}\right|^{+} + CrO_4^{2-} \rightarrow \left.\begin{matrix} M-OH_2 \\ M-CrO_4 \end{matrix}\right|^{0} + OH^- \tag{8.3}$$

$$\left.\begin{matrix} M-OH_2 \\ M-OH \end{matrix}\right|^{+} + CrO_4^{2-} \rightarrow \left.\begin{matrix} M-CrO_4 \\ M-OH \end{matrix}\right|^{-} + H_2O \tag{8.4}$$

$$\left.\begin{matrix} M-OH_2 \\ M-OH \end{matrix}\right|^{+} + CrO_4^{2-} \rightarrow \left.\begin{matrix} M-O \\ M-O \end{matrix}\right\rangle CrO_2 + OH^- + H_2O \tag{8.5}$$

$$\left.\begin{matrix} M-OH \\ M-OH \end{matrix}\right|^{0} + CrO_4^{2-} \rightarrow \left.\begin{matrix} M-O \\ M-O \end{matrix}\right\rangle CrO_2 + 2OH^- \tag{8.6}$$

Reactions (8.3) and (8.6) depict that the adsorption of chromate leads to the release of OH^-, Reaction (8.4) shows the release of water molecules as a result of the adsorption, and Reaction (8.5) indicates the release of both OH^- and water molecules caused by the adsorption.

8.1.2.2 Adsorption of Arsenic

Arsenic (As) is a toxic element to humans, animals and plants. The inorganic As is usually present as the species of arsenate and arsenite in natural environments. Sorption and desorption reactions on the surfaces of soils and oxides are two important processes influencing the mobility, toxicity, and availability of arsenic in natural environments (Violante et al. 2005; Pigna et al. 2006; Huang 2008). Jain et al. (1999) observed that the adsorption of arsenate on ferrihydrite led to surface charge reduction and net OH^- release, which suggested that the specific adsorption of arsenate occurred on the oxides. Electrophoretic mobility (EM) measurements show that arsenite adsorption does not significantly change the EM values of γ-Al_2O_3 suspension in 0.1 M $NaNO_3$ at pH 4–8, while arsenate adsorption lowered the EM values at pH 4–10 (Arai et al. 2001). This suggested that arsenate was adsorbed by γ-Al_2O_3 specifically. Amorphous $Al(OH)_3$ has greater adsorption capacity for arsenate than crystalline $Al(OH)_3$. The adsorption rate of arsenate by Fe/Al oxides is very fast (Fuller et al. 1993; Xu et al. 2008). After a reaction period of only 5 min, 52.7 and 74.9% of the arsenate added were adsorbed by the crystalline and amorphous $Al(OH)_3$, respectively (Xu et al. 2008).

It has been reported that the chemical behavior of arsenate is similar to that of phosphate in soils (Pigna et al. 2006). Both of these oxyanions are specifically adsorbed on soil minerals, mainly on Fe/Al oxides, forming inner-sphere complexes (Pigna et al. 2006). Therefore, phosphate has been used as a desorbing agent to investigate the desorption of arsenate from soils and Fe/Al oxides (O'Reilly et al. 2001; Violante et al. 2005; Pigna et al. 2006). Competitive adsorption of arsenate and phosphate on different clay minerals and soils was investigated by Violante and Pigna (2002). They found that Mn, Fe, and Ti oxides and phyllosilicates particularly rich in Fe (nontronite, ferruginous smectites) were more effective in sorbing arsenate than phosphate; the opposite was true for gibbsite, boehmite, noncrystalline Al precipitation products, allophane, kaolinite, and halloysite. Therefore, arsenate may form relatively stronger surface complexes on Fe and Mn oxides than those formed by phosphate. Kinetic investigation indicated that the rate constant of arsenate desorption from the crystalline $Al(OH)_3$ by phosphate was 3.8–15.5 times greater than that from the amorphous $Al(OH)_3$ in the fast reaction period, suggesting that, compared with the crystalline $Al(OH)_3$, the arsenate adsorbed on the amorphous $Al(OH)_3$ was much more difficult to be desorbed by phosphate (Xu et al. 2008). The rate constants increased with the increase in the concentration of phosphate (Xu et al. 2008). The adsorption of phosphate and arsenate to Al oxides proceeds through a similar mechanism to form inner-sphere surface complexes when they compete with each other for adsorption sites on Al oxides (Pigna et al. 2006). Therefore, more phosphate in the system induced more arsenate desorbed from Al hydroxides and accelerated the desorption process.

Similar with the systems of Fe/Al oxides, the release of hydroxyl was observed in variable charge soils due to arsenate adsorption. The amount of hydroxyl released from the soils increased with the increase in arsenate added because of the increase in arsenate adsorption. The release of hydroxyl from an Oxisol from Guangdong

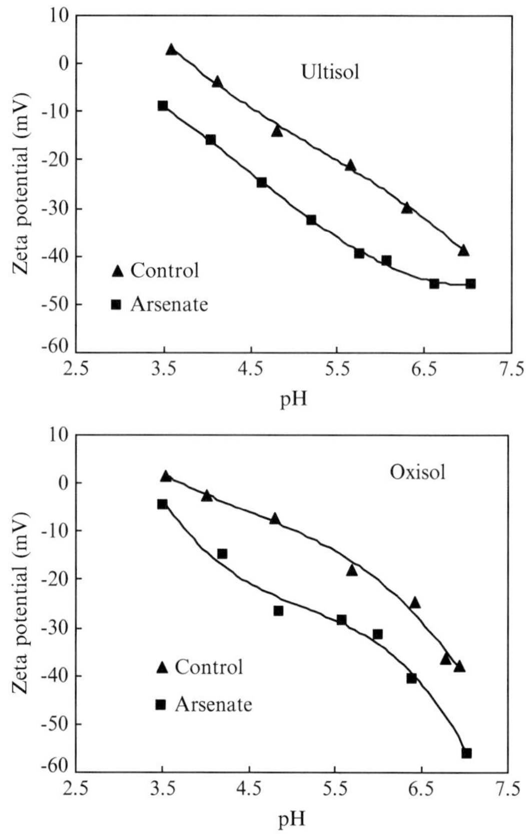

Fig. 8.3 Effect of arsenate adsorption on zeta potential of colloids of the Oxisol from Guangdong Province and the Ultisol from Jiangxi Province, China

Province, China due to arsenate adsorption increased with the increasing system pH and reached a maximum value at pH 5.4 and then decreased with the further increase in pH. For example, when arsenate added was 100 mmol/kg, the amount of hydroxyl released from the Oxisol was 9.35, 11.76, 12.71 and 12.23 mmol/kg at pH 4.0, 5.0, 5.4 and 6.0, respectively. The adsorption of arsenate also resulted in the shift of zeta potential-pH curves to the negative value direction as shown in Fig. 8.3. These results further confirmed that arsenate can be adsorbed by variable charge soils specifically.

8.1.3 Adsorption Mechanisms of Heavy Metals at Atom and Molecular Level

In the past two decades, advanced micro-focused synchrotron-based techniques have been used to probe the adsorption mechanisms of heavy metals and oxyanions on soils and minerals at molecular level. Extended X-ray absorption fine structure

(EXAFS) spectroscopy was used to investigated zinc sorption mechanisms at the clay-mineral/water interface (Nachtegaal and Sparks 2004). The results indicated that zinc partitioned to sorbents by specific chemical binding to hydroxyl surface sites at pH 5.0. While the dominant sorption mechanism changed with reaction time at pH 7.0, at the kaolinite surface, Zn was incorporated into a mixed metal Zn-Al layered double hydroxide. At the goethite surface, Zn initially formed monodentate inner-sphere adsorption complexes, with typical Zn-Fe distances of 3.18Å. However, with increasing reaction time, the major Zn sorption mechanism shifted to the formation of a zinc hydroxide surface precipitate, with characteristic Zn-Zn bond distances of 3.07Å (Nachtegaal and Sparks 2004). EXAFS was also used to investigate the characteristics of Cu(II) bonding in goethite-humate complexes (Alcacio et al. 2001). The analysis of EXAFS results suggests that Cu(II) is present in a distorted octahedral configuration containing four short equatorial (1.94–1.97 Å) and two longer axial bonds with oxygen. When the concentration of adsorbed humic acid on goethite was increased from 0 to 28 g/kg, the axial Cu-O bond length decreased to 2.24 ± 0.03 Å, which was less than for individual humic acid (2.32 ± 0.02 Å) or goethite (2.29 ± 0.03 Å) samples. The apparent decrease in the axial Cu-O bond length was attributed to a decrease in the ligand field splitting energy. When humic acid ligands replace equatorial water molecules in the Cu(II) coordination sphere, a weaker ligand field strength occurs. For absorbed humic acid up to 88 g/kg goethite, second-shell iron neighbors were observed between 3.17 and 3.20 Å, suggesting that Cu(II) was bonded equally to both inorganic (goethite) and organic (humic acid) functional groups (Alcacio et al. 2001).

The adsorption mechanisms of arsenate have been also investigated by synchrotron-based techniques. X-ray absorption fine structure spectroscopic studies indicated the formation of both inner-sphere bidentate binuclear and monodentate arsenate complexes on ferrihydrite and goethite, and bidentate binuclear complexes on γ-Al_2O_3 (Arai et al. 2001; Fendorf et al. 1997; Waychunas et al. 1993). In the study of Sherman and Randall (2003), the observed 3.26Å As-Fe distance in EXAFS spectra agreed with that predicted for the bidentate corner-sharing surface complex, no evidence for monodentate complexes. X-ray standing wave measurements also showed the formation of bridging bidentate surface arsenate complexes on hematite (Catalano et al. 2007).

8.2 Effect of Ionic Strength on Adsorption of Heavy Metal on Variable Charge Surfaces

8.2.1 Effect of Ionic Strength on Adsorption of Heavy Metal Cations

According to the traditional principles of colloidal chemistry, the formation of inner-sphere complexes is apparently unaffected by change in ionic strength, whereas the presence of outer-sphere complexes is likely to be influenced significantly by change in ionic strength because of competitive adsorption with counter-ions (Hayes et al. 1998). However, it was observed that the adsorption of some heavy metal cations on Fe and Ti oxides and soils increased with the rise of ionic strength under acidic conditions. The increase in Cu(II) adsorption on goethite with increasing ionic strength was ascribed to the depression of diffuse layer and OH^- release (Zhou et al. 1996). The increase of Cu(II) adsorption onto TiO_2 with increasing ionic strength was explained by the increase of $\exp(-F\Psi/RT)$ as a part of the electrostatic correction within a surface complexation model, when $NaClO_4$ was used as a background electrolyte (Yang et al. 2006). Barrow et al. (1980) applied an adsorption model to explain the effect of ionic strength and pH on the adsorption of phosphate on goethite. According to the model, the effect of ionic strength on the adsorption operates through its effect on electrostatic potential in the plane of adsorption, rather than through its effect on surface charge (Barrow et al. 1980). The model was also applied to interpret the effect of ionic strength on the adsorption of Cd by a variable charge soil at very low pH (Naidu et al. 1994). However, applications of such a complex model are generally not subject to direct experimental confirmation because they employed several fitting parameters that can not be analytically measured (McBride 1997). In very recent time, when the effect of ionic strength on adsorption of phosphate and arsenate by variable charge soils and adsorption of Cu(II) by Fe/Al oxides was investigated, the experimental proofs have been provided to support the hypothesis of the adsorption model through the measurement of zeta potential in the systems of variable charge soils and Fe/Al oxides (Wang et al. 2009; Xu et al. 2009; Jiang et al. 2010).

8.2.1.1 Effect of Ionic Strength on Cu(II) Adsorption by Fe/Al Oxides

The adsorption isotherms of Cu(II) by goethite at different ionic strengths at pH 4.4 were examined. The results indicated that the amount of Cu(II) adsorbed by goethite increased with the increase in its equilibrium concentration as well with the concentrations of the background electrolyte ($NaNO_3$). The similar trends for the adsorption isotherms of Cu(II) were observed for γ-Al_2O_3 system under similar conditions. As increasing the concentration of $NaNO_3$ from 0.001 and 0.1 to 1.0 M, Cu(II) adsorption onto goethite increased from 51.0 and 56.5 to 74.0 mmol/kg at Cu(II) equilibrium concentration of 0.5 mM, and from 69.5 and 86.5 to 100.5 mmol/kg at Cu(II) equilibrium concentration of 1.5 mM. Similarly, for the γ-Al_2O_3 system, the amount of Cu(II) adsorbed was increased from 43.0 and 46.0 to 61.0 mmol/kg and from 60.2 and 67.5 to 95.5 mmol/kg at its equilibrium concentration of 0.5 and 1.5 mM, respectively, for the increase in $NaNO_3$ concentrations from 0.001 and 0.1 to 1.0 M. The larger increase was observed at higher equilibrium concentration of Cu(II) for both goethite and γ-Al_2O_3 systems.

It is evident from the Cu(II)-pH adsorption curves obtained at two different background electrolyte concentration i.e., 1.0 M and 0.001 M $NaNO_3$ for these two oxides, the adsorption of Cu(II) was higher at the 1.0 M $NaNO_3$ than at the 0.001 M $NaNO_3$. These results suggested that the increasing ionic strength of a system enhanced the adsorption of Cu(II) by Fe/Al oxides under the prevailing acidic conditions. The point of zero salt effect (PZSE) is 7.92 and 8.15 for goethite and γ-Al_2O_3, respectively. Therefore, under acidic condition, both goethite and γ-Al_2O_3 carry net positive surface charge on their surfaces and the value of positive surface charge increases with increasing ionic strength. Therefore, Cu(II) was mainly adsorbed specifically by goethite and γ-Al_2O_3 under acidic condition and the change of surface charge on goethite and γ-Al_2O_3 can not be used to explain the change of Cu(II) adsorption by these Fe/Al oxides with changing ionic strength of the systems.

8.2.1.2 Interpretation of Effect of Ionic Strength on Cu(II) Adsorption

The adsorption model developed by Bowden et al. (1980), which was used for specific adsorption of anions previously, is used to interpret the mechanism involved in the adsorption of Cu(II) by goethite and γ-Al_2O_3 as a function of ionic strength (Jiang et al. 2010). According to the model, the specific adsorption of Cu(II) on the surfaces of goethite and γ-Al_2O_3 takes place in a separate plane, "a". Because Cu(II) has a high binding constant, it coordinates to the surface and is nearer the surface than electrolyte cations. The variation in the number of electrolyte ions in the diffuse layer will affect the potential (Ψa) developed in the adsorption plane (plane "a") and thus affect the adsorption of Cu(II).

Under acidic conditions, both goethite and γ-Al_2O_3 possessed net positive surface charge, and thus the surface potential and the potential in the adsorption plane were positive. With increasing ionic strength, the positive surface charge of goethite and γ-Al_2O_3 increased for the pH values less than the PZSE of the oxides. Under this condition, the counter-ion in the diffuse layer was anion (NO_3^-) and the number of anions per unit area was apparently increased with an increase in background electrolyte concentration. Therefore, the increase in the number of anions in the diffuse layer caused the decrease in potential at the adsorption plane (plane "a"), thus allowed the increase of Cu(II) adsorption by goethite and γ-Al_2O_3. The increase in Cu(II) adsorption by goethite and γ-Al_2O_3 with increasing concentration of $NaNO_3$ was attributed to the change of the potential at the adsorption plane induced by the change in ionic strength.

The zeta potential is an electrical potential at the shear plane of the electric double layer on colloid particles. Although the exact location of the shear plane in the electric double layer cannot be ascertained, it is generally considered that it is located near the interface between the Stern layer and the diffuse layer (Yu 1997). The location of the shear plane is the boundary of solid surface. Therefore, the shear plane is near the adsorption plane (plane "a"). Hence, the change trends of zeta potential with pH at different ionic strengths should be similar to that on the adsorption plane for goethite and γ-Al_2O_3. The zeta potential-pH curves for γ-Al_2O_3 at different ionic strengths are shown in Fig. 8.4. The zeta potential-pH curves at different ionic strengths intersect at one point. Similar changing trends of zeta potential with ionic strength were observed for goethite (data not shown).

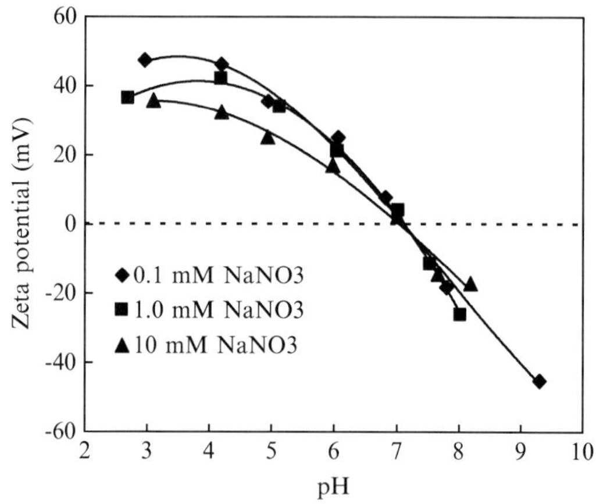

Fig. 8.4 Effect of ionic strength on zeta potential of γ-Al_2O_3

The point of intersection is obviously the isoelectric point (IEP) for the oxides. The value of IEP is 7.7 for goethite and 7.3 for γ-Al_2O_3. When pH is less than the IEP, the value of zeta potential is positive and decreases with increasing ionic strength. Therefore, when pH is less than IEP, the potential at the adsorption plane on goethite and γ-Al_2O_3 is also positive and decreases with increasing ionic strength. The change of the potential at the adsorption plane is consistent with that predicted by the adsorption model mentioned above. The results of zeta potential provide direct support to the interpretation that the change of potential at the adsorption plane with ionic strength was responsible for the effect of ionic strength on Cu(II) adsorption by goethite and γ-Al_2O_3.

8.2.2 Effect of Ionic Strength on Adsorption of Arsenate

Arsenate can be adsorbed by variable charge soils specifically. Intersects of arsenate adsorption–pH curves at different ionic strengths (a characteristic pH) are observed for an Oxisol from Guangdong and an Ultisol from Jiangxi Province, China (Fig. 8.5). It was noted that above this pH, the adsorption of arsenate was increased with the increasing ionic strength, whereas below it the reverse trend was true.

The adsorption model developed by Bowden et al. (1980) can also be used to interpret the mechanism involved in the adsorption of arsenate by variable charge soils as a function of ionic strength. According to the model, the specific adsorption of anions on variable charge soil surfaces was supposed to take place on a plane (adsorption plane) between the Stern and diffuse layers on mineral particles. The variation in the number of ions in the diffuse layer will affect the potential on the adsorption plane and thus affect the adsorption of anions (Barrow et al. 1980). For example, when pH was higher than the PZSE of a variable charge soil, the soil possessed net negative variable charge, and the surface potential and the potential

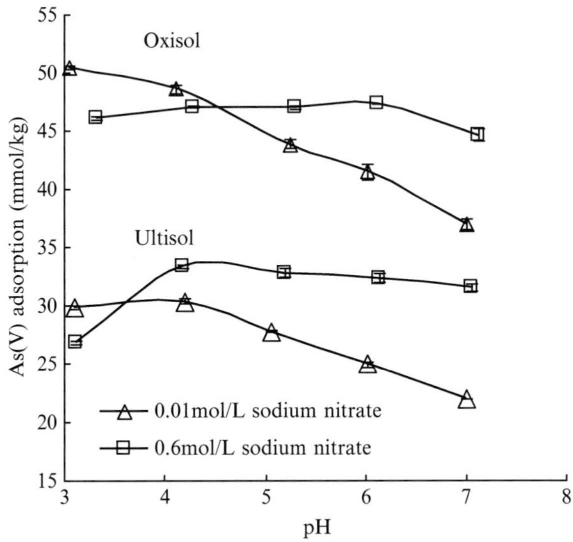

Fig. 8.5 Effects of pH and ionic strength on adsorption of arsenate by two variable charge soils (Xu et al. 2009)

in the adsorption plane were negative. Under this condition, the counter-ion in the diffuse layer is Na^+ and the number of cations per unit area increased with an increase in electrolyte concentration. Therefore, the number of cations in the adsorption plane also increased with the increase in ionic strength, which made the potential at the adsorption plane less negative and thus increased the adsorption of arsenate by the soil. Therefore, the increase of arsenate adsorption by the soil with the increasing concentration of electrolyte was attributed to the change of the potential in the adsorption plane induced by the change of ionic strength.

When the pH was lower than the PZSE of a variable charge soil, the soil possessed a net positive variable charge, and the surface potential and the potential at the adsorption plane were positive. The counter-anion in the diffuse layer was NO_3^-. The increasing number of NO_3^- ions in the adsorption plane with increasing ionic strength may cause the reduction of the potential at the adsorption plane and thus decreased the adsorption of arsenate by the soil.

The zeta potential of Oxisol colloids become less negative with the increase in ionic strength when pH is above the IEP of the soil, whereas it behaves oppositely when the pH is below the IEP of the soil (Fig. 8.6). Therefore, the potential on the adsorption plane of the soil changed similarly with the changing ionic strength. The zeta potential values provided the direct correlation to the hypothesis that the potential in the adsorption plane increased with the increasing (become less negative) ionic strength when the pH is above the PZSE of a variable charge surface and follows the reverse trend with ionic strength when pH is below the PZSE of the variable charge surface.

The change of the adsorption plane potential on variable-charge soils and Fe/Al oxides with ionic strength was responsible for the increase in the specific adsorption of anions when pH was greater than the PZSE of the adsorbents and in the specific

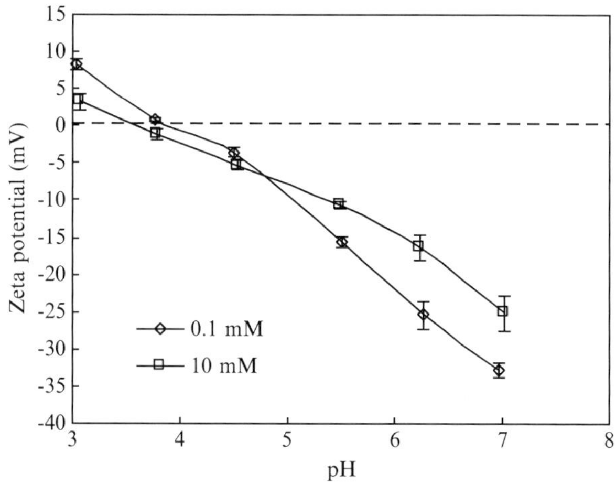

Fig. 8.6 Effect of ionic strength on zeta potential of the Oxisol from Guangdong Province, China

adsorption of heavy metal cations when pH was lower than the PZSE of the adsorbents. The zeta potential of soil colloids and Fe/Al oxides changed with ionic strength gave an experimental evidence to support the interpretation. Therefore, the principle in adsorption model developed by Bowden et al. (1980) can be used to interpret the effect of ionic strength on the specific adsorption of both cations and anions on variable charge surfaces.

8.3 Effect of Heavy Metal Adsorption on Surface Chemical Properties of Variable Charge Soils

Specific adsorption of heavy metal cations on Fe/Al oxides and variable charge soils led to the change of surface charge and IEP (Yu 1997). Specific adsorption of Zn(II), Cu(II) and Co(II) on variable charge soils induced the shifts of zeta potential of soil colloids to positive value direction, which suggested that the specific adsorption of heavy metal cations increased the positive surface charge of the soils (Yu 1997). The effects of Cd(II) and Pb(II) on zeta potential of colloids of a paddy soil derived from an Ultisol from Zhejiang Province, China were compared and the results were presented in Fig. 8.7. Both Cd(II) and Pb(II) induced the shifts of zeta potential of the soil colloids to positive values, and Pb(II) induced more shift of zeta potential than Cd(II). The effect of Cd(II) and Pb(II) on zeta potential of soil colloids increased with increasing pH (Fig. 8.7).

The effect of Cd(II) and Pb(II) on soil zeta potential is consistent with the adsorption of the two metals on the soil. The adsorption of Pb(II) on the soil is much greater than Cd(II) (Fig. 8.8), thus more adsorption of metal cations induced more

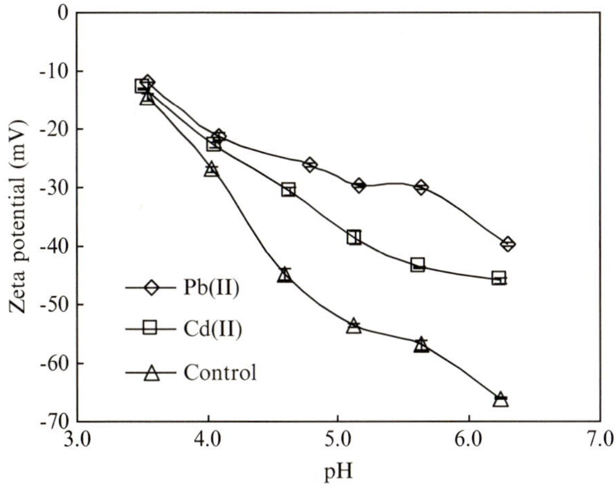

Fig. 8.7 Effect of Pb(II) and Cd(II) on zeta potential of colloids of a paddy soil derived from an Ultisol from Zhejiang Province, China

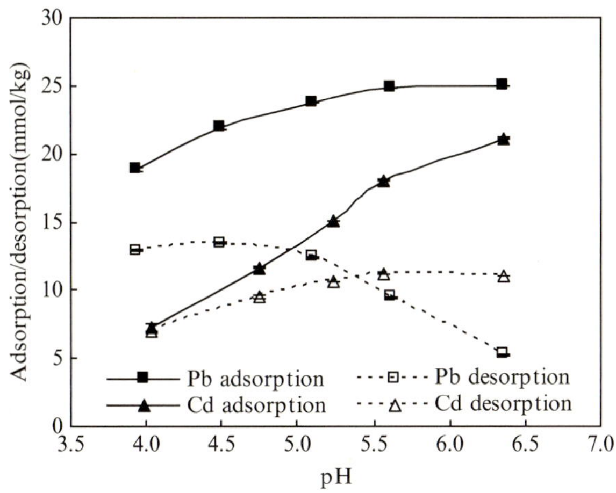

Fig. 8.8 Adsorption and desorption of Pb and Cd in the paddy soil derived from Ultisol from Zhejiang Province, China

shift of zeta potential of soil colloids. From the adsorption and desorption of metal cations, electrostatic adsorption and specific adsorption can be distinguished. The difference between adsorption and desorption was attributed to specific adsorption. From Fig. 8.8, it can be seen that specific adsorption of both Pb(II) and Cd(II) increased with the increasing pH. This is main reason for the increasing effect of Cd(II) and Pb(II) on zeta potential of soil colloids with increase of pH. This was also in agreement with the previous report (Yu 1997).

Specific adsorption induced the shift of zeta potential of soil colloids, thus the relative contribution of specific adsorption of different heavy metals can be estimated from the zeta potential of the systems of soil colloid-heavy metals. When the total concentration of Cd(II) and Pb(II) was maintained at a constant level, the

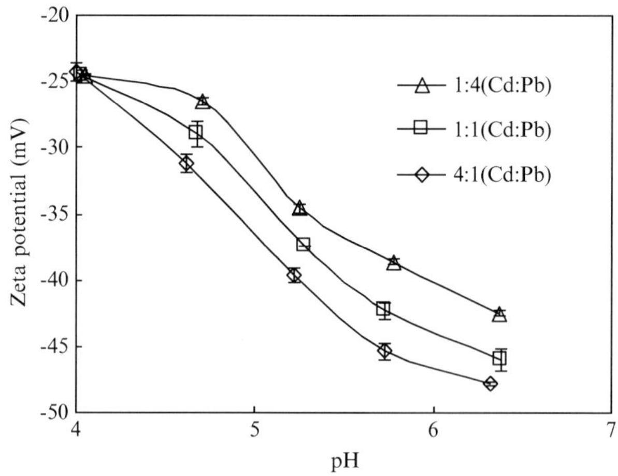

Fig. 8.9 Zeta potential of colloid systems of the paddy soil derived from Ultisol from Zhejiang Province, China containing Cd and Pb with different mole ratio of Cd/Pb

zeta potential of soil colloids containing the two heavy metals changed with the change of mole ratio of Cd(II)/Pb(II) as shown in Fig. 8.9. The zeta potential of soil colloids increased with the decrease in the mole ratio of Cd(II)/Pb(II), which suggested that the contribution of specific adsorption of Pb(II) was greater than Cd(II). This is also consistent with the results shown in Fig. 8.8.

The specific adsorption of heavy metal cations led to the increase in soil positive charge, thus should increase the electrostatic adsorption of electrolyte anions. The results in Fig. 8.10 showed that at the same initial concentration of Cl^-, the adsorption of Cl^- by the Oxisol was greatest in $PbCl_2$ system, followed by the systems of $CuCl_2$ and $CdCl_2$ and the adsorption of Cl^- in NaCl system was lowest. The specific adsorption of Pb(II), Cu(II), Cd(II) increased soil positive charge and thus increased the adsorption of Cl^- compared with Na^+. The stronger specific adsorption of Pb(II) and Cu(II) induced more increase in adsorption of Cl^- compared with Cd(II).

Opposite to the effect of heavy metal cations on surface charge, the specific adsorption of arsenate and chromate made the negative surface charge of variable charge soils more negative and thus led to the shift of zeta potential of soil colloids to negative value direction as shown in Fig. 8.3. The results in Figs. 8.11 and 8.12 indicated that the presence of arsenate decreased the net surface charge of two Oxisols and soil point of zero net charge (PZNC). The PZNC of the Oxisol from Guangdong Province, China was decreased from 6.6 in the control to 4.8 in the arsenate system, and the PZNC of the Oxisol from Yunnan Province, China was decreased from 6.3 in control to 4.7 in the arsenate system. The specific adsorption of arsenate on variable charge soils made soil negative charge more negative and thus is of benefit for the adsorption of heavy metal cations by the soils, which will be discussed below.

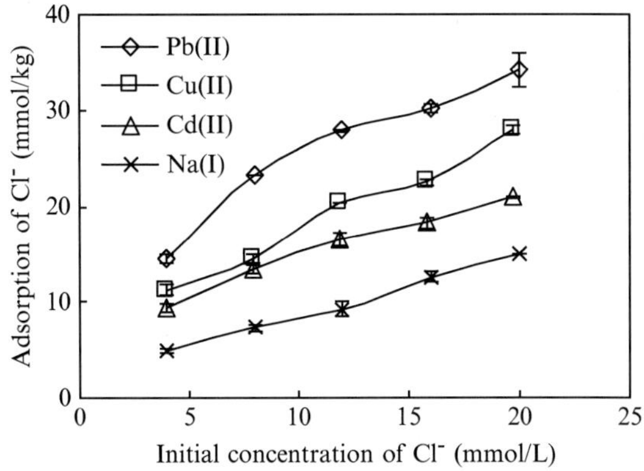

Fig. 8.10 Effect of Pb(II), Cu(II) and Cd(II) on adsorption of Cl^- by an Oxisol from Guangdong Province, China

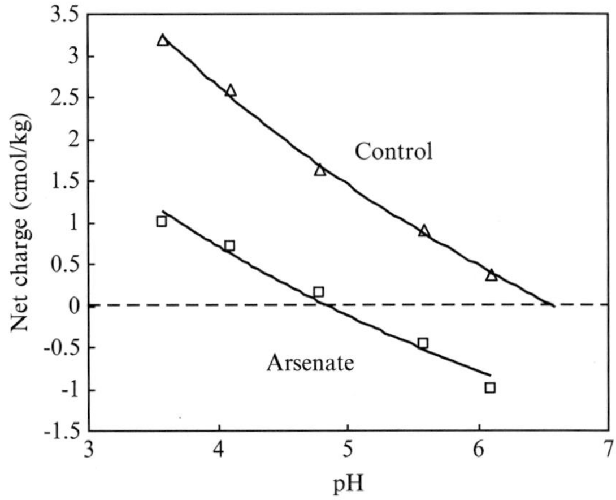

Fig. 8.11 Effect of arsenate on surface charge of the Oxisol from Guangdong Province, China at different pH (amount of arsenate added was 25.0 mmol/kg)

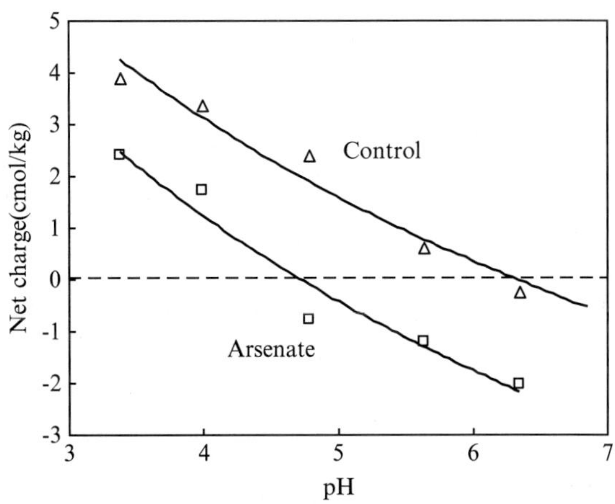

Fig. 8.12 Effect of arsenate on surface charge of the Oxisol from Yunnan Province, China at different pH (amount of arsenate added was 25.0 mmol/kg)

8.4 Competition of Heavy Metals for Adsorption Sites on Variable Charge Surfaces

8.4.1 Competition of Heavy Metal Cations

Heavy metal exists as multiple pollutants, including different metals and different forms of the same metals, in wastes (Zagury et al. 2003). When a soil system contains more than one heavy metal cation, these cations will compete with each other for adsorption sites and thus affect their adsorption with each other (Echeverría et al. 1998; Fontes and Gomes 2003; Srivastava et al. 2005). Data in Fig. 8.13 showed that the adsorption of the individual metal cations Cu(II), Pb(II) and Cd(II) on an Ultisol from Guangdong Province, China in a single system was higher than that in competitive systems, which suggested that the competition of metal cations for adsorption sites on soils led to the decrease in adsorption of individual metal cation by soils. Results in Fig. 8.13 also indicated that the adsorption of Cu(II) and Pb(II) by the soil was greater than that of Cd both in the single metal system and competitive system. This is consistent with previous report of the adsorption of these heavy metals in other variable charge soils (Fontes and Gomes 2003).

The selectivity sequence of the metals adsorbed by the Ultisol is Pb(II) > Cu(II) > Cd(II) at pH 4.2, which is also consistent with the observation of these metals adsorbed by the tropical soils from Brazil reported by Fontes and Gomes (2003). However, the selectivity sequence changed to be Cu(II) > Pb(II) > Cd(II) at pH 5.3. These results suggested that the selectivity sequence of heavy metals adsorbed by variable charge soils changed with environmental pH. In order to probe the reasons for the effect of pH on selectivity sequence of heavy metal adsorption in variable charge soils, the relationship between adsorption of heavy metals and pH in two variable charge soils was compared. Under acidic conditions, the pH-dependent adsorption of Cu(II) and Pb(II) intersected at pH 4.9 for the Ultisol from Guangdong Province, China and at pH 5.1 for the Ultisol from Jiangxi Province, China (Figs. 8.14 and 8.15). A similar phenomenon was observed in the system of goethite, the intersection was 4.4 (Fig. 8.16). Above the characteristic pH, the adsorption of Cu(II) in these soils and minerals was higher than that of Pb(II), and below which a reverse trend occurs.

Cu(II) and Pb(II) can be specifically adsorbed by variable charge soils and Fe/Al oxides. At relative low pH, both of the metal cations were adsorbed through the formation of surface complexes on the soils and oxides. Thus Pb(II) has greater ability to form surface complexes on variable charge surfaces than Cu(II) at low pH condition. At relative higher pH, Cu(II) and Pb(II) can be adsorbed by variable charge soils and minerals through hydrolysis mechanism. The soils and minerals have higher affinity for the hydrolysis products of the complexes of the metals with OH^- than the free metal cations. The hydrolysis potential of Cu(II) is greater than Pb(II), because the hydrolysis constant of Cu(II) (pK_{Cu}) is smaller than the pK_{Pb}. Therefore, more complexes of Cu(II) with OH^- were formed than Pb(II) at the same

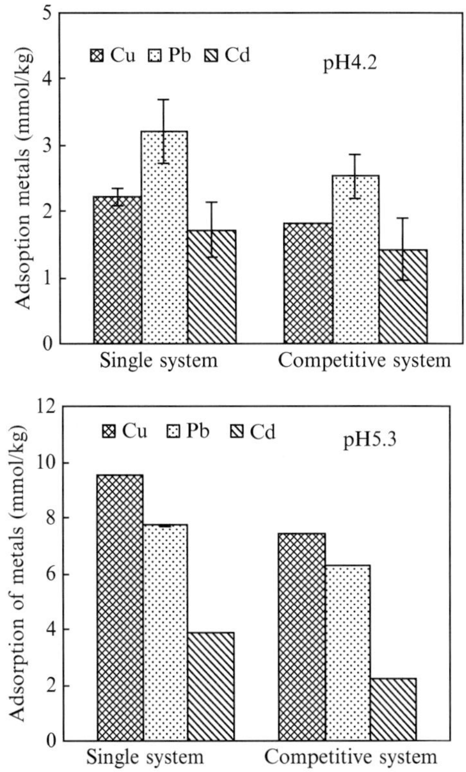

Fig. 8.13 Adsorption of Cu(II), Pb(II), Cd(II) on an Ultisol from Guangdong Province, China in single and competitive systems at pH 4.2 and 5.3, respectively

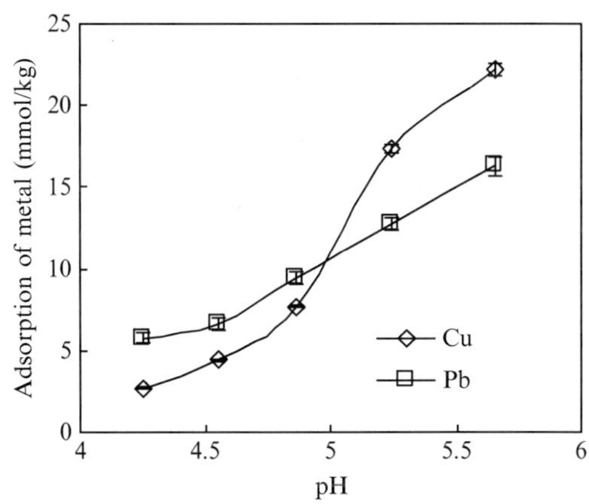

Fig. 8.14 Effect of pH on the adsorption of Cu(II) and Pb(II) on the Ultisol from Guangdong Province, China

pH and same concentration of metals, which led to more adsorption of Cu(II) with increasing pH than Pb(II) at relative higher pH. The hydrolysis constant of metals is an important factor influencing selectivity sequence of heavy metals adsorbed by variable charge soils and minerals.

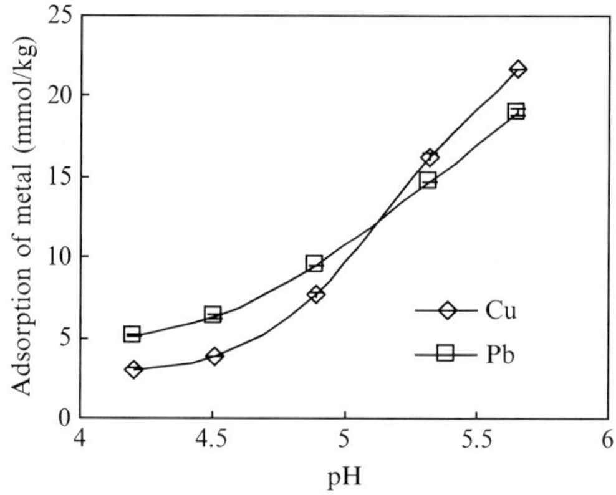

Fig. 8.15 Effect of pH on the adsorption of Cu(II) and Pb(II) on the Ultisol from Jiangxi Province, China

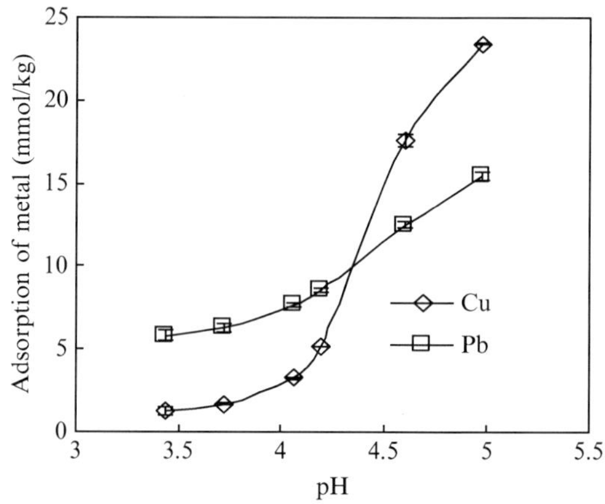

Fig. 8.16 Effect of pH on the adsorption of Cu(II) and Pb(II) on goethite

8.4.2 Competition of Arsenate and Other Anions

The competitive adsorption of arsenate with arsenite and other anions on variable charge soils and Fe/Al oxides has also been investigated. Gräfe et al. (2001) reported that humic acid (HA) reduced arsenate adsorption on goethite between pH 6 and 9 by 27%, fulvic acid (FA) inhibited the adsorption between pH 3 and 8 by a maximum of 17%. Arsenite adsorption on goethite was also decreased by both HA and FA in the range of pH 3–8 (Gräfe et al. 2002). Arsenate can also compete for adsorption sites on soils and minerals with phosphate and anions of low molecular weight organic acids (Liu et al. 2001; Violante and Pigna 2002; Zhang and Selim 2007). The rates and amounts of arsenate adsorption are significantly reduced by the presence of phosphate (Zhang and Selim 2007). Thus the presence of phosphate and organic anions decreased the adsorption of arsenate by variable charge soils and increase the mobility of arsenate in the soils.

Results from the binary system of As(III) and As(V) indicated that the two species of arsenic compete with each other for adsorption sites on variable charge

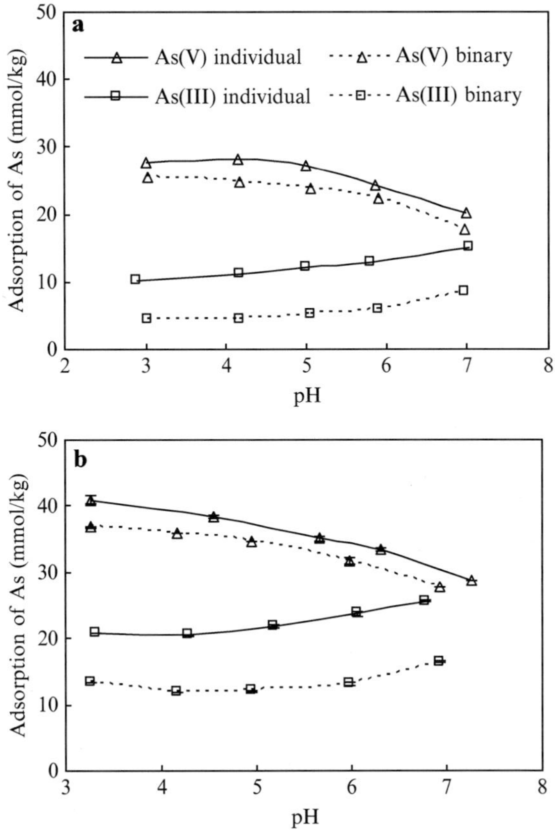

Fig. 8.17 Competitive adsorption of As(V) and As(III) on variable charge soils at different pH (**a**) Ultisol from Jiangxi province, China; (**b**) Oxisol from Guangdong Province, China

soils; and As(V) was more competitive than As(III) in acidic conditions, because the presence of As(V) depressed the adsorption of As(III) to a larger extent in the Ultisol and Oxisol (Fig. 8.17). As(III) can compete for the adsorption sites on variable charge soils with As(V), thus both arsenate and arsenite have similar adsorption mechanism on variable charge soils. Arsenate was adsorbed mainly through the formation of inner sphere surface complexes on variable charge soils and Fe/Al oxides, while both inner sphere complexes and outer sphere complexes were involved in the adsorption of arsenite. Therefore arsenate and arsenite competed for specific adsorption sites on variable charge soils each other.

8.5 Co-adsorption of Arsenate, Chromate and Heavy Metals on Variable Charge Soils

Chromated copper arsenate often used to treat the lumber for raised garden beds, which resulted in the multiple pollutions of Cr, Cu and As in nearby soil (Zagury et al. 2003; Rahman et al. 2004). Arsenic, cadmium and zinc contamination common co-occur in many contaminated environments including mining areas, soils and sediments (Liu et al. 2005; Mirlean and Roisenberg 2006). Therefore, the investigation of interactions between multiple metal pollutants in soil samples would have greater practical importance to understand the fate of these pollutants in soils.

8.5.1 Effect of Arsenate on Adsorption of Zn(II) and Cd(II)

The effect of arsenate on the adsorption of Cd(II) and Zn(II) in two variable charge soils was studied (Liang et al. 2007a, b). The batch type experiments showed, the presence of arsenate led to increase in the adsorption of both Cd(II) and Zn(II). The effect of arsenate on adsorption of two metal cations was also related to soil content of Fe/Al oxides. The larger content of Fe/Al oxides led to the more adsorption of arsenate by the soil, hence greater increase in the adsorption of Cd(II) and Zn(II). The results suggested that the adsorption of Cd(II) and Zn(II) in variable charge soils could be enhanced by arsenate through the increase in the value of negative surface charge of the soils due to the specific adsorption of arsenate. The increase in negative surface charge resulted in an apparent increase in the adsorption of Cd(II) and Zn(II) through the electrostatic attraction and the part of Cd(II) and Zn(II) adsorbed as this manner could be replaced by neutral salt (e.g, KNO_3). The presence of arsenate was found to increase not only the adsorption of Cd(II) and Zn(II), but also the desorption of pre-adsorbed Cd(II) and Zn(II). The proportion of Δdesorption in Δadsorption in Table 8.5 roughly represents the relative contribution of electrostatic adsorption to the total increment of the adsorption of Cd(II) and Zn(II) due to the presence of arsenate (Liang et al. 2007a, b). It ranged from 80.2 to 99.0% for Cd(II) and from 67.0 to 93.5% for Zn(II) in the Oxisol from Yunnan Province, China and from 69.4 to 97.1% for Cd(II) and 53.7–84.3% for Zn(II) in the Oxisol from Guangdong Province, China.

The electrostatic enhancement of Cd(II) adsorption was further supported by the results of the effect of arsenate on surface charge and the zeta potentials of the two soils. The adsorption of arsenate made the negative surface charge of the soils more negative (Figs. 8.11 and 8.12). The presence of arsenate also led to the shifts of the zeta potentials of soil suspensions to negative value direction (Fig. 8.3). These results suggested that the surface became more attractive for Cd(II) and Zn(II) due to arsenate adsorption. Since the electrostatic adsorption of ion by soils is determined by electrostatic attraction force between ion and the surface of soil colloid, it is to be believed that the change in surface charge (or surface potential) is the principal reason for the enhancement of electrostatic sorption of Cd(II) and Zn(II) in the presence of arsenate. This interpretation is in agreement with the conclusion obtained by Collins et al. (1999) who investigated the effect of phosphate and sulfate on Cd(II) sorption to goethite using EXAFS spectroscopy.

Table 8.5 Effect of initial arsenate concentration on adsorption and desorption of Zn(II) and Cd(II)

Soil	Arsenate added (mmol/kg)	Arsenate adsorbed (mmol/kg)	Zn(II)			Cd(II)		
			ΔAdsorption (mmol/kg)	ΔDesorption (mmol/kg)	ΔDesorption/ Δadsorption (%)	ΔAdsorption (mmol/kg)	ΔDesorption (mmol/kg)	ΔDesorption/ Δadsorption (%)
Oxisol from Yunnan	12.5	12.49	2.24	1.50	67.0	1.91	1.81	94.8
	20.0	19.94	3.52	3.16	89.8	2.88	2.85	99.0
	37.5	36.34	7.01	6.33	90.3	7.03	5.69	80.9
	50.0	44.53	8.63	8.06	93.4	9.13	7.32	80.2
Oxisol from Guangdong	12.5	12.47	2.05	1.72	83.9	1.49	1.10	73.8
	20.0	19.83	3.11	2.36	75.9	2.78	2.70	97.1
	37.5	34.35	6.12	4.66	76.1	6.24	4.50	72.1
	50.0	40.92	7.23	5.85	80.9	7.62	5.29	69.4

The pH of equilibrium solution after adsorption was 5.2 and the initial concentration of Zn(II) and Cd(II) was 1.0 mmol/L. ΔAdsorption is the difference in the adsorption of Zn(II) and Cd(II) between the arsenate system and control, and ΔDesorption is the difference in the desorption of Zn(II) and Cd(II) pre-adsorbed between the arsenate system and control

8.5.2 Effect of Arsenate on Cu(II) Adsorption

Cu(II) adsorption in variable charge soils could also be enhanced by arsenate through the increase in net negative surface charge of the soils due to the specific sorption of arsenate. The proportion of desorption in adsorption in Table 8.6 roughly represents the relative contribution of electrostatic adsorption to the total increments of Cu(II) sorption. It ranged from 34.7 to 46.9% in the Oxisol from Guangdong Province, China and from 12.5 to 31.5% in the Oxisol from Yunnan Province, China. The relative contribution of electrostatic adsorption to the total increment of Cu(II) adsorption by the variable charge soils due to the presence of arsenate is lower compared with Cd(II) and Zn(II).

In addition to electrostatic attraction, Cu(II) sorption can also be enhanced by arsenate through formation of soil-arsenate-Cu ternary surface complexes or surface precipitation on soil, because some adsorbed Cu(II) was not desorbed by KNO_3. When copper was added, more than 98% of added arsenate was likely to be adsorbed by the soils. The product of $[Cu^{2+}]$ and $[AsO_4{}^{3-}]$ was not more than the Ksp (8×10^{-36}) of $Cu_3(AsO_4)_2$ (the dissociation constant of H_3AsO_4, pKa_1, pKa_2 and pKa_3 was 2.1, 6.7 and 11.2, respectively). Thus, the possibility of the formation of $Cu_3(AsO_4)_2$ precipitate in soil solution can be excluded. However, both Cu^{2+} and arsenate can be concentrated on soil surface, it is possible to form surface metal-arsenate precipitate on the soils. The formation of an adamite-like surface precipitate on goethite was observed with EXAFS spectroscopy by Gräfe and co-workers when they studied the co-sorption of Zn and arsenate at the goethite-water interface (Gräfe et al. 2004). However, no ternary complexes were observed in Cd(II)-phosphate-goethite system with EXAFS spectroscopy by other researchers (Collins et al. 1999). Although, there is no surface structural evidence to support the present hypothesis of the formation of ternary surface complexes or surface precipitation in this study, these two mechanisms can not be excluded now, some further investigations need to be conducted in future.

8.5.3 Effect of Chromate on Cu(II) Adsorption

Similar with arsenate, the presence of chromate (dichromate at low pH) also enhanced the adsorption of Cu(II) by variable charge soils (Xu et al. 2005). For example, in the presence of 0.5, 0.8, 1.0 and 1.5 mmol/L of chromate, the adsorption of Cu(II) increased for the Oxisol from Guangdong Province, China by 15.3, 18.0, 19.0 and 20.2%, respectively. For the Oxisol from Yunnan Province, the corresponding figures were 11.9, 17.0, 20.3, and 26.1%, respectively. The presence of chromate also caused an increase in the desorption of pre-adsorbed Cu(II). For instance, in the presence of 0.5, 1.0, and 1.5 mmol/L of chromate, the desorption for the Oxisol from Guangdong increased by 16.9, 27.5, and 34.1%, respectively. For the Oxisol from Yunnan, the corresponding figures were 18.1, 35.6, and 51.4%, respectively. Thus the increment of the adsorption was caused

Table 8.6 Effect of arsenate on adsorption and desorption of Cu(II) by tow Oxisols (initial concentration of Cu(II) was 1.0 mmol/L)

Soil	Arsenate added (mmol/kg)	Cu(II)				
		Adsorption (mmol/kg)	ΔSorption (mmol/kg)	Desorption (mmol/kg)	ΔDesorption (mmol/kg)	ΔDesorption/Δadsorption (%)
Oxisol from Guangdong	0	8.75	–	3.79	–	–
	2.5	9.50	0.75	4.05	0.26	34.7
	12.5	11.18	2.43	4.67	0.88	36.2
	20.0	12.65	3.90	5.49	1.70	43.6
	25.0	13.28	4.53	5.86	2.07	45.7
	37.5	14.66	5.91	6.56	2.77	46.9
Oxisol from Yunnan	0	7.56	–	2.50	–	–
	2.5	8.20	0.64	2.58	0.08	12.5
	12.5	9.16	1.60	2.75	0.25	15.6
	20.0	10.69	3.13	3.05	0.47	15.0
	25.0	11.41	3.85	3.60	1.10	28.6
	37.5	14.07	6.51	4.55	2.05	31.5

Δ sorption is the difference of Cu(II) sorption between arsenate system and control; Δ desorption is the difference of the desorption of Cu(II) adsorbed in arsenate system and in control. Adsorption equilibrium pH was 4.35 for the Oxisol from Guangdong Province, China and 4.53 for the Oxisol from Yunnan Province, China

by the change in surface charge of the soils induced by the adsorption of chromate and the cooperative adsorption of chromate adsorbed and Cu(II). The increase of electrostatically adsorbed Cu(II) was responsible for the increase in the desorption of Cu(II).

8.6 Conclusions

Adsorption reactions of heavy metals are important factors controlling the mobility of these metals in variable charge soils. The adsorption of heavy metal cations involved the electrostatic and specific interactions of the soils with the metals. The relative contribution of the two mechanisms to the total adsorption varied with the type of metals and soil properties. Chromate (dichromate at low pH) was also adsorbed by variable charge soils electrostatically and specifically. However, arsenate was mainly adsorbed by the soils specifically. Analysis of EXAFS spectra indicated that heavy metal cations, chromate and arsenate can be adsorbed by variable charge soils and Fe/Al oxides specifically through the formation of inner sphere surface complexes. pH, CEC and organic matter content are main factors influencing the adsorption of metal cations. The increase in pH and content of organic matter increased not only the amount of metal cations adsorbed, but also the adsorption strength, especially for Cu(II) and Pb(II). Therefore, increasing soil pH and the content of organic matter through adding alkaline materials and organic manure can increase fixation of heavy metals in variable charge soils.

Ionic strength can affect adsorption of heavy metals by variable charge soils and Fe/Al oxides through the effect on electrostatic potential in adsorption plane. Increasing ionic strength increased specific adsorption of heavy metal cations when pH is lower than soil ZPC, and specific adsorption of arsenate and chromate when pH is higher than soil ZPC. Thus increasing ionic strength of soil solutions through fertilizing can increase the fixation of heavy metals in polluted soils.

In multiple pollutant systems, different heavy metal cations competed for adsorption sites with each other, which led to the decrease in adsorption of individual metal cation in soils. The selectivity sequence of the metals adsorbed by variable charge soils is Pb(II) > Cu(II) > Cd(II) at low pH(pH 4.2); while the selectivity sequence changed to be Cu(II) > Pb(II) > Cd(II) at relative high pH(pH 5.3). Arsenate can compete for adsorption sites with other anions. The presence of phosphate inhibited the adsorption of arsenate by variable charge soils and Fe/Al oxides obviously. Thus, phosphorus fertilizing in polluted soils may increase mobility of arsenate. When arsenate and arsenite exist in the same system, arsenate decreased the adsorption of arsenite by variable charge soils greatly, while the effect of arsenite on the adsorption of arsenate is slight. When arsenate or chromate exist with heavy metal cations in one system, co-adsorption of the anion and heavy metal cations occurred in variable charge soils. Specific adsorption of arsenate and chromate on variable charge soils made soil negative charge more negative, which induced more heavy metal cations adsorbed by the soils through the electrostatic attraction. The adsorption of heavy metal cations can also be enhanced by arsenate or chromate through formation of soil-anion-metal ternary surface complexes or surface precipitation on the soils. Co-adsorption of arsenate or chromate and heavy metal cations increased the fixation of these pollutants and decreased their mobility in polluted soils.

Acknowledgement This study was supported by National Natural Science Foundation of China (40971135).

References

Abdel-Samad H, Watson PR (1997) An XPS study of the adsorption of chromate on goethite (α-FeOOH). Appl Surf Sci 108:371–377

Agbenin JO, Olojo LA (2004) Competitive adsorption of copper and zinc by a Bt horizon of a savanna Alfisol as affected by pH and selective removal of hydrous oxides and organic matter. Geoderma 119:85–95

Aide MT, Cummings MF (1997) The influence of pH and phosphorus on the adsorption of chromium(VI) on boehmite. Soil Sci 162:599–603

Ainsworth CC, Pilon JL, Gassman PL et al (1994) Cobalt, cadmium, and lead sorption to hydrous iron oxide: residence time effect. Soil Sci Soc Am J 58:1615–1623

Alcacio TM, Hesterber GD, Chou JW et al (2001) Molecular scale characteristics of Cu(II) bonding in goethite–humate complexes. Geochim Cosmochim Acta 65:1355–1366

Arai Y, Elzinga EJ, Sparks DL (2001) X-ray absorption spectroscopic investigation of arsenite and arsenate adsorption at the aluminum oxide-water interface. J Colloid Interface Sci 235:80–88

Backes CA, McLaren RG, Rate AW et al (1995) Kinetics of cadmium and cobalt desorption from iron and manganese oxides. Soil Sci Soc Am J 59:778–785

Barrow NJ, Bowden JW, Posner AM et al (1980) Describing the effects of electrolyte on adsorption of phosphate by a variable charge surface. Aust J Soil Res 18:395–404

Bleam WF, McBride MB (1986) The chemistry of adsorbed Cu(II) and Mn(II) in aqueous titanium dioxide suspensions. J Colloid Interface Sci 110:335–346

Bowden JW, Nagarajah S, Barrow NJ et al (1980) Describing the adsorption of phosphate, citrate and selenite on a variable-charge mineral surface. Aust J Soil Res 18:49–60

Brümmer GW, Gerth J, Herms U (1986) Heavy metal species, mobility and availability in soils. Z Pflanzenernaehr Bodenk 149:382–398

Brümmer GW, Gerth J, Tiller KG (1988) Reaction kinetics of the adsorption and desorption of nickel, zinc and cadmium by goethite. I. Adsorption and diffusion of metals. J Soil Sci 39: 37–51

Catalano JG, Zhang Z, Park C, Fenter P, Bedzyk MJ (2007) Bridging arsenate surface complexes on the hematite (012) surface. Geochim Cosmochim Acta 71:1883–1897

Collins CR, Ragnarsdottir KV, Sherman DM (1999) Effect of inorganic and organic ligands on the mechanism of cadmium sorption to goethite. Geochim Cosmochim Acta 63:2989–3002

Echeverría JC, Morera MT, Mazkiarán C et al (1998) Competitive sorption of heavy metal by soils. Isotherms and factorial experiments. Environ Pollut 101:275–284

Fendorf S, Eick MJ, Grossl P et al (1997) Arsenate and chromate retention mechanisms on goethite. 1. Surface structure. Environ Sci Technol 31:315–320

Fischer L, Brümmer GW, Barrow NJ (2007) Observations and modeling of the reactions of 10 metals with goethite: adsorption and diffusion processes. Eur J Soil Sci 58:1304–1315

Fontes MPF, Gomes PC (2003) Simultaneous competitive adsorption of heavy metals by the mineral matrix of tropical soils. Appl Geochem 18:795–804

Forbes EA, Posner AM, Quirk JP (1976) The specific adsorption of divalent Cd, Co, Cu, Pb and Zn on goethite. J Soil Sci 27:154–166
Fuller CC, Davis JA, Waychunas GA (1993) Surface chemistry of ferrihydrite. Part 2. Kinetics of arsenate adsorption and coprecipitation. Geochim Cosmochim Acta 57:2271–2282
Garman SM, Luxton TP, Eick MJ (2004) Kinetics of chromate adsorption on goethite in the presence of sorbed silicic acid. J Environ Qual 33:1703–1708
Gräfe M, Eick MJ, Grossl PR (2001) Adsorption of arsenate(V) and arsenite(III) on goethite in the presence and absence of dissolved organic carbon. Soil Sci Soc Am J 65:1680–1687
Gräfe M, Eick MJ, Grossl PR, Saunders AM (2002) Adsorption of arsenate and arsenite on ferrihydrite in the presence and absence of dissolved organics carbon. J Environ Qual 31:1115–1123
Gräfe M, Nachtegaal M, Sparks DL (2004) Formation of metal-arsenate precipitates at the goethite-water interface. Environ Sci Technol 38:6561–6570
Grossl PR, Eick MJ, Sparks DL, Goldberg S, Ainsworth CC (1997) Arsenate and chromate rentention mechanisms on geothite. 2. Kinetics evaluation using a press-jump relaxation technique. Environ Sci Technol 31:321–326
Guo XY, Zhang SZ, Shan XQ et al (2006) Characterization of Pb, Cu, and Cd adsorption on particulate organic matter in soil. Environ Toxic Chem 25:2366–2373
Harter RD (1983) Effect of soil pH on adsorption lead, copper, zinc, and nickel. Soil Sci Soc Am J 47:47–51
Harter RD, Naidu R (1995) Role of metal-organic complexation in metal sorption by soils. Adv Agron 55:219–263
Hayes KF, Papelis C, Leckie JO (1998) Modeling ionic strength effects on anion adsorption at hydrous oxides/solution interfaces. J Colloid Interface Sci 125:717–726
Hodgson JF, Tiller KG, Fellows M (1964) The role of hydrolysis in the reaction of heavy metals with soil-forming materials. Soil Sci Soc Am Proc 28:42–46
Hsia T, Lo S, Lin C et al (1993) Chemical and spectroscopic evidence for specific adsorption of chromate on hydrous iron-oxide. Chemosphere 26:1897–1904
Huang PM (2008) Impacts of physicochemical-biological interactions on metal and metalloid transformations in soils: an overview. In: Violante A, Huang PM, Gadd GM (eds) Biophysico-chemical processes of heavy metals and metalloids in soil environments, vol 1, Wiley-IUPAC series biophysico-chemical processes in environment systems. Wiley, Hoboken
Jain A, Raven KP, Loeppert RH (1999) Arsenite and arsenate adsorption on ferrihydrite: surface charge reduction and net OH^- release stoichiometry. Environ Sci Technol 33:1179–1184
Jiang J, Xu RK, Wang Y et al (2008) The mechanism of chromate sorption by three variable charge soils. Chemosphere 71:1465–1471
Jiang J, Xu RK, Li SZ (2010) Effect of ionic strength and mechanism of Cu(II) adsorption by goethite and γ-Al_2O_3. J Chem Eng Data 55:5547–5552
Liang J, Xu RK, Tiwari D et al (2007a) Effect of arsenate on adsorption of Zn(II) by three variable charge soils. Aust J Soil Sci 45:465–472
Liang J, Xu RK, Jiang X et al (2007b) Effect of arsenate on adsorption of Cd(II) by two variable charge soils. Chemosphere 67:1949–1955
Liu F, De Christofaro A, Violante A (2001) Effect of pH, phosphate and oxalate on the adsorption/desorption of arsenate on/from goethite. Soil Sci 166:197–208
Liu HY, Probst A, Liao BH (2005) Metal contamination of soils and crops affected by the Chenzhou lead/zinc mine spill (Hunan, China). Sci Total Environ 339:153–166
Ma L, Xu RK, Jiang J (2010) Adsorption and desorption of Cu(II) and Pb(II) in paddy soils cultivated for different years in subtropical China. J Environ Sci 22:689–695
McBride MB (1997) A critique of diffuse double layer models applied to colloid and surface chemistry. Clays Clay Miner 45:598–608
McBride M, Sauvé S, Hendershot W (1997) Solubility control of Cu, Zn, Cd and Pb in contaminated soils. Eur J Soil Sci 48:337–346
McKenzie RM (1980) The adsorption of lead and other heavy metals on oxides of manganese and iron. Aust J Soil Res 18:61–73

McLaren RG, Backes CA, Rate AW et al (1998) Cadmium and cobalt desorption kinetics from soil clays: effect of sorption period. Soil Sci Soc Am J 62:332–337
Mirlean N, Roisenberg A (2006) The effect of emissions of fertilizer production on the environment contamination by cadmium and arsenic in Southern Brazil. Environ Pollut 143:335–340
Nachtegaal M, Sparks DL (2004) Effect of iron oxide coatings on zinc sorption mechanisms at the clay-mineral/water interface. J Colloid Interface Sci 276:13–23
Naidu R, Bolan NS, Kookana RS et al (1994) Ionic-strength and pH effects on the sorption of cadmium and the surface charge of soils. Eur J Soil Sci 26:602–617
Naidu R, Kookana RS, Sumner ME et al (1997) Cadmium sorption and transport in variable charge soils: a review. J Environ Qual 26:602–617
Naidu R, Sumner ME, Harter RD (1998) Sorption of heavy metals in strongly weathered soil: an overview. Environ Geochem Health 20:5–9
O'Reilly SE, Strawn DG, Sparks DL (2001) Residence time effects on arsenate adsorption/desorption mechanisms on goethite. Soil Sci Soc Am J 65:67–77
Padmanabham M (1983) Adsorption–desorption behaviour of copper(II) at the goethite-solution interface. Aust J Soil Res 21:309–320
Parfitt RL, Smart RSC (1978) The mechanism of sulfate adsorption on iron oxides. Soil Sci Soc Am J 42:48–50
Pigna M, Krishnamurti GSR, Violante A (2006) Kinetics of arsenate sorption-desorption from metal oxides: effect of residence time. Soil Sci Soc Am J 70:2017–2027
Rahman FA, Allan DL, Rosen CJ et al (2004) Arsenic availability from chromated copper arsenate (CCA)–treated wood. J Environ Qual 33:173–180
Rajan SSS (1978) Sulfate adsorbed on hydrous alumina, ligands displaced, and changes in surface. Soil Sci Soc Am J 42:39–44
Ross DS, Sjogren RE, Bartlett RJ (1981) Behaviour of chromium: IV. Toxicity to microorganisms. J Environ Qual 10:145–148
Selim HM, Amacher MC, Iskandar IK (1989) Modeling the transport of chromium(VI) in soil columns. Soil Sci Soc Am J 53:996–1004
Sherman DM, Randall SR (2003) Surface complexation of arsenic(V) to iron(III) (hydr)oxides: structural mechanism from ab initio molecular geometries and EXAFS spectroscopy. Geochim Cosmochim Acta 22:4223–4230
Silveira MLA, Alleoni LRF, Camargo OA et al (2002) Copper adsorption in oxidic soils after removal of organic matter and iron oxides. Commun Soil Sci Plant Anal 33:3581–3592
Srivastava P, Singh B, Angove M (2005) Competitive adsorption behavior of heavy metals on kaolinite. J Colloid Interface Sci 290:28–38
Tiller KG, Nayyar VK, Clayton PM (1979) Specific and non-specific sorption of Cd by soil clays as influenced by zinc and calcium. Aust J Soil Res 17:17–28
Violante A, Pigna M (2002) Competitive sorption of arsenate and phosphate on different clay minerals and soils. Soil Sci Soc Am J 66:1788–1796
Violante A, Pigna M, Del Gaudio S (2005) Adsorption/desorption processes of arsenate in soil environments. In: Huang PM, Violante A, Bollag J-M et al (eds) Soil abiotic and biotic interactions and impact on the human welfare. Science Publishers, Inc., Enfield
Wang Y, Jiang J, Xu RK et al (2009) Phosphate adsorption at variable charge soil/water interfaces as influenced by ionic strength. Aust J Soil Res 47:529–536
Waychunas GA, Rea BA, Fuller CC et al (1993) Surface chemistry of ferrihydrite: Part 1. EXAFS studies of the geometry of coprecipitated and adsorbed arsenate. Geochim Cosmochim Acta 57:2251–2269
Xu RK, Xiao SC, Zhao AZ et al (2005) Effect of Cr(VI) on adsorption and desorption behavior of Cu(II) in the colloidal systems of two authentic variable charge soils. J Colloid Interface Sci 284:22–29
Xu RK, Kozak LM, Huang PM (2008) Kinetics of phosphate-induced desorption of arsenate adsorbed on crystalline and amorphous aluminum hydroxides. Soil Sci 173:683–693
Xu RK, Wang Y, Tiwari D et al (2009) Effect of ionic strength on adsorption of As(III) and As(V) by variable charge soils. J Environ Sci 21:927–932

Yang JK, Lee SM, Davis AP (2006) Effect of background electrolytes and pH on the adsorption of Cu(II)/EDTA onto TiO_2. J Colloid Interface Sci 295:14–20
Yu TR (1997) Chemistry of variable charge soils. Oxford University Press, New York
Zachara JM, Ainsworth CC, Cowen CE et al (1989) Adsorption of chromate by subsurface soil horizons. Soil Sci Soc Am J 53:418–428
Zagury GJ, Samson R, Deschenes L (2003) Occurrence of metals in soil and ground water near chromated copper arsenate–treated utility poles. J Environ Qual 32:507–514
Zhang H, Selim HM (2007) Modeling competitive arsenate-phosphate retention and transport in soils: a multi-component multi-reaction approach. Soil Sci Soc Am J 71:1267–1277
Zhong K, Xu RK, Zhao AZ et al (2010) Adsorption and desorption of Cu(II) and Cd(II) in the tropical soils during pedogenesis in the basalt from Hainan, China. Carbonate Evaporite 25:27–34
Zhou DH, Xu FL, Dong YY et al (1996) Some problems relating to characterizing specific adsorption of heavy-metal ions on surface of oxide–effect of accompanying anions. Chinese Sci Bull 47:1483–1487

Part III
Adsorption and Diffusion Processes in Soil

Chapter 9
The Electrostatic Field Effect from Surface Charges on Ion Diffusion/Adsorption in Soil

Hang Li, Rui Li, Xinmin Liu, and Jie Hou

Abstract Soil is a system exhibiting strong electrostatic field in aqueous solution, even though it is electrically neutral macroscopically. Therefore, the gradient of electric potential will be the main driving forces for ions transportation, adsorption/desorption in the soil. In this chapter, we will firstly discuss the theoretical descriptions for mass diffusion in a system with multi-potential fields. Based on macroscopic thermodynamic analysis, a generalized linear theory for describing non-linear diffusion in external fields and non-ideal systems, such as soil, can be established, and this theory included: a generalized linear flux equation for non-linear diffusion; an apparent mass conservation equation and a generalized linear non-steady state equation for non-linear diffusion. This section is critical in ion diffusion/adsorption study in soil, because in this theory the coupling diffusion with multiple driving forces in soil have been unified to a single force: the apparent concentration gradient, and which changes the complex non-linear diffusion problem of ions in soil to linear. In the second section, we will discuss the electrostatic field effect from surface charges of soil particles on ion exchange equilibrium. In this section, we will show that, the electrostatic adsorption of cations in soil can be treated as a diffusion process driven by the apparent concentration gradient of ions in the external electric field from soil particle surface. Therefore the electrostatic adsorption/desorption process and the diffusion process in the electric field of soil are essentially the same process, and can be treated as one. Based on this concept, new ion equilibrium distribution equations for describing ion exchange equilibrium can be obtained, and those equation clearly show how the surface charges, the electric field strength, ionic volume and ionic interaction in soil solution influence the exchange. In the last section, we will discuss the electrostatic field

H. Li (✉) • R. Li • X. Liu • J. Hou
College of Resources and Environment, Southwest University, Chongqing, PR China
e-mail: hli22002@yahoo.com.cn

J. Xu and D.L. Sparks (eds.), *Molecular Environmental Soil Science*,
Progress in Soil Science, DOI 10.1007/978-94-007-4177-5_9,

effect from surface charges of soil particles on ion adsorption/diffusion kinetics. In this section, we will show that, the electric field in the soil exerts a significant influence on ion adsorption and diffusion kinetics. By considering the electric field from soil particle surface, new exact kinetic equations for describing cation adsorption can be developed. The new equations show that, when both strong force adsorption and weak force adsorption coexist in an exchange experiment, the initial stage of strong force adsorption can be described by the zero-order kinetics. The adsorption then shifts to the first-order kinetics of the weak force adsorption. The new kinetic equations are mathematically precise, and they were derived from the actual physical mechanism of cation adsorption. Thus, those new equations are fundamentally different from the classic apparent kinetic equations. All the parameters in the new equations have definite physical significance. Therefore, using the new equations, some important dynamic and thermodynamic parameters in cation exchange can be estimated from the experimental data.

Keywords Ion diffusion • Ion exchange • Soil electric field • Adsorption/desorption kinetics • Exchange equilibrium

Natural soil is a non-equilibrium system with multi-phase, multi-fields and multi-processes taking place simultaneously. For example, ion diffusion is a well known basic process in soil, and the driving forces involved in the diffusion may include: (1) gradient of potential coming from concentration; (2) gradient of potential coming from molecular interaction; (3) gradient of electrostatic potential producing from surface charges of soil particles combined with counter-ions in diffuse layer (electrostatic field); and (4) the temperature gradient (temperature field). Even considering a constant temperature system, the driving forces for ion diffusion include at least three aspects: electrostatic potential gradient, concentration gradient and molecular interaction potential gradient in a non-ideal system.

Under ideal condition, ion diffusion can be described by the linear Fickian diffusion equations; under a non-ideal condition, the non-linear Fickian diffusion equations should be employed. When considering the electrostatic field in the soil, however, both linear and non-linear Fickian diffusion equations are inapplicable.

Soil is a system exhibiting strong electrostatic field in aqueous solution, even though it is electrically neutral macroscopically. Figure 9.1 shows the distribution of the electrostatic field strength around a soil particle in aqueous solutions of different concentrations of 1:1 type electrolyte in bulk solution as surface charge number = 10 $cmol_c/Kg$, specific surface area = 50 m^2/g, and temperature = 298 K. The electrostatic field originating from particle surfaces may extend to deep soil solution as far as 1,000 nm, implying the driving force from the potential gradient could not be neglected in ion diffusion and adsorption studies.

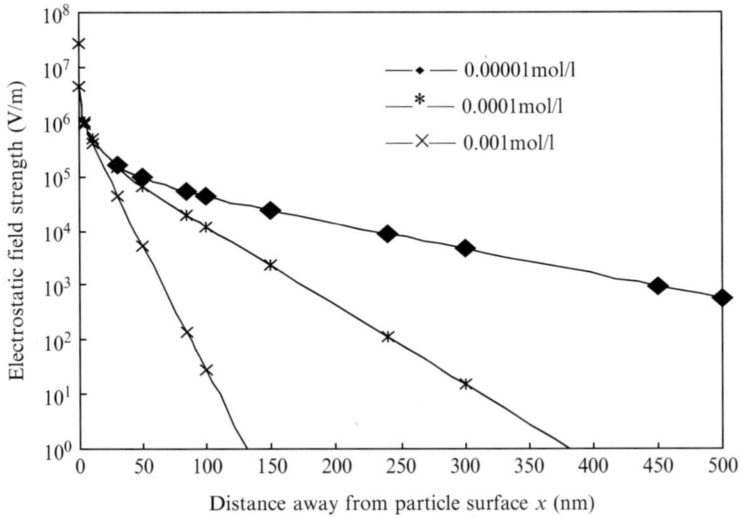

Fig. 9.1 The distribution of the electrostatic field strength around a soil particle

It is well known that, at equilibrium ion adsorption, the distribution of the adsorbed exchangeable ions will obey the Boltzmann distribution and set up a diffuse layer adjacent to the particle surface, and the concentration of counter ions adjacent particle surface is much higher than that in the bulk solution at equilibrium. This phenomenon further implies that, for the two driving forces in ion diffusion and exchange adsorption, the electric potential gradient is much more important than the concentration gradient. However, over the 100 years of studies in ion diffusion and adsorption, the Fickian laws are usually applied, in which only the concentration gradient is considered. This is done in order to describe ion diffusion and adsorption in soil with a modification of the diffusion coefficient through introducing the tortuosity factor. Therefore, the Fick's law is only applied in a system of ideal solution and external electrostatic field absence. Kemper and Van Schaik (1966) and Achari et al. (1997) used the Nernst-Planck equation to describe ion diffusion in soil. However, Kemper's study used the potential originating from ion charges, and in the study of Achari's, the electrostatic potential from surface charges was used to modify the diffusion coefficient. Recently, Chatterji (2004) also used the Nernst-Planck equation to study ionic diffusion in charged particles. Another important issue is that, in the process of ion diffusion and adsorption, both electrostatic potential gradient and chemical potential gradient are time-dependent. Therefore, the Nernst-Planck equation can not be used to describe ion diffusion or adsorption in a non-steady state process, and another more complex equation: the Fokker-Planck equation with time dependent potential must be employed (Zhang 2007).

9.1 Generalized Linear Equation for Describing Non-linear Diffusion in External Fields and Non-ideal Systems

9.1.1 Fokker-Planck Equation and Smoluchoswki Equation for Describing Mass Diffusion in External Fields and Non-ideal Systems

Soil is a system with multi-potential fields. To describe mass diffusion in a system with multi-potential fields, the Fokker-Planck equation would be employed.

Firstly, we consider the following stochastic differential equation:

$$\frac{\partial \mathbf{r}(t)}{\partial t} = \mathbf{A}[\mathbf{r}(t), t] + \mathbf{B}[\mathbf{r}(t), t] \cdot \eta(t) \tag{9.1}$$

where $\boldsymbol{A}[\mathbf{r}(t); t]$ is the drift term and $\mathbf{B}[\mathbf{r}(t); t]\ \eta(t)$ is the noise term. The drift term relates to the forces from external fields and the force from friction (dissipative force). The noise term represents a stochastic process, which relates to the stochastic force of stochastic fluctuation. $\eta(\mathbf{t})$ is the essential time dependence of the noise source in the different degrees of freedom, and the matrix $\mathbf{B}[\mathbf{r}(t); t]$ is the amplitude and the correlation of noise between the different degrees of freedom.

Corresponding to the stochastic differential Eq. (9.1), the Fokker-Planck equation may be expressed as:

$$\frac{\partial p\left(\mathbf{r}, t\right)}{\partial t} = -\nabla \cdot \mathbf{A}\left(\mathbf{r}, t\right) p\left(\mathbf{r}, t\right) + \frac{1}{2}\nabla^2 \left[\mathbf{B}\left(\mathbf{r}, t\right) \cdot \mathbf{B}^{\mathbf{T}}\left(\mathbf{r}, t\right) p\left(\mathbf{r}, t\right)\right] \tag{9.2}$$

where $p(\mathbf{r}, t)$ is the possibility density. Obviously $p(\mathbf{r}, t) \propto f(\mathbf{r}, t)$, and $f(\mathbf{r}, t)$ is the particle (molecule, ion, etc.) density or concentration.

On the other hand, there is the Langevin equation:

$$m\frac{d^2\mathbf{r}}{dt^2} = -\nabla W_{ext}(\mathbf{r}, t) - \lambda\frac{d\mathbf{r}}{dt} + \alpha\eta(t) \tag{9.3}$$

where m is the mass of a particle (molecule, ion, etc.), $W_{ext}(\mathbf{r}, t)$ is the potential energy per particle in external fields, λ is the friction coefficient, α is the amplitude of the fluctuating force.

We know that mass diffusion is often a slow process with constant velocity, because usually the friction force is strong during diffusion. Thus considering the strong friction limit, the Langevin equation can be changed to:

$$\frac{d\mathbf{r}}{dt} = -\frac{1}{\lambda}\nabla W_{ext}(\mathbf{r}, t) + \frac{\alpha}{\lambda}\eta(t) \tag{9.4}$$

Comparison between Eqs. (9.4) and (9.1), we have

$$\mathbf{A}(\mathbf{r},t) = -\frac{1}{\lambda}\nabla W_{ext}\left(\mathbf{r},t\right) \text{ and } \mathbf{B}(\mathbf{r},t) = \frac{\alpha}{\lambda} \tag{9.5}$$

Introducing (9.5) into the Fokker-Planck equation, Eq. (9.2), we have:

$$\frac{\partial p\left(\mathbf{r},t\right)}{\partial t} = \left\{\nabla\cdot\left[\frac{1}{\lambda}\nabla W_{ext}(\mathbf{r},t)\right] + \frac{1}{2}\nabla^2\frac{\alpha^2}{\lambda^2}\right\} p\left(\mathbf{r},t\right) \tag{9.6}$$

Given $D = \alpha^2/2\lambda^2$, Eq. (9.6) transformed to:

$$\frac{\partial p\left(\mathbf{r},t\right)}{\partial t} = \nabla\cdot\left[\nabla D + \frac{1}{\lambda}\nabla W_{ext}(\mathbf{r},t)\right] p\left(\mathbf{r},t\right) \tag{9.7}$$

Equation (9.7) is the Fokker-Planck equation as applied to a diffusion system of strong friction limit.

Equation (9.7) means:

$$\mathbf{j}(\mathbf{r},t) = \left[\nabla D + \frac{1}{\lambda}\nabla W_{ext}(\mathbf{r},t)\right] p\left(\mathbf{r},t\right) \tag{9.8}$$

where $\mathbf{j}(\mathbf{r}, t)$ is the flux.

Considering a stationary state, we have $\mathbf{j}(\mathbf{r}, t) = 0$. For a stationary state W is time-independent, there is $W_{ext}(\mathbf{r}, t) = W_{ext}(\mathbf{r})$; and for a stationary state, $p(\mathbf{r})$ obeys Boltzmann distribution: $p(\mathbf{r}) = p_0\exp[-W_{ext}(\mathbf{r})/kT]$, in which k is the Boltzmann constant, and T is the absolute temperature.

Thus from Eq. (9.8), we have:

$$p_0\left[\nabla D + \frac{1}{\lambda}\nabla W_{ext}(\mathbf{r})\right] e^{-\frac{W_{ext}(\mathrm{r})}{kT}} = 0 \tag{9.9}$$

where p_0 is a constant.

$$\nabla\left[De^{-\frac{W_{ext}(\mathrm{r})}{kT}}\right] = e^{-\frac{W_{ext}(\mathrm{r})}{kT}}\left[\nabla D - D\frac{1}{kT}\nabla W_{ext}(\mathbf{r})\right] \tag{9.10}$$

Introducing Eq. (9.10) into Eq. (9.9), there is:

$$e^{-\frac{W_{ext}(\mathrm{r})}{kT}}\left[\nabla D - D\frac{1}{kT}\nabla W_{ext}(\mathbf{r}) + \frac{1}{\lambda}\nabla w_{ext}(\mathbf{r})\right] = 0 \tag{9.11}$$

Equation (9.11) means:

$$\nabla D = \left(D \frac{1}{kT} - \frac{1}{\lambda} \right) \nabla W_{ext}(\mathbf{r}) \tag{9.12}$$

An identity which is known as the so-called **fluctuation-dissipation theorem**.

Since

$$\nabla \cdot \nabla D p(\mathbf{r}, t) = \nabla \cdot D \nabla p(\mathbf{r}, t) + \nabla \cdot p(\mathbf{r}, t) \nabla D \tag{9.13}$$

Introducing Eq. (9.12) into Eq. (9.13), gives

$$\nabla \cdot \nabla D p(\mathbf{r}, t) = \nabla \cdot D \nabla p(\mathbf{r}, t) + \nabla \cdot p(\mathbf{r}, t) \left(D \frac{1}{kT} - \frac{1}{\lambda} \right) \nabla W_{ext}(\mathbf{r}) \tag{9.14}$$

As diffusion occurs in a time-independent potential, Eq. (9.7) changed to:

$$\begin{aligned} \frac{\partial p\,(\mathbf{r}, t)}{\partial t} &= \nabla \cdot \left[\nabla D + \frac{1}{\lambda} \nabla W_{ext}(\mathbf{r}) \right] p\,(\mathbf{r}, t) \\ &= \nabla \cdot \nabla D p(\mathbf{r}, t) + \nabla \cdot \frac{1}{\lambda} \nabla W_{ext}(\mathbf{r}) p(\mathbf{r}, t) \end{aligned} \tag{9.15}$$

Considering Eqs. (9.14) and (9.15) becomes

$$\begin{aligned} \frac{\partial p\,(\mathbf{r}, t)}{\partial t} &= \nabla \cdot D \nabla p(\mathbf{r}, t) + \frac{1}{kT} \nabla \cdot D \nabla W_{ext}(\mathbf{r}) p(\mathbf{r}, t) - \nabla \cdot \frac{1}{\lambda} \nabla W_{ext}(\mathbf{r}) p(\mathbf{r}, t) \\ &\quad + \nabla \cdot \frac{1}{\lambda} \nabla W_{ext}(\mathbf{r}) p(\mathbf{r}, t) \\ &= \nabla \cdot D \nabla p(\mathbf{r}, t) + \frac{1}{kT} \nabla \cdot D \nabla W_{ext}(\mathbf{r}) p(\mathbf{r}, t) \end{aligned} \tag{9.16}$$

Equation (9.16) can also be rewritten as:

$$\frac{\partial p\,(\mathbf{r}, t)}{\partial t} = \nabla \cdot D \left[\nabla + \frac{1}{kT} \nabla W_{ext}(\mathbf{r}) \right] p(\mathbf{r}, t) \tag{9.17}$$

This is the famous Smoluchowski equation.

There are two important disadvantages of the Smoluchowski equation: (1) it can be applied for a time-independent potential only; (2) since D is not a constant because of the non-ideal condition, Eq. (9.17) is non-linear.

As Eq. (9.17) is applied to one mole particles, considering $p(\mathbf{r}, t) \propto f(\mathbf{r}, t)$, we have:

$$\frac{\partial f\,(\mathbf{r}, t)}{\partial t} = \nabla \cdot D \left[\nabla + \frac{1}{RT} \nabla w_{ext}(\mathbf{r}) \right] f(\mathbf{r},\, t) \tag{9.18}$$

where R is the gas constant and $R = N_0 k$, N_0 is the Avogadros constant; $w_{ext}(\mathbf{r})$ is the potential energy per mole particles and $w_{ext}(\mathbf{r}) = N_0 W_{ext}(\mathbf{r})$.
Corresponding to Eq. (9.18), the flux equation can be expressed as:

$$\mathbf{j}(\mathbf{r}, t) = D\nabla f(\mathbf{r}, t) + D\frac{1}{RT} f(\mathbf{r}, t)\nabla U(\mathbf{r}) \tag{9.19}$$

In the absence of external fields, Eq. (9.19) reduces to the Fick's first law.

$$\mathbf{j}(\mathbf{r}, t) = D\nabla f(\mathbf{r}, t) \tag{9.20}$$

As Eq. (9.19) is applied to electrostatic field for describing charged particles diffusion, since $U(\mathbf{r}) = ZF\varphi(\mathbf{r})$, we have:

$$\mathbf{j}(\mathbf{r}, t) = D\nabla f(\mathbf{r}, t) + D\frac{ZF}{RT} f(\mathbf{r}, t)\nabla \varphi(\mathbf{r}) \tag{9.21}$$

Equation (9.21) is the well known Nerst-Planck equation, in which Z is the charge number of a cation species, F is the Faraday constant, $\varphi(x)$ is the electrostatic potential.

The Smoluchowski equation was widely applied to many fields such as physics, chemistry, biology, colloid and surface science, soft matter physics, solid state physics and chemical kinetics (Hanggi et al. 2005). Hong and Noolandi (1978a, b), Ben and Chang (2002), Rice et al. (1979), Fatin-Rouge et al. (2003), Van Duijvenbode and Koper (2001), Chan and Halle (1984), Chan (1987), Chan and Hughes (1988) and Samson et al. (1999) used the Smoluchowski equation to model ion diffusion in an electric field or in charged porous media.

However, even with simple initial and boundary conditions, the solution to the classic Smoluchowski equation is difficult, especially when an exact analytical solution is desired. Researchers in different scientific fields have been seeking solutions to the equation for various initial and boundary conditions. Although some solutions under special conditions were exactly analytical (Hong and Noolandi 1978a; Chan 1987; Pechukas et al. 2001; Monchick 1956; Tachiya 1979), many more were numerical or approximately analytical (Rice et al. 1979; Jones 2003; Berezhkovskii and D'yakov 1998; Delaire et al. 1981; Montroll 1946; Mozumder 1968; Magee and Tayler 1972; Abell and Mozumber 1972; Abell et al. 1972; Butler et al. 1977; Pedersen and Freed 1977; Schulten and Schulten 1977; Zientara and Freed 1979; Sibani and Boiden 1981; Benesi 1986; Gitterman and Weiss 1993; Klik and Yao 2000; Clifford et al. 1984). Especially, if a diffusion process involves several external fields, the solution of the Smoluchowski equation becomes extremely difficult.

Unfortunately, for ion diffusion and adsorption occurring in soil, the external electrostatic field originating from surface charges together with the counter ions in the diffuse layer will be time-dependent, because: (1) in diffusion and adsorption processes, the ion density (or charge density) in the diffuse layer is time-dependent;

(2) in adsorption process, the surface charge of soil particles in both sign and quantity may be time-dependent because of specific adsorption. Therefore, the Smoluchowski equation might be not totally applicable for describing ion diffusion and adsorption in the soil.

9.1.2 Generalized Linear Theory for Non-linear Diffusion Based on the Macroscopic Thermodynamic Analysis

The new theory established by Li and Wu (2007a), however, eliminate the disadvantages of the Smoluchowski equation, and the new theory is much more simple mathematically than the Eq. (9.17).

9.1.2.1 Definition of the Apparent Variables and Derivation of the Generalized Flux Equation

It is commonly accepted that the assumption of local equilibrium is correct in a diffusion system, thus the equations of equilibrium thermodynamics should be applicable (Li 1986). As diffusion occurs in an ideal system, according to the macroscopic thermodynamic theory, the relationship between the chemical potential, $\mu(\mathbf{r}, t)$, and the concentration, $f(\mathbf{r}, t)$, can be expressed by:

$$\mu(\mathbf{r}, t) = \mu^0 + RT \ln f(\mathbf{r}, t) \tag{9.22}$$

where $\mu(\mathbf{r}, t)$ is the chemical potential of one mole of diffusion particles in an ideal system with unit J/mol, $f(\mathbf{r}, t)$ is the concentration or density with unit mol/l, and μ^0 is the reference chemical potential when $f(\mathbf{r}, t) \to 1$.

For a non-ideal system, Eq. (9.22) becomes:

$$\mu'(\mathbf{r}, t) = \mu(\mathbf{r}, t) + \mu_{mol}(\mathbf{r}, t) = \mu^0 + RT \ln A(\mathbf{r}, t) \tag{9.23}$$

where $\mu_{mol}(\mathbf{r}, t)$ is the extra potential energy arising from particles interaction with unit J/mol; $A(\mathbf{r}, t)$ is the activity (refereed as *apparent concentration* here) for a non-ideal system with unit mol/l, and $A(\mathbf{r}, t) = \gamma(\mathbf{r}, t)f(\mathbf{r}, t)$, where $\gamma(\mathbf{r}, t)$ is the activity coefficient of particles in a non-ideal system. The variables of $A(\mathbf{r}, t)$, $\gamma(\mathbf{r}, t)$ and $f(\mathbf{r}, t)$ are all defined under conditions of local equilibrium in a diffusion system.

When a non-ideal system is subjected to external fields, there is:

$$\mu(\mathbf{r}, t) + w(\mathbf{r}, t) = \mu^0 + RT \ln A(\mathbf{r}, t) \tag{9.24}$$

where $A(\mathbf{r}, t)$ is the activity (apparent concentration) for a non-ideal system in external fields, and $A(\mathbf{r}, t) = \gamma(\mathbf{r}, t)\, f(\mathbf{r}, t)$, where $\gamma(\mathbf{r}, t)$ is the activity coefficient for a non-ideal system that is subjected to external fields; $w(\mathbf{r}, t)$ is the total extra potential energy arising from particles interaction and the external fields, and there is:

$$w(\mathbf{r}, t) = w_{ext}(\mathbf{r}, t) + \mu_{mol}(\mathbf{r}, t) \tag{9.25}$$

where $w_{ext}(\mathbf{r}, t)$ is the potential energy per mole particles that comes from external fields, for ionic diffusion in electrostatic field, there is $w_{ext}(\mathbf{r}, t) = ZF\varphi(\mathbf{r}, t)$.

The following discussions will show that the definition of Eq. (9.25) could result in a non-linear diffusion process changing to a linear problem if the local equilibrium assumption is correct in the diffusion. Therefore, Eq. (9.25) is a very important definition in the new theory.

From Eqs. (9.22) and (9.24), the relationship between the activity (apparent concentration) and the real concentration is

$$A(\mathbf{r}, t) = f(\mathbf{r}, t) \exp\left[\frac{w(\mathbf{r}, t)}{RT}\right] \tag{9.26}$$

Multiplying both sides of Eq. (9.26) with a small volume element dV, we have:

$$A(\mathbf{r}, t)dV = [f(\mathbf{r}, t)dV] \exp\left[\frac{w(\mathbf{r}, t)}{RT}\right] \tag{9.27}$$

Since $m \equiv \iiint_V f(\mathbf{r}, t)dV$ is the real mass, correspondingly we can define an *apparent mass* as: $M \equiv \iiint_V A(\mathbf{r}, t)dV$.

Giving $dM(\mathbf{r}, t) = A(\mathbf{r}, t)dV$ and $dm(\mathbf{r}, t) = f(\mathbf{r}, t)dV$, the relationship between the apparent mass and the real mass in a small volume element (dV) is:

$$dM(\mathbf{r}, t) = dm(\mathbf{r}, t) \exp\left[\frac{w(\mathbf{r}, t)}{RT}\right] \tag{9.28}$$

Multiplying both sides of Eq. (9.26) by the diffusion velocity vector, $\mathbf{v}(\mathbf{r}, t)$, we have:

$$A(\mathbf{r}, t)\mathbf{v}(\mathbf{r}, t) = [f(\mathbf{r}, t)\mathbf{v}(\mathbf{r}, t)] \exp\left[\frac{w(\mathbf{r}, t)}{RT}\right] \tag{9.29}$$

since

$$\mathbf{j}(\mathbf{r}, t) = f(\mathbf{r}, t)\mathbf{v}(\mathbf{r}, t) \tag{9.30}$$

where $\mathbf{j}(\mathbf{r}, t)$ is the real mass flux.

Similar to Eq. (9.30), the *apparent mass flux*, $\mathbf{J}(\mathbf{r}, t)$, can be defined as:

$$\mathbf{J}(\mathbf{r}, t) = A(\mathbf{r}, t)\mathbf{v}(\mathbf{r}, t) \tag{9.31}$$

Thus, according to Eqs. (9.30) and (9.31), Eq. (9.29) can be expressed as:

$$\mathbf{J}(\mathbf{r},t) = \mathbf{j}(\mathbf{r},t)\exp\left[\frac{w(\mathbf{r},t)}{RT}\right] \tag{9.32}$$

Equation (9.32) reveals the relationship between the apparent mass flux and the real mass flux.

According to the assumption of local equilibrium, the diffusion velocity vector at any given time of diffusion can be expressed as (Eriksson 1952; Jacob and Reddy 1991):

$$\mathbf{v}(\mathbf{r},t) = -D_0\nabla\left[\frac{\mu(\mathbf{r},t) + w(\mathbf{r},t) - \mu_0}{RT}\right] \tag{9.33}$$

where D_0 is the diffusion coefficient.

We would like to emphasize that, in our discussions, because the molecular interaction energy has been combined with the total potential energy $w(\mathbf{r}, t)$, the D_0 in Eq. (9.33) and in the following equations will be the value in ideal condition and thus always will be a constant.

Substituting Eqs. (9.33) and (9.30) into the right side of Eq. (9.32), we obtain

$$\mathbf{J}(\mathbf{r},t) = -D_0\nabla A(\mathbf{r},t) \tag{9.34}$$

It can be demonstrated that this new flux Eq. (9.34) can be used to describe both coupling and non-coupling diffusion processes (Li and Wu 2004). Equation (9.34) implies that, for any coupling diffusion, the concentration gradient and the potential gradient can be unified into a single gradient: *the apparent concentration gradient.* Li and Wu (2004) successfully applied this generalized flux equation to a thermal diffusion system to describe mass diffusion in the thermal field, and to establish the exact relationship between the Soret coefficient and the molecular interaction energy. The following discussion will also show that D_0 in Eq. (9.34) is the diffusion coefficient in an ideal system, thus a constant. Therefore the generalized flux Eq. (9.34) is indeed a linear equation and can be used to describe the non-linear diffusion in external fields and non-ideal conditions in the classic theory.

9.1.2.2 The Apparent Mass Conservation Equation

The Conservation Equation of the Defined Total Potential Energy

If there is no convection and no work from surface tension, the energy conservation equation can be written as (Li 1986):

$$\frac{\partial\{f(\mathbf{r},t)[E + w(\mathbf{r},t) + \varepsilon_T]\}}{\partial t} = -\nabla\cdot\{[\mathbf{j}(\mathbf{r},t)\,w(\mathbf{r},t)] + \mathbf{j}_q\} \tag{9.35}$$

where E is the kinetic energy from diffusion movement, $\mathbf{j}_q$ is the heat flux through boundary, ε_T represents the internal energy. However, since the energy coming from particles interaction is included in the total extra potential energy, $w(\mathbf{r}, t)$, ε_T here should merely represents the energy coming from the thermal motion.

The strong friction limit is consistent with the local thermodynamic equilibrium (Chavanis 2003), since the diffusion process can not be accelerated under conditions of strong friction, and thus diffusion is a slow and constant velocity process. Since the diffusion velocity is much smaller than the velocity of thermal motion of particles, the statistical state of molecules will approach the Maxwell-Boltzmann equilibrium distribution locally in a diffusion process. $E \ll \varepsilon_T$. Besides, a constant diffusion velocity under the strong friction limit implies the kinetic energy does not change with time. Thus, Eq. (9.35) can be written as:

$$\frac{\partial \{f\,(\mathbf{r},t)\,[w\,(\mathbf{r},t)+\varepsilon_T]\}}{\partial t} = -\nabla \cdot \{[\mathbf{j}\,(\mathbf{r},t)\,w\,(\mathbf{r},t)] + \mathbf{j}_q\} \tag{9.36}$$

On the other hand, under those conditions, the heat flow will be the sole source of the change of ε. This means that $\partial[f(\mathbf{r}, t)\ \varepsilon_T]/\partial t = -\nabla \bullet \mathbf{j}_q$. Therefore, Eq. (9.36) can be further reduced to:

$$\frac{\partial\,[f(\mathbf{r},t)w(\mathbf{r},t)]}{\partial t} = -\nabla \cdot [\mathbf{j}(\mathbf{r},t)w(\mathbf{r},t)] \tag{9.37}$$

Equation (9.37) implies that, under the condition of local equilibrium, the total potential energy defined in Eq. (9.25) is conservative. This also shows that the total potential energy conservation is consistent with the assumption of local equilibrium and the strong friction limit. However, Eq. (9.37) does not mean that the potential energies from external fields and from particles interaction are conservative independently.

The Apparent Mass Conservation

The classic law of mass conservation can be expressed as:

$$\frac{\partial}{\partial t} f(\mathbf{r},t) = -\nabla \cdot \mathbf{j}(\mathbf{r},t) \tag{9.38}$$

Under isothermal conditions, substituting Eqs. (9.26) and (9.32) into Eq. (9.38), one obtains:

$$\frac{\partial A(\mathrm{r},\,t)}{\partial t} - \frac{A(\mathrm{r},\,t)}{RT}\frac{\partial w(\mathrm{r},\,t)}{\partial t} = -\nabla \cdot \mathrm{J}(\mathrm{r},\,t) + \frac{A(\mathrm{r},\,t)}{RT}\mathrm{v}(\mathrm{r},\,t) \cdot \nabla w(\mathrm{r},\,t) \tag{9.39}$$

From the conservation equation of the total potential energy, Eq. (9.37), we have:

$$f(\mathbf{r},t)\frac{\partial\,[w(\mathbf{r},t)]}{\partial t} + w(\mathbf{r},t)\frac{\partial\,[f(\mathbf{r},t)]}{\partial t} = -\mathbf{j}(\mathbf{r},t)\cdot\nabla w(\mathbf{r},t) - w(\mathbf{r},t)\cdot\nabla\mathbf{j}(\mathbf{r},t) \tag{9.40}$$

According to the mass conservation equation, there will be:

$$w(\mathbf{r},t)\frac{\partial\left[f(\mathbf{r},t)\right]}{\partial t}=-w(\mathbf{r},t)\nabla\mathbf{j}(\mathbf{r},t) \tag{9.41}$$

Thus Eq. (9.40) reduces to:

$$f(\mathbf{r},t)\frac{\partial\left[w(\mathbf{r},t)\right]}{\partial t}=-\mathbf{j}(\mathbf{r},t)\cdot\nabla w(\mathbf{r},t) \tag{9.42}$$

Introducing Eq. (9.30) into Eq. (9.42), we obtain:

$$\frac{\partial w(\mathbf{r},t)}{\partial t}=-\mathbf{v}(\mathbf{r},t)\cdot\nabla w(\mathbf{r},t) \tag{9.43}$$

Substituting Eq. (9.43) into Eq. (9.39), there is:

$$\frac{\partial}{\partial t}A(\mathbf{r},t)=-\nabla\cdot\mathbf{J}(\mathbf{r},t) \tag{9.44}$$

Equation (9.44) indicates that the defined apparent mass is conservative.

9.1.2.3 The Generalized Non-steady State Diffusion Equation

Substituting Eq. (9.34) into Eq. (9.44), and considering D_0 is the diffusion coefficient under ideal condition, we can get the new diffusion equation of the non-steady state in a linear form for describing non-linear diffusion in external fields and non-ideal systems:

$$\frac{\partial A(\mathbf{r},t)}{\partial t}=D_0\nabla^2A(\mathbf{r},t) \tag{9.45}$$

It can be demonstrated that, under isothermal conditions, the linear Fickian diffusion equation, and the non-linear Fickian diffusion equation of the non-steady state; the Smoluchowski equation, and the Nernst-Planck equation, are all special cases of Eq. (9.45). Therefore Eq. (9.45) is a generalized linear equation for describing non-steady state diffusion in external fields and non-ideal system.

9.1.3 Generalized Linear Theory for Non-linear Diffusion Based on the Irreversible Thermodynamic and the Microscopic Molecular Statistical Theories

In above sections, our discussions were based a macroscopic thermodynamic analysis. Here, we would like to give additional discussions on this problem based on the irreversible thermodynamic theory and the microscopic molecular statistical theory, and as the results the linear flux equation and the linear non-steady state diffusion equation for describing non-linear diffusion in external fields and non-ideal system may be obtained more consistently.

9.1.3.1 The Linear Flux Equation for Describing Non-linear Diffusion Based on the Irreversible Thermodynamic Analysis (Li et al. 2009a)

Under isothermal and isobaric conditions, considering particle diffusion in external fields and non-ideal system, the thermodynamic force can be expressed as:

$$X = -\frac{1}{T}\left[\nabla\mu(\mathbf{r},t) + \nabla\mu_{mol}(\mathbf{r},t) + \nabla w_{ext}(\mathbf{r},t)\right] = -\frac{1}{T}\nabla W(\mathbf{r},t) \tag{9.46}$$

where $W(\mathbf{r}, t)$ is the total potential energy, and $W(\mathbf{r}, t) = w(\mathbf{r}, t) + \mu(\mathbf{r}, t) = \mu_{mol}(\mathbf{r}, t) + w_{ext}(\mathbf{r}, t) + \mu(\mathbf{r}, t)$.

According to the thermodynamic theory of the local equilibrium, there is:

$$W(\mathbf{r},t) = \mu(\mathbf{r},t) + \mu_{mol}(\mathbf{r},t) + w_{ext}(\mathbf{r},t) = \mu_0 + RT\ln\left[f(\mathbf{r},t)\gamma(\mathbf{r},t)\right] \tag{9.47}$$

and

$$\mu_{mol}(\mathbf{r},t) + w_{ext}(\mathbf{r},t) = w(\mathbf{r},t) = RT\ln\gamma(\mathbf{r},t) \tag{9.48}$$

For a system containing k different component species, correspondingly the local entropy production is:

$$\sigma = -\frac{1}{T}\sum_{i=1}^{k}\mathbf{j}_i(\mathbf{r},t)\cdot\nabla W_i(\mathbf{r},t) \tag{9.49}$$

Considering the mechanical equilibrium requirement, there is (Li 1986)

$$\mathbf{j}_k(\mathbf{r},t) = -\sum_{i=1}^{k-1}\frac{M_i}{M_k}M_i\mathbf{j}_i(\mathbf{r},t) \tag{9.50}$$

where M_i is the molecular weight of i-th component species.

Under the condition of the *local equilibrium assumption*, for isothermal and isobaric conditions, from the Gibbs-Duhem relationship, we have:

$$\nabla W_k(\mathbf{r},t) = -\sum_{j=1}^{k-1}\frac{n_j}{n_k}\nabla W_j(\mathbf{r},t) \tag{9.51}$$

Introducing Eqs. (9.50) and (9.51) into Eq. (9.49), we get:

$$\sigma = -\frac{1}{T}\sum_{i=1}^{k-1}\mathbf{j}_i(\mathbf{r},t)\cdot\sum_{j=1}^{k-1}\left(\delta_{ij}+\frac{M_i n_j}{M_k n_k}\right)\nabla W_j(\mathbf{r},t) \tag{9.52}$$

where

$$\delta_{ij} = \begin{cases} 0 & i \neq j \\ 1 & i = j \end{cases}$$

From Eq. (9.52), we obtain a new thermodynamic force:

$$X_i = -\frac{1}{T}\sum_{j=1}^{k-1}\left(\delta_{ij}+\frac{M_i n_j}{M_k n_k}\right)\nabla W_j(\mathbf{r},t) \tag{9.53}$$

Correspondingly, the flux of i-th component species can be expressed as:

$$\mathbf{j}_i(\mathbf{r},t) = -\frac{1}{T}\sum_{j}^{k-1}L_{ij}\sum_{j=1}^{k-1}\left(\delta_{ij}+\frac{M_i n_j}{M_k n_k}\right)\nabla W_j(\mathbf{r},t) \tag{9.54}$$

As for two components system ($k=2$), Eq. (9.54) reduces to:

$$\mathbf{j}_1(\mathbf{r},t) = -\frac{1}{T}L_{11}\left(1+\frac{M_1 n_1}{M_2 n_2}\right)\nabla W_1(\mathbf{r},t) \tag{9.55}$$

As Eq. (9.55) applied to ideal system with external fields absence, there is $W_1 = \mu_1(\mathbf{r}, t)$, Eq. (9.55) changed to:

$$\mathbf{j}_1(\mathbf{r},t) = -\frac{1}{T}L_{11}\left(1+\frac{M_1 n_1}{M_2 n_2}\right)\nabla\mu_1(\mathbf{r},t) = -L_{11}\left(1+\frac{M_1 n_1}{M_2 n_2}\right)\frac{R}{f_1(\mathrm{r},t)}\nabla f_1(\mathbf{r},t) \tag{9.56}$$

thus we can get the diffusion coefficient under ideal conditions:

$$D_{01} = L_{11}\left(1+\frac{M_1 n_1}{M_2 n_2}\right)\frac{R}{f_1(\mathbf{r},t)} \tag{9.57}$$

where D_{01} is the diffusion coefficient of the component in an ideal system.

As Eq. (9.55) to be applied to a non-ideal system with external fields presence, considering Eqs. (9.47) and (9.48), we can get the flux equation:

$$\mathbf{j}_1(\mathbf{r},t) = -L_{11}\left(1+\frac{M_1n_1}{M_2n_2}\right)\frac{R}{f_1(\mathbf{r},t)}\left[1+\frac{f_1(\mathbf{r},t)}{RT}\frac{\partial w_1(\mathbf{r},t)}{\partial f_1(\mathbf{r},t)}\right]\nabla f_1(\mathbf{r},t)$$

$$= D_{01}\left[1+\frac{f_1(\mathbf{r},t)}{RT}\frac{\partial w_1(\mathbf{r},t)}{\partial f_1(\mathbf{r},t)}\right]\nabla f_1(\mathbf{r},t) \tag{9.58}$$

Thus the diffusion coefficient of the component is:

$$D_1 = D_{01}\left[1+\frac{f_1(\mathbf{r},t)}{RT}\frac{\partial w_1(\mathbf{r},t)}{\partial f_1(\mathbf{r},t)}\right] \tag{9.59}$$

where D_1 is the diffusion coefficient of the component in non-ideal system with external fields presence.

Considering:

$$\frac{\partial w_1(\mathbf{r},t)}{\partial f_1(\mathbf{r},t)} = \frac{\nabla w_1(\mathbf{r},t)}{\nabla f_1(\mathbf{r},t)} = \frac{RT}{e^{\frac{w_1(\mathbf{r},t)}{RT}}}\frac{\nabla e^{\frac{w_1(\mathbf{r},t)}{RT}}}{\nabla f_1(\mathbf{r},t)}$$

Equation (9.58) can be rewritten as:

$$\mathrm{j}_1(\mathrm{r},\,t) = -D_{01}\left[1+\frac{f_1(\mathrm{r},\,t)}{e^{\frac{w_1(\mathrm{r},\,t)}{RT}}}\frac{\nabla e^{\frac{w_1(\mathrm{r},\,t)}{RT}}}{\nabla f_1(\mathrm{r},\,t)}\right]\nabla f_1(\mathrm{r},\,t) \tag{9.60}$$

Equation (9.60) can be further changed to:

$$\mathrm{j}_1(\mathrm{r},t)e^{\frac{w_1(r,t)}{RT}} = -D_{01}e^{\frac{w_1(\mathrm{r},t)}{RT}}\nabla f_1(\mathrm{r},t) - D_{01}f_1(\mathrm{r},t)\nabla e^{\frac{w_1(\mathrm{r},t)}{RT}}$$

$$= -D_{01}\left[f_1(\mathrm{r},t)e^{\frac{w_1(\mathrm{r},t)}{RT}}\right] \tag{9.61}$$

According to the definition of the apparent variables, Eqs. (9.26) and (9.32), from Eq. (9.61) we get:

$$\mathrm{J}(\mathrm{r},t) = -D_{01}\nabla A(\mathrm{r},t) \tag{9.62}$$

Here the D_{10} is defined as the diffusion coefficient under ideal conditions as in Eq. (9.57), showing that, the linear flux equation for describing non-linear diffusion can also be obtained from the irreversible thermodynamic theory. Once again, we should emphasize that the introduction of the apparent variables is not merely a mathematic expedient for transforming the non-linear equations to linear as discussed above.

9.1.3.2 The Linear Non-steady State Diffusion Equation for Describing Non-linear Diffusion Based on the Microscopic Statistical Thermodynamic Analysis (Li et al. 2009a)

According to the equation of phase space density, in a mechanical system with dissipative force caused from diffusion, there is (Xiong 1981)

$$\frac{dH(\mathbf{r},\,t)}{dt} = -\sum_{i=1}^{N} \beta_i \mathbf{v}_{iD}(\mathbf{r},\,t)\cdot\mathbf{v}_i(\mathbf{r},\,t) = -\sum_{i=1}^{N} \beta_i v_{iD}^2(\mathbf{r},\,t) \tag{9.63}$$

where β_i is the damping coefficient, N is the total numbers of particles, $\mathbf{v}_{iD}$ $(\mathbf{r},\,t)$ is the diffusion velocity of i-th particle; H is the Hamilton function. In diffusion systems, the Hamilton function may be expressed as:

$$H(\mathbf{r},t) = \left[\sum_{i=1}^{N}\frac{1}{2}m_i\left[v_{iT}^2(\mathbf{r},t)+v_{iD}^2(\mathbf{r},t)\right]+\sum_{i<j}^{N}\sum_{j=1}^{N}w_{ij}(\mathbf{r},t)+\sum_{i=1}^{N}w_{(ex)i}(\mathbf{r},t)\right] \tag{9.64}$$

where v_{iT} is the velocity of thermal motion for i-th particle; w_{ij} is the interaction potential energy between i-th and j-th particles, $\mathrm{w}_{(ex)i}$ is the potential energy of i-th particle in external fields; m_i is the mass of the i-th particle.

Introducing Eq. (9.64) into Eq. (9.63), we have:

$$\frac{d}{dt}\left[\sum_{i=1}^{N}\frac{1}{2}m_i\left[v_{iT}^2(\mathbf{r},t)+v_{iD}^2(\mathbf{r},t)\right]+\sum_{i<j}^{N}\sum_{j=1}^{N}w_{ij}(\mathbf{r},t)+\sum_{i=1}^{N}w_{(ex)i}(\mathbf{r},t)\right] = -\sum_{h=1}^{N}\beta_i v_{iD}^2(\mathbf{r},t) \tag{9.65}$$

The Fokker-Planck equation could be derived from the Langiven equation as considering the *strong friction limit*. The *strong friction limit* means that, in the diffusion process, the inertial force of a particle is so small as compared with the friction force that it can be neglected. Therefore, the *strong friction limit* will ensure ***a constant and a slow diffusion*** velocity (ε_0).

On the other hand, the *local equilibrium assumption* requires the diffusion velocity to be so slow that, in diffusion processes, the statistical state of particles will approach the Maxwell-Boltzmann equilibrium distribution locally; it means:

$$|\mathbf{v}_{iD}(\mathbf{r},t)| = \varepsilon_0 << |\mathbf{v}_{iT}(\mathbf{r},t)| \tag{9.66}$$

where ε_0 would be an infinitesimal number.

We know that the *strong friction limit*, which means $m_i(dv_{iD}/dt) \to 0$, is consistent with the *local equilibrium assumption*, which means $\varepsilon_0{}^2 \to 0$ (Chavanis 2003), therefore the $m_i(dv_{iD}/dt) \to 0$ would be consistent with

$$v^2{}_{iD}(\mathbf{r},t) = \varepsilon_0{}^2 \to 0 \tag{9.67}$$

Therefore, as considering Eqs. (9.67) and (9.65) reduces to:

$$\frac{d}{dt}\left[\sum_{i=1}^{N}\frac{1}{2}m_i v_{iT}^2(\mathbf{r},t) + \sum_{i<j}^{N}\sum_{j=1}^{N} w_{ij}(\mathbf{r},t) + \sum_{i=1}^{N} w_{(ex)i}(\mathbf{r},t)\right] = 0 \tag{9.68}$$

For an isothermal condition, the sum of kinetic energy for all particles is a constant, so Eq. (9.68) can be further reduced to:

$$\frac{d}{dt}\left[\sum_{i<j}^{N}\sum_{j=1}^{N} w_{ij}(\mathbf{r},t) + \sum_{i=1}^{N} w_{(ex)i}(\mathbf{r},t)\right] = 0 \tag{9.69}$$

If N equals to the Avogadro's number, we will have:

$$\mu_{mol}(\mathbf{r},t) = \sum_{i<j}^{N}\sum_{j=1}^{N} w_{ij}(\mathbf{r},t) \tag{9.70}$$

is the molecular interaction energy per mole particles; and

$$w_{ext}(\mathbf{r},t) = \sum_{i=1}^{N} w_{(ex)i}(\mathbf{r},t) \tag{9.71}$$

is the potential energy from external fields per mole particles. Thus Eq. (9.69) becomes:

$$\frac{d\left[\mu_{mol}(\mathbf{r},t) + w_{ext}(\mathbf{r},t)\right]}{dt} = \frac{dw(\mathbf{r},t)}{dt} = 0 \tag{9.72}$$

Equation (9.72) means that the total potential energy is conservative in diffusion as considering the *strong friction limit* or the *local equilibrium assumption*. In other words, the *total potential energy conservation* may be consistent with the *strong friction limit* or the *local equilibrium assumption*.

Since:

$$\frac{dw(\mathbf{r},t)}{dt} = \frac{\partial w(\mathbf{r},t)}{\partial t} + \nabla w(\mathbf{r},t) \cdot \mathrm{v}(\mathbf{r},t)$$

from Eq. (9.72), there will be:

$$\frac{\partial w(\mathbf{r},t)}{\partial t} = -\nabla w(\mathbf{r},t) \cdot \mathrm{v}(\mathbf{r},t) \tag{9.73}$$

Equation (9.73) is identical to Eq. (9.33). Similar to the Eq. (9.34) based on the macroscopic analysis, from Eq. (9.73), we could demonstrate that the defined apparent mass is conservative, and can be expressed as (Li and wu 2007a):

$$\frac{\partial A(\mathbf{r},t)}{\partial t} = -\nabla \cdot \mathbf{J}(\mathbf{r},t) \tag{9.74}$$

As introducing the apparent linear flux equation, Eq. (9.62), into Eq. (9.73), we still can get the Eq. (9.45), but here it was derived from the irreversible thermodynamic theory and the microscopic molecular statistical theory.

However, for special cases, the local equilibrium assumption or the strong friction limit is not required for getting the Eq. (9.45). For instance, we consider the total potential to be a periodic potential as follow:

$$w(\mathbf{r},t) = w_0 \cos(\omega t - \mathbf{kr}) \tag{9.75}$$

As considering $\mathbf{v}(\mathbf{r}, t) = \omega/\mathbf{k}$, Eq. (9.75) will satisfy Eq. (9.73) naturally without the restriction of the local equilibrium assumption or the strong friction limit.

Equation (9.45) to be referred to as the generalized equation because some of the currently existed diffusion equations are the special cases of this equation.

1. *Reducing to the linear Fick's equation as Eq. (9.45) applied to an ideal system with external fields absence*

According to the definition of the total potential energy in Eq. (9.25), in an ideal system without external fields means $w(\mathbf{r}, t) = 0$, and as $w(\mathbf{r}, t) = 0$ from Eq. (9.26) there is $A(\mathbf{r}, t) = f(\mathbf{r}, t)$, thus Eq. (9.45) reduces to:

$$\frac{\partial f(\mathbf{r},t)}{\partial t} = D_0 \nabla^2 f(\mathbf{r},t) \tag{9.76}$$

This is Fick's second law, so Fick's second law is a special case of Eq. (9.45). Equation (9.76) implies that the defined D_0 is really the diffusion coefficient under an ideal condition.

2. *Reducing to the non-linear Fick's equation as Eq. (9.45) applied to non-ideal system with external fields absence*

According to the definition of the total potential energy in Eq. (9.25), in a non-ideal system without external fields, $w(\mathbf{r}, t) = \mu_{mol}(\mathbf{r}, t) = RT\ln\gamma(\mathbf{r}, t)$. In the diffusion process, $\mu_{mol}(\mathbf{r}, t)$ will change with both position (r) and time (t). By substituting Eq. (9.26) into Eq. (9.45), we obtain:

$$\begin{aligned}&\frac{\partial f(\mathbf{r},t)}{\partial t}+\frac{1}{RT}f(\mathbf{r},t)\frac{\partial \mu_{mol}(\mathbf{r},t)}{\partial t}\\&=D_0\nabla^2 f(\mathbf{r},t)+D_0\frac{1}{RT}f(\mathbf{r},t)\nabla^2\mu_{mol}(\mathbf{r},t)+D_0\frac{1}{RT}\nabla\mu_{mol}(\mathbf{r},t)\cdot\nabla f(\mathbf{r},t)\\&+\frac{1}{RT}f(\mathbf{r},t)\left[\frac{D_0}{RT}\nabla\mu_{mol}(\mathbf{r},t)+D_0\frac{1}{f(\mathbf{r},t)}\nabla f(\mathbf{r},t)\right]\cdot\nabla\mu_{mol}(\mathbf{r},t)\end{aligned}\tag{9.77}$$

Under isothermal conditions, from Eqs. (9.33) and (9.22), we get:

$$\begin{aligned}\mathbf{v}(\mathbf{r},t)&=-\frac{D_0}{RT}\nabla\left[\mu_{mol}(\mathbf{r},t)+\mu(\mathbf{r},t)-\mu_0\right]\\&=-\frac{D_0}{RT}\nabla\mu_{mol}(\mathbf{r},t)-D_0\frac{1}{f(\mathbf{r},t)}\nabla f(\mathbf{r},t)\end{aligned}\tag{9.78}$$

Substitution of Eq. (9.78) into Eq. (9.77), there will be:

$$\begin{aligned}&\frac{\partial f(\mathbf{r},t)}{\partial t}+\frac{1}{RT}f(\mathbf{r},t)\frac{\partial \mu_{mol}(\mathbf{r},t)}{\partial t}\\&=D_0\nabla^2 f(\mathbf{r},t)+D_0\frac{1}{RT}f(\mathbf{r},t)\nabla^2\mu_{mol}(\mathbf{r},t)+D_0\frac{1}{RT}\nabla\mu_{mol}(\mathbf{r},t)\cdot\nabla f(\mathbf{r},t)\\&-\frac{1}{RT}f(\mathbf{r},t)\mathbf{v}(\mathbf{r},t)\cdot\nabla\mu_{mol}(\mathbf{r},t)\end{aligned}\tag{9.79}$$

Since here $w(\mathbf{r}, \mathrm{t}) = \mu_{mol}(\mathbf{r}, \mathrm{t})$, according to Eq. (9.43), Eq. (9.79) can be changed to:

$$\begin{aligned}\frac{\partial f(\mathbf{r},t)}{\partial t}&=D_0\nabla^2 f(\mathbf{r},t)\\&+D_0\frac{1}{RT}f(\mathbf{r},t)\nabla^2\mu_{mol}(\mathbf{r},t)+D_0\frac{1}{RT}\nabla\mu_{mol}(\mathbf{r},t)\cdot\nabla f(\mathbf{r},t)\end{aligned}\tag{9.80}$$

Equation (9.80) can be rewritten as:

$$\frac{\partial}{\partial t}f(\mathbf{r},t)=D_0\nabla\cdot\left\{\left[\nabla+\frac{1}{RT}\nabla\mu_{mol}(\mathbf{r},t)\right]f(\mathbf{r},t)\right\}\tag{9.81}$$

Since $\mu_{mol}(\mathbf{r}, t) = RT \ln \gamma(\mathbf{r}, t)$, Eq. (9.81) becomes:

$$\frac{\partial}{\partial t} f(\mathbf{r}, t) = D_0 \nabla \cdot \left\{ \left[\nabla + \frac{1}{\gamma(\mathbf{r}, t)} \nabla \gamma(\mathbf{r}, t) \right] f(\mathbf{r}, t) \right\} \tag{9.82}$$

Since

$$\nabla f(\mathbf{r}, t) = \frac{\partial f}{\partial x} i + \frac{\partial f}{\partial y} i + \frac{\partial f}{\partial z} k$$

there will be:

$$\frac{1}{\nabla f(\mathbf{r}, t)} = \frac{\nabla f(\mathbf{r}, t)}{\nabla f(\mathbf{r}, t) \cdot \nabla f(\mathbf{r}, t)} = \frac{1}{\left(\frac{\partial f}{\partial x}\right)^2 + \left(\frac{\partial f}{\partial y}\right)^2 + \left(\frac{\partial f}{\partial z}\right)^2} \left(\frac{\partial f}{\partial x} i + \frac{\partial f}{\partial y} i + \frac{\partial f}{\partial z} k \right)$$

Therefore,

$$\frac{1}{\nabla f(\mathrm{r}, t)} \cdot \nabla f(\mathrm{r}, t) = 1$$

Considering the relationship:

$$\frac{\partial h}{\partial y} \frac{\partial g}{\partial x} \frac{\partial g}{\partial y} = \frac{\partial h}{\partial x} \frac{\partial x}{\partial y} \frac{\partial g}{\partial x} \frac{\partial g}{\partial y} = \frac{\partial h}{\partial x} \left(\frac{\partial g}{\partial y} \right)^2$$

Therefore we have:

$$\begin{aligned} \frac{\nabla \gamma(\mathbf{r}, t)}{\nabla f(\mathbf{r}, t)} \nabla f(\mathbf{r}, t) &= \left[\nabla \gamma(\mathbf{r}, t) \cdot \frac{1}{\nabla f(\mathbf{r}, t)}) \right] \nabla f(\mathbf{r}, t) \\ &= \nabla \gamma(\mathbf{r}, t) \left[\frac{1}{\nabla f(\mathbf{r}, t)} \cdot \nabla f(\mathbf{r}, t) \right] = \nabla \gamma(\mathbf{r}, t) \end{aligned} \tag{9.83}$$

Introducing Eq. (9.83) into Eq. (9.82), we get:

$$\frac{\partial}{\partial t} f(\mathbf{r}, t) = \nabla \cdot \left\{ \left[D_0 + D_0 \frac{f(\mathbf{r}, t)}{\gamma(\mathbf{r}, t)} \left(\frac{\nabla \gamma(\mathbf{r}, t)}{\nabla f(\mathbf{r}, t)} \right) \right] \nabla f(\mathbf{r}, t) \right\} \tag{9.84}$$

If we write the activity coefficient γ as a compound function of $f(\mathbf{r}, t)$, that is $\gamma = \gamma[f(\mathbf{r}, t)]$, Eq. (9.84) changes to:

$$\frac{\partial}{\partial t} f(\mathbf{r}, t) = \nabla \cdot \left\{ \left[D_0 + D_0 \frac{f}{\gamma(f)} \frac{\partial \gamma(f)}{\partial f} \right] \nabla f(\mathbf{r}, t) \right\} \tag{9.85}$$

Given

$$D(f) = D_0\left[1 + \frac{f}{\gamma(f)}\frac{\partial\gamma(f)}{\partial f}\right] \tag{9.86}$$

we have:

$$\frac{\partial}{\partial t} f\,(\mathrm{r}, t) = \nabla\cdot[D(f)\nabla f(\mathrm{r}, t)] \tag{9.87}$$

Equation (9.87) is the traditional non-linear Fickian diffusion equation for describing mass diffusion in a non-ideal system, which was obtained from the generalized linear Eq. (9.45). It is most important that in the obtained relationship between diffusion coefficient and concentration, the Eq. (9.87) based on the new theory, clearly shows how concentration (f) and molecular interaction (γ) affect the diffusion coefficient. From the expression of $D(f)$, if $\partial\gamma/\partial f < 0$ for a non-ideal system, there will be $D(f) < D_0$; if $\partial\gamma/\partial f > 0$, there will be $D(f) > D_0$; and as $f \to 0$ or $\partial\gamma/\partial f \to 0$, there will be $D(f) = D_0$.

3. *Reducing to the Smoluchowski equation as Eq. (9.45) applied to an ideal system with external fields presence*

When a time-independent external field is present in a ideal system, there is: $w(\mathbf{r}, t) = w_{ext}(\mathbf{r}t)$. Similar to Eqs. (9.81) and (9.45) under such conditions can be reduced to:

$$\frac{\partial}{\partial t} f\,(\mathbf{r}, t) = D_0\nabla\cdot\left\{\left[\nabla + \frac{1}{RT}\nabla w_{ext}(\mathbf{r})\right] f(\mathbf{r}, t)\right\} \tag{9.88}$$

On the other hand, Eq. (9.88) can also be derived from the Smoluchowski equation, Eq. (9.18). Thus, the Smoluchowski equation might be the special case of the new equation, Eq. (9.45).

4. *Reducing to the Nernst-Planck equation as Eq. (9.45) applied to an external electric field for describing charged particles diffusion*

Considering the diffusion of ions in an electric field: since the nature of the forces involved in the interaction between two ions, and between ion and external field is the same, these interactions can be unified to the Poisson equation at equilibrium, thus: $w(\mathbf{r}, t) = \mu_{mol}(\mathbf{r}, t) + w_{ext}(\mathbf{r}, t) = ZF\varphi(\mathbf{r}, t)$ is the solution of the Poisson equation under assumption of local equilibrium. For this case, Eq. (9.45) could be reduced to the unsteady state Nernst-Planck equation:

$$\frac{\partial f(\mathbf{r}, t)}{\partial t} = D_0\nabla^2 f(\mathbf{r}, t) + D_0\frac{ZF}{RT} f(\mathbf{r}, t)\nabla^2\phi(\mathbf{r}, t) + D_0\frac{ZF}{RT}\nabla\phi(\mathbf{r}, t)\cdot\nabla f(\mathbf{r}, t) \tag{9.89}$$

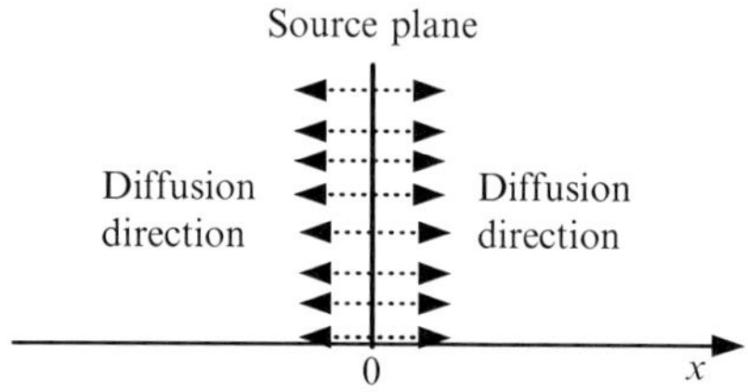

Fig. 9.2 Diffusion of an instantaneous source of cation in a negative electric field

9.1.4 Application Examples

9.1.4.1 An Analytical Solution of the New Equation for an Instantaneous Source Diffusion of Cation in a Negative Electrostatic Field

Here we consider a one-dimensional case of cation diffusion in a negative electrostatic field. Obviously, in classic theory, from Eq. (9.89), this problem will be the solution of the following equation:

$$\frac{\partial f(x,t)}{\partial t} = D_0 \frac{\partial^2 f(x,t)}{\partial x^2} + D_0 \frac{ZF}{RT} f(x,t) \frac{\partial^2 \phi(x,t)}{\partial x^2} + D_0 \frac{ZF}{RT} \frac{\partial \phi(x,t)}{\partial x} \frac{\partial f(x,t)}{\partial x} \tag{9.90}$$

For an instantaneous boundary condition with instantaneous source plane at $x=0$ as showing in Fig. 9.2, the initial and boundary conditions for solving the Eq. (9.90) are:

1. $f(x=0, t=0) = \infty$;
2. $f(x=\infty, t>0) = 0$; and
3. $f(x \neq 0, t=0) = 0$.

Obviously, even for such a simple diffusion problem, the solution of the classic Eq. (9.90) will be extremely difficult.

By our new theory, however, from the Eqs. (9.45) and (9.90) is equivalent to the following simple linear equation:

$$\frac{\partial A(x,t)}{\partial t} = D_0 \frac{\partial^2 A(x,t)}{\partial x^2} \tag{9.91}$$

in which

$$A(x,t) = f(x,t) \exp\left[\frac{ZF\phi(x,t)}{RT}\right] \tag{9.92}$$

Correspondingly, for Eq. (9.91), the initial and boundary conditions could be expressed as:

1. $A(x = 0, t = 0) = \infty$;
2. $A(x = \infty, t > 0) = 0$; and
3. $A(x \neq 0, t = 0) = 0$.

The solution for Eq. (9.91) is (Crank 2001):

$$A(x,t) = \frac{M}{2\sqrt{\pi D_0 t}} \exp\left[-\frac{x^2}{4D_0 t}\right] \tag{9.93}$$

where M is the total apparent mass with $M = m\cdot\exp[ZF\varphi(0, 0)/RT]$, and m is the real mass, and $ZF\varphi(0, 0)$ is the molar potential energy of ions at $x = 0$ and $t = 0$.

Substituting Eq. (9.92) into Eq. (9.93) and considering $M = m\cdot\exp[ZF\varphi(0, 0)/RT]$, we get

$$f(x,t) = \frac{m}{2\sqrt{\pi D_0 t}} \exp\left\{-\left[\frac{ZF\phi(x,t)}{RT} - \frac{ZF\phi(0,0)}{RT}\right] - \frac{x^2}{4D_0 t}\right\} \tag{9.94}$$

Equation (9.94) is the final form of the solution to the problem. This demonstrated that even with a time-dependent potential, the solution is simple. From Eq. (9.94), as $t \to \infty$, the concentration distribution approaches to the Boltzmann distribution.

For a non-equilibrium system of cation diffusion in diffuse double layer, by assuming local equilibrium, from the Poisson-Boltzmann equation, we suppose:

$$\begin{cases} \phi(x,t) = \phi(0,t)\exp(-\kappa x) \ldots\ldots\ldots x > 0 \\ \phi(x,t) = \phi(0,t)\exp(-\kappa x) \ldots\ldots\ldots\ldots x < 0 \end{cases} \tag{9.95}$$

where $\varphi(0, t)$ is the potential at $x = 0$ (surface potential), and it is a function of time during the diffusion process since the charge density changes with time near the plane of $x = 0$; κ is the Debye-Hückel parameter, which is a function of ionic concentration in the bulk solution. When support electrolyte is present in bulk solution, κ can be treated as a constant.

Introducing Eq. (9.95) into Eq. (9.94) and using $\chi = \kappa x$ as the dimensionless distance, $\tau = \kappa^2 D_0 t$ as the dimensionless time, $f(\chi, \tau) = f(x, t)/[m/(1^2\kappa^{-1})] = f(x, t)/m\kappa$ is the dimensionless concentration (1^2 means the per unit area), there is:

$$f(\chi,\tau) = \frac{1}{2\sqrt{\pi\tau}} \exp\left\{-\frac{ZF}{RT}\left[\phi(0,\tau)\exp(-|\chi|) - \phi(0,0)\right] - \frac{\chi^2}{4\tau}\right\} \tag{9.96}$$

Form Eq. (9.96), if $\varphi(0, \tau)$ is known, the concentration distribution $f(\chi, \tau)$ can be calculated in the diffusion process.

Correspondingly, Eq. (9.95) can be written as:

$$\phi(\chi,\tau) = \phi(0,\tau)\exp(-|\chi|) \tag{9.97}$$

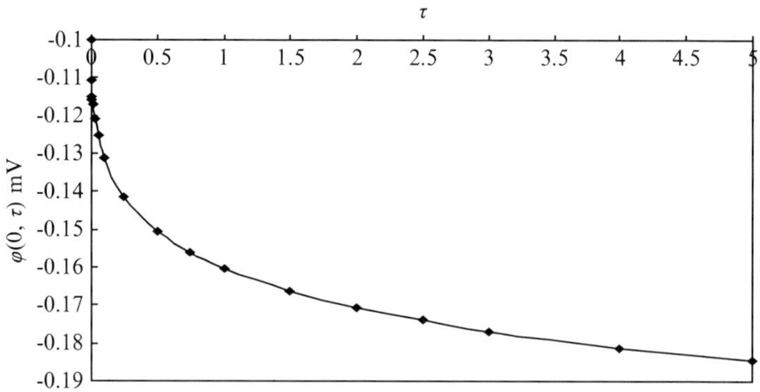

Fig. 9.3 Change in surface potential with diffusion time (τ) at $x = 0$ for $T = 298$ K and $Z = 1$

Considering the mass conservation, there is:

$$\int_{-\infty}^{0} f(-x, t)dx + \int_{0}^{\infty} f(x, t)dx = m \tag{9.98}$$

Considering $f(\chi, \tau) = f(x, t)/m\kappa$, Eq. (9.98) changes:

$$\int_{0}^{\infty} f(\chi, \tau)d\chi = \frac{1}{2} \tag{9.99}$$

Substituting Eq. (9.96) into Eq. (9.95), we have:

$$\int_{0}^{\infty} \frac{1}{\sqrt{\pi\tau}} \exp\left\{-\frac{ZF}{RT}[\phi(0, \tau)\exp(-|x|) - \phi(0, 0)] - \frac{\chi^2}{4\tau}\right\} d\chi = 1 \tag{9.100}$$

Under any given values of $\varphi(0, 0)$, the $\varphi(0, \tau)$ can be calculated using the Simpson rule (Li and Zhang 2001) for solving Eq. (9.90). Figure 9.3 is the relationship curve of $\varphi(0, \tau)$ *vs.* τ in the diffusion process as given $\varphi(0, 0) = -0.1$ mV. Figure 9.3 indicates that, through our new theory, the surface potential change with time in the diffusion process could be evaluated quantitatively.

As the $\varphi(0, \tau)$ values were obtained, introducing those value into Eq. (9.97), the relationship curve of $\varphi(\chi, \tau)$ *vs.* τ in the diffusion process could also be calculated. The results are shown in Fig. 9.4.

Introducing the obtained $\varphi(\chi, \tau)$ data into Eq. (9.96), the relationship curve of $f(\chi, \tau)$ vs. τ can be further obtained (Fig. 9.5).

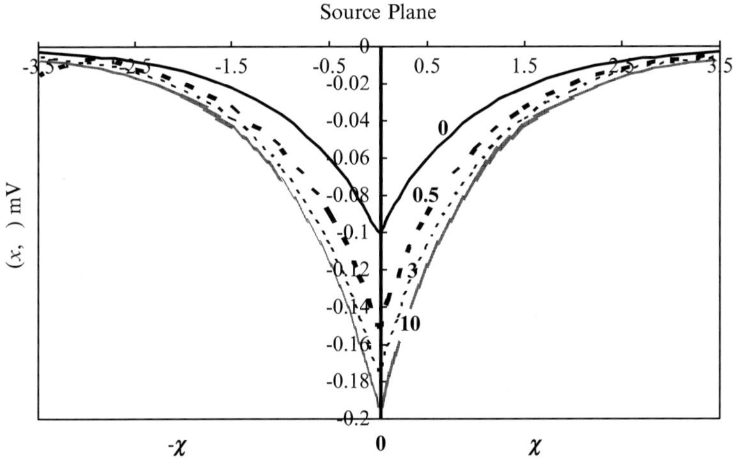

Fig. 9.4 The potential distribution as a function of time in the diffuse double layer. Numbers on the *curves* show the dimensionless time

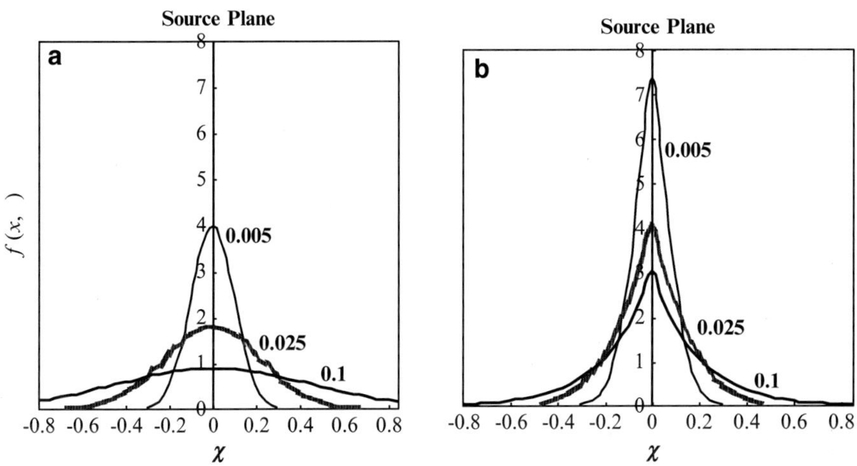

Fig. 9.5 Comparison of concentration distribution in space and time with the presence (**b**) and without the presence (**a**) of external electrical field. Numbers on the *curves* show the dimensionless time

9.1.4.2 An Application of the New Theory for Describing Possibility Evolution in Non-linear Diffusion System (Li et al. 2009a)

Supposing the diffusion particle is placed at $x = 0$ initially, and supposing the external potential is time-independent and can be expressed as:

$$w(x) = -\frac{5kT}{\pi}(\sin 2\pi x + 0.25 \sin 4\pi x) \tag{9.101}$$

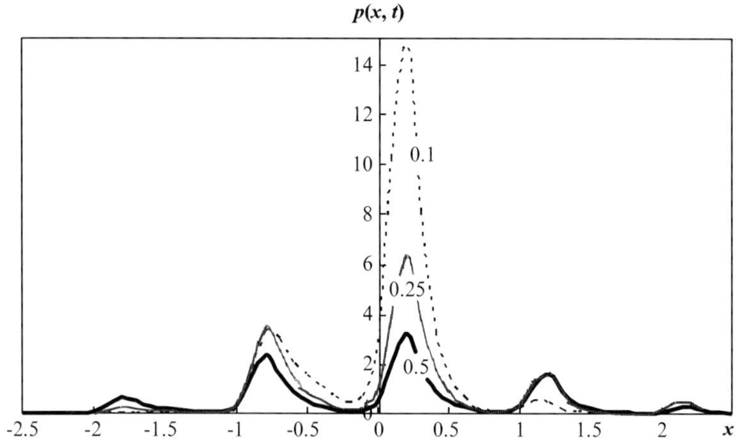

Fig. 9.6 The distribution of probability density at different times (The numbers on *curve* are the values of *Dt*)

Therefore, in the classic theory of the Fokker-Planck equation, the diffusion problem is the solution of the following equation:

$$\begin{cases} \frac{\partial}{\partial t} p\,(x,t) = D_0 \frac{\partial}{\partial x}\left\{\left[\frac{\partial}{\partial x} - \frac{5}{\pi}\frac{\partial}{\partial x}\,(\sin 2\pi x + 0.25 \sin 4\pi x)\right] p(x,t)\right\} \\ p(x,0) = \delta(x,0) \begin{cases} \infty & x = 0 \\ 0 & x \neq 0 \end{cases} \end{cases} \tag{9.102}$$

where $p(x, t)$ is the probability density.

Obviously, the exact solution of Eq. (9.102) would be extremely difficult as well as based on the classic theory. However, with our new theory, the problem could be changed to the solution of the following simple equation:

$$\begin{cases} \frac{\partial}{\partial t} A\,(x,t) = D \frac{\partial^2 A(x,t)}{\partial x^2} \\ A(x,0) = \delta(x,0) \begin{cases} \infty & x = 0 \\ 0 & x \neq 0 \end{cases} \end{cases} \tag{9.103}$$

where: $A(x, t) = p(x, t)\exp[-(5kT/\pi)(\sin 2\pi x + 0.25 \sin 4\pi x)]$.

Similar to the result of Eq. (9.94), the exact solution of that problem is:

$$p(x,t) = \frac{1}{2\sqrt{\pi D t}} e^{\frac{5}{\pi}(\sin 2\pi x + .25 \sin 4\pi x) - \frac{x^2}{4Dt}} \tag{9.104}$$

The evolution of the probability density of Eq. (9.102) can be illustrated as Fig. 9.6.

9.2 The Electrostatic Field Effect from Surface Charges on Ion Exchange Equilibrium

9.2.1 The Classic Equations for Describing Cation Exchange Equilibrium

Ion exchange is a basic process and universally exists in soil. Currently there are three different approaches to treat the ion exchange equilibrium between the solid/liquid interface: (1) treat ion exchange as an exchange reaction following the concept of thermodynamics, (2) treat ion exchange as a physicochemical process using the theory of Donnan equilibrium, and (3) treat ion exchange equilibrium using the double layer theory.

9.2.1.1 Exchange Reaction Approach to Cation Exchange Equilibrium

For *i*th cation species M_i with electric charges Z_i and *j*th cation species M_j with electric charges Z_j, the exchange reaction between M_i and M_j can be expressed as (Ogwada and Sparks 1986a; Morgan et al. 1995; Graul et al. 1999; Hui and Baker 2001; Berber-Mendoza et al. 2006):

$$Z_i(Soil - M_j)_{ex} + Z_j(M_i)_{aq} \Leftrightarrow Z_j(Soil - M_i)_{ex} + Z_i(M_j)_{aq} \tag{9.105}$$

where *Mi* and *Mj* are the exchangeable cations, and *aq* and *ex* refer to the aqueous solution and exchanger phases, respectively.

For this reaction equation, based on the mass action law of chemical reactions, the equilibrium was expressed as:

$$K_{eq} = \frac{a_{j(aq)}^{Z_i} a_{i(ex)}^{Z_j}}{a_{i(aq)}^{Z_j} a_{j(ex)}^{Z_i}} \tag{9.106}$$

where K_{eq} is the exchange equilibrium constant, a represents the activity.

The exchange equilibrium Eq. (9.106) is referred as the Kerr's equation (Tan 1982).

As $Z_i = 1$ and $Z_j = 2$, by taking the square root, Eq. (9.106) changes to (Tan 1982):

$$K_{eq} = \frac{\sqrt{a_{j(aq)}}a_{i(ex)}}{a_{i(aq)}\sqrt{a_{j(ex)}}} \tag{9.107}$$

Equation (9.107) is known as the Gapon equation.

In Eqs. (9.106) and (9.107), the difficulty in calculating K_{eq} is the estimation of the activity values of the cation species in the exchange phase. For avoiding this difficulty, Vanselow supposed the activity of adsorbed cations proportional to the mole fraction of total occupying cations, which means:

$$N_i = \frac{n_i}{n_i + n_j} \propto a_{i(ex)} \tag{9.108}$$

and

$$N_j = \frac{n_j}{n_i + n_j} \propto a_{j(ex)} \tag{9.109}$$

where N_i and N_j are the mole fraction of ith and jth cation species in the exchanger phase respectively, and n_i and n_j are the corresponding adsorbed quantity (mol/g) in exchanger phase.

Introducing Eqs. (9.108) and (9.109) into Eq. (9.106) gives:

$$K_V = \frac{a_{j(aq)}^{Z_i} n_i^{Z_j}}{a_{i(aq)}^{Z_j} n_j^{Z_i}} \frac{1}{(n_i + n_j)^{Z_j - Z_i}} \tag{9.110}$$

or

$$K_V = \frac{a_{j(aq)}^{Z_i} N_i^{Z_j}}{a_{i(aq)}^{Z_j} N_j^{Z_i}} \tag{9.111}$$

where K_V is referred as the Vanselow selectivity coefficient or the conditional equilibrium constant.

Based on the Eq. (9.111), Sposito developed a new equation for obtaining the activity coefficient of the cation species in the exchange phase and the exchange equilibrium constant through experiment.

The activity coefficient is defined as (Sposito 1981):

$$\gamma_j = \frac{a_{j(ex)}}{N_j} \text{ and } \gamma_i = \frac{a_{i(ex)}}{N_i} \tag{9.112}$$

Thus we have:

$$K_V = K_{eq} \frac{\gamma_j^{Z_i}}{\gamma_i^{Z_j}} \tag{9.113}$$

Taking the natural logarithm of both sides of Eq. (9.113), gives:

$$\ln K_V = \ln K_{eq} + Z_i \ln \gamma_j - Z_j \ln \gamma_i \tag{9.114}$$

Considering K_{eq} is always a constant, the differential of Eq. (9.114) gives:

$$d \ln K_V = Z_i d \ln \gamma_j - Z_j d \ln \gamma_i \tag{9.115}$$

An exchange phase could be regarded as a three-component system, thus according to the Gibbs-Duhem equation, we have:

$$n_i d\mu_i + n_j d\mu_j + n_w d\mu_w = 0 \tag{9.116}$$

where n_w refers to the mole number of water (mol/g), and μ refers to the chemical potential.

If the chemical potential of water does not change approximately as the exchange composition is varied, Eq. (9.116) reduces to:

$$n_i d\mu_i + n_j d\mu_j = 0 \tag{9.117}$$

Thus at constant T and P, there is

$$n_i d \ln \gamma_i + n_j d \ln \gamma_j = 0 \tag{9.118}$$

From Eqs. (9.115) and (9.118), we have:

$$d \ln K_V = Z_i d \ln \gamma_j + Z_j \frac{n_j}{n_i} d \ln \gamma_j = \frac{Z_i n_i + Z_j n_i}{n_i} d \ln \gamma_j \tag{9.119}$$

Since $\frac{Z_i n_i}{Z_i n_i + Z_j n_i} = E_i$ is the equivalent fraction of the ith cation species, Eq. (9.119) gives:

$$Z_i d \ln \gamma_j = E_i d \ln K_V \tag{9.120}$$

The corresponding solution of the ith cation species is:

$$Z_j d \ln \gamma_i = -(1 - E_i) d \ln K_V \tag{9.121}$$

Equation (9.120) can also be written as:

$$Z_i d \ln \gamma_j = d\,(E_i \ln K_V) - \ln K_V dE_i \tag{9.122}$$

Thus, as the reference state to be employed is the Gaines-Thomas reference state, the integration of Eq. (9.122) from the reference state, $\gamma_j = 1$, to the final state gives:

$$Z_i \int_0^{\ln \gamma_j} d \ln \gamma_j = \int_0^{E_i} d\,(E_i \ln K_V) - \int_0^{E_i} \ln K_V dE_i \tag{9.123}$$

In Eq. (9.123), we consider that, as $\gamma_j = 1$, $E_j = 1$ and $E_i = 0$.

The integration of Eq. (9.123) gives:

$$Z_i \ln \gamma_j = E_i \ln K_V - \int_0^{E_i} \ln K_V dE_i \tag{9.124}$$

Correspondingly, the integration of Eq. (9.121) would be expressed as:

$$Z_j \int_0^{\ln \gamma_i} d \ln \gamma_i = -\int_1^{E_i} d\left[(1 - E_i)\ln K_V\right] - \int_1^{E_i} \ln K_V dE_i \tag{9.125}$$

The result of Eq. (9.125) is:

$$Z_j \ln \gamma_i = -(1 - E_i)\ln K_V - \int_1^{E_i} \ln K_V dE_i \tag{9.126}$$

Introducing Eqs. (9.124) and (9.126) into Eq. (9.114), there is:

$$\begin{aligned} \ln K_{eq} &= \int_0^{E_i} \ln K_V dE_i - \int_1^{E_i} \ln K_V dE_i \\ &= \int_0^1 \ln K_V dE_i \end{aligned} \tag{9.127}$$

Equations (9.124), (9.126) and (9.114) show that the exchange constant and the activity coefficient of the cation species in the exchange phase could be experimentally determined.

For exchange adsorption by electrostatic force (physical adsorption), however, the adsorbed ions do not change their chemical nature. Thus, for a given cation species in exchange phase and solution phase, the reference states for the two phases can be taken as the same. For this case, the exchange equilibrium constant K_{eq} will equal to 1.

9.2.1.2 Donnan Equilibrium Approach to Cation Exchange Equilibrium

We know the distribution of the adsorbed ions as a function of distance from clay particle surface obeys Boltzmann distribution. However, as the distribution curve is very steep, one can approximate the space between the distribution curve and the particle surface as a micellar, and the remainder of the liquid is regarded as an intermicellar. Thus the micellar and the intermicellar form an Donnan system (Bolt 1955), and the exchange equilibrium of Eq. (9.105) can be expressed as (Tan 1982; Eriksson 1952; Jacob and Reddy 1991; Tanioka et al. 1998; Biesuz et al. 2001):

$$\frac{a_{j(aq)}^{Z_i} a_{i(ex)}^{Z_j}}{a_{i(aq)}^{Z_j} a_{j(ex)}^{Z_i}} = 1 \tag{9.128}$$

Thus by treating the cation exchange equilibrium as a Donnan equilibrium, the exchange equilibrium constant always equals 1. In the following we will show that, the Donnan equation is actually a natural result of the Boltzmann distribution of adsorbed ions.

Considering an exchange equilibrium between a monovalent cation species (M_i) and a divalent cation species (M_j)

$$(Soil - M_j)_{ex} + 2(M_i)_{aq} \Leftrightarrow 2(Soil - M_i)_{ex} + (M_j)_{aq} \tag{9.129}$$

Based on Eq. (9.128), Eriksson derived (Eriksson 1952):

$$\frac{a_{j(aq)} n_i^2 CEC}{a_{i(aq)}^2 n_j [n_i + 2n_j]} = \frac{\gamma_j}{\gamma_i^2} = K \tag{9.130}$$

where *CEC* is the cation exchange capacity.

9.2.1.3 Cation Exchange Equilibrium Based on the Double Layer Theory

For the exchange reaction as shown in Eq. (9.129), Eriksson established an exchange equilibrium equation based on the double layer theory, which can be expressed as (Bolt 1955; Eriksson 1952):

$$\frac{n_i}{n_i + n_j} = \frac{a_{i(aq)}}{\sqrt{\kappa' \sigma_0^2 a_{j(aq)}}} \sinh^{-1} \frac{\sqrt{\kappa' \sigma_0^2 a_{j(aq)}}}{a_{i(aq)} + 4a_{j(aq)} v_c} \tag{9.131}$$

where σ_0 is the surface charge density; $\kappa' = 8\pi v F^2 / \varepsilon RT$, v is the valence of the cation; $v_c = \cosh(ve\varphi_c/kT)$, φ_c is the potential in the central plane midway between clay plates, and for most practical purposes $v_c \approx 1$ (Eriksson 1952).

As $a_i \approx f_i$, and $a_j \approx f_j$, Eq. (9.131) can be changed to:

$$\frac{\Gamma_i}{\Gamma} = \frac{f_i / \sqrt{f_j}}{\Gamma \sqrt{\kappa'}} \sinh^{-1} \frac{\Gamma \sqrt{\kappa'}}{f_i / \sqrt{f_j} + 4v_c \sqrt{f_j}} \tag{9.132}$$

where $\Gamma_i = n_i/S$, $\Gamma = (n_i + n_j)/S$, and S is the specific surface area.

Using the experimental data of the exchange equilibrium data between Na-Ca-illite and solution, Bolt found that, Eq. (9.132) could give a quantitative description of cation exchange equilibrium, implying the cation exchange would be absolutely a Coulombic effect, as shown in Fig. 9.7.

Bolt further used Eq. (9.132) to calculate the Γ_i/Γ_j, and then, by plotting this ratio against $f_i/f_j^{0.5}$, found that the calculated result closely agree with the experimental data (Bolt 1955). From this Bolt obtained the following relationship:

$$\frac{n_i}{n_j} = 0.0125 f_i \sqrt{f_j} \tag{9.133}$$

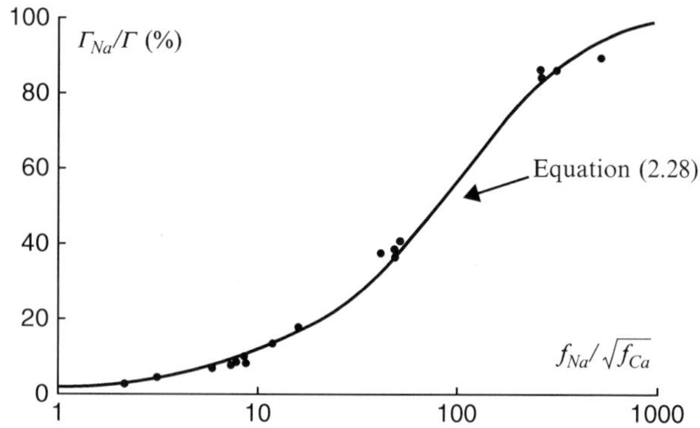

Fig. 9.7 The comparison between theoretical calculation from Eq. (9.132) and the experimental results of the exchange equilibrium between Na-Ca-illite and solution (Bolt 1955)

This equation is identical with the Gapon equation. But here the exchange equilibrium constant could be theoretically predicted.

In the above classic equations for describing cation exchange, only the Eriksson equation has a distinct physical meaning, therefore the cation distribution coefficient of each cation species between exchanger phase and solution phase could be theoretically calculated. But the Eriksson equation is too complex in mathematic formula. For other equations, this calculation could not be done because the activity coefficients of the adsorbed cations are unknown (Hui and Baker 2001; Berber-Mendoza et al. 2006; Biesuz et al. 2001; De Bokx and Boots 1989). For the Sposito equation, even though the corresponding theoretical calculation could not be done, the activity coefficient (average) of adsorbed cation species and the cation distribution coefficient or cation exchange equilibrium constant could be experimentally determined. For the Donnan equilibrium equation, just the ratio of the activity coefficient of two cation species in exchanger phase could be obtained experimentally. But for the Kerr's equation and the Gapon's equation, even the experimental determination of the exchange equilibrium constants were difficult.

In soil science, all the above classic equations for describing cation exchange equilibrium has been applied. However, the physical mechanisms of cation exchange for those equations are different. For the Kerr's equation, the Gapon's equation and the Sposito equation, the cation exchange was treated similarly to a chemical reaction. For the Donnan equation, the exchange was treated as a Donnan equilibrium; and for the Eriksson equation, the exchange was absolutely taken as a Coulombic effect. For reaction mechanisms of cation exchange, the exchange process and the diffusion process were considered as two separate processes, and usually the reaction rate is limited by diffusion (Ogwada and Sparks 1986b; Tang and Sparks 1993; Suresh et al. 2004).

It is well known that, for the exchange adsorption, the electrostatic force will be the sole source of adsorption. For a electrostatic force adsorption, the electrostatic field strength around soil particles, the charge number of cation species, the ionic diameter and the hydration diameter of cation species will have significant effects on the cation exchange equilibrium. Unfortunately, at present, researches seldom pay much attention to these important issues in cation exchange equilibrium.

9.2.2 The Identity of Exchange Adsorption and Diffusion in the External Electric Field of Cations

Based on the generalized linear equation of diffusion in external fields and non-ideal system, from Eq. (9.34) the flux equation of diffusion in an external electrostatic field of diffuse layer of soil particles can be expressed as:

$$J(x,t) = -D\nabla\left[f(x,t)e^{\frac{ZF\phi(x,t)}{RT}}\right] \tag{9.132a}$$

In the absence of an external electric field, $j(x, t) = 0$ means $\nabla f(x, t) = 0$, this is an equilibrium state of diffusion under an isothermal condition, as indicated by Fick's law. While in the presence of an external field, $J(x, t) = 0$ also indicates an equilibrium state if $\nabla\left[f(x,t)e^{\frac{ZF\phi(x,t)}{RT}}\right] = 0$ from Eq. (9.132), even if $\nabla f(x, t) \neq 0$. Because the Fick's flux equation is a special case of Eq. (9.132) with $\phi(x, t) = 0$, a general criterion for judging the equilibrium state of diffusion is $\nabla A\,(x, t = \infty) = \nabla\left[f(x,t)e^{\frac{ZF\phi(x,\infty)}{RT}}\right] = 0$. Consequently, the criterion for judging the equilibrium of diffusion for a multi-components system containing different types of cations is:

$$\nabla A_i\,(x,\infty) = \nabla\left[f_i(x,t)e^{\frac{Z_i F\phi(x,\infty)}{RT}}\right] = 0 \tag{9.133a}$$

It is well known that, at the equilibrium state of exchange adsorption, the ionic distribution in the diffuse layer will obey the Boltzmann equation:

$$f_i(x,\infty) = f_i^0 \cdot e^{-\frac{Z_i F\phi(x,\infty)}{RT}} \tag{9.134}$$

where f_i^0 is the concentration of ith ion species in bulk solution.

Rearranging Eq. (9.134) and then differentiating it, we have:

$$\nabla\left[f_i(x,\infty)\cdot e^{\frac{Z_i F\phi(x,\infty)}{RT}}\right] = \nabla A_i(x,\infty) = \nabla f_i^0 = 0 \tag{9.135}$$

Equations (9.133) describes the diffusion equilibrium and (9.135) describing the equilibrium of exchange adsorption. The identity of the two equations implies that the ion exchange of electrostatic adsorption occurring in the diffuse layer may be essentially a diffusion process. Therefore, the diffusion process and exchange process of *electrostatic adsorption* in the diffuse layer can be combined and treated as one process by using the apparent concentration, in which the driving force is the combination of the concentration gradient and potential gradient. Based on this concept, the kinetic equation of ion exchange in the diffuse layer might be theoretically derived directly from the diffusion equation of ions in an electric field, and this will be further discussed in the section of cation exchange kinetics.

If ion distribution in adsorption phase obeys the Boltzmann distribution equation, Eq. (9.134), the above demonstrate the exchange adsorption process is essentially a diffusion process of ions in electrostatic field. And here we will show that, the Boltzmann distribution equation actually implies the reference states for a given ion species in exchange phase and solution phase are really the same reference states.

The chemical potential of ith ion species in solution phase under isobaric and isothermal conditions is:

$$\mu_{i\,(\textbf{solution})} = \mu_{i\,0(solution)} + RT\ln f_i{}^0$$

where $\mu_{i(\text{solution})}$ is the chemical potential of the ith ion species in solution phase, $\mu_{i0(\text{solution})}$ is the chemical potential of the ith ion species at reference state in solution phase.

Correspondingly, the chemical potential of ith ion species in exchange phase under isobaric and isothermal conditions is:

$$\mu_{i\,(\textbf{exchange})} = \mu_{i\,0(\textbf{exchange})} + RT\ln f_i(x) + Z_i F\phi(x)$$

where $\mu_{i(\text{exchange})}$ is the chemical potential of the ith ion species in exchange phase, $\mu_{i0(\text{exchange})}$ is the chemical potential of the ith ion species at reference state in exchange phase.

At equilibrium, there is $\mu_{i(\text{exchange})} = \mu_{i(\text{solution})}$. Thus, if the Boltzmann distribution equation, Eq. (9.134), is correct, there must be $\mu_{i0(\text{solution})} = \mu_{i0(\text{exchange})}$.

9.2.3 Cation Exchange Equilibrium as Considering the Electrostatic Field in the Soil

9.2.3.1 Calculation of the Mean Activity Coefficient of Ions in Exchanger Phase

The mean concentration of ith cation species in the diffuse layer can be defined as (Li et al. 2004):

$$\tilde{f_i} = \frac{n_i}{S(1/\kappa)} = \kappa\int_0^{\frac{1}{\kappa}} f_i(x)dx \tag{9.136}$$

where $1/\kappa$ is the effective thickness of the diffuse layer, n_i is the adsorbed quantity of ith cation species in the diffuse layer.

According to the chemical thermodynamic theory, at equilibrium, we have:

$$f_i(x) = \frac{1}{\gamma_i(x)} e^{\frac{\mu_i - \mu_0}{RT}} \tag{9.137}$$

Where μ_0 is the reference potential of ith cation species. At equilibrium, both μ_i and $\mu_i{}^0$ are constant, therefore introducing (9.137) into (9.136) we have:

$$\tilde{f_i} = \kappa e^{\frac{\mu_i - \mu_{i0}}{RT}} \int_0^{\frac{1}{\kappa}} \frac{1}{\gamma_i(x)} dx \tag{9.138}$$

Here we define an average activity coefficient $\tilde{\gamma}_i$ as:

$$\frac{1}{\tilde{\gamma}_i} = \kappa \int_0^{\frac{1}{\kappa}} \frac{1}{\gamma_i(x)} dx \tag{9.139}$$

Thus Eq. (9.138) can be rewritten as:

$$\mu_i = \mu_i^0 + RT \ln \tilde{f_i} \tilde{\gamma}_i \tag{9.140}$$

Obviously, in the electrostatic field of diffuse layer, the mean activity coefficient of the ith cation species can also be written as:

$$\frac{1}{\tilde{\gamma}_i} = \kappa \int_0^{\frac{1}{\kappa}} \frac{1}{\gamma_i(x)} dx = \kappa \int_0^{\frac{1}{\kappa}} e^{-\frac{Z_i F \phi(x)}{RT}} dx \tag{9.141}$$

On the other hand, introducing Eq. (9.134) into Eq. (9.136), there is:

$$\tilde{f_i} = f_i{}^0 \kappa \int_0^{\frac{1}{\kappa}} e^{-\frac{Z_i F \phi(x)}{RT}} dx \tag{9.142}$$

From Eqs. (9.141) and (9.142), we get:

$$\tilde{f_i} \tilde{\gamma}_i = f_i^0 \tag{9.143}$$

The comparison of Eqs. (9.137) and (9.138) gives:

$$f_i(x)\gamma_i(x) = \tilde{f_i} \tilde{\gamma}_i = f_i^0 \tag{9.144}$$

According to Eq. (9.143), for a solution mixture of two types of electrolytes, when the diffusion reaches equilibrium, we have:

$$\begin{cases} \tilde{f_1} \tilde{\gamma}_1 = f_1^0 \\ \tilde{f_2} \tilde{\gamma}_2 = f_2^0 \end{cases} \tag{9.145}$$

where $i = 1$ or 2 represents 1th cation species or 2th cation species respectively.

Equation (9.145) implies that, for theoretical calculation of cation exchange, the key is how to calculate the mean activity coefficient of the ith cation species in the diffuse layer: $\tilde{\gamma}_i$.

It has been demonstrated that, as charged density of colloidal particles not exceeding 0.2–0.3 C/m^2, the simple Gouy-Chapman theory could give fairly reliable results (Sparks 1998a), which means the simple Gouy-Chapman theory could be applied to describe the diffuse layer of most types of colloidal particle in the soil quantitatively, because the surface charge density of soil particles is often lower than 0.3 C/m^2. For example, for kaolinite the surface charge density is about ~0.16 C/m^2, for montmorrilonite the surface charge density is about ~0.15 C/m^2, and for illite the surface charge density is about ~0.3 C/m^2. Bolt has demonstrated that, Na/Ca exchange in illite was totally a Coulombic effect. So, the Gouy-Chapman theory could be used to describe the equilibrium of cation's exchange adsorption in the soil.

1. *The calculation of $\tilde{\gamma}_i$ in an exchange equilibrium between two symmetric types of electrolytes.*

For an exchange equilibrium between two symmetric electrolytes, e.g. NaCl, KCl, based on the Poisson-Boltzmann equation, the expression $\varphi(x)$ can be written as (Sposito 1984):

$$\phi(x) = \frac{4RT}{ZF}\tanh^{-1}\left(he^{-\kappa x}\right) \tag{9.146}$$

where κ could be expressed as:

$$\kappa = \sqrt{\frac{8\pi F^2\left(\frac{1}{2}\sum_{i=1}^{2} f_{i0} Z_i^2\right)}{\varepsilon RT}} \tag{9.147}$$

and h is a constant expressed as

$$h = \tanh\left(\frac{ZF\phi_0}{4RT}\right) \tag{9.148}$$

Introducing Eq. (9.146) into Eq. (9.141), we get:

$$\frac{1}{\tilde{\gamma}_i} = \kappa\int_0^{\frac{1}{\kappa}}\left(\frac{1-he^{-\kappa x}}{1+he^{-\kappa x}}\right)^2 dx \tag{9.149}$$

Giving $y = e^{-\kappa x}$, Eq. (9.149) changes to

$$\frac{1}{\tilde{\gamma}_i} = -\int_1^{e^{-1}}\frac{(1-hy)^2}{y(1+hy)^2}dy \tag{9.150}$$

The integration of Eq. (9.150) gives:

$$\frac{1}{\tilde{\gamma}_i} = 1 + \frac{4}{1+h} - \frac{1}{1+e^{-1}h} \tag{9.151}$$

On the other hand, as $|\varphi_0|$ is large enough (usually $|Z\varphi_0| > 100$ mV (Li et al. 2004)), from Eq. (9.148) we may have $h \to -1$. Thus Eq. (9.151) can be simplified to:

$$\frac{1}{\tilde{\gamma}_i} = \frac{4}{1+h} \tag{9.152}$$

On the other hand, Eq. (9.148) can be written as

$$\phi_0 = -\frac{2RT}{ZF}\ln\frac{1-h}{1+h} \approx -\frac{2RT}{ZF}\ln\frac{2}{1+h} \tag{9.153}$$

Thus the combination of Eqs. (9.152) and (9.153) gives

$$\tilde{\gamma}_i = \frac{1}{2}\exp\left[\frac{Z_i F\phi_0}{2RT}\right] \tag{9.154}$$

Showing that, under an isothermal condition, the surface potential and the valence of the cation species determine the average activity coefficient of the adsorbed cations in the diffuse layer.

2. *The calculation of* $\tilde{\gamma}_i$ *in an exchange equilibrium between two 1:2 types of electrolytes*

For an exchange equilibrium between two 1:2 types of electrolytes, e.g. Na_2SO_4, K_2SO_4, based on the Poisson-Boltzmann equation, the expression of $\varphi(x)$ can be written as (Sposito 1984):

$$\phi(x) = -\frac{RT}{F}\ln\left[1 + \frac{6be^{\kappa x}}{(6be^{\kappa x}-1)^2}\right] \tag{9.155}$$

where

$$b = \frac{\sqrt{1+2e^{-\frac{F\phi_0}{RT}}}+3}{\sqrt{1+2e^{-\frac{F\phi_0}{RT}}}-3} \tag{9.156}$$

If a cation is positively adsorbed, introducing Eq. (9.155) into Eq. (9.141), we get:

$$\frac{1}{\tilde{\gamma}_i} = \kappa\int_0^{\frac{1}{\kappa}}\left[1 + \frac{6be^{\kappa x}}{(6be^{\kappa x}-1)^2}\right]dx \tag{9.157}$$

Taking the same procedure as obtaining Eq. (9.151), the integration of Eq. (9.157) gives:

$$\frac{1}{\tilde{\gamma}_i} = 1 + \frac{6}{b-1} - \frac{6}{eb-1} \tag{9.158}$$

On the other hand, Eq. (9.156) can be rewritten as

$$\phi_0 = -\frac{RT}{F} \ln \frac{b^2 + 4b + 1}{(b-1)^2} \tag{9.159}$$

Similarly, as $Z\varphi_0 > 100$ mV, the combination of Eqs. (9.158) and (9.160) gives

$$\tilde{\gamma}_i = \frac{1}{\sqrt{6}} \exp\left[\frac{F\phi_0}{2RT}\right] \tag{9.160}$$

3. *The calculation of* γ_i *in an exchange equilibrium between two 2:1 types of electrolytes*

For an exchange equilibrium between two 2:1 types of electrolytes, e.g. $CaCl_2$ and $MgCl_2$, based on the Poisson-Boltzmann equation, the expression of $\varphi(x)$ can be written as (Sposito 1984):

$$\phi(x) = \frac{RT}{F} \ln\left[\frac{3}{2}\tanh^2\left(\frac{1}{2}\kappa x + s\right) - \frac{1}{2}\right] \tag{9.161}$$

where

$$s = \tanh^{-1}\sqrt{\frac{1}{3}\left(1 + 2e^{\frac{F\phi_0}{RT}}\right)} \tag{9.162}$$

If cation is positively adsorbed, introducing Eq. (9.161) into Eq. (9.141), we get:

$$\frac{1}{\tilde{\gamma}_i} = \kappa \int_0^{\frac{1}{\kappa}} \left(\frac{e^{2s}e^{2\kappa x} + 2e^{\kappa x} + e^{-2s}}{e^{2s}e^{2\kappa x} - 4e^{\kappa x} + e^{-2s}}\right)^2 dx \tag{9.163}$$

Given $y = e^{\kappa x}$, Eq. (9.163) changes to

$$\frac{1}{\tilde{\gamma}_i} = \int_1^e \left[\frac{1}{y} + \frac{12}{e^{2s}y^2 - 4y + e^{-2s}} + \frac{36y}{(e^{2s}y^2 - 4y + e^{-2s})^2}\right] dy \tag{9.164}$$

Since

$$\int \frac{1}{y} dy = \ln y \tag{9.165}$$

$$\begin{aligned}
&\int \frac{12}{e^{2s}y^2 - 4y + e^{-2s}}dy \\
&= \frac{12}{e^{2s}} \int \frac{dy}{\left[y - \left(2 + \sqrt{3}\right)e^{-2s}\right]\left[y - \left(2 - \sqrt{3}\right)e^{-2s}\right]} \\
&= \frac{6}{\sqrt{3}} \int \frac{dy}{y - \left(2 + \sqrt{3}\right)e^{-2s}} - \frac{6}{\sqrt{3}} \int \frac{dy}{y - \left(2 - \sqrt{3}\right)e^{-2s}} \\
&= 2\sqrt{3}\ln \frac{y - \left(2 + \sqrt{3}\right)e^{-2s}}{y - \left(2 - \sqrt{3}\right)e^{-2s}}
\end{aligned} \tag{9.166}$$

and

$$\begin{aligned}
&\int \frac{36y}{\left(e^{2s}y^2 - 4y + e^{-2s}\right)^2}dy \\
&= \frac{36}{e^{4s}} \int \frac{ydy}{\left[y - \left(2 + \sqrt{3}\right)e^{-2s}\right]^2\left[y - \left(2 - \sqrt{3}\right)e^{-2s}\right]^2} \\
&= \frac{36}{e^{4s}} \left\{ \int \frac{-\sqrt{3}e^{4s}dy}{18\left[y - \left(2 + \sqrt{3}\right)e^{-2s}\right]} + \int \frac{\left(2 + \sqrt{3}\right)e^{-2s}dy}{12\left[y - \left(2 + \sqrt{3}\right)e^{-2s}\right]^2} \right. \\
&\quad \left. + \int \frac{\sqrt{3}e^{4s}dy}{18\left[y - \left(2 - \sqrt{3}\right)e^{-2s}\right]} + \int \frac{\left(2 - \sqrt{3}\right)e^{-2s}dy}{12\left[y - \left(2 - \sqrt{3}\right)e^{-2s}\right]^2} \right\} \\
&= -2\sqrt{3}\ln \frac{y - \left(2 + \sqrt{3}\right)e^{-2s}}{y - \left(2 - \sqrt{3}\right)e^{-2s}} - \frac{3\left(2 + \sqrt{3}\right)}{e^{2s}\left[y - \left(2 + \sqrt{3}\right)e^{-2s}\right]} \\
&\quad - \frac{3\left(2 - \sqrt{3}\right)}{e^{2s}\left[y - \left(2 - \sqrt{3}\right)e^{-2s}\right]}
\end{aligned} \tag{9.167}$$

the integration of Eq. (9.164) gives

$$\begin{aligned}
\frac{1}{\tilde{\gamma}_i} = 1 &- \frac{3\left(2 + \sqrt{3}\right)}{e^{2s+1} - \left(2 + \sqrt{3}\right)} + \frac{3\left(2 + \sqrt{3}\right)}{e^{2s} - \left(2 + \sqrt{3}\right)} \\
&- \frac{3\left(2 - \sqrt{3}\right)}{e^{2s+1} - \left(2 - \sqrt{3}\right)} + \frac{3\left(2 - \sqrt{3}\right)}{e^{2s} - \left(2 - \sqrt{3}\right)}
\end{aligned} \tag{9.168}$$

On the other hand, Eq. (9.162) can be rewritten as

$$\phi_0 = -\frac{RT}{F} \ln \frac{\left(e^{2s}+1\right)^2}{\left(e^{2s}\right)^2 - 4e^{2s} + 1} \tag{9.169}$$

Similarly, as $|Z\varphi_0| > 100$ mV, the combination of Eqs. (9.168) and (9.169) gives

$$\tilde{\gamma}_i = \frac{1}{\sqrt{3}} \exp\left[\frac{Z_i F \phi_0}{2RT}\right] \tag{9.170}$$

In the derivation of Eqs. (9.154), (9.160) and (9.170), charged cations were considered as point charges, which implies that different ions with equivalent charges (e.g. Cl^- and $NO_3{}^-$, or K^+ and Na^+) were treated equally.

Those equations are applicable for calculation of the mean activity coefficient of cations in the exchanger phase for a solution mixture of two 1:1 types of electrolytes, two 2:2 types of electrolytes, two 1:2 types of electrolytes, or two 2:1 types of electrolytes, respectively. On the other hand, because Eq. (9.154) is applicable for any symmetric electrolyte solution, we can expect the mathematical form will be the same as Eq. (9.155) for calculating the mean activity coefficient of each cation species for a solution mixture of two different types of symmetric electrolytes (e.g. a mixture of 1:1 and 2:2 electrolytes).

The exact solutions for the calculation of the mean activity coefficient of each cation species in the exchange phase for a solution mixture of two different types of electrolytes are difficult to obtain. Considering a solution mixture of 1:2 (e.g. K_2SO_4) and 2:1 types of electrolytes (e.g. $MgCl_2$): even though Eqs. (9.160) and (9.170) are respectively applicable only to the two separate systems of 1:2 and 2:1 electrolytes, we can use them to define an arithmetic mean of $\tilde{\gamma}_i$ if the values of $Z_i\varphi_0$ are the same for the two separate systems:

$$\tilde{\gamma}_i = \frac{f_i^0}{\tilde{f}_i} = \frac{1}{2}\left(\frac{1}{\sqrt{3}} + \frac{1}{\sqrt{6}}\right) \exp\left[\frac{Z_i F \phi_0}{2RT}\right] \approx \frac{1}{2} \exp\left[\frac{Z_i F \phi_0}{2RT}\right] \tag{9.171}$$

When the difference of mole number is ***not very significant*** for the two types of electrolytes (K_2SO_4 and $MgCl_2$), this solution mixture can also be approximately taken as a solution mixture of 1:1 type (KCl) and 2:2 type ($MgSO_4$) electrolytes. Therefore Eq. (9.154) for symmetric electrolytes is also approximately applicable for this solution mixture of 1:2- and 2:1-type electrolytes. A comparison between Eqs. (9.154) and (9.170) shows that the defined arithmetic mean of $\tilde{\gamma}_i$ can be used to calculate the mean activity coefficient of cation in the exchanger phase for a solution mixture of two different types of electrolytes.

The above discussion indicates that a generalized expression for calculating the mean activity coefficient of cation in exchange phase in any solution mixture can be approximately expressed as (Li and Wu 2007b):

$$\tilde{\gamma}_i = \frac{f_i^0}{\tilde{f}_i} = \frac{1}{2} A_{ij} \exp\left[\frac{Z_i F \phi_0}{2RT}\right] \tag{9.172}$$

Table 9.1 Values of A_{ij} for common solution mixtures containing different types of electrolytes (Li and Wu 2007b)

Types of solutions	Two types of symmetric electrolytes	1:1 and 2:1	1:1 and 1:2	1:2 and 2:2
A_{ij}	$\frac{1}{2}+\frac{1}{2}=1$	$\frac{1}{2}+\frac{1}{\sqrt{3}}$	$\frac{1}{2}+\frac{1}{\sqrt{6}}$	$\frac{1}{2}+\frac{1}{\sqrt{6}}$
Types of solutions	**2:1 and 2:2**	**1:2 and 2:1**	**1:2 and 1:2**	**2:1 and 2:1**
A_{ij}	$\frac{1}{2}+\frac{1}{\sqrt{3}}$	$\frac{1}{\sqrt{3}}+\frac{1}{\sqrt{6}}$	$\frac{1}{\sqrt{6}}+\frac{1}{\sqrt{6}}=\frac{2}{\sqrt{6}}$	$\frac{1}{\sqrt{3}}+\frac{1}{\sqrt{3}}=\frac{2}{\sqrt{3}}$

where A_{ij} is a constant. Based on the above discussion, a list of A_{ij} values for different solution mixtures of two types of electrolytes is presented in Table 9.1.

9.2.3.2 Cation Exchange Equilibrium Equation Based on the Diffusion Mechanism in External Electric Field

Considering the case of cation diffusion equilibrium in a solution mixture of two types of electrolytes, introducing Eq. (9.172) into Eq. (9.145), we have:

$$\frac{\left(f_1^0\right)^{Z_2}\cdot \tilde{f}_2^{Z_1}}{\tilde{f}_1^{Z_2}\cdot \left(f_2^0\right)^{Z_1}}=\left(\frac{2}{A_{ij}}\right)^{Z_1-Z_2} \tag{9.173}$$

According to the definition of average concentration in diffuse layer, Eq. (9.173) can also be written as:

$$\frac{\left(f_1^0\right)^{Z_2}{n_2}^{Z_1}}{n_1^{Z_2}\left(f_2^0\right)^{Z_1}}\left(\frac{\kappa}{S}\right)^{Z_1-Z_2}=\left(\frac{2}{A_{ij}}\right)^{Z_1-Z_2} \tag{9.174}$$

Considering: $S\sigma_0 = Z_1n_1 + Z_2n_2$
and

$$\frac{1}{\kappa}=\sqrt{\frac{\varepsilon RT}{8\pi F^2\left(\frac{1}{2}\sum f_i^0 Z_i^2\right)}}$$

Equation (9.174) changes to:

$$\frac{{n_2}^{Z_1}}{n_1^{Z_2}(Z_1n_1+Z_2n_2)^{Z_1-Z_2}}=\left(\frac{2}{A_{ij}}\right)^{Z_1-Z_2}\left(\frac{\varepsilon RT}{8\pi F^2{\sigma_0}^2}\right)^{\frac{Z_1-Z_2}{2}} \cdot\left[\frac{\left(f_2^0\right)^{Z_1}}{\left(f_1^0\right)^{Z_2}\left(\frac{1}{2}\sum f_i^0 Z_i^2\right)^{\frac{Z_1-Z_2}{2}}}\right] \tag{9.175}$$

The exchange equilibrium equation, Eq. (9.175), was derived from the diffusion equation of cations in the electric field of the diffuse layer. This equation shows that, if the surface charge density of solid particles is known in advance, the adsorbed quantities of cations or the cation distribution between exchanger phase and solution phase can be theoretically calculated.

From Eqs. (9.173) and (9.174), we also have:

$$\frac{\left(a_1^0\right)^{Z_2} \cdot \tilde{a}_2^{Z_1}}{\tilde{a}_1^{Z_2} \cdot \left(a_2^0\right)^{Z_1}} = 1 \tag{9.176}$$

Here we consider $a_i^0 = 1 \cdot f_i^0$ for an ideal bulk solution. Equation (9.176) is identical to the Donnan equilibrium equation, but here it was obtained by treating cation exchange as a diffusion process in the external electric field of diffuse layer. This might imply that the cation exchange kinetics occurring in the diffuse layer could indeed be treated as a diffusion process.

For different cation exchange system of two types of electrolytes, introducing the values of Z_1, Z_2 and A_{ij} into Eq. (9.175), the corresponding ion distribution equations for cation exchange equilibrium can be obtained.

9.2.3.3 An Application of Equation (9.175)

The cation exchange between $Ca^{2+}(CaCl_2)$ and $Na^+(NaCl)$ in a solid/solution suspension, the equilibrium Eq. (9.175) of cation exchange between 2:1 and 1:1 types electrolytes and in a dilute bulk solution and well dispersed clay system is (Li and Wu 2007b):

$$\frac{{n_{Na}}^2}{n_{Ca}\left(2n_{Ca} + n_{Na}\right)} = \frac{4\sqrt{3}}{2+\sqrt{3}}\frac{1}{\sigma} \cdot \left[\left(\frac{\varepsilon RT}{8\pi F^2}\right)^{\frac{1}{2}} \frac{\left(f_{Na}^0\right)^2}{\left(f_{Ca}^0\right)\left(3f_{Ca}^0 + f_{Na}^0\right)^{\frac{1}{2}}}\right] \tag{9.177}$$

where ${f_{Ca}}^0$ and ${f_{Na}}^0$ are the concentrations of Ca^{2+} and Na^+ (mol/l), respectively, in the bulk solution; n_{Na} and n_{Ca} are the respective numbers of mole of the total adsorbed Na^+ and Ca^{2+} per gram of solid. At $T = 298$ K, from Eq. (9.177) there is

$$\frac{{n_{Na}}^2}{n_{Ca}\left(n_{Na} + 2n_{Ca}\right)} = \frac{4\sqrt{3}}{2+\sqrt{3}}\frac{1}{\sigma} \cdot \left[3 \cdot 10^{-9} \frac{\left(f_{Na}^0\right)^2}{f_{Ca}^0\sqrt{f_{Na}^0 + 3f_{Ca}^0}}\right] \tag{9.178}$$

By taking the common logarithm of both sides of Eq. (9.178), there is:

$$\log\left[\frac{{n_{Na}}^2}{n_{Ca}\left(n_{Na} + 2n_{Ca}\right)}\right] = \log\left(\frac{4\sqrt{3}}{2+\sqrt{3}}\frac{1}{\sigma}\right) + \log\left[3 \cdot 10^{-9} \frac{\left(f_{Na}^0\right)^2}{f_{Ca}^0\sqrt{f_{Na}^0 + 3f_{Ca}^0}}\right] \tag{9.179}$$

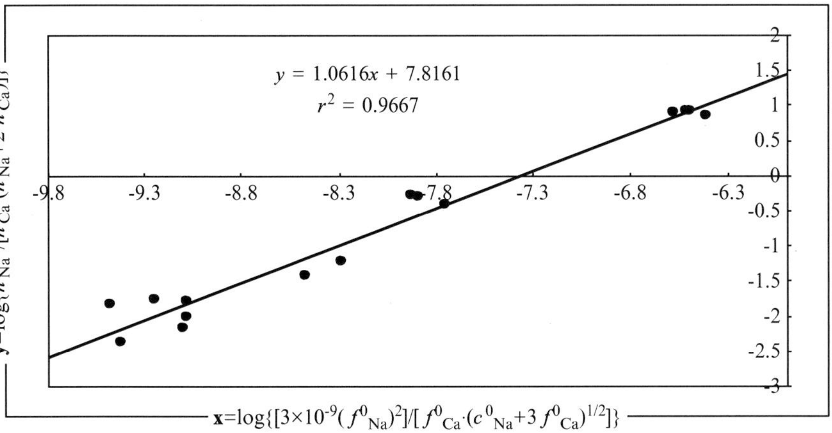

Fig. 9.8 The experimental data points (Bolt 1955) and the linear regression (*line*) of the relationship between $\lg\{n_{\mathrm{Na}}{}^2/[n_{\mathrm{Ca}}\cdot(n_{\mathrm{Na}}+2\,n_{\mathrm{Ca}})]\}$ and $\lg\{[3\times10^{-9}(f_{\mathrm{Na}}{}^0)^2]/[f_{\mathrm{Ca}}{}^0\cdot(f_{\mathrm{Na}}{}^0+3f_{\mathrm{Ca}}{}^0)^{1/2}]\}$ for cation exchange at equilibrium

From Eq. (9.179), there will be: (1) if we determine the adsorbed quantity of Na^+ and Ca^{2+} at different concentrations of Na^+ and Ca^{2+} in the bulk solution at equilibrium, a linear relationship between $\log\{n_{Na}{}^2/[n_{Ca}\,(n_{Na}+2n_{Ca})]\}$ and $\log\{3\cdot10^{-9}\left(f_{Na}^0\right)^2/[f_{Ca}^0\sqrt{f_{Na}^0+3f_{Ca}^0}]\}$ can be obtained; (2) the slope of the straight line would be equal to 1; and (3) the surface charge density of the solid particles can be calculated from the intercept obtained from the linear regression.

Figure 9.8 shows the results using the experimental data of cation exchange equilibrium between Na-Ca-illite and solution of NaCl and $CaCl_2$ by Bolt (1955), and the linear regression line obtained by fitting Eq. (9.179) to the data. It shows that: (1) the experimental data fits a straight line, the value of r^2 is 0.9667; (2) the slope of the straight line is 1.0616, which is close to the theoretically value of 1.0000; and (3) the value of the intercept "$\log[(4\cdot3^{1/2})/(2+3^{1/2})\sigma]$" is 7.816, which means the surface charge density of the illite is $\sigma = 2.84\times10^{-7}$mmol$_c$/cm^2. This value of surface charge density was close to the value (3.0×10^{-7} mmol$_c$/cm^2) that Bolt determined independently by a negative adsorption method (Bolt 1955).

Equation (9.179) was based on the cation diffusion analyses in an electrostatic field. The experimental result showed that the cation exchange adsorption occurring in the diffuse layer could be treated as a diffusion process driven by the apparent concentration gradient of ions in the external electric field from the solid particle surface. Therefore the exchange process of electrostatic adsorption and the diffusion process in the external electric field are essentially the same process, and can be treated as one.

Here we should emphasize that the above approach was based on two important assumptions: (1) the exchange process of cations is merely an electrostatic adsorption process occurring in the diffuse layer, and (2) because the Poisson-Boltzmann equation was used to describe the concentration and potential distribution in the diffuse layer, the assumptions for this equation must be valid. However, when the Poisson-Boltzmann equation is to be used to describe cation exchange, there are two defects: (1) in the adopted Boltzmann distribution equation of the Gouy-Chapman theory, the ionic interaction energy in bulk solution is not considered, and (2) all the ionic species were assumed to be point charges. Unfortunately this assumption is incorrect in cation exchange. For example K^+ and Na^+ have the same charge number, and there were not any different in the Poisson-Boltzmann equation. However, in cation exchange equilibrium, we have known that, the adsorption selectivity of K^+ is greatly superior to Na^+, because the hydration radius of Na^+ is larger than that of K^+.

9.2.3.4 Cation Exchange Equilibrium Equation as Considering the Effects of Ionic Hydration and Ionic Interaction

As cation exchange adsorption is essentially an electrostatic interaction, cation exchange rate and exchange equilibrium will be controlled by the electrostatic interaction energies of the participating exchange cation species in the electrostatic field of soil. In a given strength of electrostatic field of cation exchange systems, two factors will influence the electrostatic interaction energy: the charge number of a cation species and the hydration diameter of the cation species. At present, however, all the exchange equilibrium equations mentioned above did not consider the hydration effect of cation species in cation exchange.

As early as 1932, Jenny found the lyotropic series of the selectivity of a purified swelling clay mineral for alkali cations (Jenny 1932, 1936):

$$Cs^+ > Rb^+ > K^+ > Na^+ > Li^+$$

and the explanations for this ranking are myriad. The most accepted explanation for the series is that cation exchange selectivity results from cation hydrated radius (Pauley 1953; Shainberg and Kemper 1966). A larger hydrated radius of cation means that the cationic center of charge is farther from the clay surface so the clay-cation electrostatic interaction is weaker (Pauley 1953; Shainberg and Kemper 1966).

For two cation species with the same charge number, the cation species with a larger hydration radius must have weaker electrostatic adsorption force in the electrostatic field from the charged particle surface, because the charge density of this cation species is lower. Also, because all the adsorbed cations are randomly distributed in the diffuse layer, for a cation species with larger hydration radius the minimum distance between the cationic center and the particle surface is larger, and considering the minimum distance corresponding to a maximum electrostatic force for the cation species to be adsorbed on particle surface, the larger the minimum distance means the lower probability to be adsorbed on the particle

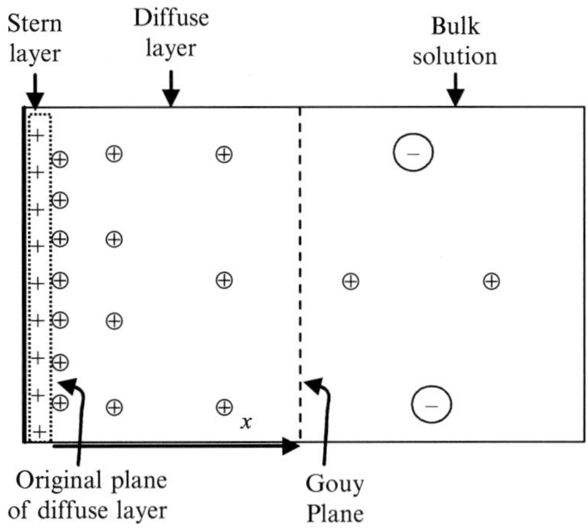

Fig. 9.9 A conceptual diagram of the diffuse layer for a charged particle in solution

surface. Therefore, the relative electrostatic force between cation and soil particle surface for two cation species with the same charge number could be applied to the explanation of the lyotropic series of the selectivity for alkali cations.

In the above section, we established the cation exchange equilibrium equation through the application of the Poisson-Boltzmann equation and treating the cation exchange process as a diffusion process. Also we are aware that, there were two defects as the Poisson-Boltzmann equation to be used to describe cation exchange: neglecting the ionic interaction energy in bulk solution and the hydration effect of cation species in aqueous solution. Thus, if the Poisson-Boltzmann equation is to be used to describe cation exchange equilibrium, the corresponding two modifications should be made.

1. *Non-ideal Modification for the Poisson-Boltzmann Equation and Its Solutions*

When cation exchange is taken as an electrostatic interaction, the non-linear Poisson-Boltzmann equation could be used to describe the exchange equilibrium as (Bolt 1955; Eriksson 1952). In order to get more reliable results in the theoretical analyses of cation exchange as mentioned above, the ionic interaction in aqueous solution (non-ideal condition) will be taken into consideration in the application of the Poisson-Boltzmann equation (Li et al. 2011a).

At equilibrium, between the diffuse layer and the bulk solution (shown in Fig. 9.9), for ith cation species, there is:

$$\mu_{i(DL)} = \mu_{i(Bulk)} \tag{9.180}$$

where $\mu_i(_{DL})$ is the total potential of ith ion species in the diffuse layer (DL), $\mu_{i(Bulk)}$ is the chemical potential of the ith ion species in bulk solution.

In the diffuse layer, there is:

$$\mu_{i(DL)} = \mu_{chemical} + \mu_{electric} = \mu_{i0} + RT \ln f_i(x) + Z_i F\phi(x) \tag{9.181}$$

where $\mu_{chemical}$ is the chemical potential of the ith ion species in diffuse layer; $\mu_{electric}$ is the electric potential of the ith ion species in the diffuse layer, which comes from the electrostatic interaction of all charges including the charges at the original plane of the diffuse layer and the counter ions in the diffuse layer; μ_{i0} is the standard chemical potential of ith ion species; $f_i(x)$ (mol/l) is the concentration of the ith ion species at position x in the diffuse layer.

As considering the ionic interaction (non-ideal condition), in the bulk solution there is:

$$\mu_{i(Bulk)} = \mu_{i0} + RT \ln f_i^0 + RT \ln \gamma_i^0 \tag{9.182}$$

where $\gamma_i{}^0$ is the activity coefficient of ith cation species in bulk solution.

Introducing Eqs. (9.181) and (9.182) into Eq. (9.180), the Boltzmann distribution equation as considering ionic interaction can be expressed as:

$$f_i(x) = f_i{}^0 e^{-\frac{Z_i F\phi(x) - RT \ln \gamma_i}{RT}} = f_i{}^0 \gamma_i^0 e^{-\frac{Z_i F\phi(x)}{RT}} = a_i^0 e^{-\frac{Z_i F\phi(x)}{RT}} \tag{9.183}$$

where $a_i{}^0$ (mol/l) is the activity of the ith cation species in the bulk solution.

Correspondingly, the one-dimensional Poisson-Boltzmann equation could be expressed as:

$$\frac{d^2\phi(x)}{dx^2} = -\frac{4\pi F}{\varepsilon} \sum_i a_i^0 Z_i e^{-\frac{Z_i F\phi(x)}{RT}} \tag{9.184}$$

The average concentration of the adsorbed cation species is still defined as Eq. (9.136), nevertheless the Debye-Huckel parameter is changed to:

$$\kappa = \sqrt{\frac{8\pi F^2 \left(\frac{1}{2} \sum_i a_{i0} Z_i^2\right)}{\varepsilon RT}} \tag{9.185}$$

Therefore the solutions of Eq. (9.184) for two symmetric electrolytes, two 1:2 types of electrolytes and two 2:1 types of electrolytes of cation exchange may be expressed respectively:

1. for solution of two symmetric electrolyte types:

$$\tilde{\gamma}_i = \frac{1}{2\gamma_i^0} \exp\left[\frac{Z_i F\phi_0}{2RT}\right] \tag{9.186}$$

2. for solution of two 1:2 types of electrolytes

$$\tilde{\gamma}_i = \frac{1}{\sqrt{6}\gamma_i^0} \exp\left[\frac{F\phi_0}{2RT}\right] \tag{9.187}$$

3. for solution of two 2:1 types of electrolytes

$$\tilde{\gamma}_i = \frac{1}{\sqrt{3}\gamma_i^0} \exp\left[\frac{Z_i F\phi_0}{2RT}\right] \tag{9.188}$$

Equations (9.186), (9.187) and (9.188) show that, for solutions with two symmetric types of electrolytes, two 1:2 types of electrolytes and two 2:1 types of electrolytes, the average activity coefficient of the adsorbed cation species in diffuse layer could be expressed as:

$$\tilde{\gamma}_i = \frac{1}{m\gamma_i^0} \exp\left[\frac{Z_i F\phi_0}{2RT}\right] \tag{9.189}$$

For system containing a single or two different 1:1 types of electrolytes, there is $m = m_{1:1} = 2$; for system containing a single or different 2:2 types of electrolyte, also there is $m = m_{2:2} = 2$; for system containing a single or different 2:1 types of electrolytes, there is $m = m_{2:1} = 3^{0.5} \approx 1.732$; for a single or different 1:2 types of electrolytes, there is $m = m_{1:2} = 6^{0.5} \approx 2.449$.

Unfortunately, for a mixture containing two different types of electrolytes, e.g. one 1:1 and one 2:1 type of electrolytes, the exact solution like Eqs. (9.186) and (9.187) or Eq. (9.188) can not be currently obtained. However, based on the above discussions, here we can make a reasonable assumption that, for a mixture solution containing two different types of electrolytes, the expression of Eq. (9.187) is still correct with an unknown constant m. Li and Wu has demonstrate that this assumption is correct for a mixture solution containing one 1:1 type and one 2:1 type of electrolytes (Li and Wu 2007b), and found that if the difference of mole numbers of the 1:1 and 2:1 type of electrolytes in the mixed solution were not significant (which means $c^0{}_{NaCl}/c^0{}_{CaCl2} \rightarrow 1$), there might be $m \approx (\mathrm{m}_{1:1} + \mathrm{m}_{2:1})/2 = (2+\sqrt{3})/2 = 1.856$ (Li and Wu 2007b).

2. *Modification of Cation Exchange equation as considering the hydration effect of cations*

Firstly, we consider a cation exchange equilibrium between K^+ and Li^+ of KCl and LiCl electrolytes. According to Eq. (9.186), we have (here $m = 2$ in Eq. (9.189)):

$$\begin{cases} \tilde{f}_K = 2f_K^0\gamma_K^0 \exp\left(-\dfrac{Z_K F\phi_0}{2RT}\right) = 2a_K^0 \exp\left(-\dfrac{Z_K F\phi_0}{2RT}\right) & \text{(9.190a)} \\ \tilde{f}_{Li} = 2f_{Li}^0\gamma_{Li}^0 \exp\left(-\dfrac{Z_{Li} F\phi_0}{2RT}\right) = 2a_{Li}^0 \exp\left(-\dfrac{Z_{Li} F\phi_0}{2RT}\right) & \text{(9.190b)} \end{cases}$$

where $\tilde{f}_K$ and $\tilde{f}_{\mathrm{Li}}$ are the average concentration of adsorbed K^+ and Li^+ in diffuse layer respectively, $f_K{}^0$ and $f_{Li}{}^0$ are the concentration of K^+ and Li^+ in bulk solution respectively, $\gamma_K{}^0$ and $\gamma_{Li}{}^0$ are the activity coefficient of K^+ and Li^+ in bulk solution respectively, $a_K{}^0$ and $a_{Li}{}^0$ are the activity of K^+ and Li^+ in bulk solution respectively, Z_K and Z_{Li} are the charge numbers of K^+ and Li^+ respectively.

From Eqs. (9.190a) and (9.190b), we may obtain:

$$\begin{cases} \dfrac{n_K}{a_K^0} = 2\dfrac{S}{\kappa}\exp\left(-\dfrac{Z_K F\phi_0}{2RT}\right) & (9.191\text{a}) \\ \dfrac{n_{Li}}{a_{Li}^0} = 2\dfrac{S}{\kappa}\exp\left(-\dfrac{Z_{Li} F\phi_0}{2RT}\right) & (9.191\text{b}) \end{cases}$$

where n_K and n_{Li} are the adsorbed quantities of K^+ and Li^+ in the diffuse layer (mol/g).

Thus we may further obtain:

$$K_{K/Li} = \frac{a_{Li}^0 n_K}{n_{Li} a_K^0} = \exp\left[-\frac{(Z_K - Z_{Li})\, F\phi_0}{2RT}\right] \tag{9.192}$$

where $K_{K/Li}$ is the selectivity coefficient between K^+ and Li^+.

If the K^+ and Li^+ are taken as point charge, there is $Z_K = Z_{\mathrm{Li}} = 1$. From Eq. (9.192), we will have $K_{K/Li} = 1$, which means K^+ and Li^+ have the same selectivity, and if $a_K{}^0 = a_{Li}{}^0$ there will be $n_K = n_{Li}$. It is well known, however, as $a_K{}^0 = a_{Li}{}^0$ there will be $N_K > N_{Li}$ according to the lyotropic series. If it is the hydration effect that results in the lyotropic series, a modification factor would be introduced for the charge number of each cation species, and the modification factor (β) reflects the ***relative effective charges*** of cation species i as comparison to the cation species j in i/j exchange. Furthermore, considering the electric neutrality law, the total charge number of the adsorbed cations still must be $Z_i + Z_j$ as introducing the modification factor β, so we should have: $\beta_i Z_i + \beta_j Z_j = Z_i + Z_j$. For K^+ and Li^+, since $Z_i = 1$ and $Z_j = 1$, we have $\beta_i + \beta_j = 2$. Considering the introduced factors can only be applied for the given two cation species, $\beta_i + \beta_j = 2$ means if the electrostatic potential energy decrease a given amount for the ith cation species, another cation species (jth) must increase the same amount correspondingly.

Thus, considering the modification of hydration effect of cations, Eq. (9.192) could be rewritten as (Liu et al. 2012):

$$K_{K/Li} = \frac{a_{Li}^0 n_K}{n_{Li} a_K^0} = \exp\left[-\frac{(\beta_K Z_K - \beta_{Li} Z_{Li})\, F\phi_0}{2RT}\right] \tag{9.193}$$

where β_K and β_{Li} are the effective charge coefficient for cation species K^+ and Li_+ respectively.

Table 9.2 The exchange equilibrium between K^+/Na^+-montmorillonite and solutions of KCl and LiCl (Liu et al. 2012)

Equilibrium activity (mol/l)		Equilibrium conc. (mol/l)		Equilibrium absorbance (mol/g)		Selectivity coefficient
$c_K{}^0$	$c_{Li}{}^0$	$a_K{}^0$	$a_{Li}{}^0$	N_K	N_{Li}	$K_{K/Li}$
0.005776	0.01596	0.004968	0.01373	0.0004152	0.0002975	3.858
0.005750	0.01568	0.004951	0.01350	0.0004056	0.0002870	3.854
0.005141	0.01461	0.004450	0.01264	0.0004149	0.0002986	3.948
0.004904	0.01404	0.004256	0.01218	0.0004119	0.0002969	3.970
0.004622	0.01304	0.004028	0.01136	0.0004084	0.0002898	3.976
0.004526	0.01264	0.003951	0.01104	0.0004059	0.0002887	3.928
0.004949	0.01296	0.004309	0.01129	0.0003956	0.0002733	3.791
0.003885	0.01118	0.003417	0.009834	0.0004165	0.0002960	4.050
0.004609	0.01179	0.004035	0.01032	0.0003935	0.0002700	3.726
0.003827	0.01057	0.003375	0.009324	0.0004119	0.0002906	3.916
0.003987	0.01043	0.003516	0.009197	0.0003974	0.0002764	3.761
0.003840	0.01018	0.003392	0.008991	0.0003983	0.0002760	3.826
0.003346	0.009393	0.002971	0.008340	0.0004169	0.0002984	3.921

Equation (9.193) indicates that there are three factors determining the distribution of the two cation species between the adsorption phase and the solution phase: (1) the surface potential of the charged particles, and the higher the surface potential value, the more quantity of the cation species with smaller hydration radius will be adsorbed; (2) the relative effective charge coefficient, which depends on the relative hydration radius of the two cation species; (3) the charge number of the cation species.

Correspondingly, Eqs. (9.191a) and (9.191b) change to

$$\begin{cases} \dfrac{n_K}{a_K^0} = 2\dfrac{S}{\kappa}\exp\left(-\dfrac{\beta_K Z_K F\phi_0}{2RT}\right) & (9.194a) \\[2ex] \dfrac{n_{Li}}{a_{Li}^0} = 2\dfrac{S}{\kappa}\exp\left(-\dfrac{\beta_{Li} Z_{Li} F\phi_0}{2RT}\right) & (9.194b) \end{cases}$$

Table 9.2 lists the exchange equilibria between K^+/Na^+-montmorillonite and solutions of KCl and LiCl. Because the specific surface area of the montmorillonite is $S = 650.5\ m^2/g$, the κ value for each exchange equilibrium experiment can be calculated from Eq. (9.185). By considering $\beta_K + \beta_{Li} = 2$, the β_K, β_{Li} and φ_0 values can be obtained by introducing the corresponding data shown in Table 9.2 into Eqs. (9.194a) and (9.194b).

Theoretically, both β and φ_0 values are ionic strength dependent for a given exchange system. Figures 9.10 and 9.11 show the $\beta \sim I^{0.5}$ and $\varphi_0 \sim I^{0.5}$ relationships based on the experimental data. From the $\beta \sim I^{0.5}$ relationship, we can see that, (1) in the domain of the ionic strength in the experiment, the β_K and β_{Li} values could be taken as constants; (2) $\beta_K = 1.249$ and $\beta_{Li} = 0.751$, thus $\beta_K > \beta_{Li}$, indicating K^+ has higher selectivity than Li^+ in K^+/Li^+ exchange.

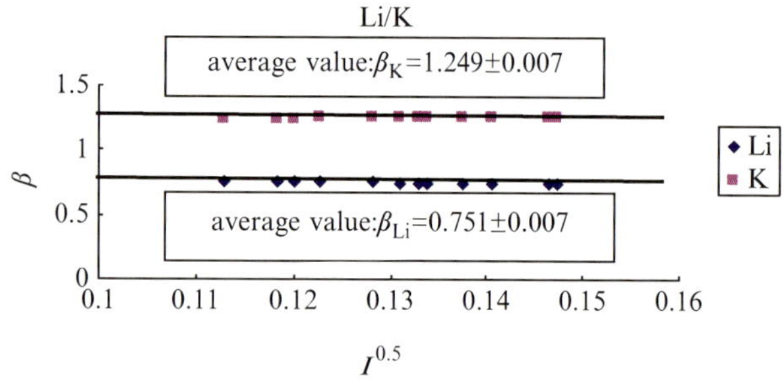

Fig. 9.10 The relationship of β vs. $I^{0.5}$ in Li^+/K^+ exchange (I is the ionic strength and $I=0.5\Sigma c_i{}^0$) (Liu et al. 2012)

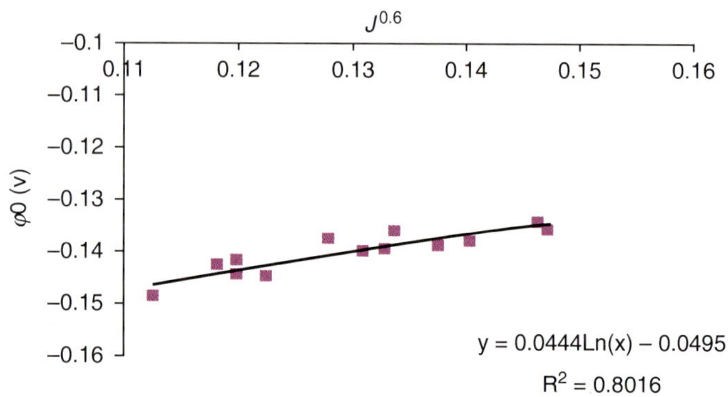

Fig. 9.11 The relationship of φ_0 vs. $I^{0.5}$ in Li^+/K^+ exchange (I is the ionic strength and $I=0.5\Sigma c_i{}^0$) (Liu et al. 2012)

From Fig. 9.11 we can see that the surface potential φ_0 was sensitive to the ionic strength change in the system, and with the increase of ionic strength the negative value of the surface potential decreases.

Because β_K and β_{Li} are constants as discussed as above, from Eq. (9.193), theoretically we may have for K^+/Li^+ exchange:

$$K_{K/Li} = \frac{a_{Li}^0 N_K}{N_{Li} a_K^0} = e^{-9.697\phi_0} \tag{9.195}$$

Equation (9.195) shows that $K_{K/Li}$ will be the sole function of surface potential φ_0 under an isothermal condition. Equation (9.195) also indicates that with the increase of the negative value of the surface potential, the defined $K_{K/Li}$ increases, and as $\varphi_0 \to 0$ there will be $K_{K/Li} \to 1$. Figure 9.12 shows the comparison between the theoretical calculation from Eq. (9.195) and the experimental data.

Secondly, we consider an exchange equilibrium between 1:1 (e.g. NaCl) and 2:1(e.g. CaCl2) types of electrolytes. From Eq. (9.189), introducing the relative effective charge coefficient, we may have:

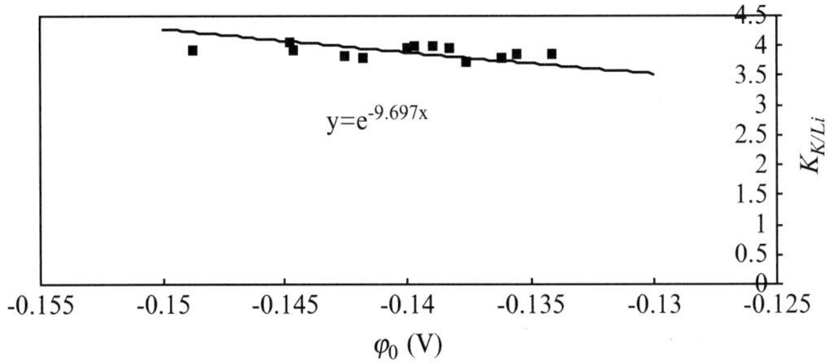

Fig. 9.12 The theoretical curve (*solid line*) and the experimental data (*dots*) of the relationship $K_{K/Li}$ vs. φ_0 (Liu et al. 2012)

$$\begin{cases} \dfrac{n_{Na}}{a^0_{Na}} = m\dfrac{S}{\kappa}\exp\left(-\dfrac{\beta_{Na}Z_{Na}F\phi_0}{2RT}\right) & (9.196a) \\ \dfrac{n_{Ca}}{a^0_{Ca}} = m\dfrac{S}{\kappa}\exp\left(-\dfrac{\beta_{Ca}Z_{Ca}F\phi_0}{2RT}\right) & (9.196b) \end{cases}$$

Here m is an unknown constant, but as $f_{Na}{}^0 \approx f_{Ca}{}^0$, there may be $m \approx 1.856$ theoretically (Li and Wu 2007b).

Thus we may further have:

$$K_{Ca/Na} = \frac{a^0_{Na}n_{Ca}}{n_{Na}a^0_{Ca}} = \exp\left[-\frac{(\beta_{Ca}Z_{Ca} - \beta_{Na}Z_{Na})\,F\phi_0}{2RT}\right] \quad (9.197)$$

Here we still use the experimental data of the exchange equilibrium between Na^+/Ca^{2+}-illite and solutions of NaCl and $CaCl_2$ by Bolt (1955).

In Bolt's experiment, the specific surface area, surface charge density and the surface charge number independently determined, and they are 74.3 m^2/g, 3×10^{-3} $mmol_{(-)}/m^2$ (or 0.0028946 C/dm^2) and 0.223 $mmol_{(-)}/g$ respectively.

On the other hand, the activity of Na^+ and Ca^{2+} in the bulk solution could be obtained based on the calculated activity coefficients from the modified Debye-Huckel equation (as $T = 298$ K) (Su 1986):

$$\lg \gamma_i = -\frac{0.509Z_i^2\sqrt{I}}{1+\sqrt{I}} \quad (9.198)$$

Thus the experimental data of the exchange equilibrium between Na^+/Ca^{2+}-illite and solutions of NaCl and $CaCl_2$ by Bolt are shown in Table 9.3.

As considering $\beta_{Na} + \beta_{Ca} = 2$, using the data shown in Table 9.3, the β_{Na}, β_{Ca} and m values can be obtained. Here, because the m value was not known in advance, another equation to be combined with Eqs. (9.196a) and (9.196b) for solving the β_{Na}, β_{Ca} and m values would be introduced.

Table 9.3 The exchange equilibrium between Na/Ca-illite and solutions of NaCl and $CaCl_2$

Equilibrium solution (mol/l)					Adsorbed ions in DL (mol/g) (n_i)	
Concentration (c_i^0)		Ionic strength	Activity (a_i^0)			
Na	Ca	$I^{0.5}$	Na	Ca	Na	Ca
0.0007	0.0002	0.03605	0.000672	0.0001699	0.00000540	0.000111
0.001	0.0002	0.04	0.0009559	0.000167	0.0000100	0.000105
0.0032	0.00015	0.06042	0.002993	0.0001148	0.0000290	0.0000985
0.0032	0.00056	0.06986	0.002964	0.0004123	0.0000120	0.0001005
0.0051	0.0002	0.0755	0.004697	0.0001439	0.0000340	0.0000910
0.006	0.0013	0.0995	0.005396	0.0008505	0.0000150	0.000104
0.009	0.0023	0.1261	0.007893	0.001361	0.0000190	0.000100
0.029	0.0008	0.1772	0.02431	0.000395	0.0000770	0.0000700
0.019	0.009	0.2145	0.01545	0.003933	0.0000190	0.0000985
0.056	0.0029	0.2544	0.04415	0.001121	0.0000820	0.0000660
0.078	0.00017	0.2802	0.06035	0.00006093	0.000169	0.0000195
0.095	0.0068	0.3397	0.07058	0.002071	0.0000840	0.0000640
0.034	0.0294	0.3496	0.0251	0.008729	0.0000180	0.000101

As considering the non-ideal modification and the relatively effective charge modification, the classic equation for the calculation of surface charge density would be changed to (Sposito 1984; Li et al. 2011a):

$$\sigma_0 = \text{sgn}(\phi_0)\sqrt{\frac{\varepsilon RT}{2\pi}\left[a^0_{Na}\left(e^{-\frac{\beta_{Na}Z_{Na}F\phi_0}{RT}}-1\right)+a^0_{Ca}\left(e^{-\frac{\beta_{Ca}Z_{Ca}F\phi_0}{RT}}-1\right)\right]} \tag{9.199}$$

Introducing the corresponding data shown in Table 9.3 into Eqs. (9.196a), (9.196b) and (9.199), the combined solution of those three equations with the equation of $\beta_{Na}+\beta_{Ca}=2$, the β_{Na}, β_{Ca} and m values can be obtained. Theoretically, m will be the function of $c^0{}_{NaCl}/c^0{}_{CaCl2}$ as discussed as above. Therefore, the empirical relationship between β and $I^{0.5}$, or between m and $c^0{}_{NaCl}/c^0{}_{CaCl2}$ can be obtained, and the results are shown in Figs. 9.13 and 9.14 respectively.

Figure 9.13 indicates that, β_{Na} and β_{Ca} still could be taken as constant in the Ca^{2+}/Na^+ exchange. Figure 9.14 showed that as $c^0{}_{Na}/c^0{}_{Ca}=1$, there was $m=1.992$ experimentally, which was close to the theoretical prediction of $m=1.856$ (Li and Wu 2007b).

By using the obtained β_{Na} and β_{Ca} values, according to Eq. (9.197), the $K_{Ca/Na}$ value can be theoretically calculated from the following equation:

$$K_{Ca/Na} = \frac{a^0_{Na}N_{Ca}}{N_{Na}a^0_{Ca}} = e^{-35.59\phi_0} \tag{9.200}$$

The surface potential dependence of $K_{Ca/Na}$ as shown in Fig. 9.15 once again indicates that the cation adsorption is simply the Coulombic effect.

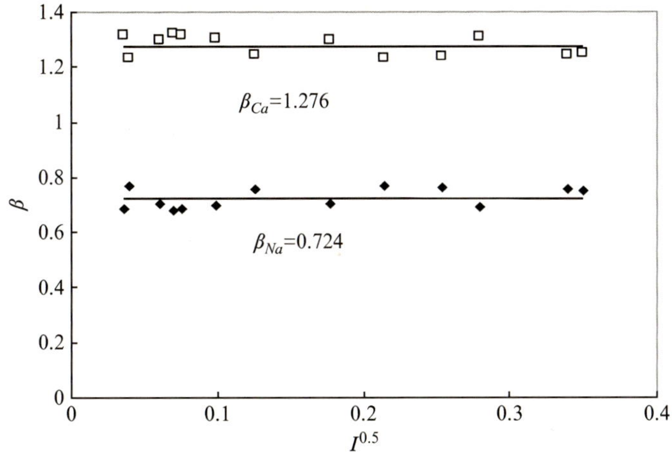

Fig. 9.13 The empirical relationship between β and $I^{0.5}$ (Li et al. 2011a)

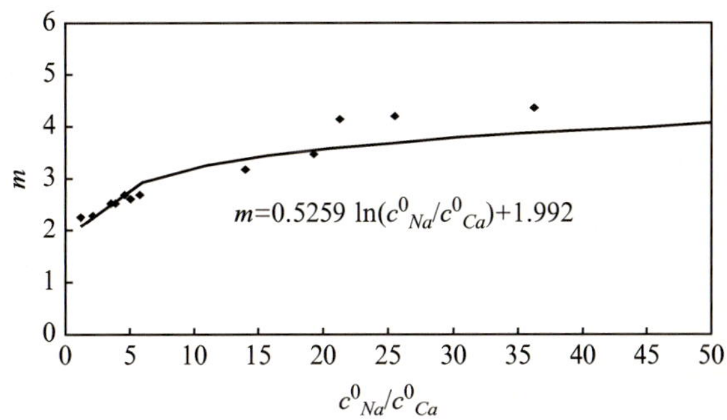

Fig. 9.14 The empirical relationship between m and c^0_{Na}/c^0_{Ca} (Li et al. 2011a)

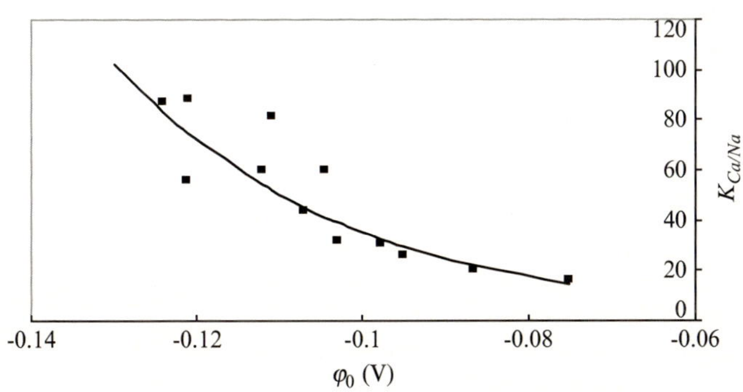

Fig. 9.15 The theoretical curve (*solid line*) and the experimental data (*dots*) of the relationship $K_{Ca/Na}$ vs. φ_0

9.3 The Electrostatic Field Effect from Surface Charges on Ion Adsorption/Diffusion Kinetics

Many researchers have devoted great efforts to study the ion diffusion/adsorption processes in soil (Sparks 1989, 1998a, b; Porter et al. 1960; Olsen et al. 1962; Van Schaik et al. 1966; Sparks et al. 1980; Rao et al. 1980; Nkedi-Kizza et al. 1982; Aharoni et al. 1991). Generally, it was considered that, ion adsorption at soil particle surface contained the following processes (Sparks 1989): (1) ion diffusion in bulk solution, (2) ion diffusion in solution film surrounding particles, (3) ion diffusion through a hydrated interlayer space of particles, (4) adsorption. Thus, generally ion adsorption includes two separate processes: diffusion and adsorption, and usually the apparent adsorption rate is limited by ion diffusion (Ogwada and Sparks 1986a; Tang and Sparks 1993; Suresh et al. 2004). For describing ion diffusion in the soil, however, Fick's first and second laws are directly employed (Sparks 1989; Aharoni et al. 1991), for which only ion movement caused by the concentration gradient was considered. Even though soil is electrically neutral macroscopically, ion diffusion and exchange occur in the diffuse layer adjacent to the solid/liquid interface where the electric potential gradient is not zero. Thus, the driving forces for ion movement must include both the concentration gradient and the electric potential gradient in order to explain the spontaneous flow of ions from lower concentrations (in bulk solution) to higher concentrations (in the diffuse layer), and to explain the Boltzmann distribution at equilibrium of adsorption between the diffuse layer and bulk solution. Kemper and Van Schaik (1966) used the Nernst-Plank equation that includes both the concentration gradient and the potential gradient to describe ion diffusion in soils. However, they only considered the potential gradient caused by charged ions and neglected the electrical potential caused by the charged clay surface in their approach.

We know that, for exchange adsorption, all the adsorbed ions are located in the diffuse layer and all the adsorbed ions in the diffuse layer obey the Boltzmann distribution law. We have show that, all the charges including the surface charges of particles and the counter ions in the diffuse layer could set up a strong electrostatic field in the soil solution around soil particles. When considering the influence of electric field in the diffuse layer, ions diffusion in the soil is a process occurring in an external electric field, and the electric field is time-dependent, because the density of ions in diffuse layer changes with time (Li et al. 2009b). Classically, mass diffusion in a time-dependent potential can be described by the well-known Fokker-Planck equation. However, because of the inherent complexity, the exact solution of the Fokker-Planck equation with a time-dependent potential is difficult to obtain. Li and Wu (2007a) established a simple theory as discussed in the first section. In this theory, the complex non-linear Fokker-Planck equation is essentially a simple linear equation as described by equation (9.45) under the condition of strong friction limit, which is consistent with the local equilibrium assumption

As discussed above, the application of Eqs. (9.34) and (9.45) means the exchange adsorption process (electrostatic adsorption) and the diffusion process of ions in

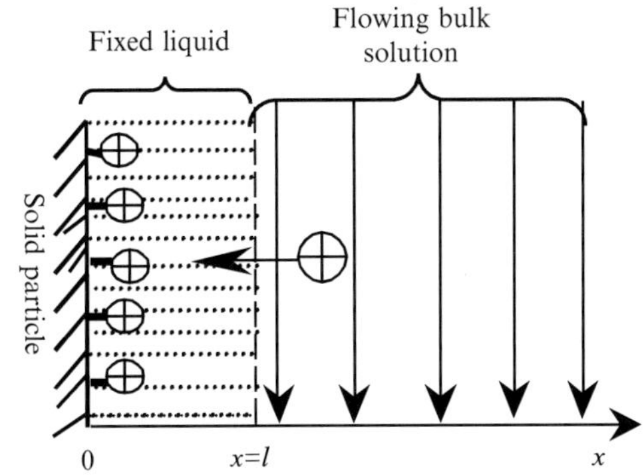

Fig. 9.16 A conceptual diagram of ion adsorption/diffusion from bulk solution to EDL

the electrostatic field of the diffuse layer may be essentially the same process. On the other hand, in Eqs. (9.34) and (9.45), there are three forces driving cation adsorption/diffusion: (1) the force coming from the chemical potential, (2) the force coming from the ionic interaction potential and (3) the force coming from the electrostatic potential. Thus, this equation might unify the three diffusion processes (diffusion in bulk solution, in solution film, and in hydrated interlayer space) and the adsorption (Coulumbic) process to one. Therefore, Eqs. (9.34) and (9.45) should be the basic equations for describing the kinetics of Coulombic adsorption of cations.

9.3.1 *Kinetic Analysis of Cation Exchange Adsorption*

9.3.1.1 The Dynamic Distribution of Cations in the Diffuse Layer in an Adsorption Process

When the miscible displacement technique is employed to study ion adsorption under the conditions of constant flow, the process of ion diffusion/adsorption can be illustrated as in Fig. 9.16.

From Fig. 9.16, the initial and boundary conditions for solving Eq. (9.45) can be expressed as:

$$\begin{cases} A_i(l,t) \approx f_{i0} \\ A_i(0,t) = A_i(t) \\ A_i(x,0) = 0 \end{cases} \tag{9.201}$$

where f_{i0} is the concentration of the ith ion species in bulk (flowing) solution, A_i $(x, 0) = 0$ means the initial concentration of ith ion species in the domain $0 \leq x \geq 1$ is zero.

To obtain the solution of Eq. (9.45) by considering the initial and boundary conditions shown in Eq. (9.201) (Li et al. 2010) and applying the Laplace transform for t, from Eq. (9.45), we have:

$$pF(x,p) = D_i \frac{\partial F^2(x,p)}{\partial x^2} \tag{9.202}$$

Here, $A_i(x, t)$ is the original function of $F(x, p)$. The corresponding transformation results for the boundary conditions as shown in Eq. (9.201) are:

$$L\left[A_i\ (0,t)\right] = L\left[A_i(t)\right] = F(p) \tag{9.203}$$

$$L\left[A_i\ (l,t)\right] = F(l,p) = \frac{f_{i0}}{p} \tag{9.204}$$

We search the Laplace transform for variable x for Eq. (9.202). Since the domain of x is $x = [0, l]$, we must extend the domain of the x to $x = [0, \infty]$, thus $x = [0, l]$ belongs to $x = [0, \infty]$. Considering Eq. (9.203), the corresponding transformation result for Eq. (9.202) is:

$$pG(s,p) = D_i s^2 G(s,p) - D_i s F(p) - 0 \tag{9.205}$$

where $F(x, p)$ is the original function of $G(s, p)$. From this algebra equation, we have:

$$G(s,p) = \frac{D_i s F(p)}{D_i s^2 - p} \tag{9.206}$$

It can also be written as:

$$G(s,p) = F(p) \frac{s}{s^2 - \left(\sqrt{\frac{p}{D_i}}\right)^2} \tag{9.207}$$

To search for the inversion of the Laplace transform of Eq. (9.207), we obtain the original function $F(x, p)$ of $G(s, p)$:

$$F(x,p) = F(p)\cosh\left(\sqrt{\frac{p}{D_i}}x\right) \tag{9.208}$$

Based on Eq. (9.208), using the boundary condition Eq. (9.204), we have:

$$\frac{f_{i0}}{p} = F(p)\cosh\left(\sqrt{\frac{p}{D_i}}l\right)$$

Thus there is:

$$F(p) = \frac{f_{i0}}{p} \frac{1}{\cosh\left(\sqrt{\frac{p}{D_i}} l\right)}$$

Introducing the above equation into Eq. (9.208) gives:

$$F(x, p) = \frac{f_{i0}}{p} \frac{\cosh\left(\sqrt{\frac{p}{D_i}} x\right)}{\cosh\left(\sqrt{\frac{p}{D_i}} l\right)} \tag{9.209}$$

The extreme points of Eq. (9.209) are $p = 0$ and $p = -(D_i \pi^2/l^2)(n-1/2)^2 = p_n$, each of them belongs to the first order extreme points of $F(x, p)$, in which $n = 1$, 2, 3, ... According to the universal inversion formula for Laplace transform, the original function $A_i(x, t)$ of $F(x, p)$ can be expressed as:

$$A_i(x, t) = \mathrm{Re}\left[F(x, p)e^{pt}\right] + \sum_{n=1}^{\infty} \mathrm{Re}\left[F(x, p_n)e^{p_n t}\right] \tag{9.210}$$

The first item in the above equation is for searching the residue of $F(x, p)e^{pt}$ with respect to the extreme point $p = 0$; the second item is for searching the residue of $F(x, p)e^{pt}$ with respect to the extreme point $p = p_n$, $n = 1, 2, 3, \ldots$, where "Re" is a symbol for searching the residue.

From Eq. (9.209), we can obtain the residue of the first term in Eq. (9.210):

$$\mathrm{Re}\left[F(x, p)e^{pt}\right] = \lim_{p \to 0} pF(x, p)e^{pt} == \lim_{p \to 0} f_{i0} \frac{\cosh\left(\sqrt{\frac{p}{D_i}} x\right)}{\cosh\left(\sqrt{\frac{p}{D_i}} l\right)} e^{pt} = f_{i0} \tag{9.211}$$

From Eq. (9.209), we can obtain the residue of the second term in Eq. (9.210):

$$\begin{aligned}
&\sum_{n=1}^{\infty} \mathrm{Re}[F(x, p_n)e^{p_n t}] \\
&= \sum_{n=1}^{\infty} f_{i0} \frac{\cosh\left(\sqrt{\frac{p_n}{D_i}} x\right)}{\frac{d\left[p_n \cosh\left(\sqrt{\frac{p_n}{D_i}} l\right)\right]}{dp_n}} e^{p_n t} \\
&= \sum_{n=1}^{\infty} \frac{f_{i0} \cos\left[\left(n - \frac{1}{2}\right) \frac{\pi}{l} x\right] e^{-\frac{D_i \pi^2}{l^2}\left(n - \frac{1}{2}\right)^2 t}}{(-1)^n \left(n - \frac{1}{2}\right) \frac{\pi}{2}} \\
&= f_{i0} \sum_{n=0}^{\infty} \frac{(-1)^{n+1} 4}{(2n+1)\pi} e^{-\frac{D_i \pi^2}{4l^2}(2n+1)^2 t} \cos\left[\frac{x}{l}(2n+1)\frac{\pi}{2}\right]
\end{aligned} \tag{9.212}$$

Considering:

$$\cos\left[\frac{x}{l}(2n+1)\frac{\pi}{2}\right] = (-1)^n \sin\left[\frac{l-x}{l}(2n+1)\frac{\pi}{2}\right]$$

Equation (9.212) can be written as:

$$\sum_{n=1}^{\infty} \mathrm{Re}[F(x,p_n)e^{p_n t}] = -f_{i0}\sum_{k=0}^{\infty}\frac{4}{(2n+1)\pi}e^{-\frac{D_i\pi^2}{4l^2}(2n+1)^2 t}\sin\left[\frac{l-x}{l}(2n+1)\frac{\pi}{2}\right] \tag{9.213}$$

Introducing Eq. (9.211) and Eq. (9.213) into Eq. (9.210), we get:

$$A_i(x,t) = f_{i0} - f_{i0}\sum_{n=0}^{\infty}\frac{4}{(2n+1)\pi}e^{-\frac{D_i\pi^2}{4l^2}(2n+1)^2 t}\sin\left[\frac{l-x}{l}(2n+1)\frac{\pi}{2}\right] \tag{9.214}$$

Introducing the definition of $A_i(x, t)$ into Eq. (9.214), we get

$$f_i(x,t) = f_{i0}e^{-\frac{Z_i F\phi(x,t)}{RT}}\left\{1-\sum_{n=0}^{\infty}\frac{4}{(2n+1)\pi}e^{-\frac{\pi^2 D_i}{4l^2}(2n+1)^2 t}\sin\left[\frac{(l-x)}{l}(2n+1)\frac{\pi}{2}\right]\right\} \tag{9.215}$$

Therefore, if $\varphi(x, t)$ is known, the dynamic distribution of the ith cation species in the diffuse layer can be obtained. From Eq. (9.215) we can see that, as $t \to \infty$, the Boltzmann distribution equation is obtained:

$$f_i(x) = f_{i0}e^{-\frac{Z_i F\phi(x)}{RT}} \tag{9.216}$$

For one particle diffusion, the time scale for the ion movement in diffuse layer would be much longer than the *configuration* adjustment of the diffuse layer back to the equilibrium state, according to the Instantaneous Relaxation Approximation (Chan 1987; Åkesson et al. 1986). Therefore when many ions diffuse in the system simultaneously, and as long as the diffuse layer keeps electro-neutrality in the diffusion process, it is reasonable to further assume that the *configuration of the diffuse layer* can reach the final equilibrium state quickly for the miscible displacement experiment. During experiments, because the ionic concentration in bulk solution (flowing liquid) keeps approximately constant, the diffuse layer can reach the final thickness quickly. As a result, the charge density in the diffuse layer can approach the final density quickly. This assumption may be illustrated in Fig. 9.17, which means $\varphi(x, t) \approx \varphi(x, \infty)$ after a short time in the experiment.

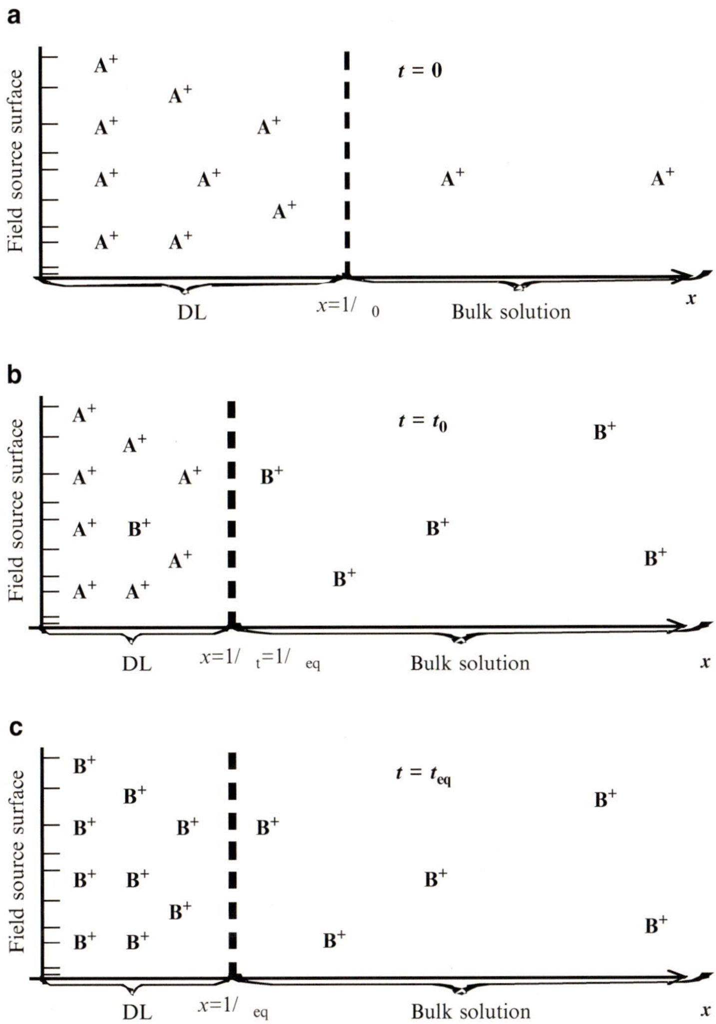

Fig. 9.17 The conceptual diagram of diffuse layer in the ion adsorption(exchange) process. (**a**) The initial state of diffuse layer (*DL*). (**b**) As ion exchange occurs within a short time $t = t_0$, the DL configuration approaches to the equilibrium state, in which the thickness of DL and the charge density in DL approaches to final state. But at this moment the exchange process just begins. (**c**) The equilibrium state of exchange as $t = t_{eq}$ ($t_{eq} >> t_0$)

Based on the above assumption, Eq. (9.215) can be transformed to:

$$f_i(x,t) = f_{i0} e^{-\frac{Z_i F \phi(x,\infty)}{RT}} \left\{ 1 - \sum_{n=0}^{\infty} \frac{4}{(2n+1)\pi} e^{-\frac{\pi^2 D_i}{4l^2}(2n+1)^2 t} \sin\left[\frac{(l-x)}{l} (2n+1) \frac{\pi}{2} \right] \right\} \tag{9.217}$$

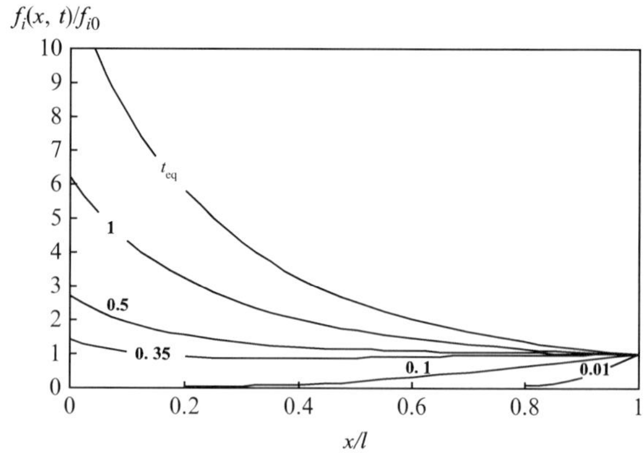

Fig. 9.18 The dynamic distribution patterns of the ion in diffuse layer for an adsorption process in the experiment of the miscible displacement ($T = 298$ K, $Z = 1, \varphi_0 = -0.1$ V, $l = 1/\kappa$. Numbers on curves are dimensionless time ($(\pi^2 D/4\,l^2)t$)

Figure 9.18 gives the conceptual dynamic distribution patterns of the adsorption ion in the diffuse layer based on Eq. (9.217) by assuming:

$$\phi(x,\infty) = \phi_0(\infty)e^{-\kappa(\infty)x}]$$

where $\varphi_0(\infty)$ is the surface potential at equilibrium of exchange, $\kappa(\infty)$ is the Debye-Hückel parameter at equilibrium of adsorption.

9.3.1.2 Theoretical Analysis of Cation Adsorption Kinetics Based on the Diffusion Mechanism of Cation in External Electric Field

According to Eq. (9.217) and Fig. 9.18, from $t = 0$ to $t = t$, the total number of ith cation species adsorption/diffusion from bulk solution (flowing liquid) to the diffuse layer is:

$$n_i(t) = Sf_{i0}\int_0^l e^{-\frac{Z_i F\phi(x,\infty)}{RT}}\left\{1-\sum_{n=0}^{\infty}\frac{4}{(2n+1)\pi}e^{-\frac{\pi^2 D_i}{4l^2}(2n+1)^2 t}\sin\left[\frac{(l-x)}{l}(2n+1)\frac{\pi}{2}\right]\right\}dx \tag{9.218}$$

where $n(t)$ is the total adsorption quantity from $t = 0$ to t, S is the specific surface area of the soil sample.

Equation (9.218) seems similar to the equation through solving the Fick's second law mathematically (Aharoni et al. 1991), but the two are quite different. In Eq. (9.218), it contains the Boltzmann factor: $\exp[ZF\varphi(x)/RT]$.

Since the series in Eq. (9.218) converges rapidly as t is large (tens of minutes), the only term in the series needing to be considered is when $n = 0$. Thus Eq. (9.218) becomes:

$$n_i(t) = Sf_{i0}\int_0^l e^{-\frac{Z_i F\phi(x,\infty)}{RT}}\left\{1-\frac{4}{\pi}e^{-\frac{\pi^2 D_i}{4l^2}t}\sin\left[\frac{(l-x)}{l}\frac{\pi}{2}\right]\right\}dx \tag{9.219}$$

When adsorption reaches equilibrium, from Eq. (9.219) there is:

$$n_i(\infty) = S f_{i0} \int_0^l e^{-\frac{Z_i F\phi(x,\infty)}{RT}} dx \tag{9.220}$$

where the unit of $n_i(\infty)$ is mol/g.

Obviously, the Eq. (9.220) can also be obtained directly from the Boltzmann distribution equation.

According to Eq. (9.219), the flux of the ith cation species adsorption (diffusion) through the plane of $x = l$ into the diffuse layer can be expressed as:

$$j_i\,[\phi(x) \neq 0] = \frac{dn_i(t)}{Sdt} = \frac{\pi D_i f_{i0}}{l^2} e^{-\frac{\pi^2 D_i}{4l^2}t} \left\{ \int_0^l e^{-\frac{ZF\phi(x)}{RT}} \sin\left[\frac{(l-x)}{l}\frac{\pi}{2}\right] dx \right\} \tag{9.221}$$

In the mean time, if the electric field in the diffuse layer is not considered, the solution from the Fick's diffusion equations:

$$f_i(x,t) = f_{i0} \left\{ 1 - \sum_{n=0}^{\infty} \frac{4}{(2n+1)\pi} e^{-\frac{\pi^2 D_i}{4l^2}(2n+1)^2 t} \sin\left[\frac{(l-x)}{l}(2n+1)\frac{\pi}{2}\right] \right\} \tag{9.222}$$

Equation (9.222) means that the equilibrium distribution of the cation species i in diffuse layer is $f_i(x, \infty) = f_{i0}$, evidently which is incorrect for a charged soil. It is well known that ion distribution in diffuse layer follows the Boltzmann distribution when adsorption reaches equilibrium. The discussion implies that correct result could not be obtained for the kinetic study of ion adsorption/diffusion if the Fick's diffusion equation directly employed, for which the effect of the electric field in the soil was neglected.

From Eq. (9.222), similar to Eq. (9.218), we can further obtain:

$$n_i(t) = S f_{i0} \int_0^l \left\{ 1 - \sum_{n=0}^{\infty} \frac{4}{(2n+1)\pi} e^{-\frac{\pi^2 D_i}{4l^2}(2n+1)^2 t} \sin\left[\frac{(l-x)}{l}(2n+1)\frac{\pi}{2}\right] \right\} dx \tag{9.223}$$

When the equilibrium of adsorption is reached, Eq. (9.223) becomes:

$$n_i'(\infty) = Sl f_{i0}$$

Obviously, this result is also incorrect for a charged soil. For a charged soil $n_i(\infty)$ will be much larger than Slf_{i0}.

Similar to Eq. (9.221), from Eq. (9.223), we can get the flux as neglecting the electric field:

$$j_i\left[\phi_0(x)=0\right]=\frac{dn_i(t)}{Sdt}=\frac{\pi D_i f_{i0}}{l^2}e^{-\frac{\pi^2 D_i}{4l^2}t}\left\{\int_0^l \sin\left[\frac{(l-x)}{l}\frac{\pi}{2}\right]dx\right\}=\frac{2f_{i0}D_i}{l}e^{-\frac{\pi^2 D_i}{4l^2}t} \tag{9.224}$$

Therefore:

$$\frac{j_i\left[\phi(x)\neq 0\right]}{j_i\left[\phi(x)=0\right]}=\frac{\pi}{2}\left[\frac{1}{l}\int_0^l e^{-\frac{ZF\phi(x)}{RT}}\sin\left(\frac{l-x}{l}\frac{\pi}{2}\right)dx\right] \tag{9.225}$$

Equation (9.225) implies that, there will be $j_i[\varphi(x)\neq 0] >> j_i[\varphi(x)=0]$, which means the electrostatic field in the soil will increase the adsorption rate of cations remarkably.

The above theoretical analyses showed that: (1) in the study of ion adsorption/diffusion in charged soil, the electric field in diffuse layer must be considered; (2) since $j_i[\varphi(x)\neq 0] >> j_i[\varphi(x)=0]$, the electric field around soil particles may strongly influence ion adsorption/diffusion rates; and (3) the term in the square brackets of the right side of Eq. (9.225) implies that it is the average effect of the electric field from 0 to l that influences the cation adsorption/diffusion. Since a smaller value of l means a larger value of $(1/l)\int_0^l \exp[-ZF\varphi(x)/RT]dx$, the flux will be larger as the penetrating plane ($x=l$) of ion adsorption/diffusion is closer to the particle surface.

When the electrostatic field is considered, from Eq. (9.219), we have:

$$\frac{dn_i(t)}{dt}=\left(\frac{\pi^2 D_i}{4l^2}\right)\left[Sf_{i0}\int_0^l e^{-\frac{Z_i F\phi(x)}{RT}}dx\right]-\frac{\pi^2 D_i}{4l^2}n_i(t) \tag{9.226}$$

As considering

$$Sf_{i0}\int_0^l e^{-\frac{Z_i F\phi(x,\infty)}{RT}}dx\approx\frac{CEC}{10^5 Z_i} \tag{9.227}$$

where the unit of CEC is $\mathrm{cmol}_{(+)}/\mathrm{Kg}$.

$$\frac{dn_i(t)}{dt}=\left(\frac{\pi^2 D_i}{4l^2}\right)\frac{CEC}{10^5 Z_i}-\frac{\pi^2 D_i}{4l^2}n_i(t) \tag{9.228}$$

It is obvious that Eq. (9.227) or Eq. (9.228) is mathematically similar to the widely applied apparent first order kinetic equation in cation exchange (Sparks et al. 1980; Sparks 1989; Jardine and Sparks 1984). However, the nature between the classic apparent first order kinetic equation and the Eq. (9.226) or Eq. (9.228) is quite different. Because each parameter in Eq. (9.226) or Eq. (9.228) has its defined physical meanings, the rate coefficient in the new kinetic equation could be theoretically predicted.

According to the new kinetic equation, we can see:

1. If we define the rate coefficient k as:

$$k = \frac{\pi^2 D_i}{4l^2} S f_{i0} \int_0^l e^{-\frac{Z_i F \phi(x)}{RT}} dx \tag{9.229}$$

From Eq. (9.226), there is:

$$\frac{dn_i(t)}{dt} = k\left[1 - \frac{n_i(t)}{n_i(\infty)}\right] \tag{9.230}$$

In order to using the experimental data directly, Eq. (9.230) can be changed to the discrete form:

$$\frac{n_i(t_{m+1}) - n_i(t_m)}{S\,(t_{m+1} - t_m)} = k - \frac{k}{n_i(\infty)} n_i\left(t_{m+1/2}\right) \tag{9.231}$$

where $n_i(t_{m+1/2}) = n_i(t_m) + 0.5[n_i(t_{m+1}) - n_i(t_m)]$, $m = 1, 2, 3, \ldots$, and here k_2 is:

Equation (9.231) shows that, for the electrostatic force adsorption, using experimental data to plot a relationship curve of $[n_i(t_{m+1}) - n_i(t_m)]/S(t_{m+1} - t_m)$ *vs.* $n_i(t_{m+1/2})$, a straight line could be obtained, and from the straight line we may deduce that: (1) the intercept of the straight line will be the rate coefficient k in the adsorption process; (2) the slope of the straight line will be $\pi^2 D_i/4l^2 S$; (3) the total adsorption quantity $n_i(\infty)$ at the equilibrium can be obtained directly by using the intercept and the slope values.

When Eq. (9.231) is directly used to deal with the experimental kinetic data, the value of N_∞ is not required in advance. Previously, the integral form was used deal with the experiment data, and a straight line of $\ln(1 - n_i(t)/n_i(\infty))$ *vs.* t was expected to be obtained as the first order process existed (Aharoni et al. 1991; Jardine and Sparks 1984; Sparks and Jardine 1984; Li and Xue 1996). Obviously, if the integral form to be used to deal with experimental data, the $n_i(\infty)$ value must be known in advance. Unfortunately, it will be very difficult to obtain the exact $n_i(\infty)$ value. Theoretically, the directly determined $n_i(\infty)$ within a finite range of time will always be lower than the actual value, because the actual value could be obtained just as $t = \infty$. If the adopted $n_i(\infty)$ is lower than the actual value, in dealing with the experimental data, the slope of the fitted curve of $\ln(1 - n_i(t)/n_i(\infty))$ *vs.* t must tend to infinity as time t increase, which is similar to the dashed line shown in Fig. 9.19. Theoretically, the slope of the curve of $\ln(1 - n_i(t)/n_i(\infty))$ *vs.* t is the $\pi^2 D_i/4l^2 S$, so if the experimental result of the slope is infinite, this result must be considered incorrect.

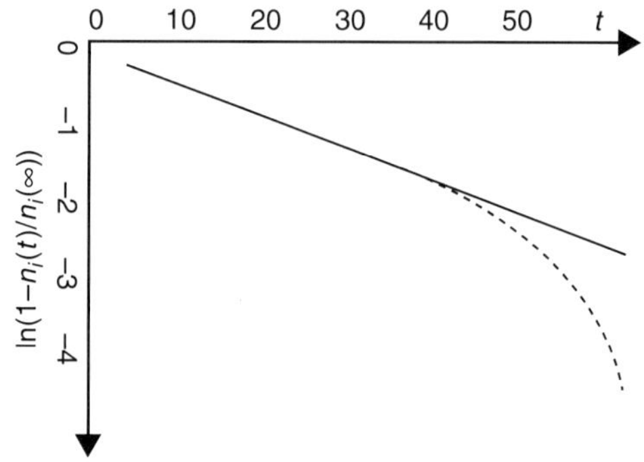

Fig. 9.19 The fitted result for the kinetic data as using the integral form of the first-order kinetic equation—as the adopted $n_i(\infty)$ to be the true value, ·····as the adopted $n_i(\infty)$ value is lower than the true value

2. If the electrostatic field in the diffuse layer influences the cation adsorption rate, from Eq. (9.228), theoretically we should have:

$$\left|\frac{\textbf{intercept}}{\textbf{slope}}\right| = \frac{CEC}{10^5 Z_i} \tag{9.232}$$

Therefore, from the experimentally obtained | intercept/slope | ratio value, the CEC value can be estimated. On the other hand, if the theoretically estimated CEC value from Eq. (9.232) based on the kinetic data corresponds to the CEC value obtained from a classic measurement method, we may conclude that the obtained new rate equation for describing cation adsorption is correct.

3. For the same soil, if the experiment is carried out at two different cation concentrations in bulk solution, e.g. $f^{(1)}{}_{i0}$ and $f^{(2)}{}_{i0}$, the surface potential of the soil under those two different concentrations will be different. As a result, the potential distribution ($\varphi(x)$) in space 0 to l will be different as well. However, for a constantly charged material:

$$CEC = 10^5 Z_i S \left(f_{i0}^{(1)} \int_0^l e^{-\frac{Z_i F \phi^{(1)}(x)}{RT}} dx \right) = 10^5 Z_i S \left(f_{i0}^{(2)} \int_0^l e^{-\frac{Z_i F \phi^{(2)}(x)}{RT}} dx \right)$$

Therefore, based on Eq. (9.229), the theoretical value of $k^{(1)}(f_{i0})/k^{(2)}(f_{i0})$ will be:

$$\frac{k^{(1)}(f_{i0})}{k^{(2)}(f_{i0})} = \frac{\frac{\pi^2 D_i^{(1)}}{4l^2} f_{i0}^{(1)} \int_0^l e^{-\frac{Z_i F \phi^{(1)}(x)}{RT}} dx}{\frac{\pi^2 D_i^{(2)}}{4l^2} f_{i0}^{(2)} \int_0^l e^{-\frac{Z_i F \phi^{(2)}(x)}{RT}} dx} = \frac{\frac{\pi^2 D_i^{(1)}}{4l^2}}{\frac{\pi^2 D_i^{(2)}}{4l^2}} \cdot \frac{CEC}{CEC} = \frac{D_i^{(1)}}{D_i^{(2)}} \tag{9.233}$$

Comparison between the theoretical and experimental values of $k^{(1)}(f_{i0})/k^{(2)}(f_{i0})$ provides another means of quantitative verification of the suggested new rate equation.

9.3.1.3 Theoretical Analyses of Cation Adsorption Kinetics as Considering the Strong and Weak Electrostatic Force Adsorptions

The New Flux Equations for Describing Ion Adsorption/Diffusion in an Electric Field

Because the series in Eq. (9.217) is rapidly convergent, as the t is large (often larger than several ten minutes depending on the value of $\pi^2 D_i/4l^2$), only the item $n = 0$ may be considered, thus there is:

$$f_i(x,t) = f_{i0} e^{-\frac{Z_i F\phi(x,t)}{RT}} \left[1 - \frac{4}{\pi} e^{-\frac{\pi^2 D_i}{4l^2} t} \sin \frac{l-x}{l} \frac{\pi}{2} \right] \tag{9.234}$$

Correspondingly, the flux of ith cation species, the diffusion/adsorption through the plane of $x = l$ into the space $x = 0 \sim l$ can be expressed as:

$$j_i = \frac{dn_i(t)}{Sdt} = D_i \left\{ \int_0^l \frac{\pi f_{i0}}{l^2} e^{-\frac{\pi^2 D_i}{4l^2} t} e^{-\frac{ZF\phi(x)}{RT}} \sin \left[\frac{(l-x)}{l} \frac{\pi}{2} \right] dx \right\} \tag{9.235}$$

According to the definition of the apparent concentration or activity as described above, Eq. (9.234) changes to:

$$A_i(x,t) = f_i(x,t) e^{\frac{Z_i F\phi(x,t)}{RT}} = f_{i0} \left[1 - \frac{4}{\pi} e^{-\frac{\pi^2 D_i}{4l^2} t} \sin \frac{l-x}{l} \frac{\pi}{2} \right] \tag{9.236}$$

where $A_i(x, t)$ is the apparent concentration or activity of ith cation species at the distance x away from particle surface and time t.

Introducing Eq. (9.236) into Eq. (9.235), gives (Li et al. 2011b)

$$j_i = \frac{dn_i(t)}{Sdt} = \frac{\pi^2 D_i}{4l} \left\{ \frac{1}{l} \int_0^l e^{-\frac{ZF\phi(x)}{RT}} \left[f_{i0} - A_i(x,t) \right] dx \right\} \tag{9.237}$$

Equation (9.237) is the flux equation for describing ion diffusion/adsorption by considering the electric field from soil particle surfaces under the miscible displacement experiment.

The Adsorption Rate Equation as Strong Electrostatic Adsorption Force Presence

When the strong electrostatic adsorption force is present, it means a strong negative energy area or a deep potential well near the soil particle surface exists. Once a cation falls into (adsorbed) the potential well, the cation will be strongly adsorbed

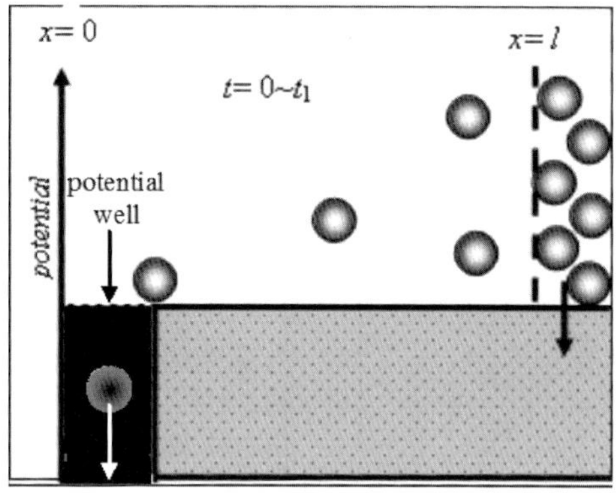

Fig. 9.20 A conceptual diagram of ion adsorption as strong adsorption presence at particle surface

and thus will be very difficult to escape from the potential well. So, it could be taken that this strongly adsorbed cation "disappears" immediately from the solution adjacent particle surface. Here, "disappear" means that the activity of the cation in the solution near the soil particle surface (in the potential well) is approximately zero. Obviously, this phenomenon will only appear in the initial stage of the cation adsorption experiment. This process could be illustrated in Fig. 9.20.

Thus, there will be:

$$f_{i0} - A_i(x,t) \approx f_{i0} \tag{9.238}$$

For this case, the flux Eq. (9.237) changes to:

$$j_i = \frac{dn_i(t)}{Sdt} = \frac{\pi^2 D_i}{4l^2} f_{i0} \int_0^l e^{-\frac{Z_i F\phi(x)}{RT}} dx \tag{9.239}$$

If we define the rate coefficient k_1 as:

$$k_1 = \frac{\pi^2 D_i}{4l^2} f_{i0} \int_0^l e^{-\frac{Z_i F\phi(x)}{RT}} dx \tag{9.240}$$

The rate equation for describing strong electrostatic adsorption can be expressed as:

$$\frac{dn_i(t)}{Sdt} = k_1 = k_1[n_i(t)]^0 \tag{9.241}$$

Equation (9.241) indicates that, the strong electrostatic force adsorption will appear to be a zero order rate process.

Correspondingly, the discrete form of Eq. (9.241) is:

$$\frac{n_i(t_{m+1}) - n_i(t_m)}{S\,(t_{m+1} - t_m)} = k_1[n_i(t_{k+1/2})]^0 \tag{9.242}$$

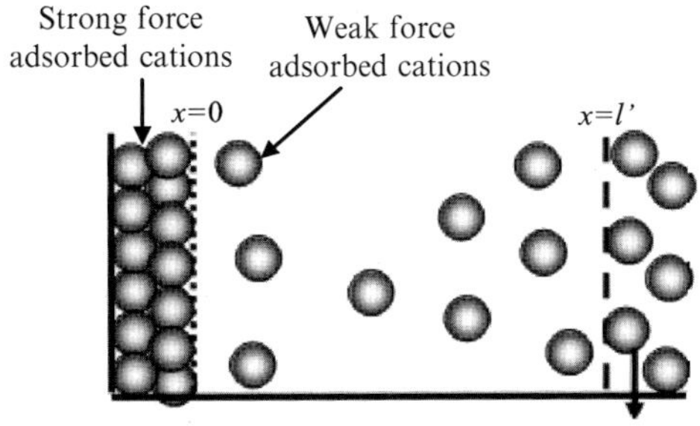

Fig. 9.21 A conceptual diagram of ion adsorption for the weak force adsorption

From Eq. (9.242), if strong adsorption forces are present, using experimental data to plot a relationship curve of $[n_i(t_{m+1}) - n_i(t_m)]/S(t_{m+1} - t_m)$ *vs.* $n_i(t_{m+1/2})$, a straight line which is parallel to the abscissa axis could be obtained, and the intercept of the straight line will be the rate coefficient k_1 of ion adsorption for the strong force adsorption.

As the strong forces adsorption is completed (i.e. the strong adsorption sites have been saturated by the adsorbed cations), then the driving force for cation adsorption will change to the weak electrostatic force, which can be illustrated by Fig. 9.21.

For the weak force adsorption, the rate equation will be the first order equation and can be expressed as Eq. (9.230) or Eq. (9.231). However, in this case, as strong force adsorption exist, for the weak force adsorption, the adsorption or diffusion space decreases from $0 \to l$ to $0 \to l'$ as shown in Fig. 9.21. Thus for the weak force adsorption, the rate coefficient and the adsorbed quantity of the weak force at equilibrium would be expressed respectively as:

$$k_2 = \frac{\pi^2 D_i}{4l'^2} f_{i0} \int_0^{l'} e^{-\frac{Z_i F\phi(x)}{RT}} dx \tag{9.243}$$

$$n_i'(\infty) = S f_{i0} \int_0^{l'} e^{-\frac{Z_i F\phi(x)}{RT}} dx \tag{9.244}$$

Correspondingly, the Eqs. (9.230) and (9.231) change respectively to:

$$\frac{dn_i(t)}{dt} = k_2 \left[1 - \frac{n_i(t)}{n_i(\infty)} \right] \tag{9.245}$$

$$\frac{n_i(t_{m+1}) - n_i(t_m)}{S(t_{m+1} - t_m)} = k_2 - \frac{k_2}{n'_i(\infty)} n_i\left(t_{m+1/2}\right) \tag{9.246}$$

If there is the strong force adsorption in cation exchange experiments, usually both strong and weak force adsorption will appear in the experiment, and in the initial stage of adsorption there will be the strong force adsorption, then the adsorption will change to the weak force adsorption. Thus using experimental data to plot the relationship curves of $[n_i(t_{m+1}) - n_i(t_m)]/S(t_{m+1} - t_m)$ *vs.* $n_i(t_{m+1/2})$, two

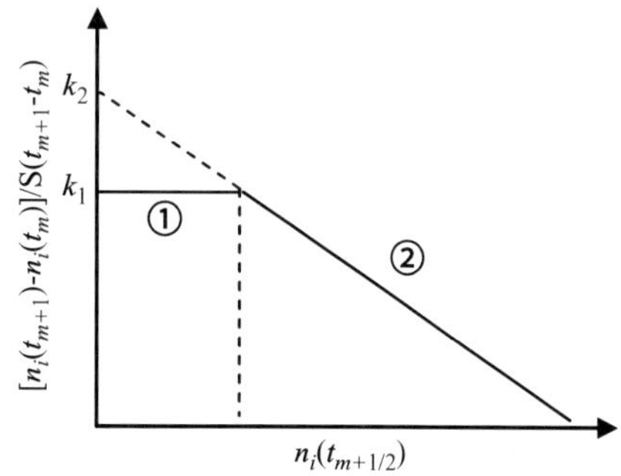

Fig. 9.22 The adsorption rate curves as both strong and weak force adsorption presence

straight lines could be obtained. In the initial stage of adsorption, the straight line will be parallel to the abscissa axis, and the intercept of the straight line will be the rate coefficient k_1. This is the zero order rate process. Then, at a given time moment, the adsorption changes to the first order rate process, and the relationship curve will transfer to another straight line, the intercept and the slope of the straight line will be k_2 and $\pi^2 D_i/4\,l'^2 S$ respectively. The turning point of the two straight lines corresponds to the time when the strong force adsorption transfers to the weak force adsorption. Considering $l' < l$, there must be $k_2 > k_1$. Thus, if both strong and weak forces adsorptions occur, the two straight lines could be illustrated in Fig. 9.22.

9.3.1.4 Some Experimental Results of Cation Adsorptions

Mg^{2+} Adsorption in a Ca^{2+} Saturated Montmorrilonite

Figure 9.23 gives the experimental results of Mg^{2+} adsorption in a Ca^{2+} saturated montmorrilonite under two different Mg^{2+} concentrations in flowing liquid. The experimental results show that, the relationship between $[n_i(t_{m+1})-n_i(t_m)]/S(t_{m+1}-t_m)$ and $n_i(t_{m+1/2})$ were indeed straight lines for the two concentrations. According to the theoretical analysis, the intercept of the straight lines represents the adsorption rate coefficient k. As the Mg^{2+} concentrations were 8.833×10^{-5} mol/l and 5.000×10^{-5} mol/l, the corresponding k values were 2.882×10^{-7} and 1.527×10^{-7} mol/ming respectively.

According to the theoretical analysis, from the ratio of the intercept and the slope shown in Fig. 9.23, the CEC value can be calculated. Conversely, if the CEC value of the montmorrilonite was obtained in advance, this ratio value can be theoretically predicted. The CEC value of the montmorrilonite was determined in advance by using a cation adsorption method, it was: 51.7cmol$_+$/Kg. Thus the theoretically predicted ratio value of $|$ intercept/slope $|$ was 25.9×10^{-5} mol/g. Based on the experimental results shown in Fig. 9.23, for the two concentrations of 8.833×10^{-5} mol/l and 5.000×10^{-5} mol/l, the experimental ratio values were 28.9×10^{-5} and 29.3×10^{-5} mol/g respectively, thus the average value is $(29.1 \pm 0.2) \times 10^{-5}$ mol/g. This shows that the experimental result agree with the

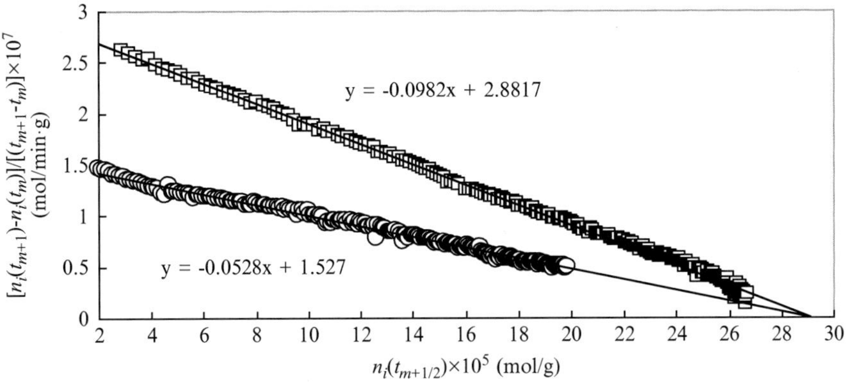

Fig. 9.23 Experimental results of $[n_i(t_{k+1})-n_i(t_k)]/[(t_{k+1}-t_k)]$ *vs.* $n_i(t_{k+1/2})$ for the motmorillonite □□□: $f_{i0} = 8.833 \times 10^{-5}$ mol/l; ooo: $f_{i0} = 5.000 \times 10^{-5}$ mol/l

theoretical prediction well. However, if the electrostatic field is not considered, correspondingly the theoretical prediction can also be made, and the result is 0.00595×10^{-5} mol/g in average for the two concentrations (Li et al. 2010), which is much lower than the experimental result.

Mg^{2+} Adsorption in a K^+ Saturated Montmorrilonite

Here, montmorillonite was still used as the experimental material. The surface charge number was 62.2cmol$_c$/kg, and the specific surface area was 650.5 m^2/g. Theoretically, the monovalent cations, e.g., K^+, Na^+, etc., have the weaker effect in shielding the negative electrostatic field from soil particle surface than that of the divalent cations, e.g., Ca^{2+}, Mg^{2+}, etc. Therefore, the electric field strength of the monovalent cations saturated sample will be much larger than that of the divalent cations saturated sample under a given electrolyte concentration and temperature conditions, this can be shown in Fig. 9.24.

Figure 9.24 shows that, as the negative charges on particle surface were saturated by K^+, a strong electrostatic field may possibly be created possibly, and the strong adsorption force may exist for Mg^{2+} adsorption in the Mg^{2+}/K^+ exchange experiment. On the other hand, as the negative charges on particle surfaces were saturated by Ca^{2+}, the strong electrostatic field may not be set up possibly, and the strong force adsorption may not exist for Mg^{2+} adsorption in the Mg^{2+}/Ca^{2+} exchange experiment.

The experimental results of $[N_i(t_{m+1})-N_i(t_m)]/S(t_{m+1}-t_m)$ *vs.* $N_i(t_{m+1/2})$ for Mg^{2+} adsorption in Mg^{2+}/K^+ and Mg^{2+}/Ca^{2+} Exchanges are shown in Figs. 9.25 and 9.26 respectively.

Figure 9.25 showed that: (1) the zero-order kinetic process was present in the initial stage of adsorption for $t = 0$–405 min. Based on the theoretical analysis, there

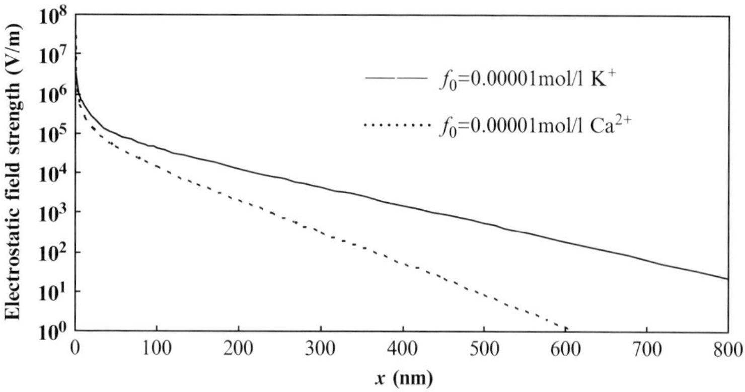

Fig. 9.24 The distribution of the electrostatic field strength away from soil particle surfaces, supposing surface charge number = 10 cmol$_c$/Kg, specific surface area = 50 m^2/g, T = 298 K

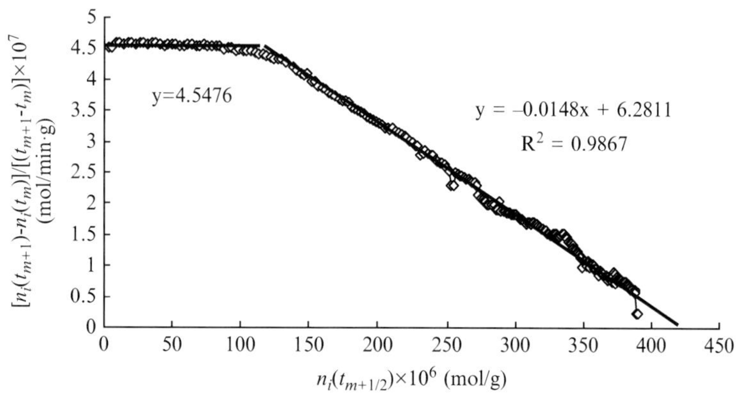

Fig. 9.25 The experimental results of the relationship between $y = [N_i(t_{m+1}) - N_i(t_m)]/S(t_{m+1} - t_m)$ and $x = N_i(t_{m+1/2})$ of Mg^{2+} adsorption for Mg^{2+}/K^+ Exchange ◇, experimental data;——, *fitting line*

would be the strong force adsorption for Mg^{2+} in the experiment of Mg^{2+}/K^+ exchange. Because the chemical bond might not exist between Mg^{2+} and the particle surface for the sample, the strong force would be the electrostatic force. (2) After a given time moment of $t = 405$ min in the experiment, the adsorption process turned to the first-order kinetic process, and the turning point from the zero-order to the first-order was very distinct.

From Fig. 9.26, however, for Mg^{2+} adsorption in Mg^{2+}/Ca^{2+} exchange, the zero-order kinetic process was not appeared yet as discussed as above. Therefore, for Mg^{2+} adsorption in Mg^{2+}/Ca^{2+} exchange, only weak force adsorption process was present.

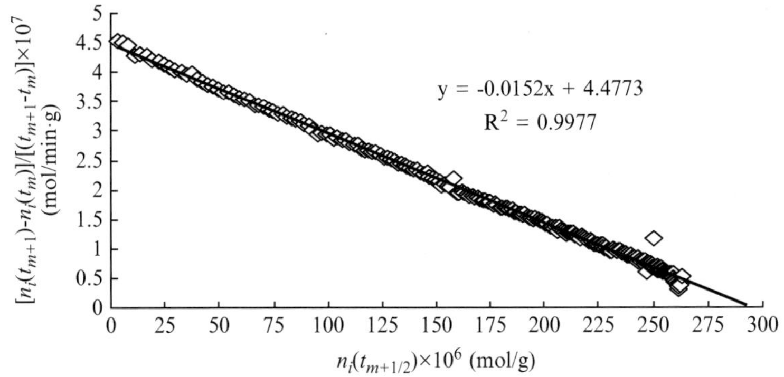

Fig. 9.26 The experimental results of the relationship between $y=[N_i(t_{m+1})-N_i(t_m)]/S(t_{m+1}-t_m)$ and $x=N_i(t_{m+1/2})$ of Mg^{2+} adsorption for Mg^{2+}/Ca^{2+} Exchange ◇, experimental data;——, *fitting line*

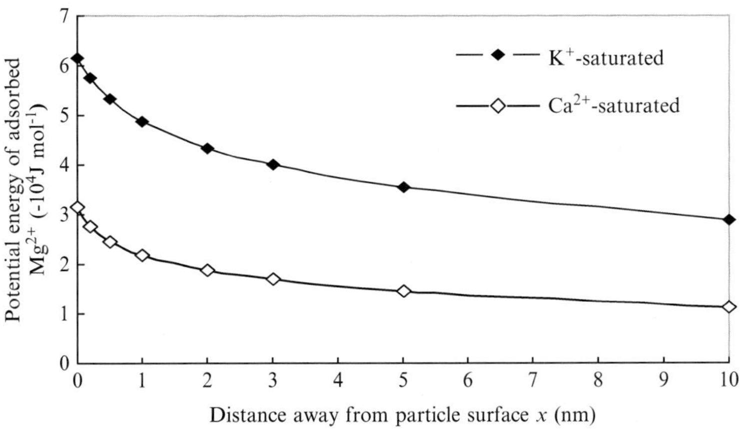

Fig. 9.27 The distributions of the potential energy of Mg^{2+} in the space adjacent montmorillonite particle surface for the Ca^{2+} and K^+-saturated samples respectively

As discussed above, the electrostatic field strength around a particle in solution for Ca^{2+}-saturated sample will be much weaker than that for the K^+-saturated samples. Using the surface charge number and the specific surface area values of the montmorrilonite, the potential energy $E(x)=ZF\varphi(x)$ of Mg^{2+} change with the distance away from the montmorrilonite surface could be calculated in the K- and Ca-saturated samples (Li et al. 2004), and they are shown in Fig. 9.26.

Figure 9.27 indicates that, at a given position in the space adjacent particle surface, the potential energy (negative) of Mg^{2+} in the K^+-saturated sample was much lower than that in the Ca^{2+}-saturated sample. For example, at a distance of 1 nm away from the particle surface, for the K^+-saturated sample, the potential energy of adsorbed Mg^{2+} was $E_{Mg}=-48803\text{Jmol}^{-1}$; correspondingly, for the Ca^{2+}-saturated sample, the potential energy of adsorbed Mg^{2+} was merely $E_{Mg} = -21{,}806$ Jmol^{-1}. Therefore, Mg^{2+} would be adsorbed much stronger in the K^+-saturated montmorillonite than in the Ca^{2+}-saturated montmorillonite in the exchange process. On the other hand, at a distance of 1 nm away from the particle surface, the difference of potential energy between E_{Mg} and E_{Ca} was $\Delta E_{Mg/Ca} = E_{Mg} - E_{Ca} = -21{,}806 + 21{,}806 = 0$ J mol^{-1} for the Mg/Ca exchange; correspondingly, the difference of potential energy between E_{Mg} and E_K was $\Delta E_{Mg/K} = E_{Mg} - E_K = -48{,}803 + 24{,}401 = -24402$Jmol^{-1} for the Mg/K exchange. This shows that, the Mg^{2+} exchange K^+ was much easier than Mg^{2+} exchange Ca^{2+}. Therefore, the Mg/K exchange could occur by the zero-order kinetic process, but the Mg/Ca exchange could not.

According to the theoretical analysis, using the slope and the intercept of the straight line shown in Fig. 9.24 or 9.25, the totally adsorbed quantities N_∞ can be calculated. (1) *As strong and weak force adsorption coexisted for* Mg^{2+}*adsorption in the Mg/K exchange*, the calculated value was $N_\infty = 424.4$ mmol kg^{-1}. Here the totally adsorbed quantities: N_∞ included the adsorbed quantities of the strong force and weak force adsorptions. The adsorption total quantity of Mg^{2+} at the turning point was the maximum quantity of the strong force adsorption at equilibrium. According to the fitted equations shown in Fig. 9.24, at the turning point we must have:

$$4.5476 = -0.0148x + 6.2811 \tag{9.247}$$

where x is the adsorption total quantity of Mg^{2+} at the turning point.

From Eq. (9.246) we could get the quantity of the strong force adsorption of Mg^{2+} at equilibrium, it was 117.1 mmol kg^{-1}. Thus we can calculate the quantity of the weak force adsorption of Mg^{2+} at equilibrium Mg/K exchange, given a value of 424.4–117.1 = 307.3 mmol/kg.

As only weak force adsorption existed for Mg^{2+} adsorption in the Mg/Ca exchange, from Fig. 9.26, the calculated adsorption quantity of Mg^{2+} at equilibrium was 4.4773/0.0152 = 294.6 mmol kg^{-1}. This shows that, the obtained weak force adsorption quantity for Mg^{2+} in Mg^{2+}/Ca^{2+} exchange equals to that quantity of Mg^{2+} in Mg^{2+}/K^+ exchange approximately (307.3 mmol/kg). This result also indicated that, because Ca^{2+} has a stronger adsorption force in a given electrostatic field than does Mg^{2+}, a portion of the strongly adsorbed Ca^{2+} could not be exchanged by Mg^{2+}.

The previous study by Li et al. showed that (2010), for Mg^{2+} adsorption in the Mg^{2+}/Ca^{2+} exchange in an Inceptisols soil sample (the neutral purple soil in Chinese soil taxonomy), there was merely the first order kinetic process present. If using the same type of Inceptisols soil, and the Inceptisols soil 90% + montmorillonite 10% in mass percent as the experimental materials, however, the Mg^{2+} and Ca^{2+} adsorptions in Mg^{2+}/K^+ and Ca^{2+}/K^+ exchange respectively under different background electrolyte composition and concentration conditions in bulk solution, both the zero order and the first order kinetic processes occur, as shown in Fig. 9.28 (Li and Li 2010).

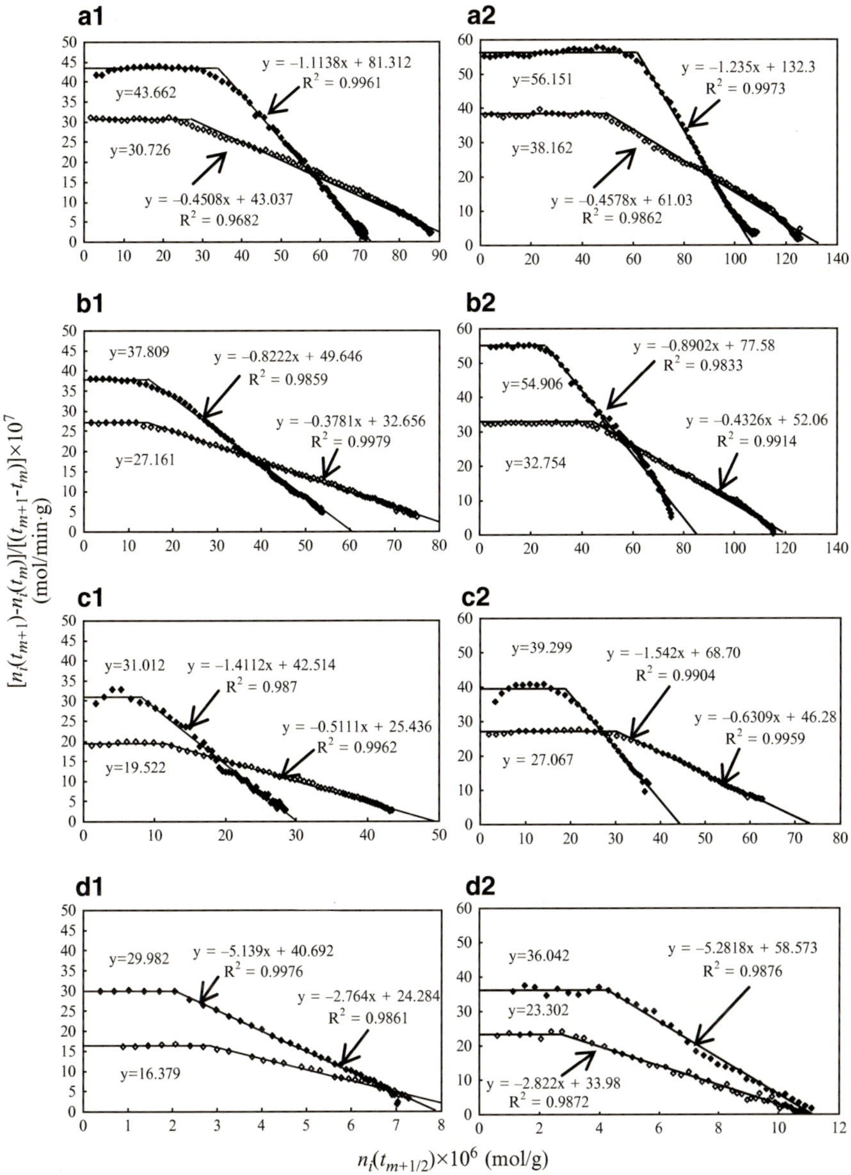

Fig. 9.28 The experimental results of the relationship between $[n_i(t_{m+1})-n_i(t_m)]/S(t_{m+1}-t_m)$ and $n_i(t_{m+1/2})$ o Mg^{2+} and Ca^{2+} adsorptions in Mg^{2+}/K^+ and Ca^{2+}/K^+ exchange respectively (**a₁**): 10^{-4} mol/L(Mg^{2+}); (**a₂**):10^{-4} mol/L(Ca^{2+}) (**b₁**): 10^{-4} mol/L(Mg^{2+}) + 10^{-4} mol/L(K^+); (**b₂**): 10^{-4} mol/L(Ca^{2+}) + 10^{-4} mol/L(K^+) (**c₁**): 10^{-4} mol/L(Mg^{2+}) + 10^{-3} mol/L(K^+); (**c₂**): 10^{-4} mol/L(Ca^{2+}) + 10^{-3} mol/L(K^+) (**d₁**): 10^{-4} mol/L(Mg^{2+}) + 10^{-2} mol/L(K^+); (**d₂**): 10^{-4} mol/L(Ca^{2+}) + 10^{-2} mol/L(K^+) —♦—soil100%;—◇—soil 90% + montmorillonite10%

Thus, the new rate equations and experimental results show that, when strong and weak force adsorption coexist in the exchange process, in the initial stage of strong force adsorption, the adsorption process will appear as zero-order kinetics, and then transfer to the first-order kinetics of the weak force adsorption. Because K^+-saturated samples could form a much stronger electrostatic field adjacent particle surface than Ca^{2+}-saturated samples did, Mg^{2+} or Ca^{2+} adsorption in K^+-saturated sample appeared zero-order kinetic process in the initial stage of the strong force adsorption, and then the adsorption process transfer to the first-order kinetics of the weak force adsorption. However, for the Ca^{2+}-saturated sample, because only weak electrostatic field presence, merely the first-order kinetic process appeared for Mg^{2+} adsorption.

References

Abell GC, Mozumber A (1972) Application of diffusion model for recombination of isolated pairs in condensed media. J Chem Phys 56:4079

Abell GC, Mozumder A, Magee JL (1972) Laplace transformation method in theory of ion neutralization–Application to scavenging probability in gamma-radiolysis of dielectric liquids. J Chem Phys 56:5422

Achari G, Chatterji S, Joshi RC, Bentley LR (1997) Effect of ion concentration on diffusion through clay. Soil Sci 162:51

Aharoni C, Sparks DL, Levinson L, Ravina I (1991) Kinetics of soil chemical reactions: relationships between empirical equations and diffusion models. Soil Sci Soc Am J 55:1307–1312

Åkesson B, Jőnsson B, Chan DYC (1986) Ion diffusion at charged interfaces – A stochastic dynamics simulation test of the Smoluchowski-Poisson-Boltzmann approximation. Mol Phys 57:1105

Ben Y, Chang HC (2002) Nonlinear Smoluchowski slip velocity and micro-vortex generation. J Fluid Mech 461:229

Benesi AJ (1986) Diffusion in potentials-A method for solving the Smoluchowski equation. J Chem Phys 85:374

Berber-Mendoza MS, Leyva-Ramos R, Alonso-Davila P, Fuentes-Rubio L, Guerrero-Coronado RM (2006) Comparison of isotherms for the ion exchange of Pb(II) from aqueous solution onto homoionic clinoptilolite. J Colloid Interf Sci 301:40–45

Berezhkovskii AM, D'yakov YuA (1998) Smoluchowski equation with a sink term: Analytical solutions for the rate constant and their numerical test. J Chem Phys 109:4182

Biesuz R, Zagorodni AA, Muhammed M (2001) Estimation of deprotonation coefficients for chelating ion exchange resins: comparison of different thermodynamic models. J Phys Chem B 105:4721–4726

Bolt GH (1955) Ion adsorption by clay. Soil Sci 79:267–276

Butler PR, Pilling MJ, Rice SA, Stone TJ (1977) Numerical investigation of tunneling contributions to electron scavenging reactions in liquids at short times. Can J Chem 55:2124

Chan DYC (1987) Ion diffusion near charged surfaces – Exact analytical solutions. J Chem Soc Faraday Trans 83:2271

Chan DYC, Halle B (1984) The Smoluchowski-Poisson-Boltzmann description of ion diffusion at charged interface. Biophys J 46:387

Chan DYC, Hughes BD (1988) Ion diffusion in Coulombic field. J Stat Phys 52:383

Chatterji S (2004) Ionic diffusion through thick matrices of charged particles. J Colloid Interface Sci 269:186–191

Chavanis PH (2003) Generalized thermodynamics and Fokker-Planck equations: Applications to stellar dynamics and two-dimensional turbulence. Phys Rev E 68:036108

Clifford P, Green NJB, Pilling MJ (1984) Analysis of the Debye-Smoluchowski equation – Approximation for high-permittivity solvents. J Phys Chem 88:4171

Crank J (2001) The mathematics of diffusion. Oxford Science, New York

De Bokx PK, Boots HMJ (1989) The ion-exchange equilibrium. J Phys Chem 93:8243–8248

Delaire JA, Croc E, Cordier P (1981) J Phys Chem 85:1549

Eriksson E (1952) Cation-exchange equilibrium on clay minerals. Soil Sci 74:103–111

Fatin-Rouge N, Milon A, Buffle J, Goulet RR, Tessier A (2003) Diffusion and partitioning of solutes in agarose hydrogels: The relative influence of electrostatic and specific interactions. J Phys Chem B 107:12126

Gitterman M, Weiss GH (1993) Small-noise approximations to the solution of the Smoluchowski equation. Phys Rev E 47:976

Graul TW, Li M, Schlenoff JB (1999) Ion exchange in ultrathin films. J Phys Chem B 103:2718–2723

Hanggi P, Luczka J, Talkner P (2005) Statistics of transition times, phase diffusion and synchronization in periodically driven bistable systems. New J Phys 7

Hong KM, Noolandi J (1978a) Solution of Smoluchowski equation with a Coulomb potential. 1. General results. J Chem Phys 68:5163

Hong KM, Noolandi J (1978b) Solution of Smoluchowski equation with a Coulomb potential. 2. Application to fluorescence quenching. J Chem Phys 68:5172

Hui TW, Baker MD (2001) Ion exchange and electron transport at methyl viologen Y modified electrodes. J Phys Chem B 105:3204–3210

Jacob AM, Reddy MM (1991) Employment of gibbs-donnan-based concepts for interpretation of the properties of linear polyelectrolyte solutions. J Phys Chem 95:10208–10214

Jardine PM, Sparks DL (1984) Potassium-calcium exchange in multireactive soil system: I. Kinetics. Soil Sci Soc Am J 48:39–45

Jenny H (1932) Studies on the mechanism of ion exchange in colloidal aluminum silicates. J Phys Chem 36:2217–2258

Jenny H (1936) Simple kinetic theory of ion exchange. I. Ions of equal valency. J Phys Chem 40:501–517

Jones RB (2003) Adiabatic change in the Smoluchowski equation: Orientational diffusion of polar particles. J Chem Phys 119:1517

Kemper WD, Van Schaik JC (1966) Diffusion of salts in clay-water systems. Soil Sci Soc Am Proc 30:534–540

Klik I, Yao YD (2000) Solution of a separable Smoluchowski equation in one spatial dimension. Phys Rev E 62:4469

Li R (1986) Non-equilibrium thermodynamics and dissipative structure. Qinghua University Press, Beijing (In Chinese)

Li R, Li H (2010) Comparison study between Mg/K and Ca/K exchange kinetics under electric fields at the solid/liquid interface of soil. Acta Phys-Chim Sin 26(3):552–560

Li H, Wu L (2004) On the relationship between thermal diffusion and molecular interaction energy in binary mixtures. J Phys Chem B 108:13821

Li H, Wu L (2007a) A generalized linear equation for non-linear diffusion in external fields and non-ideal systems. New J Phys 9:357

Li H, Wu L (2007b) A new approach to estimate ion distribution between the exchanger and solution phases. Soil Sci Soc Am J 71:1694–1698

Li H, Xue JH (1996) Study on ion diffusion kinetics in soil systems. Acta Pedologica Sinica 33(4):327–336 (in Chinese)

Li H, Zhang H (2001) Mathematics in soil science. Agricultural Press, Beijing

Li H, Qing CL, Wei SQ, Jiang XJ (2004) An approach to the method for determination of surface potential on solid/liquid interface: theory. J Colloid Interface Sci 275:172–176

Li H, Zhu HL, Jia MY, Wu LS (2009a) Some new features of linear theory for describing non-linear diffusion. Commun Theor Phys 52:269–273

Li H, Laosheng Wu, Zhu H, Hou J (2009b) Ion diffusion in the time-dependent potential of the dynamic electric double layer. J Phys Chem C 113:13241–13248
Li H, Li R, Zhu H, Wu L (2010) Influence of electrostatic field from soil particle surfaces on ion adsorption–diffusion. Soil Sci Soc Am J 74:1129–1138
Li H, Hou J, Liu X, Li R, Zhu H, Wu W (2011a) Combined determination of specific surface area and surface charge properties of charged particles from a single experiment. Soil Sci Soc Am J 75(6):2128–2135
Li R, Li H, Zhu H, Wu L (2011b) Kinetics of cation adsorption on charged soil mineral as strong electrostatic force presence or absence. J Soils Sedim 11(1):53–61
Liu X, Li H, Li R, Hou J (2012) A new theory on cation exchange equilibrium of hydrated. J Soil Sediments (accepted)
Magee JL, Tayler AB (1972) Recombination of ions in a Coulomb field in presence of a scavenger. J Chem Phys 56:3061
Monchick L (1956) Note on the theory of diffusion controlled reactions – Application to photodissociation in solution. J Chem Phys 24:381
Montroll EW (1946) A note of the theory of diffusion controlled reactions with application of the quenching of fluorescence. J Chem Phys 14:202
Morgan JD, Napper DH, Warr GG (1995) Thermodynamics of ion exchange selectivity. J Phys Chem 99:9458–9465
Mozumder A (1968) Theory of neutralization of an isolated ion pair – Application of method of prescribed diffusion to random walk in Coulomb field. J Chem Phys 48:1659
Nkedi-Kizza P, Rao PSC, Jessup RE, Davidson JM (1982) Ion exchange and diffusive mass transfer during miscible displacement. Soil Sci Soc Am J 46:471–4476
Ogwada RA, Sparks DL (1986a) A critical evaluation on the use of kinetics for determining thermodynamics of ion exchange in soils. Soil Sci Soc Am J 50:300–305
Ogwada RA, Sparks DL (1986b) Kinetics of ion exchange on clay minerals and soil: I. Evaluation of methods. Soil Sci Soc Am J 50:1162–1166
Olsen SR, Kemper WD, Jackson RD (1962) Phosphate diffusion to plant roots. Soil Sci Soc Am Proc 26:222–226
Pauley JL (1953) Prediction of cation-exchange equilibria. J Am Chem Soc 76:1422–1425
Pechukas P, Ankerhold J, Grabert H (2001) Quantum Smoluchowski equation II: The overdamped harmonic oscillator. J Phys Chem B 105:6638
Pedersen JB, Freed JH (1973) Theory of chemically induced dynamic electron polarization. J Chem Phys 58:2746
Porter LK, Kemper WD, Jackson RD, Stewart BA (1960) Chloride diffusion in soils as influenced by moisture content. Soil Sci Soc Am Proc 24:460–463
Rao PSC, Rolston DE, Jessup RE, Davidson JM (1980) Solute transport in aggregated porous media: theoretical and experimental evaluation. Soil Sci Soc Am J 44:1139–1145
Rice SA, Butler R, Pilling MJ, Baird JK (1979) Solution of the Debye-Smoluchowski equation for the rate of reaction of ions in dilute solution. J Chem Phys 70:4001
Samson E, Marchand J, Robert JL, Bournazel JP (1999) Modelling ion diffusion mechanisms in porous media. Int J Numer Meth Eng 46:2043
Schulten Z, Schulten K (1977) General, diffusion, spin motion, and recombination of redical pairs in solution in nano second time domain. J Chem Phys 66:4616
Shainberg I, Kemper WD (1966) Hydration status of adsorbed cations. Soil Sci Soc Am Proc 30:707–713
Sibani P, Boiden JP (1981) A perturbation solution of the smoluchowski equation in the weak potential region. J Chem Phys 74:6934
Sparks DL (1989) Kinetics of soil chemical processes. Academic Press, Inc., San Diego/New York/Berkeley/Boston/London/Sydney/Tokyo/Toronto
Sparks DL (1998a) Soil physical chemistry, 2nd edn. CRC Press, Boca Raton/Boston/London/New York/Washington, DC, p 17
Sparks DL (1998b) Chapter 11. In: Huang PM, Senesi N, Buffle J (eds) Structure and surface reactions of soil particles. Wiley, New York

Sparks DL, Jardine PM (1984) Comparison of kinetics equations to describe K-Ca exchange in pure and in mixed systems. Soil Sci 138:115–122
Sparks DL, Zalazny LW, Martens DC (1980) Kinetics of potassium desorption in soil using miscible displacement. Soil Sci Soc Am J 44:1205–1208
Sposito G (1981) The thermodynamics of soil solution. Oxford University Press, New York, pp 126–150
Sposito G (1984) The surface chemistry of soils. Oxford University Press, New York
Su WD (1986) Electro-chemistry. In: Jilin University (ed) Physical chemistry. Education Press, Beijing, pp 212–318
Suresh G, Scindia YM, Pandey AK, Goswami A (2004) Isotopic and ion-exchange kinetics in the Nafion-117 membrane. J Phys Chem B 108:4104–4110
Tachiya M (1979) Recombination of geminate species in the presence of a scavenger. J Chem Phys 70:238
Tan KH (1982) Principle of soil chemistry. Marcel Dekker Inc., New York/Basel, pp 162–171
Tang LY, Sparks DL (1993) Cation-exchange kinetics on montmorillonite using pressure-jump relaxation. Soil Sci Soc Am J 57:42–46
Tanioka A, Kawaguchi M, Hamada M, Yoshie K (1998) Dissociation constant of a weak electrolyte in charged membrane. J Phys Chem B 102:1730–1735
Van Duijvenbode RC, Koper GJM (2001) Fast adsorption on nonideal surfaces. J Phys Chem B 105:11729
Van Schaik JC, Kemper WD, Olsen SR (1966) Contribution of adsorbed cations to diffusion in clay-water systems. Soil Sci Soc Am Proc 30:17–22
Xiong YT (1981) Statistic physics. Peoples Educational Press, Beijing (in Chinese)
Zhang T (2007) Statistical dynamics and its applications. Metallurgical Industry Press, Beijing
Zientara GP, Freed JH (1979) J Chem Phys 70:2587

Chapter 10
Effects of "Aging" on Bioreactive Chemical Retention, Transformation, and Transport in Soil

William C. Koskinen and H.H. Cheng

Abstract The purpose of this paper is to illustrate how aging can affect the behavior of bioreactive chemicals in the soil. This paper will focus on the sorption process, which can directly or indirectly control transformation and transport processes. Most commonly, aging effects have been characterized by batch equilibration and batch sequential extraction methods, in which changes in sorption coefficients (i.e. K_d) are determined for the chemical remaining after a given equilibration or incubation period. Alternative innovative methods to determine aging effects include isotopic exchange and supercritical fluid extraction methods. K_d values have generally been shown to increase with aging time. The mechanisms that result in aging effects on sorption in soils can be artifacts of the methodology used to characterize sorption, or changes in chemical or physical processes, or a combination of these processes, all of which may affect the retention or sorption/desorption of the chemical in soil, which in turn affects pesticide transport and transformation. Regardless of the mechanisms involved in aging, observed increases in sorption during pesticide aging should be taken into account during characterization of the sorption process for mathematical models of pesticide transport. Potential transport of the pesticide would be over predicted if freshly treated soil K_d values were used to predict transport. The question of whether increases in sorption with aging affect degradation is still unanswered.

Keywords Sorption • Desorption • Aged residues • Bioavailability

W.C. Koskinen (✉)
USDA-Agricultural Research Service, St. Paul, MN 55108, USA
e-mail: koskinen@umn.edu

H.H. Cheng
Department of Soil, Water, & Climate, University of Minnesota, St. Paul, MN 55108, USA

J. Xu and D.L. Sparks (eds.), *Molecular Environmental Soil Science*, Progress in Soil Science, DOI 10.1007/978-94-007-4177-5_10,

10.1 Introduction

The "aging" effect on the process of transformation of bioreactive chemicals in soil systems has been studied extensively. For instance, in quantitative studies of nitrogen transformations in soils, partitioning soil nitrogen components into bioreactive pools of various transformation rates has greatly improved the effectiveness of model predictions (e.g., Cheng et al. 1994). Similar approaches have been applied to study the transformation of components of a single chemical using radioisotope tracing techniques (e.g., Cheng et al. 1978). However, less attention has been given to the aging effect on the processes of retention of chemicals in soil and on the effect of changing retention on transformation and transport process. Since the fate of a chemical in the soil environment is affected by retention, transformation, as well as transport processes and by their interactions, research methods adopted for studying these processes need to consider the aging effect in characterizing these processes. The purpose of this paper is to illustrate how aging can affect the behavior of bioreactive chemicals in the soil. Since many studies on aging were on the fate of pesticides, they serve as useful models for studying other bioreactive chemicals in the soil environment. This paper will focus on the sorption process, which can directly or indirectly control transformation and transport processes.

10.2 Methods to Characterize Aging Effects

Aging, which can range from hours to days to several months, and its effects on sorption have been characterized by a variety of methods. Most commonly, aging effects have been characterized by batch equilibration methods, in which changes in sorption coefficients (K_d) or Freundlich isotherm coefficients (K_f) are determined for the chemical remaining after a given equilibration or incubation period. Traditional sorption-desorption characterizations typically measure changes in solution concentrations during 24-h equilibrations for both sorption and desorption. However, for some bioreactive chemicals, sorption equilibrium is not reached for weeks to months (i.e. Mamy and Barriuso 2007; Sander and Pignatello 2009). To characterize longer term aging effects by the batch equilibration method, the parent chemical is sequentially extracted from soil, first with aqueous $CaCl_2$ to give the solution phase concentration after each aging period, and then by organic solvents to give the sorbed phase concentration. Sorption K_d values can be calculated as in traditional batch sorption studies.

Effects of aging on sorption-desorption have been shown in both short-term and long-term batch studies. Numerous studies using the traditional 24-h batch equilibration method have shown a hysteresis effect during desorption of the pesticide from soil. Hysteresis has also been shown to increase with aging (Koskinen et al. 1979) (Fig. 10.1). Using the batch sequential extraction method for long-term aged pesticide residues, increases in the sorption coefficient K_d with incubation

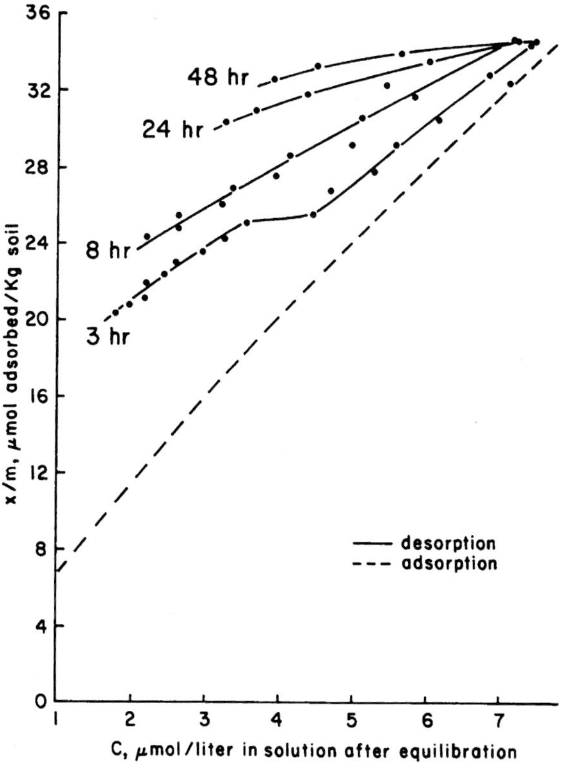

Fig. 10.1 Desorption of 2,4,5-T from silt loam soil, initial concentration = 39 μmol L^{-1}. Effect of equilibration time on hysteresis (From Koskinen et al. 1979)

time have generally been observed for diverse pesticides such as triazines (Boesten and Van der Pas 1983; Pignatello and Huang 1991), substituted ureas (Walker 1987; Blair et al. 1990), pyridine carboxylic acids (McCall and Agin 1985), amides (Walker 1987), carbofuran (Shelton and Parkin 1991), substituted ureas (Walker 1987; Blair et al. 1990; Gaillardon and Sabar 1994), nitroguanidines (Cox et al. 1998; Oi 1999), acetanilides (Pignatello and Huang 1991), imidazolinones (Bresnahan et al. 2000, 2002) sulfonylaminocarbonyltriazolinones (Koskinen et al. 2002, 2003), and sulfonylureas (Regitano and Koskinen 2008) (Fig. 10.2). The increase in sorption K_d in aged soil has been reported to be up to a factor of 8 as compared to that in freshly treated soil (Table 10.1).

Effects of aging on sorption-desorption have also been studied using alternative innovative methods. These methods have primarily been developed to determine mechanisms of aging effects. For instance, an isotopic exchange technique using ^{12}C and ^{14}C labeled pesticides was developed to determine the nature of the equilibrium of the pesticide between solution and sorbed phases (Celis and Koskinen 1999a, b; Sander and Pignatello 2005). These methods were used to determine the amounts of irreversibly bound pesticides in soil to distinguish hysteresis due to experimental artifacts from true hysteresis due to thermodynamically irreversible processes.

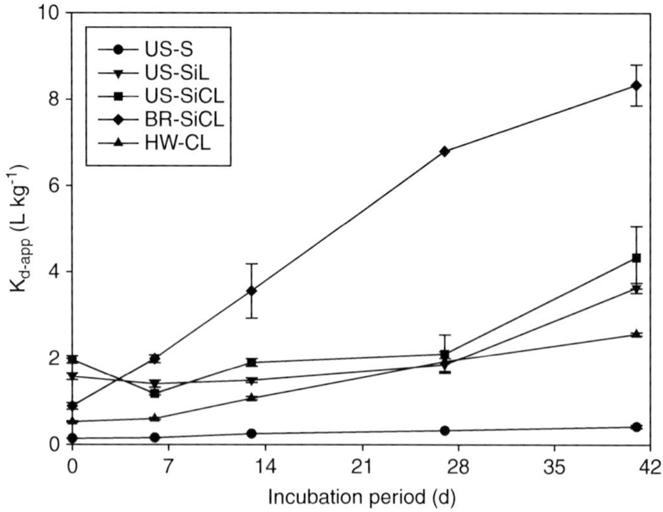

Fig. 10.2 Effect of aging on nicosulfuron sorption coefficients in two Brazilian (BR), two Midwest US (US), and a Hawaiian soil (From Regitano and Koskinen 2008)

Table 10.1 Increase in sorption K_d values as determined by the batch sequential solvent extraction method after aging selected pesticides for 4–8 week as compared to freshly treat soil

Pesticide	Soil	Increase in sorption
Imazethapyr	SiCL	2.0X
Dicamba	SL	7.0X
Flucarbazone	CL, LS	4.0X, 6.8X
Propoxycarbazone	CL, LS	5.0X, 6.1X
Benzensulfonamide	CL, LS	3.5X, 5.4X
Triazolinone	CL, LS	3.7X, 6.5X
Imidacloprid	3 soils	2.8X
I.-urea degradate	3 soils	2.0X
DKN degradate	SiCL, SL	7.5X, 2.0X

Another alternative technique to characterize aging effects on sorption-desorption is the use of supercritical fluid extraction (SFE) to selectively remove the pesticide from the soil-solution surrounding the soil particle without extracting it from soil itself. Above a certain temperature and pressure, the critical point, a substance will be in its supercritical state where there is no difference between the gas and liquid states. For instance, for carbon dioxide the critical point is reached at a modest temperature (31.1°C) and pressure (73.8 bar). The selectivity of the extraction by the supercritical fluid can be controlled by varying its density (i.e. temperature and pressure). SFE has been used to selectively extract atrazine from the soil solution of field-moist soil without also extracting residues bound to the soil particles (Rochette and Koskinen 1996) thus allowing calculation of sorption coefficients at low water contents. Multiple extractions after varying equilibration times allows characterization of sorption-desorption and calculation of sorption coefficients in aged field-moist soils. While there was no effect of

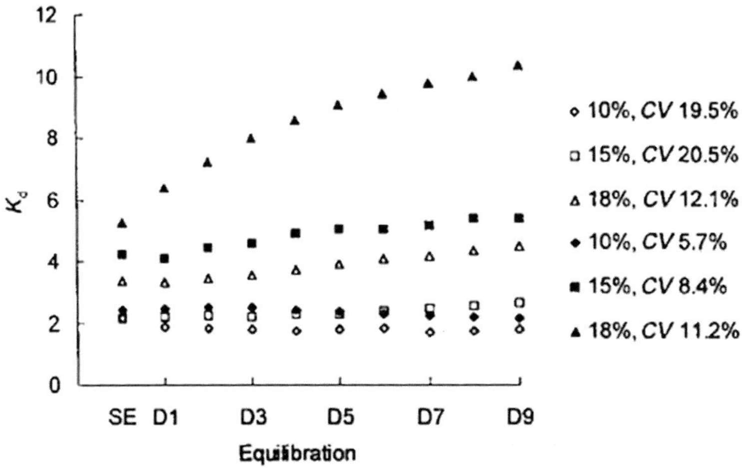

Fig. 10.3 Effect of increased equilibration on sorption of triadimefon in sandy loam soil (*solid symbols*) and silt loam soil (*open symbols*) at three initial moisture contents (From Berglöf et al. 2000b)

short-term aging on linuron sorption-desorption in soil, sorption of triadimefon increased with aging in soil at 18% soil moisture, whereas there was no aging effect at a moisture content of 10% (Fig. 10.3) (Berglöf et al. 2000a, b).

The SFE technique has also been used to determine sorption coefficients of weakly acidic herbicides (Berglöf et al. 2003). Experiments were performed using low density (i.e. 0.3 g mL^{-1}) SF-CO_2 to convert anionic metsulfuron methyl to the molecular species and remove it from the soil water phase only, thus allowing calculation of sorption coefficients (Kd) at low water contents. Kd values for sorption of the metsulfuron methyl molecular species on a clay loam soil at 11% water content increased with aging (Fig. 10.4). Another application of the SFE method is by using the neutral species Kd value, the pKa of metsulfuron methyl, and the pH of the soil we could successfully predict the Kd values obtained using the batch slurry technique, which typically has a predominance of anionic species in solution during the sorption characterization.

10.3 Mechanisms of Aging Effects

The mechanisms that result in aging effects on sorption in soils are not well understood, but can be possible artifacts of the methodology used to characterize sorption, or changes in chemical or physical processes, or a combination of these processes, all of which may affect the retention or sorption/desorption of the chemical in soil, which in turn affects pesticide transformation and transport (Cheng 1990; Cheng et al. 1994; Cheng and Koskinen 2002). For instance, increases in sorption may

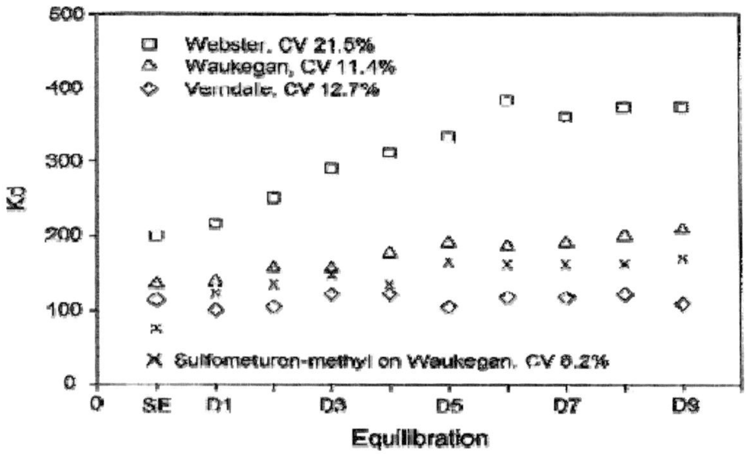

Fig. 10.4 Metsulfuron methyl sorption-desorption coefficients (Kd) for sorption (SE) and desorption equilibrations (D1-D9) for three soils at 11% soil moisture (From Berglöf et al. 2003)

be due to rates of degradation in solution and on labile sites being faster than rates of desorption from the soil particles, diffusion to less accessible or stronger binding sites with time, and/or a combination of the two processes. Degradation of the pesticide during the sorption equilibration, or during the aging periods between sorption characterizations, can significantly affect calculated sorption coefficients.

Several studies have shown that some pesticides may be significantly degraded during the time frame of the experiment, even the 24-h sorption-desorption equilibrations. For instance, 63% of the observed hysteresis during the desorption of 2,4,5-T from a silt loam, and 35% from a clay loam, could be eliminated by taking microbial degradation into account (Koskinen et al. 1979) (Fig. 10.5). Several studies have suggested that only pesticide in solution, or that which is readily desorbed from soil, is available for degradation (i.e. Anderson 1981; Shelton and Parkin 1991); therefore the sorption coefficient is overestimated due to underestimation of the solution concentration.

Pesticides that are sorbed to soil particles are not instantaneously degradable, and must first be desorbed from the soil into solution. Thus, assuming the presence and activity of pesticide degrading microorganisms, degradability would be directly related to their ability to be desorbed from soil. Weakly sorbed and easily desorbed pesticides would be readily degradable, whereas pesticides that are strongly sorbed and hysteretic during desorption would be slowly degradable over time. Conversely, very strongly sorbed pesticides, i.e. bound residues, would be unavailable for biodegradation.

Pesticides that persist in soils often become increasingly less bioavailable, as indicated by markedly declining rates of biodegradation with aging. This has been incorporated into models of biodegradation of pesticides coupled to diffusion-limited sorption-desorption (Scow and Hutson 1992; Scow and Alexander 1992;

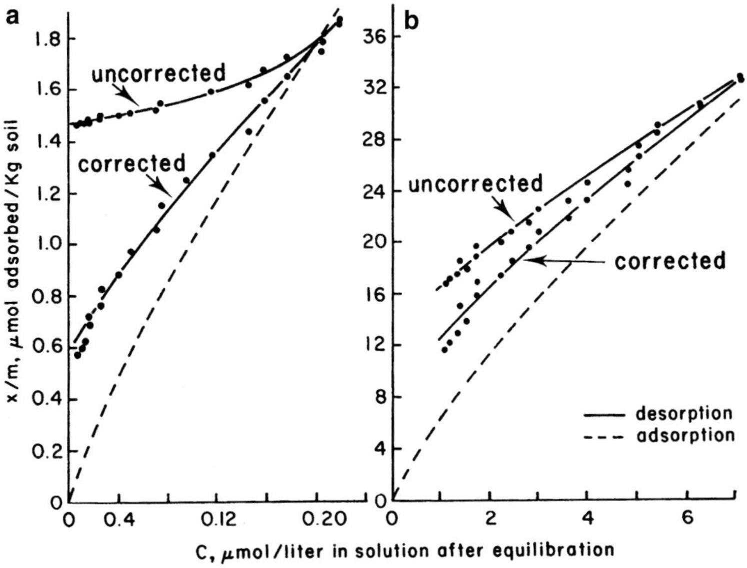

Fig. 10.5 2,4,5-T desorption in silt loam soil. Impact of correction for 2,4,5-T degradation at two concentrations: (**a**) 2 μmol L^{-1}; (**b**) 39 μmol L^{-1} (From Koskinen et al. 1979)

Shelton and Doherty 1997a, b). Coupling a two-site nonequilibrium adsorption model to two first-order degradation terms for the solution phase and sorbed phase pesticide, has indicated that nonequilibrium sorption will initially favor degradation. However, in aged samples, degradation will decrease when desorption kinetics become the limiting factor in the degradation process (Guo et al. 2000). When there is degradation during retention characterization, there could be two possible consequences. Firstly, if the degradation rate is known and the sorption is concentration dependent, i.e. Freundlich slope (1/n) <1, there would be greater sorption as amount of pesticide decreased. Secondly, if degradation occurred but was not known, sorption would overpredicted and observed hysteresis during desorption would be an experimental artifact.

Rapid formation of strongly bound residues would also exhibit a hystersis effect indistinguishable from that of degradation occurring during equilibration. The irreversibility of pesticide sorption-desorption by soil observed in batch equilibration experiments has been characterized using an isotopic exchange technique (Celis and Koskinen 1999a). Quantitative estimation of the irreversible and reversible components of sorption, experimentally derived from isotopic exchange experiments, indicated increased irreversibility with increased preequilibration time and at lower pesticide concentration. For instance, the quantitative estimation of the irreversible and reversible components of sorption, experimentally derived from isotopic exchange experiments, indicated degree of irreversibility (% irreversibly bound) in the order: imidacloprid-SiCL soil (6–32%) < imidacloprid urea-SiCL soil (15–23%) < imidacloprid guanidine-LS soil (32–51%), with greater irreversibility

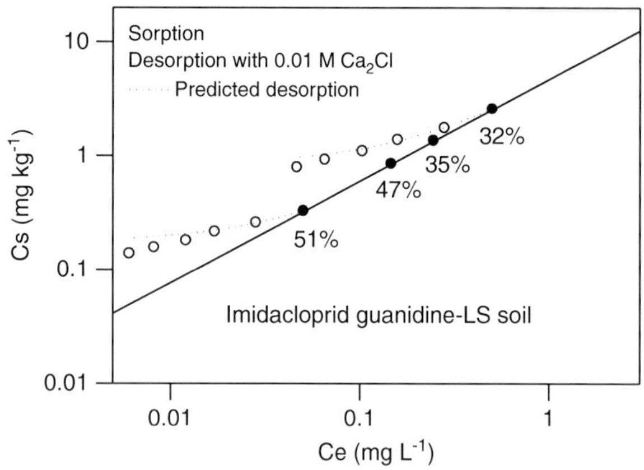

Fig. 10.6 Predicted batch desorption from amount of chemical irreversibly sorbed as determined by the isotopic exchange method (From Celis and Koskinen 1999b)

at lower pesticide concentration (Celis and Koskinen 1999b) (Fig. 10.6). The isotopic exchange of pesticides and metabolites in soil can be described by a two compartment model in which the fraction of strongly bound pesticide increases with time and is accompanied by an equal and opposite decrease in the magnitude of the reversibly bound fraction. This model closely predicted the hysteresis observed in the desorption isotherms from batch experiments.

Aging may result in the formation of a stronger bond or a change in the binding mechanism between pesticide and soil. For instance, for protonatable pesticides such as triazines and imidazolidones, the cationic species may be irreversibly sorbed by cation exchange, thereby removing it from the equilibrating solution. As more cationic species are formed, they are continually sorbed, resulting in an increased sorption coefficient. For instance, imazamox K_d increased by 50% after a 30-days incubation in a soil at pH 5.4, presumably due to cation exchange mechanisms (Bresnahan et al. 2002). The increase in sorption was reversible if the soil pH was raised.

In contrast to cationic pesticides, anionic chemicals are not sorbed by soils, yet the sorption coefficient is the net result of the equilibrium between sorbed neutral chemical and both neutral and anionic species in solution. If the K_d of the molecular species only is determined, it is much greater than the composite K_d value determined in the batch method (Berglöf et al. 2003). In aged soils, the anion may be differentially degraded as opposed to the neutral species. It may also diffuse into microsites. Both of these could result in an observed aging effect.

Chelate formation has also been shown to affect sorption-desorption equilibria. Desorption isotherms indicated that sorption of the biologically active diketonitrile degradate of isoxaflutole, DKN, on some organoclays was irreversible (Carrizosa et al. 2004). Infrared studies showed that DKN enters the interlamellar space of the organoclay and dissociates into the anion, which then forms a very stable chelate

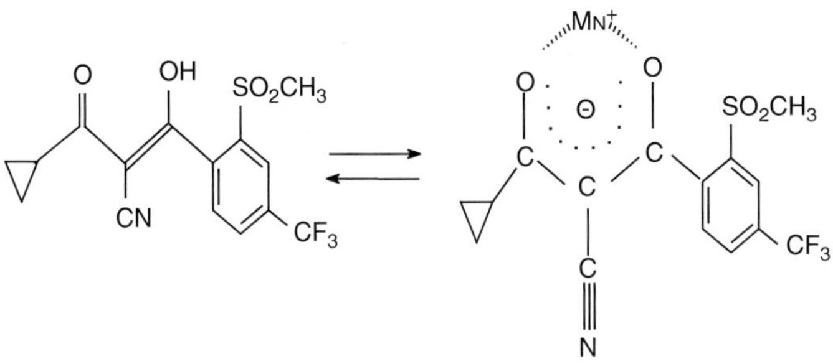

Fig. 10.7 Chelate formation of the biologically active diketonitrile degradate of isoxaflutole (From Carrizosa et al. 2004)

complex with the clay's residual cations and/or partially-coordinated structural cations (Fig. 10.7). The strong binding results in the irreversibility of the sorptive process.

Slow diffusion within small pores of soil aggregates, hydrophobic partitioning into solid humic materials, entrapment into the hydrophobic surface nanopores, and sorption into non-desorbable sites of soil organic matter have all been proposed as possible mechanisms involved in the aging process (Pignatello and Huang 1991; Pignatello et al. 1993). Aging effects for some bioreactive chemicals have been attributed to result from matrix deformation, which can affect diffusion of a pesticide into and out of remote sorption/binding sites (i.e. microsites in soil micropores) resulting in sequestration of the pesticide in soil organic matter or clays (Sander and Pignatello 2009 and the references cited therein). Other mechanisms of hysteresis have also been based on alteration of soil organic matter (OM), including physical rearrangement of the OM during desorption and equilibration resulting in physical entrapment of the pesticide.

10.4 Impacts of Aging Effects on Pesticide Behavior

Screening and transport models use sorption coefficients (K_d and K_{oc} values) to describe pesticide retention by soil for prediction of offsite movement though leaching and runoff. These models are very sensitive to sorption coefficients. The accuracy of the sorption estimates can be more important than the choice of transport model in correctly simulating pesticide leaching, emphasizing the need to understand the sorption-desorption process and the effects of soil and environmental factors on the process. For instance, a generic screening model characterizes pesticides with K_{oc} values of 0–50 as 'very high mobility', those with K_{oc} 50–150 as 'high mobility', and those with K_{oc} 150–500 as 'medium mobility'. Increasing K_{oc} by a factor of 3 would change the mobility classification to a significantly less mobile chemical. We have shown that during aging in soils of numerous types of pesticides, and some of their metabolites, sorption coefficients increased by a factor >3. For instance, K_d values for flucarbazone and propoxycarbazone and their benzensulfonamide and triazolinone metabolites increased by a factor of 3.5–6.8 within 2 weeks after application to clay loam and loamy sand soils (Koskinen et al. 2002). The K_d of imidacloprid (Cox et al. 1998) and two metabolites (Koskinen et al. 2001) increased by a factor of 2–3 within 2 weeks of application to three different soils and nicosulfuron K_d increased by a factor of 2–3 in Mollisols from the United States and by a factor of 5–9 in Oxisols from Hawaii and Brazil after 41 days as compared to freshly treated soils (Regitano and Koskinen 2008). In all these cases, potential mobility would change from high to medium mobility.

Transport models often overpredict the potential leaching in field soils compared to the leaching that is observed in the field, presumably because the models use K_d values for freshly-treated soils, as opposed to aged sorption K_d values. In addition, decreased pesticide leaching observed in aged soil in column studies has been attributed to an increase in diffusion-limited sorption (Pignatello et al. 1993; Renaud et al. 2004; van Beinum et al. 2006).

Recent reports also question the importance of sorption parameters as transport model input parameters. It is argued that the large sensitivity of models to sorption parameters disappears if the degradation rate of the pesticide is assumed to be proportional to the concentration in the liquid phase as opposed to proportional to the total soil concentration. Nonlinearity of sorption predicts an increased travel time of a pesticide pulse through the soil, however a decreased degradation rate of pesticide pulse (Beltman et al. 2008). These two effects would compensate each other exactly, if the mass fraction leached beyond a certain depth is considered.

Recently there has been renewed interest in predicting bioavailability of pesticides in aged soils. However, pesticide bioavailability in aged soils has been characterized by a variety of methods with only limited success, due in part to methodological limitations. The amount of a bioavailable pesticide in soil has been commonly estimated by the determination of aqueous 0.01 M $CaCl_2$ extractable pesticide, and in some instances, bioavailability may be predicted from soil properties or by a mild solvent extraction, but not with vigorous and more harsh solvent extraction (Alexander and Alexander 1999, 2000; Chung and Alexander 1998). However, there are conflicting results on the efficacy of using mild organic solvent extractants to estimate the amount of bioavailable pesticide in soil.

In recent studies, we have found that the bacterium *Pseudomonas* sp. strain ADP, which is capable of rapidly mineralizing triazines (atrazine, simazine) (Sadowsky and Wackett 2000) could be used to characterize the degradability of aged triazine residues, without the contribution of triazine desorption from soil (Barriuso et al. 2004). We have found that although triazine sorption to soil increased with aging, the amounts of triazine in aged soils extracted by 0.01 M $CaCl_2$ and aqueous methanol were highly correlated to amounts of triazine mineralized by *Pseudomonas* sp. strain ADP (Barriuso et al. 2004; Regitano et al. 2006; Cabrera et al. 2008) (Fig. 10.8). Consequently, 0.01 M $CaCl_2$/methanol extractable triazine in soils can be used to estimate degradable residues, even in aged soils. These results, however, question whether the increased sorption K_d values obtained by

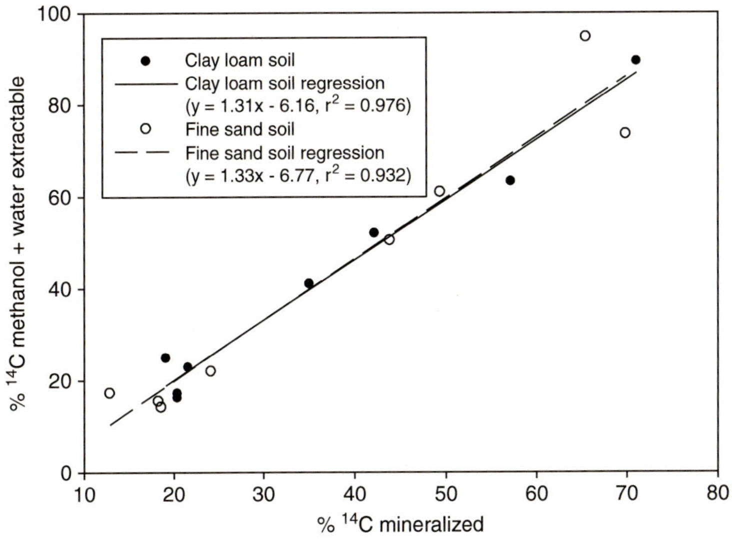

Fig. 10.8 Correlation of methanol: water extraction of aged atrazine residues from two soils as compared the aged atrazine mineralized by *Pseudomonas* ADP (From Barriuso et al. 2004)

the sequential extraction (H_20 followed by MeOH) batch equilibration method are indicative of decreased degradability since there was an excellent correlation between H_2O:MeOH extractable and mineralizable aged triazine residues.

10.5 Conclusions

Regardless of the mechanisms involved in aging, these results are further evidence that increases in sorption during pesticide aging should be taken into account during characterization of the sorption process for mathematical models of pesticide degradation and transport. These data show the importance of characterization of sorption-desorption in aged herbicide residues in soil, particularly in the case of prediction of herbicide transport in soil. In this case, potential transport of the pesticide would be over predicted if freshly treated soil Kd values were used to predict transport.

References

Alexander RR, Alexander M (1999) Genotoxicity of two polycyclic aromatic hydrocarbons declines as they age in soil. Environ Toxicol Chem 18:1140–1143

Alexander RR, Alexander M (2000) Bioavailability of genotoxic compounds in soils. Environ Sci Technol 34:1589–1593

Anderson JPE (1981) Soil moisture and the rates of biodegradation of diallate and triallate. Soil Biol Biochem 13:155–161

Barriuso E, Koskinen WC, Sadowsky MJ (2004) Solvent extraction characterization of bioavailability of atrazine residues in soil. J Agric Food Chem 52:552–6556

Beltman WHJ, Boesten JJTI, van der Zee SEATM (2008) Spatial moment analysis of transport of nonlinearly adsorbing pesticides using analytical approximations. Water Resour Res 44(5), Article Number:W05417

Berglöf T, Koskinen WC, Brucher J, Kylin H (2000a) Linuron sorption-desorption in field moist soils. J Agric Food Chem 48:3718–3721

Berglöf T, Koskinen WC, Kylin H, Moorman TB (2000b) Characterization of triadimefon sorption in soils using supercritical fluid (SFE) and accelerated solvent (ASE) extraction techniques. Pest Manag Sci 56:927–931

Berglöf T, Koskinen WC, Duffy MJ, Norberg KA, Kylin H (2003) Metsulfuron sorption-desorption in field-moist soils. J Agric Food Chem 51:3598–3603

Blair AM, Martin TD, Walker A, Welch SJ (1990) Measurement and prediction of isoproturon movement and persistence in three soils. Crop Prot 9:89–294

Boesten JJTI, Van der Pas LJT (1983) Test of some aspects of a model for the adsorption/desorption of herbicides in soil. Aspects Appl Biol 4:495–501

Bresnahan GA, Koskinen WC, Dexter AG, Lueschen WE (2000) Influence of soil pH sorption interactions on imazethapyr carryover. J Agric Food Chem 48:1929–1934

Bresnahan GA, Dexter AG, Koskinen WC, Lueschen WE (2002) Influence of soil pH-sorption interactions on the carryover of fresh and aged soil residues of imazamox. Weed Res 42:45–51

Cabrera A, Cox L, Koskinen WC, Sadowsky MJ (2008) Availability of triazine herbicides in aged soils amended with olive oil mill waste. J Agric Food Chem 56:4112–4119

Carrizosa MJ, Rice PJ, Koskinen WC, Carrizosa I, Hermosin MC (2004) Sorption of isoxaflutole and DKN on organoclays. Clays Clay Miner 52:341–349

Celis R, Koskinen WC (1999a) An isotopic exchange method for the characterization of the irreversibility of pesticide sorption-desorption in soil. J Agric Food Chem 47:782–790

Celis R, Koskinen WC (1999b) Characterization of pesticide desorption from soil by the isotopic exchange technique. Soil Sci Soc Am J 63:1659–1666

Cheng HH (1990) Pesticides in the soil environment – an overview. In: Cheng HH (ed) Pesticides in the soil environment. Soil Science Society of America, Madison, pp 1–5

Cheng HH, Koskinen WC (2002) Interactions of minerals-organic matter-living organisms on the fate of allelochemicals and xenobiotics in soil: a methodological evaluation. In: Violante A, Huang PM, Bollag J-M, Gianfreda L (eds) Soil mineral-organic matter-microorganism interactions and ecosystem health: ecological significance of the interactions among clay minerals, organic matter and soil biota. Developments in soil science, vol 28B. Elsevier, Amsterdam, pp 135–145

Cheng HH, Fuehr F, Jarczyk HJ, Mittelstaedt W (1978) Degradation of methabenzthiazuron in the soil. J Agric Food Chem 26:595–599

Cheng HH, Koskinen WC, Molina JAE (1994) Effects of soil interactions on the bioreactivity of organic chemicals in soils. In: Interactions of soil components, Agricultural ecosystems and health issues. Transact. XV. International Congress of Soil Science, Symposium, vol 3. ID20, Acapulco, Mexico, pp 638–655

Chung NH, Alexander M (1998) Differences in sequestration and bioavailability of organic compounds aged in dissimilar soils. Environ Sci Technol 32:855–860

Cox L, Koskinen WC, Yen PY (1998) Changes in sorption of imidacloprid with incubation time. Soil Sci Soc Am J 62:342–347

Gaillardon P, Sabar M (1994) Changes in the concentrations of isoproturon and its degradation products in soil and soil solution during incubation at two temperatures. Weed Res 34:243–251

Guo L, Wagenet RJ, Jury WA, Flury M (2000) Dependence of pesticide degradation on sorption: nonequilibrium model and application to soil reactors. J Contam Hydrol 43:45–62

Koskinen WC, O'Connor GA, Cheng HH (1979) Characterization of hysteresis in the desorption of 2,4,5-T from soils. Soil Sci Soc Am J 43:871–874

Koskinen WC, Cox L, Yen PY (2001) Changes in sorption/bioavailability of imidacloprid metabolites in soil with incubation time. Biol Fertil Soils 33:546–550

Koskinen WC, Rice PJ, Anhalt JA, Sakaliene O, Moorman TB, Arthur EL (2002) Sorption-desorption of "aged" sulfonylaminocarbonyl-triazolinone herbicides in soil. J Agric Food Chem 50:5368–5372

Koskinen WC, Anhalt JA, Sakaliene O, Rice PJ, Moorman TB, Arthur EL (2003) Sorption-desorption of two "aged" sulfonylaminocarbonyltriazolinone herbicide metabolites in soil. J Agric Food Chem 50:3604–3608

Mamy L, Barriuso E (2007) Desorption and time-dependent sorption of herbicides in soils. Eur J Soil Sci 58:174–187

McCall PJ, Agin GL (1985) Desorption kinetics of picloram as affected by residence time in the soil. Environ Toxicol Chem 4:37–44

Oi M (1999) Time-dependent sorption of imidacloprid in two different soils. J Agric Food Chem 47:327–332

Pignatello JJ, Huang LQ (1991) Sorptive reversibility of atrazine and metolachlor residues in field soil samples. J Environ Qual 20:222–228

Pignatello JJ, Ferrandino FJ, Huang LQ (1993) Elution of aged and freshly added herbicides from a soil. Environ Sci Technol 27:1563–1571

Regitano JB, Koskinen WC (2008) Characterization of nicosulfuron availability in aged soils. J Agric Food Chem 56:5801–5806

Regitano JB, Koskinen WC, Sadowsky MJ (2006) Influence of soil aging on sorption and bioavailability of simazine. J Agric Food Chem 54:1373–1379

Renaud FG, Brown CD, Fryer CJ, Walker A (2004) Lysimeter experiment to investigate changes with time in availability of pesticide for leaching. Environ Pollut 1331:81–91

Rochette EA, Koskinen WC (1996) Supercritical carbon dioxide for determining atrazine sorption by field-moist soils. Soil Sci Soc Am J 60:453–460

Sadowsky MJ, Wackett LP (2000) Genetics of atrazine and s-triazine degradation by *Pseudomonas* sp. strain ADP and other bacteria. In: Hall JC, Hoagland RE, Zablotowicz RM (eds) Pesticide biotransformations in plants and microorganisms, vol 777, ACS symposium series. American Chemical Society, Washington, D.C., pp 268–282

Sander M, Pignatello JJ (2005) An isotope exchange technique to assess mechanisms of sorption hysteresis applied to naphthalene in kerogenous organic matter. Environ Sci Technol 39:7476–7484

Sander M, Pignatello JJ (2009) Sorption irreversibility of 1,4-dichlorobenzene in two natural organic matter-rich geosorbents. Environ Toxicol Chem 28:447–457

Scow KM, Alexander M (1992) Effect of diffusion on the kinetics of biodegradation: experimental results with synthetic aggregates. Soil Sci Soc Am J 56:128–134

Scow KM, Hutson J (1992) Effect of diffusion and sorption and biodegradation: theoretical considerations. Soil Sci Soc Am J 56:119–127

Shelton DR, Doherty MA (1997a) A model describing pesticide bioavailability and biodegradation in soil. Soil Sci Soc Am J 61:1078–1084

Shelton DR, Doherty MA (1997b) Estimating losses of efficacy due to pesticide degradation in soil – model simulations. Soil Sci Soc Am J 61:1085–1090

Shelton DR, Parkin TB (1991) Effect of moisture on sorption and biodegradation of carbofuran in soil. J Agric Food Chem 39:2063–2068

Van Beinum W, Beulke S, Fryer C, Brown C (2006) Lysimeter experiment to investigate the potential influence of diffusion-limited sorption on pesticide availability for leaching. J Agric Food Chem 54:9152–9159

Walker A (1987) Evaluation of a simulation model for prediction of herbicide movement and persistence in soil. Weed Res 27:143–152

Chapter 11
Sorption Comparison between Pharmaceuticals and Hydrophobic Organic Chemicals in Soils and Sediments

Bo Pan and Baoshan Xing

Abstract The risks of organic contaminants are controlled by their sorption behavior. Hydrophobic organic contaminants (HOCs) have been studied for decades for their sorption properties in soils/sediments, while the relevant study on pharmaceuticals was started very recently. A systematic comparison of the sorption properties between HOCs and pharmaceuticals may facilitate our understanding on the sorption mechanisms of both HOCs and pharmaceuticals in soils/sediments. The studies on correlation between HOC sorption and soil/sediment properties have achieved some important conclusions, but a general method to predict sorption characteristics through soil/sediment properties is not available yet. The concept of relating the operational fractions, chemical compositions, and physical conformations of soil organic matter to HOC sorption might be referential to pharmaceutical studies. However, because of the comparable contribution of organic and inorganic soil/sediment fractions to pharmaceutical sorption and the various chemical properties of pharmaceuticals, the attempt to establish a common model to predict their environmental behavior, or more specifically, their sorption characteristics, may not be of great significance. A proper classification of pharmaceuticals according to their properties and functional groups may be very useful for pharmaceutical environmental fate studies. A very good start to investigate organic chemical sorption in soils/sediments may be to separate soils/sediments into individual fractions. This line of study revealed unique roles of organo-mineral complexes in controlling organic contaminant sorption. Carbon fractions are always the center of discussion on organic contaminant sorption. This chapter also summarized and compared the roles of carbon nanomaterials, activated carbon, and biochar to the

B. Pan (✉)
Faculty of Environmental Science & Engineering, Kunming University of Science and Technology, Kunming, China
e-mail: panbocai@gmail.com

B. Xing
Stockbridge School of Agriculture, University of Massachusetts, Amherst, MA, USA
e-mail: bx@pssci.umass.edu

J. Xu and D.L. Sparks (eds.), *Molecular Environmental Soil Science*, Progress in Soil Science, DOI 10.1007/978-94-007-4177-5_11,

sorption of both HOCs and pharmaceuticals. HOC and pharmaceutical sorption characteristics as affected by environmental conditions were discussed for dissolved organic matter, pH, cations, as well as other organic contaminants. The most distinct difference between HOCs and pharmaceuticals is that pharmaceuticals may present as different species depending on environmental conditions, while HOCs not. Traditional sorption studies incorporating pharmaceutical species analysis may be of 35 great importance. Several promising research directions were also discussed.

Keywords Anthropogenic pollutants • Degradation • Dual mode model • Rigidity • Risk assessment

The sorption behavior of organic chemicals in soils/sediments is a key process controlling their environmental fate and risk. This type of study has been a research focus for decades. It is widely accepted that chemicals with different properties behave differently and the risk assessments for these chemicals should consider different processes. Hydrophobic organic contaminants (HOCs) and pharmaceuticals are two different types of organic contaminants with distinct physicochemical properties. Separate reviews on sorption of HOCs (Pignatello and Xing 1996; Weber et al. 1999; Cornelissen et al. 2005; Pan et al. 2008) or pharmaceuticals (Daughton and Ternes 1999; Sarmah et al. 2006; Pan et al. 2009) are available. A systematic comparison between their sorption properties in soils/sediments could provide valuable information to understanding their sorption mechanisms and consequently improving our knowledge on their exposure and risks. Therefore, this chapter will summarize the sorption studies of HOCs and pharmaceuticals in soils/sediments as well as their components, with an emphasis on the different sorption behavior of these two classes of chemicals.

11.1 Sources and Risks of Pharmaceuticals and HOCs

HOCs are known as organic contaminants of low solubilities and include organochlorine pesticides (OCPs), polyaromatic hydrocarbons (PAHs), polybrominated diphenylethers (PBDEs) and polychlorinated biphenyls (PCBs). These chemicals are known to be mutagenic, carcinogenic and/or co-carcinogenic, and they have the tendency to be accumulated by organisms and could be magnified through food chain. Therefore, their risks have been a hot research topic for many years. Some other hydrophobic compounds, such as n-alkanes or paraffin, will not be included in this chapter because they have limited environmental risks (Pesticide Action Network North America, http://www.pesticideinfo.org). Most of the above mentioned HOCs are man-made unintentionally or specifically for agricultural and engineering purpose. For example, OCPs were mass produced to kill insects, such as pest in crops, ants, ticks, and mosquitoes indoors. It was reported that OCPs have a wide range of both acute and chronic health effects, including cancer, neurological damage, and birth defects. Therefore, the application and production of OCPs were banned in the 1970s. However, these chemicals break down slowly and can remain in the environment or in organisms for a long time. Their health risks are currently still an issue world-wide. PAHs are chemicals that contain more than one fused benzene ring. They are formed in volcano events, natural fires, and thermal geological reactions and are commonly found in petroleum fuels, coal products, and tar. Human activities, such as fuel operations, incineration, coal gasification, and heat and power generation, also promoted the release or production of PAHs. On the contrary, PCBs are man-made chemicals that never existed in the environmental until 100 years ago. These chemicals were manufactured for engineering application such as dielectric fluids, plasticizers, insulators for transformers and capacitors. PCBs contain 209 individual congeners. PCB flash point, stability, and resistance to biodegradation increase with the number of chlorines in the PCB molecules. Therefore, PCBs with large numbers of chlorines are more favorable for production. These chemicals also have greater environmental and health risks. PBDEs are organobromine compounds that are used as flame retardants. This type of chemicals could reduce human fertility and impair development of the nervous system at concentration range found in households. PBDEs are widely used in building materials, electronics, furnishings, motor vehicles, airplanes, plastics, polyurethane foams, and textiles and thus their environmental behavior and toxicity are affected by the release from their product matrices in addition to environmental factors.

Pharmaceuticals include human and veterinary drugs. These chemicals are mostly synthesized and contain various functional groups. Many widely used pesticides also contain different functional groups and may have the similar properties to pharmaceuticals. However, because their applications and pathways to enter the environment are different, different physico-chemical processes should be emphasized when examining the environmental fate and risks of pesticides and pharmaceuticals. Therefore, pesticides are not included in this chapter. The readers are suggested to refer to several review papers for pesticide interactions with soils/sediments (Gevao et al. 2000; Arias-Estevez et al. 2008).

The global pharmaceutical market reached $712 billion in 2007 (Seget and Pharma 2008). Because of the wide use of these chemicals, pharmaceuticals were discharged into the environment through various pathways. For example, the absorption efficiencies of pharmaceuticals by humans or livestock are very low and thus most of the applied pharmaceuticals may be excreted as the original compounds or the metabolites (Fig. 11.1). Another pathway is the disposal of the expired pharmaceuticals. The pharmaceutical-containing water may enter wastewater treatment plants. However, conventional methods applied in wastewater treatment plant may not be able to completely remove pharmaceuticals (Göbel et al. 2005). Therefore, pharmaceuticals that were not removed from the wastewater may enter surface water, ground water, or even tap water with the discharge of wastewater treatment plant (Yin et al. 2010; Yiruhan et al. 2010). The important pharmaceutical inputs in soils/sediments are the irrigation with wastewater, land application of sludge from

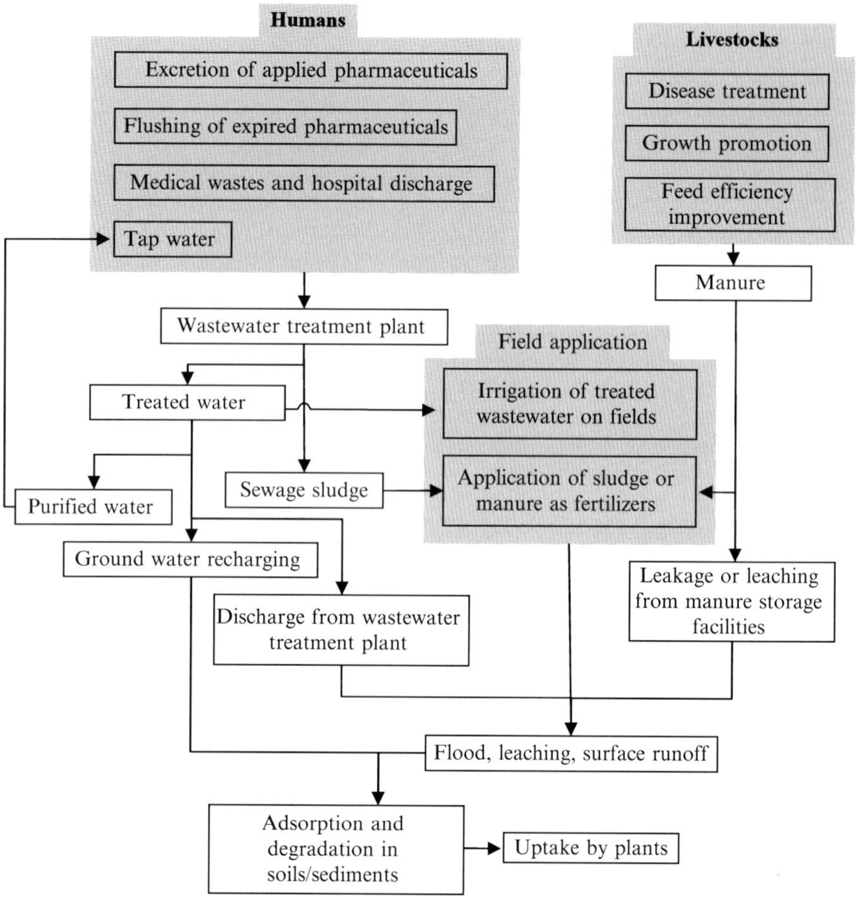

Fig. 11.1 Applications and sources of pharmaceuticals. In wastewater treatment plants, pharmaceuticals are not efficiently removed using current technologies, thus the water and sewage sludge discharged from wastewater treatment plants and tap water purified from wastewater could contain considerable amounts of pharmaceuticals. Application of treated water and sewage sludge for agricultural purposes may have environmental risks. Contamination by veterinary drugs is mainly from application of manure as fertilizer and/or leaking from manure storage facilities (Modified from Xing et al. 2011)

wastewater treatment plants and/or manure of livestock as fertilizers, and possible discharge from wastewater treatment plants (Scheytt et al. 2007; Chee-Sanford et al. 2009).

Unfortunately, pharmaceuticals were reported to pose adverse stress on wildlife and human being (Daughton and Ternes 1999). For example, some of the pharmaceuticals synthesized to adjust endocrine system for human body. The presence of hormonal pharmaceuticals or their metabolites in the environment could unexpectedly interfere with natural hormones of human and wildlife. This type of chemicals is also known as endocrine disrupting chemicals (EDCs). Another example is antibiotics or antibacterials. The occurrence of antibiotics in the environment provides an extended "training" environment for microorganisms and thus the microorganisms could develop drug-resistant genes. Virus or bacteria may adopt drug-resistant genes through gene transfer. Therefore, the environmental behavior of pharmaceuticals is important for assessing their environmental risks. Investigators have summarized the occurrence, detection, and toxicology of EDCs (Hollert et al. 2005; Ying et al. 2002), veterinary pharmaceuticals (VP) (Boxall et al. 2004; Sarmah et al. 2006; Tolls 2001) and antibiotics (Thiele-Bruhn 2003).

Identifying the concentrations of HOCs enables evaluation of their environmental risks. The parameters of EC50 and LC50 are all dependent on HOC concentrations. However, to assess the risks of pharmaceuticals, a different concept may be required. Pharmaceuticals promote the development of drug-resistant genes after long-time exposure of microorganisms at low-concentrations. Because pharmaceuticals may degrade quickly, their risks could not be evaluated through their concentrations at a single time point. Therefore, analyzing their concentrations at a certain time point may not directly reflect their environmental risks. The history of pharmaceuticals existing in the studied area is the key. However, this type of information is not easily available based on current technologies.

11.2 Sorption in Soils/Sediments

11.2.1 Apparent Nonideal Sorption in Soils/Sediments

The sorption of HOCs in soils/sediments was mostly attributed to hydrophobic interactions. Therefore, earlier studies related HOC sorption in soils/sediments to partitioning in organic solvents (Chiou et al. 1979). The ideal partitioning concept was proposed and the sorption was related with HOC hydrophobic properties (such as aqueous solubility and octanol-water partitioning coefficient) only. Based on this modeling concept, linear sorption is expected for HOC sorption in soils/sediments. However, amounts of studies reported nonlinear sorption with nonlinear factor, n, as low as 0.5 (Young and Weber 1995). Thus, new conceptual models, such as dual-mode model and multi-domain reactive model (or more practically, dual reactive model), were proposed to describe the nonideal sorption of HOCs in soils/sediments. These models consistently consider soil/sediment organic matter (SOM) as an adsorbent with sorption sites of distributed energy. In another word, both linear and nonlinear sorption sites exist in SOM and interact with organic chemicals simultaneously, therefore the overall sorption shows nonlinear isotherms. In addition to nonlinear feature of the sorption isotherms, HOC-soil/sediment interactions also showed competitive sorption with other chemicals, sorption/desorption hysteresis, and multi-stage sorption kinetics. The SOM was conceptualized as glassy and rubbery sorption domains (Xing et al. 1996). Rubbery sorption domain presents linear sorption while glassy domain contains both linear and nonlinear sorption regions. Although no method is available to quantitatively predict HOC sorption based HOC and soil/sediment properties, the proposed dual-mode model and multi-domain reactive model explains well the failure of certain soil remediation approaches, and pointed out an important research direction to improve the efficiency of environmental remediation.

Following the experiences on HOC sorption studies, investigators emphasized nonideal sorption behaviors of pharmaceuticals in soils/sediments in the very first batch of studies, such as nonlinear sorption (Li and Lee 1999; Thiele-Bruhn et al. 2004; Yu et al. 2004, 2006; Loffredo and Senesi 2006; Zeng et al. 2006; Gu et al. 2007), desorption hysteresis (Li and Lee 1999; During et al. 2002; Zeng et al. 2006; Williams et al. 2006; Gu et al. 2007), concentration-dependent sorption kinetics (Yu et al. 2004), two-stage sorption kinetics (Loffredo and Senesi 2006; Zhou 2006), and three-site sorption (Wehrhan et al. 2007). The success of normalizing HOC sorption coefficient by organic carbon content promoted the application of this concept in pharmaceutical sorption studies. The investigator observed positive relationships between Freundlich sorption coefficient (K_F) and total organic carbon (TOC) as well as single-point sorption coefficient (K_d) and TOC (During et al. 2002; Bowman et al. 2002; Holthaus et al. 2002; Loffredo and Senesi 2006; Maskaoui et al. 2007; Uslu et al. 2008). Therefore, organic content in soils/sediments greatly control the environmental fate of pharmaceuticals, such as enhancing sorption, reducing bioavailability (Jacobsen et al. 2005) and thus reducing their environmental risk (Fan et al. 2007). The amendment of soils by manure could significantly increase soil organic content, sulfadiazine sorption and sorption/desorption hysteresis (Sukul et al. 2008).

However, normalization of pharmaceutical sorption coefficients by TOC seems not applicable because of the large variation (Tolls 2001; Zhou et al. 2007; Holbrook et al. 2004; Liu et al. 2005; Patrolecco et al. 2006). The phenomenon could be understood from pharmaceutical amphiphilic characteristic, which leads to their sorption on both polar and nonpolar surfaces (Suntisukaseam et al. 2007). Strong evidence has indicated that HOC sorption in soils/sediments is mostly controlled by organic fraction, and the contribution of inorganic minerals could be neglected as long as the TOC content is higher than 0.1% (Schwarzenbach and Westall 1981). On the contrary, pharmaceutical sorption by inorganic particles accounted for 10–40% of the overall sorption (Lai et al. 2000; Schafer et al. 2002). Jones et al. (2005) investigated the sorption of oxytetracycline in 30 soils. They reported that soil texture, cation exchange capacity, and iron oxide content were the most important factors in soils with TOC between 0 and 4%. TOC content influenced oxytetracycline sorption only in a soil with 9% TOC (Jones et al. 2005). Hou et al. (2010) also observed that a positive relationship between sorption coefficient and TOC observed only at TOC higher than 4%. Therefore, both organic and inorganic fractions are important in controlling pharmaceutical sorption in soils/sediments (Lai et al. 2000; Schafer et al. 2002; Strock et al. 2005). More importantly, organic matter may block sorption sites in clays and reduce pharmaceutical sorption (Pils and Laird 2007). The competition between organic matter andpharmaceuticals on mineral particles was also reported (Hou et al. 2010). Clearly, normalization of sorption coefficient by TOC content neglects the contribution by inorganic fractions, which causes uncertainties in predicting pharmaceutical sorption.

Hydrophobic interaction was reported not sufficient to explain pharmaceutical sorption in soils/sediments. The mechanisms other than hydrophobic interaction discussed widely in literature include hydrogen bond, ion exchange, and electrostatic interactions. For example, Vasudevan et al. (2009) observed pH-dependent K_d values of ciprofloxcin in 30 soils. They stated that cation exchange capacity was a key soil parameter influencing the apparent sorption. An important observation is that soils with high effective CEC exhibited distinct pH edges and the highest sorption at all pHs in comparison to soils with low CEC, indicating cation exchange was the dominant mechanisms of sorption (Vasudevan et al. 2009). The above mentioned sorption mechanisms indicated that the concept of glassy and rubbery sorption domains of SOM for HOCs was not applicable for pharmaceuticals. Clearly, a much more complicated modeling concept should be applied for pharmaceutical sorption in soils/sediments.

11.2.2 Correlation Between Soil/Sediment Organic Matter Physicochemical Properties and Sorption

A group of investigators attributed HOC nonideal behavior in soils/sediments to the presence of black carbon (Accardi-Dey and Gschwend 2002). Carbonaceous geosorbents, such as coal and kerogen also have been reported to have strong nonlinear sorption with HOCs (Allen-King et al. 2002; Cornelissen et al. 2005; Ran et al. 2007). However, strong evidence also indicates that the SOM samples free of black carbon (such as extracted HA) exhibited distinct nonideal sorption for HOCs (Lu and Pignatello 2004; Wang and Xing 2005b; Pignatello et al. 2006; Jeong et al. 2008). Therefore, the properties of natural organic fractions of soils/sediments are the primary factor controlling HOC fate (Weber et al. 1999; Kottler and Alexander 2001). For pharmaceuticals, various studies have concluded that both organic and inorganic fractions are important as stated earlier. No detailed investigation is available yet, thus no explicit conclusion was reached for soil/sediment physicochemical properties controlling pharmaceutical sorption. The most widely available soil/sediment physicochemical property is on their organic fractions. Therefore, this section will mostly focus on soil/sediment organic matter (SOM) properties in relation with organic chemical sorption, mostly HOCs. We will highlight the possible application of the methods and concepts used for HOC sorption behaviors to pharmaceutical studies.

11.2.2.1 Operational Fractionation of SOM

SOM are operationally separated as fulvic acid (FA), humic acid (HA) and humin (HM) according to their solubilities (Weber et al. 1999). Briefly, FA and HA are dissolvable in alkaline solution. HM is insoluable at any pHs. The dissolved HA could be precipitated when pH was adjusted below 2, while FA could not. HM was also further separated into lipid, bound-HA (BHA) and insoluble residue (IR) (Rice and Maccarthy 1989). Based on the extended study on HOC sorption, the investigators provided a general sequence that the aliphaticity, molecular weight, carbon content, density, cross-linking, glass-to-rubbery transition temperature, HOC-HS interaction nonlinearity and sorption/desorption hysteresis followed an order of FA < HA < HM (Xing 2001a; Pan et al. 2006). If the BHA and IR were included in this sequence, the nonlinearity followed the order of lipid < HA < BHA < IR (Xie et al. 1997; Kohl and Rice 1998; Doick et al. 2005; Wen et al. 2007). HM fraction is considered the primary domain for irreversible binding, non-extractable, and sequestration of HOCs (Nam and Kim 2002; Macleod and Semple 2003). Dissolved organic matter was also operationally separated into seven fractions and hydrophobic fractions showed higher sorption to HOCs (Ilani et al. 2005; Chefetz et al. 2006).

The study on pharmaceutical sorption in separated SOM fractions is rare. Because this separation results in soil/sediment fractions with a same origin, extended study in this line for pharmaceuticals may provide valuable information for sorption mechanism considerations. The separated fractions should be characterized in more detail other than physical conformation alone for detailed pharmaceutical sorption study. The chemical composition and the physical locations of functional groups may determine the extent and the availability of sorption sites for pharmaceuticals. However, it should be noted that operational definition always involves poor reproducibility. For example, different extraction methods produced HM fractions with significantly different chemical composition and sorption properties (Wang and Xing 2005a). This uncertainty limits the application of operational fractions in predicting organic chemical-SOM interactions.

11.2.2.2 SOM Chemical Properties in Relation to Organic Chemical Sorption

The most widely discussed topic is on the contribution of aromatic and aliphatic fractions to HOC sorption in soils/sediments. Increased nonlinear was observed to be concurrent with aromaticity. Investigators indicated that diagenetic maturity and condensed nature of aromatic fractions could be well adopted to explain their strong and nonlinear sorption with HOCs (Xing 2001b; Ding et al. 2002; Wang et al. 2005). High sorption capacity was also observed to increase with aromaticity (Gauthier et al. 1986; Chin et al. 1997; Perminova et al. 1999). To further demonstrate the above observations, SOM sorption properties were compared before and after chemical treatment. Aromatic carbon was removed through bleaching and the sorption to phenanthrene showed more linear feature. On the other hand, when carbohydrate was removed by hydrolysis, the sorption was more nonlinear (Gunasekara et al. 2003). In addition, phenanthrene was strongly retained by aromatic fractions as indicated by aromatic carbon-related desorption-resistant phenanthrene (Golding

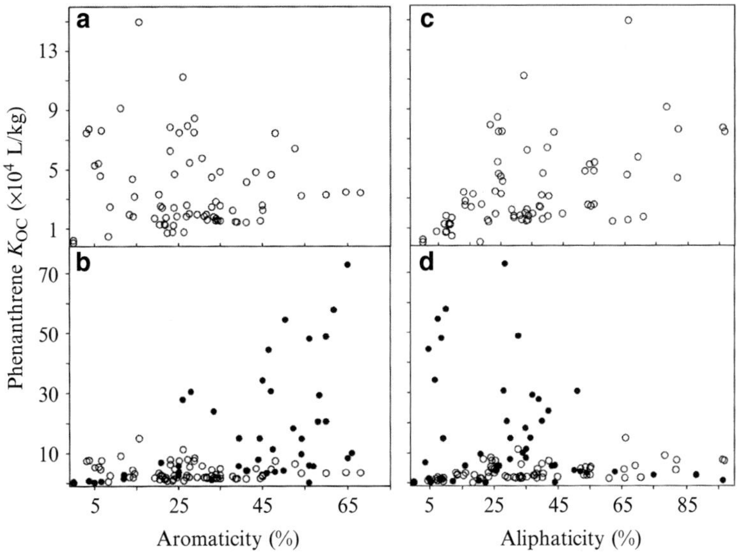

Fig. 11.2 Relationship between phenanthrene K_{OC} values (at an equilibrium concentration of 10 μg/L) and the aromaticity (panels *A* and *B*) or the aliphaticity (panels *C* and *D*) level of sorbents. *Open circles* are sorption data for natural sorbents and *filled circles* are for modified sorbents (Chefetz and Xing 2009 with permission from ACS)

et al. 2005). Another group of studies emphasized the importance of aliphatic fractions to HOC sorption. The most condensed fraction, HM, was widely reported to be important for HOC sorption and sequestration (Chefetz et al. 2000; Mao et al. 2002; Salloum et al. 2002; Simpson et al. 2003; Sun et al. 2008). The aliphatic fraction in HM was much higher than that in HA, indicating that aliphatic carbon is a major composition in HM (Kang and Xing 2005; Xing et al. 2005). Therefore, investigators proposed that both aromatic and aliphatic carbons are important to the overall sorption (Kang and Xing 2005; Chefetz and Xing 2009) (Fig. 11.2).

Molecular weight also has a positive relationship with HOC K_{OC} (Chin et al. 1994, 1997; Hur and Schlautman 2004). In addition to some specific chemical compositions, the overall properties of SOM were also proposed to be related with their sorption features to HOCs. Polarity, which is defined as (N + O)/C, was shown to be negatively related with isotherm linearity, single-point sorption coefficient (K_{OC}), and sorption/desorption hysteresis (Weber et al. 1999). The importance of SOM polarity to HOC-SOM interactions was also reported in several other studies (Xing et al. 1994; Xing 1997, 2001b; Chen and Xing 2005).

The essential ideas of relating HOC sorption and soil/sediment properties are based on hydrophobic interactions, but without considering inorganic fractions. Thus, the case of pharmaceutical sorption should be different. Because of their various functional groups and different species, pharmaceuticals may specifically interact with explicit chemical compositions, such as hydroxyl and carboxyl groups. Therefore, the application of the above mentioned discussion for pharmaceutical sorption studies is limited.

π-π Electron donor-acceptor (EDA) interaction is discussed extensively for HOC sorption in literature (Zhu et al. 2004). This theory applies not only to adsorbents, but also to adsorbates. Although how to quantitatively apply this theory in predicting organic chemical sorption in solid particles is still unknown, the basic concept is that the interaction between a donor and an acceptor is much stronger than donor-donor or acceptor-acceptor pairs. The diverse functional groups on pharmaceuticals enable them to form strong EDA pairs with soils/sediments. The quantitative description of EDA mechanisms controlling organic chemical sorption is not available yet. However, the parameters of donating electrons and accepting electrons may greatly improve our understanding on the comparison of different pharmaceuticals in forming EDA pairs (Yang et al. 2010). π-π EDA complexes often show a charge-transfer band in the ultraviolet–visible region. Zhu et al. (2004) measured the order of interaction strength of π-acceptors and π-donor based on an ultraviolet–visible spectra test (Fig. 11.3). This method was also applied by other researchers (Wang et al. 2010).

11.2.2.3 Physical Conformation of Solid SOM and Sorption Characteristics

Uncertainties were observed in establishing a general model to predict SOM sorption characteristics from any individual chemical properties (Salloum et al. 2001a; Thomsen et al. 2002; Hur and Schlautman 2003; Bonin and Simpson 2007). One of the primary reasons is that SOM of different physical forms may have different sorption properties (Pan et al. 2007). For example, the properties of SOM rigidity play an important role in predicting HOC sorption properties, because rigidity is related with the abundance of hydrophobic regions or pores. Aromatic carbons are essentially condensed and could be easily related with HOC nonlinear and strong sorption; whereas aliphatic carbons could be selectively adsorbed by mineral particles and form compact structures (Kang and Xing 2005).

In polymer science, the concept of glassy and rubbery states was applied to describe the physical conformation of a solid particle and glassy-rubbery transition temperature (T_g) was adopted to compare the rigidity. T_g could be visualized as an inflection in the curve of thermal capacity vs. temperature (Weber et al. 1999; Zhang et al. 2007) (Fig. 11.4). Solid particles with higher T_g was believed to be more rigid with a high extent of cross-linking and attractive force between molecules. In sorption studies, T_g was well applied to discuss with nonideal sorption behavior. Glassy polymers and SOM, which showed nonideal sorption, have measurable T_gs. Rubbery polymers have linear sorption and do not have T_gs. The nonlinearity of HOC sorption in SOM decreased when SOM was heated (Xing and Chen 1999). The T_g values for SOM were measured to be 43–79 °C (LeBoeuf and Weber 1997; Weber et al. 2001; Zhang et al. 2007). The analogous concept of SOM to polymers promoted the establishment of modern sorption theories for HOCs in SOM, such as dual mode model (Xing and Pignatello 1997), distributed reactive model or dual

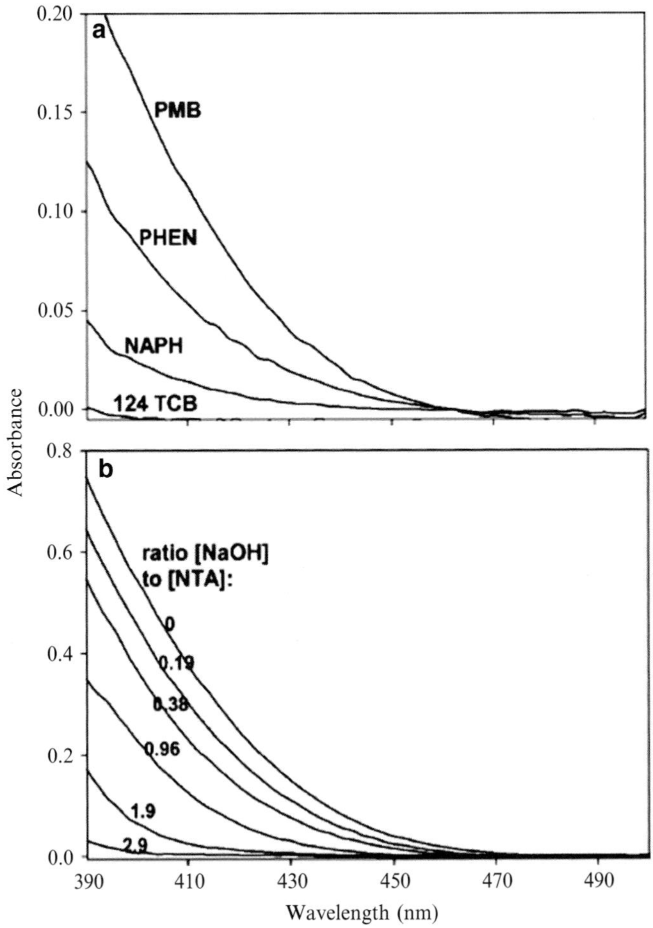

Fig. 11.3 UV/vis difference spectra in methanol showing the charge-transfer absorption band of π-π complexes and its dependence on the degree of protonation of the acceptor: (**a**) 0.0013 M 1,4,5,8-naphthalene tetracarboxylic acid (NTA) and 0.01 M π-donors, pentamethylbenzene (PMB), phenanthrene (PHEN), and naphthalene (NAPH), and a nondonor, 1,2,4-trichlorobenzene (124-TCB), which shows no significant complexation; (**b**) 0.0013 M NTA and 0.28 M PMB at different NaOH/NTA ratios. Single components have no significant absorbance in the region (Zhu et al. 2004 with permission from ACS)

reactive model (Weber et al. 1992 and the papers in the same series), and extended dual mode model (Xia and Pignatello 2001). Basically, these models emphasize the physical conformation of SOM (glassy and rubbery domains). Rubbery domain contributes to linear sorption while glassy domain to both linear and nonlinear sorption. However, for pharmaceuticals, whose sorption is not mostly controlled by hydrophobicity, the concepts in defining rigidity (such as glass-to-rubbery transition temperature) cannot completely help understand pharmaceutical sorption.

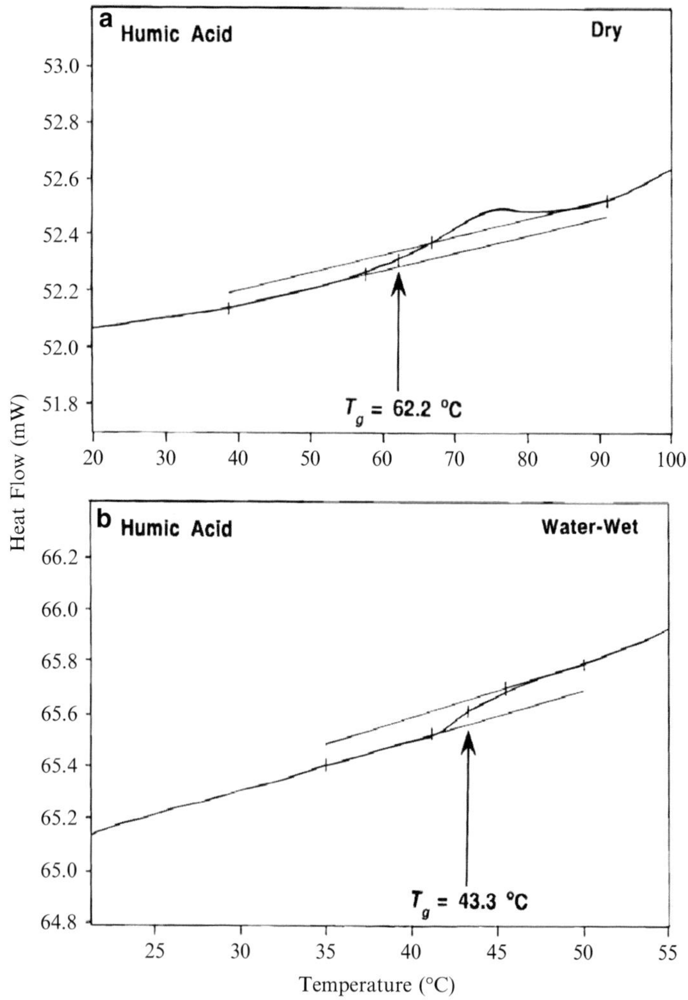

Fig. 11.4 Calorimetric analysis of the glass transitions of humic acid. (**a**) Desiccator-dried humic acid with significantly reduced over-relaxation response and broad glass transition peak at 62°C. (**b**) Water-wet specimen (equilibrated at pH 7.0 for 7 days) with little over-relaxation response and lowered glass transition of 43°C (Weber et al. 1999 with permission from Elsevier)

The effect of physical conformation on HOC sorption is also noted in sorption studies of separated SOM fractions. The sum of HOC sorption in individual fractions is normally higher than that in the integral soil/sediment particles (Salloum et al. 2001b; Bonin and Simpson 2007). Therefore, fractionation procedure may have re-organized SOM physical conformation, and exposed more sorption sites for HOCs.

Clearly, the above mentioned discussion on HOC sorption as affected by physical conformation is mostly on hydrophobic regions. Again, this concept should be carefully applied for pharmaceuticals. The same experimental design may be applied for pharmaceuticals. But the focus should be the availability of the specific sorption sites for pharmaceuticals. Therefore, different characterization of adsorbents should be used. For example, HA precipitated by water evaporation, hydrogen ions and heavy metals may be compared for HA conformation and composition effect on organic chemical sorption. For HOCs, the wettability, surface hydrophobicity and rigidity may provide information for mechanistic consideration. However, for pharmaceuticals, the abundance and distribution of functional groups may be a major controlling factor for their sorption in soils/sediments.

11.2.3 Correlation Between Organic Chemical Properties and Sorption

The sorption of organic chemicals in soils/sediments is greatly controlled by soil/sediment properties as discussed in the last section. The discussion on organic chemical properties alone may not provide enough valuable information on their fate modeling in the real environment. However, this line of study could shed light on organic chemical sorption mechanisms in soils/sediments. Therefore, this section summarizes research progress on relating organic chemical properties and sorption coefficients. This type of study for HOCs has been conducted for decades. Various linear free energy models were developed to establish the relationship between sorption coefficient (K_{OC}) and sorbate properties, such as K_{OW} (Burkhard 2000), S_W (Chiou et al. 1979), molecular connectivity index (Bahnick and Doucette 1988), or a combination of polarity, polarizablility, overall hydrogen-bond acidity and basicity (Nguyen et al. 2005). A novel correlating parameter could be derived from the molecular connectivity which provides topological description in terms of branching index and fundamental treatment of the molecular structure based on a valence weighted graph of the structural formula of a compound (Saikia and Dutta 2008; Saikia 2008).

Pharmaceuticals have various functional groups and these functional groups may play roles simultaneously on pharmaceutical sorption in soils/sediments. In order to reveal pharmaceutical sorption mechanisms, a simplification of the sorption system may provide useful information. For example, Mackay and Seremet (Mackay and Seremet 2008) used flumequine (–COOH) and phenylpiperazine ($-NH^{3+}$) to study Ciprofloxacin (CIP)-soil sorption interactions through the contribution of surface complexation (via –COOH group) and cation exchange ($-NH^{3+}$ group). Their results indicated that the adsorption of ciprofloxacin to soil could be estimated based on the contribution of these two mechanisms.

Realizing the complexity of pharmaceutical properties, artificial neural networks were adopted to comprehensively analyze the sorption behavior of a large selection of prescribed pharmaceuticals in soils and sludges (Barron et al. 2009). The results indicated that Moriguchi log P (vapor pressure) contributed most to the predicted sorption coefficient, $K_d^{predicted}$. Other less important parameters are pKa, the number of sulfur atoms, the number of double bonds, and the hydrophilic factor. This type of comprehensive computational methods could facilitate the prediction of pharmaceuticals in the environment with a lesser dependence on expensive chemical analysis.

11.3 Sorption on Soil/Sediment Fractions

Soils/sediments are the most complicated environmental media. The data explanation encountered many problems because of the complicated interactions of soil/sediment fractions. Therefore, it is useful to investigate the sorption mechanisms in the simplified systems and the results could be applied to explain the sorption behavior in real system.

11.3.1 Adsorption to Inorganic Particles

As discussed in previous sections, sorption of HOCs on inorganic particles is very low because water molecules could be strongly adsorbed on this hydrophilic surface. Thus, it has been well documented that SOM controls the behavior of HOCs, such as sorption, mobility, extractability, solubility, and bioavailability (Chiou et al. 1979). Sorption by inorganic fractions could be neglected (Kile et al. 1995; Huang and Weber 1997).

On the contrary, the adsorption of pharmaceuticals on inorganic mineral particles is important in controlling pharmaceutical environmental behavior. Thus, investigating pharmaceutical sorption on pure mineral particles could provide valuable information to understand pharmaceutical sorption mechanisms in soils/sediments.

The layer structure of clay mineral particles is critical for their high sorption capacity and strong desorption hysteresis for pharmaceuticals as compared among goethite, illite, kaolinite, and montmorillonite (Van Emmerik et al. 2003; Shareef et al. 2006). Tetracycline could intercalate into the interlayer space of montmorillonite and thus increase d-spacing. X-ray diffraction measurement indicated that montmorillonite basal spacing increased from 1.26 to 1.95 nm after tetracycline adsorption (Parolo et al. 2008) (Fig. 11.5). Crystalline tetracycline decomposition temperature increased from 230 to 410°C after tetracycline intercalated in the interlayers of rectorite (Chang et al. 2009). Tetracycline infrared spectrum bands at 1228, 1311, 1356, 1375, and 1450 shifted to higher frequencies, indicating tetracycline molecules interacted strongly with rectorite surface and thus the stability of tetracycline was increased (Chang et al. 2009).

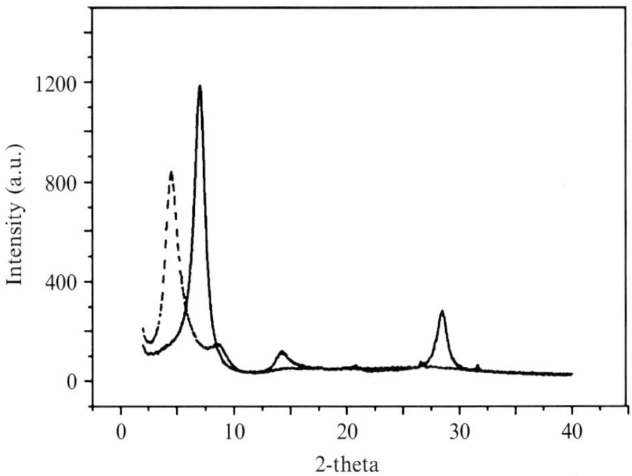

Fig. 11.5 XRD patterns of montmorillonite (*solid line*) and of montmorillonite with adsorbed TC (*dashed line*) at pH 3. The primary reflection at 7.00° 2θ corresponds to a basal spacing of 1.26 nm while reflection shifts to 4.43° 2θ corresponds to a basal spacing of 1.95 nm (Parolo et al. 2008 with permission from Elsevier)

Some other properties of inorganic mineral particles, such as pore distribution (Tsai et al. 2006a, b), surface charge density (Strock et al. 2005) and particle size (Casey et al. 2003), were all reported to affect pharmaceutical sorption. Various conclusions were reached in these correlation studies. This line of study provides only a framework of pharmaceutical sorption mechanisms.

Understanding the mineral composition of soil/sediment particles is vital in understanding pharmaceutical sorption behavior. For example, Bao et al. (2009) compared the sorption of tetracycline in a clay-rich soil and an iron oxide-rich soil. The adsorption kinetics of tetracycline indicated that longer equilibrium time (11 h) is requested for clay-rich soil because of the penetration of tetracycline into the interlayers between clays and organic matter. But the adsorption on oxide-rich soil was rather fast (5 h) because the dominant mechanism is surface complexation (Bao et al. 2009).

Identifying the specific sites for pharmaceutical sorption is a promising research direction to understand pharmaceutical sorption mechanisms. Although overall principles of pharmaceuticals occupying specific sites are not summarized, some investigators already conducted good research and provided useful information. For example, the pyrimidine N and/or the –SO_2- group, pyrimidine in sulfamethazine could complex with exchangeable cations through inner and outer sphere complexes (Akyuz and Akyuz 2003). The keto group of ciprofloxacin could form hydrogen-bonding with the protonated surface hydroxyl group of Aluminum oxides, and ciprofloxacin keto O and one of the oxygen in the carboxylate group could form mononuclear bidentate complex with an Fe atom on iron oxide surface (Gu and Karthikeyan 2005). Combining FTIR and Raman spectroscopy,

Peterson et al. (2009) proposed that cephapirin bonds to quartz surfaces by bidentate complexation between one oxygen of –COO- and oxygen from carbonyl of an acetoxymethyl group. On a feldspar surface, cephapirin adsorbs through monodentate C=O complexation. It should be noted that the above mentioned studies identified the possible interaction mechanisms between pharmaceuticals and inorganic particles. But this information is not substantial for predicting their interaction strength and extent. For this prediction goal, quantifying the contribution of individual mechanisms to the overall sorption will be useful.

11.3.2 Adsorption to Organo-Mineral Complex

Numerous studies suggested that organic matter can form stable complexes with clay minerals and oxides. The organic matter would thus be preserved because the organic-mineral complexes are highly resistant to degradation and oxidation (Rice 2001; Kalbitz et al. 2005; Tharayil et al. 2006). Thus, these organo-mineral complexes play critical roles in preserving organic carbon from biodegradation, increasing soil water content and soil fertility (Kalbitz et al. 2005). Organo-mineral complexes also have special sorption properties with organic chemicals.

The inorganic mineral particles showed linear and low sorption with HOCs. However, the organic matter-coated mineral particles, showed stronger nonlinear and higher sorption to HOCs. Two major reasons were discussed in literature: (1) the complexed organic matter formed condensed region on mineral particles, which contributed to the apparent nonlinear and strong sorption with HOCs (Wang and Xing 2005b). (2) Organic matter was fractionated during the sorption on mineral particles. The fractions with strong and nonlinear sorption to HOCs were selectively sorbed on particle surface (Jones and Tiller 1999).

Pan et al. (2007) treated soil particles with H_2O_2 for 40 days. The resulted particles contained 0.08% organic carbon, which was the highly condensed carbon fraction. The H_2O_2 treated particles showed strong nonlinear sorption to phenanthrene with n values of 0.67–0.69 (Fig. 11.6). When these particles were coated with organic matter, the apparent sorption was more linear. This result indicated that organic matter form condensed complexes with mineral particles. As organic carbon content increased, the organic matter-mineral particle complexation is more like a physical coating process. This process resulted in a loose conformation and presented relatively linear sorption to HOCs.

Because of the significant contribution of inorganic minerals to pharmaceutical sorption, the relative importance of organic and inorganic fractions to the apparent pharmaceutical sorption is the discussion focus. The importance of organic fractions to pharmaceutical sorption was emphasized in several studies. For example, carbamazepine could be adsorbed on a sandy soil because of the presence of allophone which could accumulate organic matter. This result indicated that the adsorbed organic matter play the major role in the removal of carbamazepine

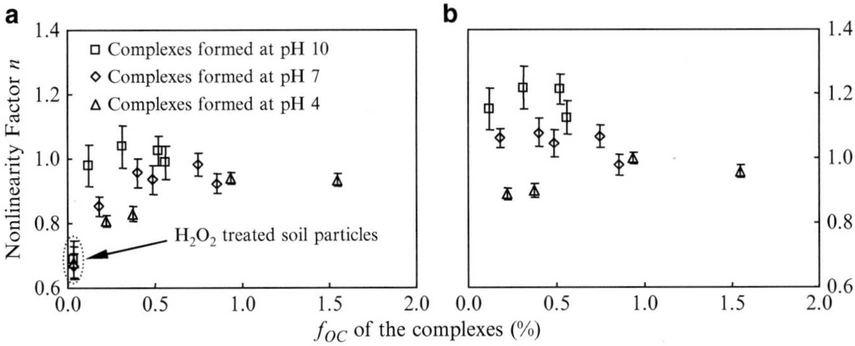

Fig. 11.6 Nonlinearity factor *n* for phenanthrene sorption on organo-mineral complexes (**a**) and the coated humic acid (**b**). The sorption by coated humic acid was calculated by deducting the maximum contribution of H_2O_2-treated soil particles. All the *n* values increased after subtracting the contribution of H_2O_2-treated soil particles (Pan et al. 2007 with permission from Elsevier)

(Gielen et al. 2009). Because the conventional sewage treatment could not efficiently remove pharmaceuticals, the authors proposed that irrigation of the effluent of the wastewater treatment plant could be beneficial for pharmaceutical removal. Sulfathiazole sorption to inorganic sorbents was more than one order of magnitude lower than that to organic sorbents (Kahle and Stamm 2007b). The authors concluded that the contribution of inorganic sorbents to sulfathiazole sorption was minor in many top soils and sediments (Kahle and Stamm 2007b). Sulfamethazine adsorption was increased after humic acid adsorption on smectite (Gao and Pedersen 2010) and the increase is consistent with the abundance of carboxyl (and possibly other O- and N-containing) moieties and aliphatic carbon content. However, it should be noted that the sorption equilibrium time may be extended because of the coating of organic matter on mineral particles. For example, pharmaceutical sorption kinetics study indicated that soil samples with higher organic carbon content have longer equilibrium time (Fan et al. 2006). The result may be understood from two aspects: (1) the diffusion of pharmaceutical in organic matter is the key process controlling kinetics. (2) The inorganic fraction provides important adsorption sites for pharmaceuticals. Pharmaceuticals need to diffuse through the coated organic matter to reach these sorption sites. Therefore, Fan's study could not exclude the possible roles of inorganic fraction to pharmaceutical sorption in soils/sediments. In addition, removing SOM from soil shortened the equilibrium time and increased the overall adsorption of tetracycline (Bao et al. 2009), indicating exposed inorganic minerals have stronger sorption to pharmaceuticals. Therefore, both inorganic and organic fractions control the sorption of pharmaceuticals on organo-mineral complexes (Gao and Pedersen 2010). Hou et al. (2010) also observed increased sulfamethoxazole sorption after removing the organic fraction from sediment particles. Interestingly, they collected the literature data on SMX sorption, and compared the apparent sorption with f_{OC}. At f_{OC} lower than 2%, a negative relationship between f_{OC} and

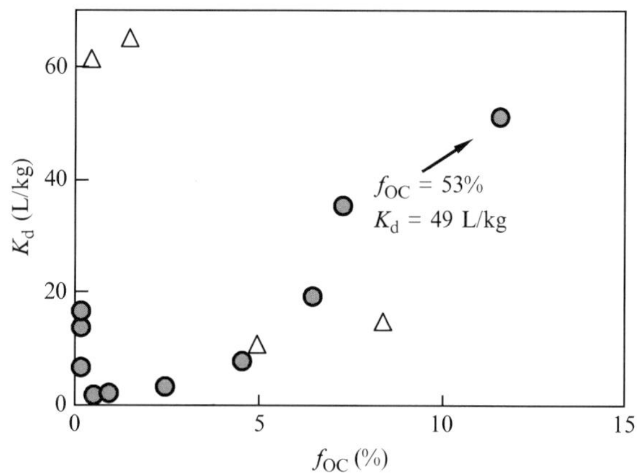

Fig. 11.7 K_d vs. f_{OC} for sulfamethoxazole adsorption in soils/sediments. A positive relationship between K_d and f_{OC} could be observed for sulfamethoxazole sorption on soils/sediments with f_{OC} higher than 2%. For adsorbents with f_{OC} lower than 2%, decreased f_{OC} may result in increased sorption (Hou et al. 2010 with permission from Elsevier)

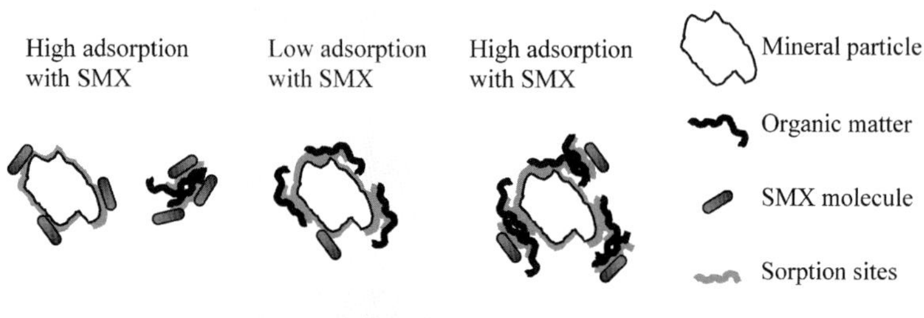

Fig. 11.8 Schematic diagram for SMX adsorption as affected by organic carbon content on solid particles. Although both mineral particle and organic matter have high adsorption with SMX, the coating of organic matter on mineral surface may occupy some adsorption sites for SMX and decrease the apparent SMX adsorption. As organic content further increases and multi-layer coating of organic matter was formed on mineral particles surface, the adsorption of SMX could be increased (Hou et al. 2010 with permission from Elsevier)

K_d could be noticed, while at f_{OC} higher than 2%, a positive relationship between f_{OC} and K_d was observed (Fig. 11.7). Based this observation, a schematic diagram was proposed to describe SMX sorption on organo-mineral complexes (Fig. 11.8), which may be applicable to other pharmaceutical sorption system. The sorption of SMX on mineral particles and organic matter are equally strong, and SMX and organic matter may share similar sorption sites, such as polar functional groups. Therefore, the apparent SMX sorption decreased with increased f_{OC} because of the competition between SMX and organic matter. However, further coating of organic matter on mineral particles resulted in multi-layer coating and more exposed polar functional groups of organic matter, which consequently increased SMX sorption.

11.3.3 Adsorption to Carbon-Based Adsorbents (Carbon Nanomaterials, Activated Carbon, and Biochar)

Carbon-based adsorbents, such as carbon nanomaterials, activated carbon, and biochar, usually have highly hydrophobic surface, and thus their sorption with HOCs was widely discussed. These adsorbents usually showed nonlinear and high sorption and strong sorption/desorption hysteresis. Activated carbons (ACs) showed high sorption because of their porous structure and high specific surface areas. These adsorbents are widely available and relatively cheap and thus are often used for water treatment. Carbon nanomaterials (CNMs) are emerging materials which have unique and explicit surface properties, and have many applications (Mu et al. 2006; Esawi and Farag 2007; Mauter and Elimelech 2008; Pan and Xing 2008). Because of their small sizes, CNMs form aggregates under most conditions. Their surface area could not be fully accessed unless they are well dispersed. Therefore, the mass-based sorption of HOCs on CNMs was lower than that on ACs. Once dispersed, CNMs are expected to have much higher sorption to organic chemicals, and thus CNM dispersion is an attractive research direction. The accessibility of sorbent surface area to organic chemicals should consider their molecular size. For example, chemicals with different molecular sizes could be compared for their adsorption on CNTs and ACs. Small-sized sulfamethoxazole had higher adsorption on ACs than other carbonaceous adsorbents. But for bulky TC and tylosin, their adsorption on AC was much lower. This result indicated that the adsorption on AC is highly molecular size selective (Ji et al. 2009). Another research also reported that solution pH and ionic strength did not have a major effect on organic chemical adsorption on ACs, indicating that the principal interactions involved in the adsorption of these compounds are non-electrostatic (Rivera-Utrilla et al. 2009). Therefore, the accessibility of the surface area or the inner pores is the essential controlling factor regarding organic chemical adsorption on ACs. But for CNTs, most surface area was accessible for the adsorption. This was confirmed by the comparable surface area – normalized adsorption between CNTs and nonporous graphite (Ji et al. 2010). The more accessible surface area of CNTs in comparison to ACs was also confirmed by their faster adsorption of organic chemicals (Ji et al. 2010). The adsorption difference between CNTs and ACs could also be understood from the adsorption mechanisms. The lower adsorption of tetracycline to AC indicates that for bulky adsorbates adsorption affinity is greatly affected by the accessibility of available adsorption sites. The remarkably strong adsorption of tetracycline to the carbon nanotubes and to graphite can be attributed to the strong adsorptive interactions (van der Waals forces, pi-pi electron-donor-acceptor interactions, cation-pi bonding) with the graphene surface (Ji et al. 2009). The above discussion indicated that for both HOCs and pharmaceuticals, the hydrophobicity of the chemicals and the accessibility of the surface area on solid particles are the critical factors controlling their adsorption on ACs and CNTs. The chemicals with pi-system should be examined with carefulness because of the abundance of pi-system in CNTs and ACs. The EDA theory discussed in the previous section applies here for pharmaceuticals.

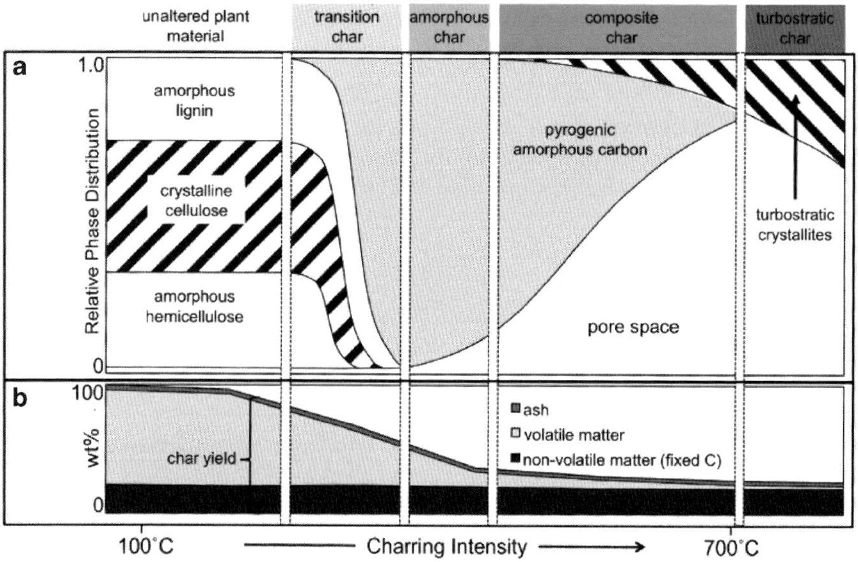

Fig. 11.9 Dynamic molecular structure of plant biomass-derived black carbon (biochar) across a charring gradient and schematic representation of the four proposed char categories and their individual phases. (*A*) Physical and chemical characteristics of organic phases. Exact temperature ranges for each category are controlled by both charring conditions (i.e., temperature, duration, and atmosphere) and relative contents of plant biomass components (i.e., hemicellulose, cellulose, and lignin). (*B*) Char composition as inferred from gravimetric analysis. Yields, VM, fixed C, and ash contents are averaged across wood and grass chars. Relative contributions above 700°C are estimates (Keiluweit et al. 2010 with permission from ACS)

Biochar is a new term for pyrolysis products of waste biomass and plants, and is likely to be widely used in near future as agricultural soil amendments. The adsorption of HOCs on biochars increased with the pyrolytic temperature (Chen et al. 2008). The application of biochar in the open environment will definitely alter the environmental behavior of organic chemicals. Therefore, the sorption of organic chemicals on biochar has attracted increasing research attention. The sorption of pharmaceuticals on biochar deserves special attention. In addition, the comparative study of HOC and pharmaceutical sorption in biochar is not available yet.

In comparison to ACs, biochars may contain noncarbonized matter because of the lower pyrolysis temperature. Recalcitrant nature of biochars enable them to release carbon slowly as carbon sources for organisms in soil system and thus facilitate their application as soil amendment to mitigate greenhouse effect, while the fraction of biochars with polar functional groups can provide cation exchange sites to retain plant nutrients. The wood and grass biomass-derived biochars experience phase transitions of plant material: transition char, amorphous chars, composite chars, and turbostratic char (Fig. 11.9) (Keiluweit et al. 2010).

11.4 Influence of Environmental Conditions on Sorption

11.4.1 Dissolved Organic Matter

Dissolved organic matter (DOM) is ubiquitous in the environment. The strong interactions between DOM and HOCs have been reported widely. Studies indicated that the dissolved HAs have higher binding capacities with HOCs than their solid state (Pan et al. 2007). Therefore, DOM could influence HOC environmental behavior greatly. The readers are suggested to read relevant review papers for the interaction between HOCs and DOM (Chefetz et al. 2006; Pan et al. 2008). This current section intends to discuss the roles of DOM in HOC-soil/sediment sorption system.

The following equation could be applied to describe the sorption system in the presence of DOM as illustrated in Fig. 11.10. This model considers organic chemical sorption on particles, DOM sorption on particles, and organic chemical sorption by DOM. The overall sorption coefficient could be calculated as:

$$K_p = \frac{S_b + S}{C_b + C_f} \tag{11.1}$$

$$\text{or } K_p = \frac{K_b \cdot K_{DOC} \cdot [\text{DOM}] + K_d}{K_{DOC} \cdot [\text{DOM}] + 1} \tag{11.2}$$

where C_b and C_f are concentrations of HOCs bound with DOM left in aqueous phase and of free dissolved HOCs, respectively. S_b and S are concentrations of HOCs bound with DOM coated on solid particles and sorbed directly by solid

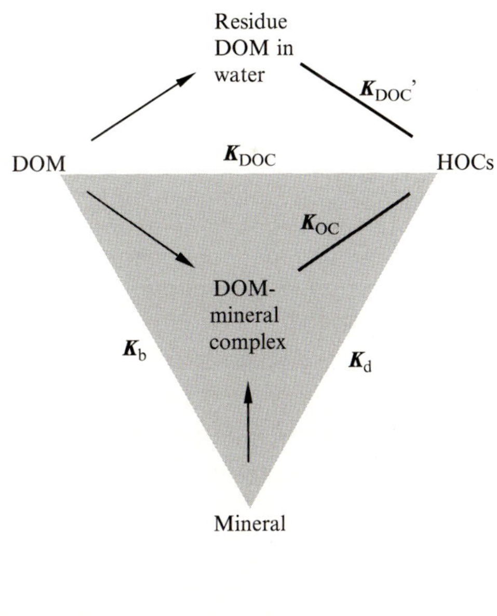

Fig. 11.10 HOC sorption in a three-phase-system. HOCs can be sorbed by both DOM and mineral particles. DOM sorption on mineral particles results in its fractionation. K_b: sorption coefficient for DOM sorption on mineral particles; K_d: sorption coefficient for HOCs on mineral particles; K_{DOC}: sorption coefficient for HOCs by bulk DOM; K_{OC}: sorption coefficient of HOCs on DOM-mineral complex; K_{DOC}': sorption coefficient of HOCs on residue DOM

particles, respectively. K_p is the overall sorption coefficient for HOCs sorption in the three-phase system, K_b is DOM sorption coefficient on particles, K_d is HOC sorption coefficient by particles, and K_{DOC} is HOC sorption coefficient by DOM.

Equation 11.2 is actually a simplified form of Eq. 11.1 based on the assumption that all the processes involved in Fig. 11.10 are linear. It is easy to notice that nonlinear interactions are more common conditions in the real system. However, incorporating all the nonlinear processes in the overall equation makes the modeling impractical, because the increased parameter number and the nonlinear expression will greatly increase the uncertainty of the overall model prediction. Therefore, although the concept of Eq. 11.2 is widely applied in current studies, it should be realized that this equation is a simple expression of the real environment and extended study is needed to verify this simplification or propose more environmental-relevant simplified concept. In addition, DOM was reported to be fractionated when adsorbed on solid particles. Namely, DOM molecules with certain chemical properties may be selectively adsorbed, while other molecules are left in the solution. This process could not be investigated based on TOC analysis. Up to date, no quantitative model is available to describe DOM fractionation after adsorption.

DOM is also reported to alter pharmaceutical environmental fate. Investigators estimated that 10–29% of pharmaceuticals in the environment is associated with colloids, and the percentage increases to 70% in aqueous phase (Zhou et al. 2007). DOM effect on pharmaceutical sorption behavior could also be noted as third phase effect or solid concentration effect on apparent sorption coefficient (Bowman et al. 2002; Zhou 2006). The above discussed modeling concept of the effect of DOM on the apparent behavior of HOCs in soil/sediment system applies for pharmaceutical sorption. However, because pharmaceuticals may exist as different species, different aspects should be considered. For example, the sorption of SMX on mineral particles decreased with the addition of DOM (Hou et al. 2010) (Fig. 11.11). The authors conducted complementary experiments to investigate the interactions between SMX and DOM using dialysis equilibrium system, and no significant interaction was observed. They concluded that at the experimental pH (pH 7), both SMX and DOM were negatively charged. The apparent decreased SMX sorption should be resulted from the competition between SMX and DOM to the sorption sites on mineral particles. Zhang and Dong (2008) also reported that low-molecular-weight (LMW) organic acids, including citric acid, malic acid and salicylic acid could inhibit NOR adsorption at pH 4.5.

11.4.2 pH

HOCs do not have pKa and thus their physicochemical properties are hardly affected by pH change. The apparent sorption as affected by pH was explained by the change of soil/sediment properties. However, pharmaceuticals have pK_a values of the comparable range of soil/sediment pHs, and thus they may exist as different species, such as cation, zwitterion, and anion (Figueroa et al. 2004; MacKay and

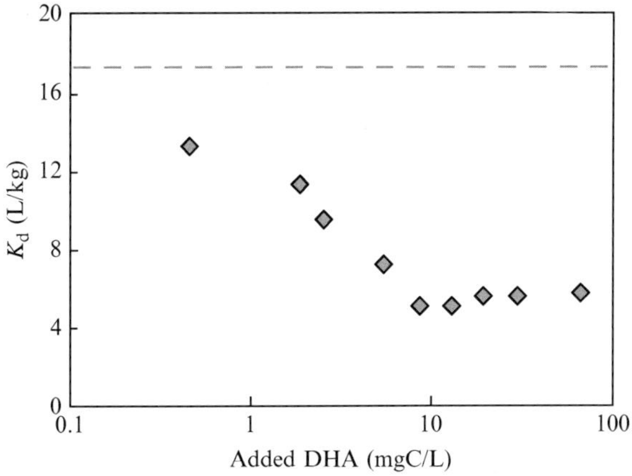

Fig. 11.11 Sulfamethoxazole K_d on IM as affected by the addition of dissolved humic acid (DHA). The *dashed line* indicates sulfamethoxazole sorption coefficient without DHA. A significant decrease of sulfamethoxazole sorption was observed with the addition of DHA (Hou et al. 2010 with permission from Elsevier)

Canterbury 2005; Gu et al. 2007). Identifying the functioning of a certain speciated form in the overall sorption facilitated the discussion on sorption mechanisms of pharmaceuticals in soils/sediments. Therefore, many studies have been conducted along this line.

For example, pharmaceutical cation species could be adsorbed through cation exchange, while zwitterion species could interact with soils/sediment through surface complexation and hydrophobic interactions (Figueroa et al. 2004; Kulshrestha et al. 2004; Sibley and Pedersen 2008). Kahle and Stamm (2007b) examined the adsorption of sulfathiazole to clay minerals (montmorillonite and illite) and ferrihydrite (anion sorbent). Cationic sulfathiazole was the most important species for the adsorption to clay minerals, followed by neutral species (Kahle and Stamm 2007b). At acidic condition ($pH < 4$), cationic tetracycline was the prevailing species. The adsorption was high reaching cation exchange capacity of the mineral (Parolo et al. 2008). After modeling of the adsorption data, the authors indicated that neutral and monoanionic species of tetracycline adsorbed to the negatively charged montmorillonite surface, indicating that non-electrostatic interaction exists in addition to electrostatic interactions (Parolo et al. 2008). Richter et al. (2009) highlighted cation binding between sulfathiazole$^+$ and anionic sites of humic acid. This process could be well modeled by NICA-Donnan cation-binding model. Sorption of sulfathiazole on organic materials followed the order of K_{OC} (cation) > K_{OC} (neutral) > K_{OC} (anion) (Kahle and Stamm 2007a). Ketonic, carboxylic, and phenolic C as well as aromatic C-H and methoxyl/N-alkyl C are important functional groups controlling sulfathiazole sorption as provided by NMR. Therefore, pharmaceutical sorption models should prudently consider species-specific sorption processes.

Although the pH effect on pharmaceutical sorption in soils/sediments is widely reported, a problem could not be completely avoided is that the physical conformation or even chemical composition may be altered with pH. Therefore, a stable surrogate adsorbent is needed to conduct experiments regarding pH effects. Zhang et al. (2010) successfully applied CNTs as model adsorbents to study the species-specified sorption of SMX on solid particles. They observed that at $pH < 3.5$, the apparent Kd of SMX increased with increasing pH, while at $pH > 3.5$, an opposite trend was observed. They compared the pHzpc of CNTs and pKa of SMX. The species-specific sorption model was proposed and applied to fit the apparent data from the adsorption edge study. Based on this modeling, the authors were able to conclude about the sorption coefficients for all the species and the contribution of certain species to the overall sorption. Thus, discussion on sorption mechanisms is easy and the possible involved mechanisms are illustrated in Fig. 11.12.

11.4.3 Effect of Cations

Pharmaceuticals may form stable complexes with cations. Consequently, pharmaceutical species alter with cation addition and the sorption could be enhanced or inhibited. If ternary complexation of pharmaceutical-metal-soil/sediment formed, pharmaceutical sorption capacity and strength could be enhanced greatly (MacKay and Canterbury 2005; Gu et al. 2007). The study on the coadsorption of Cu and tetracyclines indicated that the adsorption on montmorillonite of both Cu and tetracyclines was increased in comparison to their individual adsorption. The authors calculated that TC-Cu complexes had higher sorption coefficients than that of the corresponding TC species (Wang et al. 2008). Choi et al. (2008) studied the removal of sulfonamide and tetracycline antibiotics by AC. TAs were more hydrophilic than SAs, but the removal was higher for TAs than for SAs. The authors stated that the metal and metal oxide on the surface of AC may form complex with TAs and thus contribute to the high adsorption. Surface complexation was also attributed to the enhanced tetracycline binding by calcium (Basakcilardan-Kabakci et al. 2007). On the contrary, cations may compete with positively charged pharmaceuticals and thus decrease their sorption in soils/sediments (Sibley and Pedersen 2008; Bai et al. 2008). Therefore, the apparent effect of cations on pharmaceutical sorption is greatly dependent on pharmaceutical species and sorption mechanisms.

The hydration shell of cations alters the availability of organic chemicals to sorption sites on solid particles. Cu^{2+} and Pb^{2+} diminished the sorption of 2,4,6-trichlorophenol on humic acids because the hydration shells of adsorbed Cu^{2+} and Pb^{2+} decreased the contact of 2,4,6-trichlorophenol with available surface (Wang et al. 2009). The same reason resulted in inhibiting adsorption of atrazine by Cu^{2+}, Pb^{2+} and Cd^{2+} on carbon nanotubes (Chent et al. 2008).

11.4.4 Effect of Other Organic Contaminants

The existence of multi-pollutants is a common situation in the real environment. Therefore, the sorption behavior of organic chemicals in multi-pollutant systems is of a wide research interest. This type of sorption system is known as binary

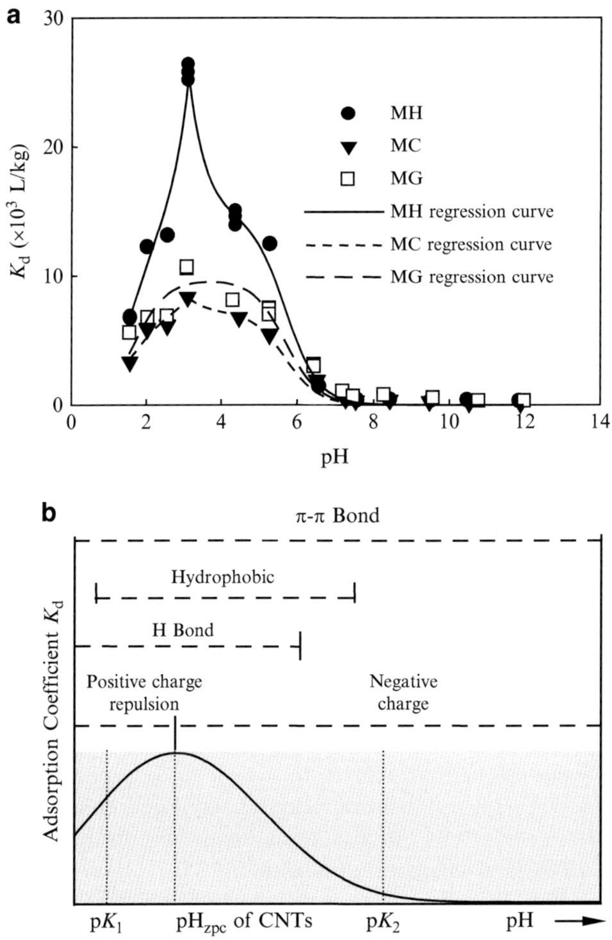

Fig. 11.12 pH-dependent SMX adsorption on CNTs. (**a**) Adsorption coefficient (K_d) of SMX on CNTs at different pHs. (**b**) Possible adsorption mechanisms of SMX on CNTs at different pHs. The experiments were carried out at SMX initial concentration of 9.5 mg/L. K_d decreased with increasing pH at $pH > pH_{zpc}$ and decreased with decreased pH at $pH < pH_{zpc}$ because of electrostatic repulsions. At pH around pHzpc, various adsorption mechanisms (such as hydrophobic interaction, π-π and hydrogen bonds) contribute to the overall SMX adsorption (Zhang et al. 2010 with permission from ACS)

sorption, co-sorption, or competitive sorption system. Two lines of results were reported: (1) cooperative adsorption. For example, 4-nitrophenol adsorption was enhanced by the presence of basic yellow 28 dye in the mixture and increased with dye concentrations (Zermane et al. 2010). This type of cooperative sorption was described using Sheindorf-Rebhun-Sheintuch model. This case is rare in literature. However, it deserves more attention, especially for pharmaceuticals. (2) A more widely reported results in the inhibited sorption of the primary sorbate in the presence of the secondary sorbate. The competitive sorption between HOCs could be easily found in literature. Pharmaceuticals could also compete with each other (Lai et al. 2000; Sanders et al. 2008) or with their metabolites (Casey et al. 2004; Strock et al. 2005), HOCs (Yu and Huang 2005), and DOM (Graham et al. 2000) for sorption sites. The apparent sorption was compared based on the sorption isotherms or sorption coefficients of the primary sorbate in the absence or presence of the secondary sorbate (Sander and Pignatello 2005; Yang et al. 2006; Wang et al. 2008, 2009).

A recent study indicated that chemicals with different properties may complimentarily occupy different sorption sites, which results in an accumulated sorption phenomenon (Pan and Xing 2010). The primary sorbate and secondary sorbate have their different preferred sorption sites as well as shared sorption sites. Therefore, both competitive and complementary sorption can occur in multi-pollutant systems, especially when these pollutants have distinct properties. From this viewpoint, the co-sorption of HOCs and pharmaceuticals in soils/sediments deserves more research attention.

The application of another organic chemical in the sorption system may also have indirect interactions with the primary sorbate. Verma et al. (2007) observed that the addition of EDTA could promote tetracycline release. The direct interaction between EDTA and tetracycline was not likely to occur. The authors proposed that a portion of tetracycline was bound with metal ions. The strong complexation between EDTA and metal ions decreased tetracycline-metal binding and thus decreased tetracycline apparent sorption (Verma et al. 2007).

Surfactants play an important role in organic chemical sorption on solid particles. The discussion on organic chemical sorption behavior in DOM-particle-water system applies in surfactant-particle-water system. Specifically, different roles of surfactants on HOC and pharmaceutical solubility should be noticed. The presence of surfactants could increase HOC solubility in aqueous phase and thus increase their "tendency" to stay in the aqueous phase. However, pharmaceuticals may form ion pairs with surfactant and thus reduce the solubilities of pharmaceuticals (Hari et al. 2005; Kibbey et al. 2007). The ionic states of surfactants (e.g. cation, anion, and neutral states) make the explanation of pharmaceutical sorption behavior in surfactant-particle-water system complicated and thus special care should be put in this aspect.

11.5 Summary and Perspectives

Both pharmaceuticals and HOCs are mostly derived from anthropogenic activities. Because they involve with human life differently, the pathways that they enter the environment are different. In addition, HOCs could pose chemical and biological stresses directly to organisms (such as mutagenic, carcinogenic and/or co-carcinogenic toxicities), while pharmaceuticals can pose risks to human indirectly (such as promoting the development of drug-resistant genes). Therefore, the environmental risks of HOCs and pharmaceuticals should be evaluated using different methodologies. A typical situation is that the risks of HOCs could be assessed by their concentrations, but pharmaceuticals risks could not be assessed by their instant concentrations without knowing pollution history because of their relatively rapid degradation and indirect environmental risks.

The physical conformation (such as rigidity) and the overall chemical properties (such as aromatic, aliphatic and polarity) of soils/sediments are the controlling factors for HOC sorption. Although quantitative prediction method has not been developed yet, the modeling concept regarding HOC sorption in organic fractions was rather clear, such as dual mode model and dual reactive model. Pharmaceuticals containing different functional groups interact with soils/sediments through different mechanisms. Thus the sorption modeling/prediction is only applicable when a proper classification of pharmaceuticals is available. This type of classification has to be comprehensive and needs extensive studies.

The sorption of HOCs in soils/sediments is mostly controlled by hydrophobic properties of both adsorbates and adsorbents and the contribution of inorganic mineral particles could be neglected. But the organic and inorganic fractions of soils/sediments have the same importance regarding pharmaceutical sorption and thus the contribution of inorganic fractions to pharmaceutical sorption could not be ignored. The most distinct different property for pharmaceuticals is their pK_as, which result in different species distributions at different water chemistry conditions (such as pH and ionic strength). Identifying the contribution of different species to the overall sorption is important to understand pharmaceutical sorption mechanisms.

Because of their different sorption behaviors in soils/sediments, direct comparative studies on pharmaceutical and HOC sorption could provide valuable insight to their sorption mechanisms. Knowing the different sorption properties of HOCs and pharmaceuticals enable us to develop chemical-specified technologies to remediate the polluted sites and assess their environmental exposure and risk properly.

Acknowledgments This work was supported by National Scientific Foundation of China (40803034 and 40973081), Scientific Research Foundation for the Returned Overseas Chinese Scholars (State Education Ministry), Research Grant for Future Talents of Yunnan Province, and USDA-AFRI (2009-35201-05819).

References

Accardi-Dey A, Gschwend PM (2002) Assessing the combined roles of natural organic matter and black carbon as sorbents in sediments. Environ Sci Technol 36:21–29

Akyuz S, Akyuz T (2003) FT-IR spectroscopic investigation of adsorption of pyrimidine on sepiolite and montmorillonite from Anatolia. J Incl Phenom Macrocycl Chem 46:51–55

Allen-King RM, Grathwohl P, Ball WP (2002) New modeling paradigms for the sorption of hydrophobic organic chemicals to heterogeneous carbonaceous matter in soils, sediments, and rocks. Adv Water Resour 25:985–1016

Arias-Estevez M, Lopez-Periago E, Martinez-Carballo E, Simal-Gandara J, Mejuto JC, Garcia-Rio L (2008) The mobility and degradation of pesticides in soils and the pollution of groundwater resources. Agric Ecosyst Environ 123:247–260

Bahnick DA, Doucette WJ (1988) Use of molecular connectivity indexes to estimate soil sorption coefficients for organic-chemicals. Chemosphere 17:1703–1715

Bai Y, Wu F, Liu C, Guo J, Fu P, Li W, Xing B (2008) Interaction between carbamazepine and humic substances: a fluorescence spectroscopy study. Environ Toxicol Chem 27:95–102

Bao YY, Zhou QX, Wang YY (2009) Adsorption characteristics of tetracycline by two soils: assessing role of soil organic matter. Aust J Soil Res 47:286–295

Barron L, Havel J, Purcell M, Szpak M, Kellehera B, Paull B (2009) Predicting sorption of pharmaceuticals and personal care products onto soil and digested sludge using artificial neural networks. Analyst 134:663–670

Basakcilardan-Kabakci S, Thompson A, Cartmell E, Le Corre K (2007) Adsorption and precipitation of tetracycline with struvite. Water Environ Res 79:2551–2556

Bonin JL, Simpson MJ (2007) Variation in phenanthrene sorption coefficients with soil organic matter fractionation: the result of structure or conformation? Environ Sci Technol 41:153–159

Bowman JC, Zhou JL, Readman JW (2002) Sediment-water interactions of natural oestrogens under estuarine conditions. Mar Chem 77:263–276

Boxall ABA, Fogg LA, Blackwell PA, Kay P, Pemberton EJ, Croxford A (2004) Veterinary medicines in the environment. Rev Environ Contam Toxicol 180:1–91

Burkhard LP (2000) Estimating dissolved organic carbon partition coefficients for nonionic organic chemicals. Environ Sci Technol 34:4663–4668

Casey FXM, Larsen GL, Hakk H, Šimůnek J (2003) Fate and transport of 17 beta-estradiol in soil-water systems. Environ Sci Technol 37:2400–2409

Casey FXM, Hakk H, Šimůnek J, Larsen GL (2004) Fate and transport of testosterone in agricultural soils. Environ Sci Technol 38:790–798

Chang PH, Jean JS, Jiang WT, Li ZH (2009) Mechanism of tetracycline sorption on rectorite. Coll Surf A-Physicochem Eng Asp 339:94–99

Chee-Sanford JC, Mackie RI, Koike S, Krapac IJ, Lin YF, Yannarell AC, Maxwell S, Aminov RI (2009) Fate and transport of antibiotic residues and antibiotic resistance genes following land application of manure waste. J Environ Qual 38:1086

Chefetz B, Xing BS (2009) Relative role of aliphatic and aromatic moieties as sorption domains for organic compounds: a review. Environ Sci Technol 43:1680–1688

Chefetz B, Deshmukh AP, Hatcher PG, Guthrie EA (2000) Pyrene sorption by natural organic matter. Environ Sci Technol 34:2925–2930

Chefetz B, Illani T, Schulz E, Chorover J (2006) Wastewater dissolved organic matter: characteristics and sorptive capabilities. Water Sci Technol 53:51–57

Chen BL, Xing BS (2005) Sorption and conformational characteristics of reconstituted plant cuticular waxes on montmorillonite. Environ Sci Technol 39:8315–8323

Chen BL, Zhou DD, Zhu LZ (2008) Transitional adsorption and partition of nonpolar and polar aromatic contaminants by biochars of pine needles with different pyrolytic temperatures. Environ Sci Technol 42:5137–5143

Chent GC, Shan XQ, Wang YS, Pei ZG, Shen XE, Wen B, Owens G (2008) Effects of copper, lead, and cadmium on the sorption and desorption of atrazine onto and from carbon nanotubes. Environ Sci Technol 42:8297–8302

Chin YP, Aiken G, Oloughlin E (1994) Molecular-weight, polydispersity, and spectroscopic properties of aquatic humic substances. Environ Sci Technol 28:1853–1858

Chin YP, Aiken GR, Danielsen KM (1997) Binding of pyrene to aquatic and commercial humic substances: the role of molecular weight and aromaticity. Environ Sci Technol 31:1630–1635

Chiou CT, Peters LJ, Freed VH (1979) Physical concept of soil-water equilibria for non-ionic organic-compounds. Science 206:831–832

Choi KJ, Kim SG, Kim SH (2008) Removal of tetracycline and sulfonamide classes of antibiotic compound by powdered activated carbon. Environ Technol 29:333–342

Cornelissen G, Gustafsson O, Bucheli TD, Jonker MTO, Koelmans AA, Van Noort PCM (2005) Extensive sorption of organic compounds to black carbon, coal, and kerogen in sediments and soils: mechanisms and consequences for distribution, bioaccumulation, and biodegradation. Environ Sci Technol 39:6881–6895

Daughton CG, Ternes TA (1999) Pharmaceuticals and personal care products in the environment: agents of subtle change? Environ Health Perspect 107:907–938

Ding GW, Novak JM, Herbert S, Xing BS (2002) Long-term tillage effects on soil metolachlor sorption and desorption behavior. Chemosphere 48:897–904

Doick KJ, Burauel P, Jones KC, Semple KT (2005) Distribution of aged C-14-PCB and C-14-PAH residues in particle-size and humic fractions of an agricultural soil. Environ Sci Technol 39:6575–6583

During RA, Krahe S, Gath S (2002) Sorption behavior of nonylphenol in terrestrial soils. Environ Sci Technol 36:4052–4057

Esawi AMK, Farag MM (2007) Carbon nanotube reinforced composites: potential and current challenges. Mater Des 28:2394–2401

Fan ZS, Casey FXM, Larsen GL, Hakk H (2006) Fate and transport of 1278-TCDD, 1378-TCDD, and 1478-TCDD in soil-water systems. Sci Total Environ 371:323–333

Fan ZS, Casey FXM, Hakk H, Larsen GL (2007) Persistence and fate of 17 beta-estradiol and testosterone in agricultural soils. Chemosphere 67:886–895

Figueroa RA, Leonard A, Mackay AA (2004) Modeling tetracycline antibiotic sorption to clays. Environ Sci Technol 38:476–483

Gao J, Pedersen JA (2010) Sorption of sulfonamide antimicrobial agents to humic acid-clay complexes. J Environ Qual 39:228–235

Gauthier TD, Shane EC, Guerin WF, Seitz WR, Grant CL (1986) Fluorescence quenching method for determining equilibrium-constants for polycyclic aromatic-hydrocarbons binding to dissolved humic materials. Environ Sci Technol 20:1162–1166

Gevao B, Semple KT, Jones KC (2000) Bound pesticide residues in soils: a review. Environ Pollut 108:3–14

Gielen G, van den Heuvel MR, Clinton PW, Greenfield LG (2009) Factors impacting on pharmaceutical leaching following sewage application to land. Chemosphere 74:537–542

Göbel A, Thomsen A, McArdell CS, Joss A, Giger W (2005) Occurrence and sorption behavior of sulfonamides, macrolides, and trimethoprim in activated sludge treatment. Environ Sci Technol 39:3981–3989

Golding CJ, Smernik RJ, Birch GF (2005) Investigation of the role of structural domains identified in sedimentary organic matter in the sorption of hydrophobic organic compounds. Environ Sci Technol 39:3925–3932

Graham MR, Summers RS, Simpson MR, MacLeod BW (2000) Modeling equilibrium adsorption of 2-methylisoborneol and geosmin in natural waters. Water Res 34:2291–2300

Gu C, Karthikeyan KG (2005) Sorption of the antimicrobial ciprofloxacin to aluminum and iron hydrous oxides. Environ Sci Technol 39:9166–9173

Gu C, Karthikeyan KG, Sibley SD, Pedersen JA (2007) Complexation of the antibiotic tetracycline with humic acid. Chemosphere 66:1494–1501

Gunasekara AS, Simpson MJ, Xing BS (2003) Identification and characterization of sorption domains in soil organic matter using structurally modified humic acids. Environ Sci Technol 37:852–858

Hari AC, Paruchuri RA, Sabatini DA, Kibbey TCG (2005) Effects of pH and cationic and nonionic surfactants on the adsorption of pharmaceuticals to a natural aquifer material. Environ Sci Technol 39:2592–2598

Holbrook RD, Love NG, Novak JT (2004) Sorption of 17-beta-estradiol and 17 alpha-ethinylestradiol by colloidal organic carbon derived from biological wastewater treatment systems. Environ Sci Technol 38:3322–3329

Hollert H, Durr M, Holtey-Weber R, Islinger M, Brack W, Farber H, Erdinger L, Braunbeck T (2005) Endocrine disruption of water and sediment extracts in. A non-radioactive dot blot/RNAse protection-assay using isolated hepatocytes of rainbow trout – deficiencies between bioanalytical effectiveness and chemically determined concentrations and how to explain them. Environ Sci Pollut Res 12:347–360

Holthaus KIE, Johnson AC, Jurgens MD, Williams RJ, Smith JJL, Carter JE (2002) The potential for estradiol and ethinylestradiol to sorb to suspended and bed sediments in some English rivers. Environ Toxicol Chem 21:2526–2535

Hou J, Pan B, Niu XK, Chen JZ, Xing BS (2010) Sulfamethoxazole sorption by sediment fractions in comparison to pyrene and bisphenol A. Environ Pollut 158:2826–2832

Huang WL, Weber WJ (1997) A distributed reactivity model for sorption by soils and sediments. 10. Relationships between desorption, hysteresis, and the chemical characteristics of organic domains. Environ Sci Technol 31:2562–2569

Hur J, Schlautman MA (2003) Using selected operational descriptors to examine the heterogeneity within a bulk humic substance. Environ Sci Technol 37:880–887

Hur J, Schlautman MA (2004) Influence of humic substance adsorptive fractionation on pyrene partitioning to dissolved and mineral-associated humic substances. Environ Sci Technol 38:5871–5877

Ilani T, Schulz E, Chefetz B (2005) Interactions of organic compounds with wastewater dissolved organic matter: role of hydrophohic fractions. J Environ Qual 34:552–562

Jacobsen AM, Lorenzen A, Chapman R, Topp E (2005) Persistence of testosterone and 17 beta-estradiol in soils receiving swine manure or municipal biosolids. J Environ Qual 34:861–871

Jeong S, Wander MM, Kleineidam S, Grathwohl P, Ligouis B, Werth CJ (2008) The role of condensed carbonaceous materials on the sorption of hydrophobic organic contaminants in subsurface sediments. Environ Sci Technol 42:1458–1464

Ji LL, Chen W, Duan L, Zhu DQ (2009) Mechanisms for strong adsorption of tetracycline to carbon nanotubes: a comparative study using activated carbon and graphite as adsorbents. Environ Sci Technol 43:2322–2327

Ji L, Liu F, Xu Z, Zheng S, Zhu D (2010) Adsorption of pharmaceutical antibiotics on template-synthesized ordered micro- and mesoporous carbons. Environ Sci Technol 44:3116–3122

Jones KD, Tiller CL (1999) Effect of solution chemistry on the extent of binding of phenanthrene by a soil humic acid: a comparison of dissolved and clay bound humic. Environ Sci Technol 33:580–587

Jones AD, Bruland GL, Agrawal SG, Vasudevan D (2005) Factors influencing the sorption of oxytetracycline to soils. Environ Toxicol Chem 24:761–770

Kahle M, Stamm C (2007a) Sorption of the veterinary antimicrobial sulfathiazole to organic materials of different origin. Environ Sci Technol 41:132–138

Kahle M, Stamm C (2007b) Time and pH-dependent sorption of the veterinary antimicrobial sulfathiazole to clay minerals and ferrihydrite. Chemosphere 68:1224–1231

Kalbitz K, Schwesig D, Rethemeyer J, Matzner E (2005) Stabilization of dissolved organic matter by sorption to the mineral soil. Soil Biol Biochem 37:1319–1331

Kang SH, Xing BS (2005) Phenanthrene sorption to sequentially extracted soil humic acids and humins. Environ Sci Technol 39:134–140

Keiluweit M, Nico PS, Johnson MG, Kleber M (2010) Dynamic molecular structure of plant biomass-derived black carbon (biochar). Environ Sci Technol 44:1247–1253

Kibbey TCG, Paruchuri R, Sabatini DA, Chen L (2007) Adsorption of beta blockers to environmental surfaces. Environ Sci Technol 41:5349–5356

Kile DE, Chiou CT, Zhou HD, Li H, Xu OY (1995) Partition of nonpolar organic pollutants from water to soil and sediment organic matters. Environ Sci Technol 29:1401–1406

Kohl SD, Rice JA (1998) The binding of contaminants to humin: a mass balance. Chemosphere 36:251–261

Kottler BD, Alexander M (2001) Relationship of properties of polycyclic aromatic hydrocarbons to sequestration in soil. Environ Pollut 113:293–298

Kulshrestha P, Giese RF, Aga DS (2004) Investigating the molecular interactions of oxytetracycline in clay and organic matter: insights on factors affecting its mobility in soil. Environ Sci Technol 38:4097–4105

Lai KM, Johnson KL, Scrimshaw MD, Lester JN (2000) Binding of waterborne steroid estrogens to solid phases in river and estuarine systems. Environ Sci Technol 34:3890–3894
LeBoeuf EJ, Weber WJ (1997) A distributed reactivity model for sorption by soils and sediments. 8. Sorbent organic domains: discovery of a humic acid glass transition and an argument for a polymer-based model. Environ Sci Technol 31:1697–1702
Li H, Lee LS (1999) Sorption and abiotic transformation of aniline and alpha-naphthylamine by surface soils. Environ Sci Technol 33:1864–1870
Liu RX, Wilding A, Hibberd A, Zhou JL (2005) Partition of endocrine-disrupting chemicals between colloids and dissolved phase as determined by cross-flow ultrafiltration. Environ Sci Technol 39:2753–2761
Loffredo E, Senesi N (2006) Fate of anthropogenic organic pollutants in soils with emphasis on adsorption/desorption processes of endocrine disruptor compounds. Pure Appl Chem 78:947–961
Lu YF, Pignatello JJ (2004) Sorption of apolar aromatic compounds to soil humic acid particles affected by aluminum(III) ion cross-linking. J Environ Qual 33:1314–1321
MacKay AA, Canterbury B (2005) Oxytetracycline sorption to organic matter by metal-bridging. J Environ Qual 34:1964–1971
Mackay AA, Seremet DE (2008) Probe compounds to quantify cation exchange and complexation interactions of ciprofloxacin with soils. Environ Sci Technol 42:8270–8276
Macleod CJA, Semple KT (2003) Sequential extraction of low concentrations of pyrene and formation of non-extractable residues in sterile and non-sterile soils. Soil Biol Biochem 35:1443–1450
Mao JD, Hundal LS, Thompson ML, Schmidt-Rohr KS (2002) Correlation of poly(methylene)-rich amorphous aliphatic domains in humic substances with sorption of a nonpolar organic contaminant, phenanthrene. Environ Sci Technol 36:929–936
Maskaoui K, Hibberd A, Zhou JL (2007) Assessment of the interaction between aquatic colloids and pharmaceuticals facilitated by cross-flow ultrafiltration. Environ Sci Technol 41:8038–8043
Mauter MS, Elimelech M (2008) Environmental applications of carbon-based nanomaterials. Environ Sci Technol 42:5843–5859
Mu SC, Tang HL, Qian SH, Pan M, Yuan RZ (2006) Hydrogen storage in carbon nanotubes modified by microwave plasma etching and Pd decoration. Carbon 44:762–767
Nam K, Kim JY (2002) Role of loosely bound humic substances and humin in the bioavailability of phenanthrene aged in soil. Environ Pollut 118:427–433
Nguyen TH, Goss KU, Ball WP (2005) Polyparameter linear free energy relationships for estimating the equilibrium partition of organic compounds between water and the natural organic matter in soils and sediments. Environ Sci Technol 39:913–924
Pan B, Xing BS (2008) Adsorption mechanisms of organic chemicals on carbon nanotubes. Environ Sci Technol 42:9005–9013
Pan B, Xing BS (2010) Competitive and Complementary Adsorption of Bisphenol A and 17a-Ethinyl Estradiol on Carbon Nanomaterials. J Agr Chem 58: 8338–8343
Pan B, Xing BS, Liu WX, Tao S, Lin XM, Zhang XM, Zhang YX, Xiao Y, Dai HC, Yuan HS (2006) Distribution of sorbed phenanthrene and pyrene in different humic fractions of soils and importance of humin. Environ Pollut 143:24–33
Pan B, Xing BS, Tao S, Liu WX, Lin XM, Xiao Y, Dai HC, Zhang XM, Zhang YX, Yuan H (2007) Effect of physical forms of soil organic matter on phenanthrene sorption. Chemosphere 68:1262–1269
Pan B, Ning P, Xing BS (2008) Humic substances. Review series. Part IV-sorption of hydrophobic organic contaminants. Environ Sci Pollut Res 15:554–564
Pan B, Ning P, Xing BS (2009) Part V-sorption of pharmaceuticals and personal care products. Environ Sci Pollut Res 16:106–116
Parolo ME, Savini MC, Valles JM, Baschini MT, Avena MJ (2008) Tetracycline adsorption on montmorillonite: pH and ionic strength effects. Appl Clay Sci 40:179–186

Patrolecco L, Capri S, De Angelis S, Pagnotta R, Polesello S, Valsecchi S (2006) Partition of nonylphenol and related compounds among different aquatic compartments in Tiber river (central Italy). Water Air Soil Pollut 172:151–166

Perminova IV, Grechishcheva NY, Petrosyan VS (1999) Relationships between structure and binding affinity of humic substances for polycyclic aromatic hydrocarbons: relevance of molecular descriptors. Environ Sci Technol 33:3781–3787

Peterson JW, O'Meara TA, Seymour MD, Wang W, Gu BH (2009) Sorption mechanisms of cephapirin, a veterinary antibiotic, onto quartz and feldspar minerals as detected by Raman spectroscopy. Environ Pollut 157:1849–1856

Pignatello JJ, Xing BS (1996) Mechanisms of slow sorption of organic chemicals to natural particles. Environ Sci Technol 30:1–11

Pignatello JJ, Lu YF, LeBoeuf EJ, Huang WL, Song JZ, Xing BS (2006) Nonlinear and competitive sorption of apolar compounds in black carbon-free natural organic materials. J Environ Qual 35:1049–1059

Pils JRV, Laird DA (2007) Sorption of tetracycline and chlortetracycline on K- and Ca-saturated soil clays, humic substances, and clay-humic complexes. Environ Sci Technol 41:1928–1933

Ran Y, Sun K, Yang Y, Xing BS, Zeng E (2007) Strong sorption of phenanthrene by condensed organic matter in soils and sediments. Environ Sci Technol 41:3952–3958

Rice JA (2001) Humin. Soil Sci 166:848–857

Rice J, Maccarthy P (1989) Isolation of humin by liquid-liquid partitioning. Sci Total Environ 81–2:61–69

Richter MK, Sander M, Krauss M, Christl I, Dahinden MG, Schneider MK, Schwarzenbach RP (2009) Cation binding of antimicrobial sulfathiazole to leonardite humic acid. Environ Sci Technol 43:6632–6638

Rivera-Utrilla J, Prados-Joya G, Sanchez-Polo M, Ferro-Garcia MA, Bautista-Toledo I (2009) Removal of nitroimidazole antibiotics from aqueous solution by adsorption/bioadsorption on activated carbon. J Hazard Mater 170:298–305

Saikia MD (2008) Revisiting adsorption of biomolecules on polymeric resins. Coll Surf A-Physicochem Eng Asp 315:196–204

Saikia MD, Dutta NN (2008) Adsorption affinity of certain biomolecules onto polymeric resins: effect of solute chemical nature. React Funct Polym 68:33–38

Salloum MJ, Dudas MJ, McGill WB (2001a) Sorption of 1-naphthol to soil and geologic samples with varying diagenetic properties. Chemosphere 44:779–787

Salloum MJ, Dudas MJ, McGill WB (2001b) Variation of 1-naphthol sorption with organic matter fractionation: the role of physical conformation. Org Geochem 32:709–719

Salloum MJ, Chefetz B, Hatcher PG (2002) Phenanthrene sorption by aliphatic-rich natural organic matter. Environ Sci Technol 36:1953–1958

Sander M, Pignatello JJ (2005) Characterization of charcoal adsorption sites for aromatic compounds: insights drawn from single-solute and bi-solute competitive experiments. Environ Sci Technol 39:1606–1615

Sanders SM, Srivastava P, Feng Y, Dane JH, Basile J, Barnett MO (2008) Sorption of the veterinary antimicrobials sulfadimethoxine and ormetoprim in soil. J Environ Qual 37:1510–1518

Sarmah AK, Meyer MT, Boxall ABA (2006) A global perspective on the use, sales, exposure pathways, occurrence, fate and effects of veterinary antibiotics (VAs) in the environment. Chemosphere 65:725–759

Schafer AI, Mastrup M, Jensen RL (2002) Particle interactions and removal of trace contaminants from water and wastewaters. Desalination 147:243–250

Scheytt TJ, Mersmann P, Rejman-Rasinski E, These A (2007) Tracing pharmaceuticals in the unsaturated zone. J Soils Sediments 7:75–84

Schwarzenbach RP, Westall J (1981) Transport of non-polar organic-compounds from surface-water to groundwater – laboratory sorption studies. Environ Sci Technol 15:1360–1367

Seget S, Pharma D (2008) Pharmaceutical market trends, 2008–2012: key market forecasts and growth opportunities. URCH Publishing, London

Shareef A, Angove MJ, Wells JD, Johnson BB (2006) Sorption of bisphenol A, 17 alpha-ethynylestradiol and estrone to mineral surfaces. J Colloid Interface Sci 297:62–69

Sibley SD, Pedersen JA (2008) Interaction of the macrolide antimicrobial clarithromycin with dissolved humic acid. Environ Sci Technol 42:422–428

Simpson MJ, Chefetz B, Hatcher PG (2003) Phenanthrene sorption to structurally modified humic acids. J Environ Qual 32:1750–1758

Strock TJ, Sassman SA, Lee LS (2005) Sorption and related properties of the swine antibiotic carbadox and associated N-oxide reduced metabolites. Environ Sci Technol 39:3134–3142

Sukul P, Lamshoft M, Zuhlke S, Spiteller M (2008) Sorption and desorption of sulfadiazine in soil and soil-manure systems. Chemosphere 73:1344–1350

Sun K, Ran Y, Yang Y, Xing BS (2008) Sorption of phenanthrene by nonhydrolyzable organic matter from different size sediments. Environ Sci Technol 42:1961–1966

Suntisukaseam U, Weschayanwiwat P, Sabatini DA (2007) Sorption of amphiphile pharmaceutical compounds onto polar and nonpolar adsorbents. Environ Eng Sci 24:1457–1465

Tharayil N, Bhowmik PC, Xing BS (2006) Preferential sorption of phenolic phytotoxins to soil: implications for altering the availability of allelochemicals. J Agric Food Chem 54:3033–3040

Thiele-Bruhn S (2003) Pharmaceutical antibiotic compounds in soils – a review. J Plant Nutr Soil Sci-Zeitschrift Fur Pflanzenernahrung Und Bodenkunde 166:145–167

Thiele-Bruhn S, Seibicke T, Schulten HR, Leinweber P (2004) Sorption of sulfonamide pharmaceutical antibiotics on whole soils and particle-size fractions. J Environ Qual 33:1331–1342

Thomsen M, Dobel S, Lassen P, Carlsen L, Mogensen BB, Hansen PE (2002) Reverse quantitative structure-activity relationship for modelling the sorption of esfenvalerate to dissolved organic matter – a multivariate approach. Chemosphere 49:1317–1325

Tolls J (2001) Sorption of veterinary pharmaceuticals in soils: a review. Environ Sci Technol 35:3397–3406

Tsai WT, Hsu HC, Su TY, Lin KY, Lin CM (2006a) Adsorption characteristics of bisphenol-A in aqueous solutions onto hydrophobic zeolite. J Colloid Interface Sci 299:513–519

Tsai WT, Lai CW, Su TY (2006b) Adsorption of bisphenol-A from aqueous solution onto minerals and carbon adsorbents. J Hazard Mater 134:169–175

Uslu MO, Yediler A, Balcioglu IA, Schulte-Hostede S (2008) Analysis and sorption behavior of fluoroquinolones in solid matrices. Water Air Soil Pollut 190:55–63

Van Emmerik T, Angove MJ, Johnson BB, Wells JD, Fernandes MB (2003) Sorption of 17 beta-estradiol onto selected soil minerals. J Colloid Interface Sci 266:33–39

Vasudevan D, Bruland GL, Torrance BS, Upchurch VG, MacKay AA (2009) pH-dependent ciprofloxacin sorption to soils: interaction mechanisms and soil factors influencing sorption. Geoderma 151:68–76

Verma B, Headley JV, Robarts RD (2007) Behaviour and fate of tetracycline in river and wetland waters on the Canadian Northern Great Plains. J Environ Sci Health A Tox Hazard Subst Environ Eng 42:109–117

Wang KJ, Xing BS (2005a) Chemical extractions affect the structure and phenanthrene sorption of soil humin. Environ Sci Technol 39:8333–8340

Wang KJ, Xing BS (2005b) Structural and sorption characteristics of adsorbed humic acid on clay minerals. J Environ Qual 34:342–349

Wang XP, Shan XQ, Luo L, Zhang SZ, Wen B (2005) Sorption of 2,4,6-trichlorophenol in model humic acid-clay systems. J Agric Food Chem 53:3548–3555

Wang YJ, Jia DA, Sun RJ, Zhu HW, Zhou DM (2008) Adsorption and cosorption of tetracycline and copper(II) on montmorillonite as affected by solution pH. Environ Sci Technol 42:3254–3259

Wang YS, Shan XQ, Feng MH, Chen GC, Pei ZG, Wen B, Liu T, Xie YN, Owens G (2009) Effects of copper, lead, and cadmium on the sorption of 2,4,6-trichlorophenol onto and desorption from wheat ash and two commercial humic acids. Environ Sci Technol 43:5726–5731

Wang F, Yao J, Sun K, Xing BS (2010) Adsorption of dialkyl phthalate esters on carbon nanotubes. Environ Sci Technol 44(18):6985–6991

Weber WJ, McGinley PM, Katz LE (1992) A distributed reactivity model for sorption by soils and sediments. 1. Conceptual basis and equilibrium assessments. Environ Sci Technol 26:1955–1962
Weber WJ, Huang WL, LeBoeuf EJ (1999) Geosorbent organic matter and its relationship to the binding and sequestration of organic contaminants. Colloids Surf, A Physicochem Eng Asp 151:167–179
Weber WJ, Leboeuf EJ, Young TM, Huang WL (2001) Contaminant interactions with geosorbent organic matter: insights drawn from polymer sciences. Water Res 35:853–868
Wehrhan A, Kasteel R, Simunek J, Groeneweg J, Vereecken H (2007) Transport of sulfadiazine in soil columns – experiments and modelling approaches. J Contam Hydrol 89:107–135
Wen B, Zhang JJ, Zhang SZ, Shan XQ, Khan SU, Xing BS (2007) Phenanthrene sorption to soil humic acid and different humin fractions. Environ Sci Technol 41:3165–3171
Williams CF, Williams CE, Adamsen EJ (2006) Sorption–desorption of carbamazepine from irrigated soils. J Environ Qual 35:1779–1783
Xia GS, Pignatello JJ (2001) Detailed sorption isotherms of polar and apolar compounds in a high-organic soil. Environ Sci Technol 35:84–94
Xie H, Guetzloff TF, Rice JA (1997) Fractionation of pesticide residues bound to humin. Soil Sci 162:421–429
Xing BS (1997) The effect of the quality of soil organic matter on sorption of naphthalene. Chemosphere 35:633–642
Xing BS (2001a) Sorption of anthropogenic organic compounds by soil organic matter: a mechanistic consideration. Can J Soil Sci 81:317–323
Xing BS (2001b) Sorption of naphthalene and phenanthrene by soil humic acids. Environ Pollut 111:303–309
Xing BS, Chen ZQ (1999) Spectroscopic evidence for condensed domains in soil organic matter. Soil Sci 164:40–47
Xing BS, Pignatello JJ (1997) Dual-mode sorption of low-polarity compounds in glassy poly(vinyl chloride) and soil organic matter. Environ Sci Technol 31:792–799
Xing BS, McGill WB, Dudas MJ (1994) Cross-correlation of polarity curves to predict partition-coefficients of nonionic organic contaminants. Environ Sci Technol 28:1929–1933
Xing BS, Pignatello JJ, Gigliotti B (1996) Competitive sorption between atrazine and other organic compounds in soils and model sorbents. Environ Sci Technol 30:2432–2440
Xing BS, Liu JD, Liu XB, Han XZ (2005) Extraction and characterization of humic acids and humin fractions from a black soil of China. Pedosphere 15:1–8
Xing BS, Senesi N, Huang PM (2011) Biophysico-chemical processes of anthropogenic organic compounds in environmental systems. Wiley, Hoboken. ISBN 978-0-470-53963-7
Yang K, Wang XL, Zhu LZ, Xing BS (2006) Competitive sorption of pyrene, phenanthrene, and naphthalene on multiwalled carbon nanotubes. Environ Sci Technol 40:5804–5810
Yang K, Wu WH, Jing QF, Jiang W, Xing BS (2010) Competitive adsorption of naphthalene with 2,4-dichlorophenol and 4-chloroaniline on multiwalled carbon nanotubes. Environ Sci Technol 44:3021–3027
Yin JF, Meng ZH, Du MJ, Liu C, Song MY, Wang HL (2010) Pseudo-template molecularly imprinted polymer for selective screening of trace beta-lactam antibiotics in river and tap water. J Chromatogr A 1217:5420–5426
Ying GG, Kookana RS, Ru YJ (2002) Occurrence and fate of hormone steroids in the environment. Environ Int 28:545–551
Yiruhan WQJ, Mo CH, Li YW, Gao P, Tai YP, Zhang Y, Ruan ZL, Xu JW (2010) Determination of four fluoroquinolone antibiotics in tap water in Guangzhou and Macao. Environ Pollut 158:2350–2358
Young TM, Weber WJ Jr (1995) A distributed reactivity model for sorption by soils and sediments. 3. Effects of diagenetic processes on sorption energetics. Environ Sci Technol 29:92–97
Yu ZQ, Huang WL (2005) Competitive sorption between 17 alpha-ethinyl estradiol and naphthalene/phenanthrene by sediments. Environ Sci Technol 39:4878–4885

Yu ZQ, Xiao BH, Huang WL, Peng P (2004) Sorption of steroid estrogens to soils and sediments. Environ Toxicol Chem 23:531–539

Yu XY, Ying GG, Kookana RS (2006) Sorption and desorption behaviors of diuron in soils amended with charcoal. J Agric Food Chem 54:8545–8550

Zeng GM, Zhang C, Huang GH, Yu J, Wang Q, Li JB, Xi BD, Liu HL (2006) Adsorption behavior of bisphenol A on sediments in Xiangjiang River, central-south China. Chemosphere 65:1490–1499

Zermane F, Bouras O, Baudu M, Basly JP (2010) Cooperative coadsorption of 4-nitrophenol and basic yellow 28 dye onto an iron organo-inorgano pillared montmorillonite clay. J Colloid Interface Sci 350:315–319

Zhang JQ, Dong YH (2008) Effect of low-molecular-weight organic acids on the adsorption of norfloxacin in typical variable charge soils of China. J Hazard Mater 151:833–839

Zhang L, Leboeuf EJ, Xing BS (2007) Thermal analytical investigation of biopolymers and humic- and carbonaceous-based soil and sediment organic matter. Environ Sci Technol 41:4888–4894

Zhang D, Pan B, Zhang H, Ning P, Xing BS (2010) Contribution of different sulfamethoxazole species to their overall adsorption on functionalized carbon nanotubes. Environ Sci Technol 44:3806–3811

Zhou JL (2006) Sorption and remobilization behavior of 4-tert-octylphenol in aquatic systems. Environ Sci Technol 40:2225–2234

Zhou JL, Liu R, Wilding A, Hibberd A (2007) Sorption of selected endocrine disrupting chemicals to different aquatic colloids. Environ Sci Technol 41:206–213

Zhu DQ, Hyun SH, Pignatello JJ, Lee LS (2004) Evidence for pi-pi electron donor-acceptor interactions between pi-donor aromatic compounds and pi-acceptor sites in soil organic matter through pH effects on sorption. Environ Sci Technol 38:4361–4368

Chapter 12
Adsorption of Dissolved Organic Compounds by Black Carbon

Joseph J. Pignatello

Abstract Black carbon is the carbonaceous product of pyrolysis or incomplete combustion of biomass or fossil fuels. Due to its widespread occurrence in atmospheric aerosols, soils and sediments, and its intrinsic strength as an adsorbent, black carbon potentially plays an important role in the partitioning of organic pollutants from water and air to natural solids, especially at low pollutant concentration. The adsorptive strength of black carbon depends greatly on pyrolysis time, temperature and other formation conditions, as well as subsequent weathering in the environment. The predominant property of black carbon governing its adsorptive strength is its small-pore porosity. The filling of micropores and mesopores of molecular dimensions eliminates the need for the cavity penalty that otherwise accompanies partitioning of molecules into bulk nonporous phases such as organic liquids, polymers and natural organic matter to accommodate incoming molecules. However, the filling of these small pores exhibits a steric effect due to size exclusion at pore throats. The polyaromatic (graphite-like) surface of black carbon serves as a strong π-electron donor in the formation of π-π electron donor-acceptor complexes with strong π-deficient aromatic systems, for example, polynitroaromatics and charged aromatic amines. Polar functional groups on the rims of polyaromatic sheets attract water clusters that crowd out adsorbates, regardless of polarity. Nevertheless, compounds such as phenols, carboxylic acids and others that are capable of forming especially strong hydrogen bonds with carboxyl or phenoxyl groups on the surface may interact strongly. In the environment, the adsorptive strength of black carbon becomes quickly attenuated by fouling with humic substances, which block pores and compete for adsorption sites. Quantifying the contribution of native black carbon to retention of a contaminant in a given natural sample is a challenge due to the lack of reliable methods for determining black carbon content in geosolids, the difficulty of separating black carbon particles from the sample, the absence in

J.J. Pignatello (✉)
Department of Environmental Sciences, Connecticut Agricultural Experiment Station, 123 Huntington St, New Haven, CT 06504-1106, USA

J. Xu and D.L. Sparks (eds.), *Molecular Environmental Soil Science*,
Progress in Soil Science, DOI 10.1007/978-94-007-4177-5_12,

most cases of a basis for choosing an appropriate reference standard, and a poor quantitative understanding of the weathering process. Adding to the challenge is the strong hysteresis that is often seen to accompany adsorption to black carbon materials, but that is unpredictable and poorly understood mechanistically.

Keywords Charcoal • Soot • Biochar • Activated carbon • Graphite • Graphene • Soil • Sediment • Atmospheric aerosols • Natural organic matter • Polyaromatic hydrocarbons • Hydrophobic effect • Sorption hysteresis • Pi-pi interactions • Competitive sorption • Irreversible sorption

12.1 Introduction—Sources and Environmental Significance of Black Carbon

Black carbon (BC) is the carbonaceous product of organic substances heated in a deficiency of oxygen. It includes char (or charcoal) and soot. Charcoal is formed by pyrolytic condensation processes in the solid state, while soot is formed in the gas phase after a complex series of free radical reactions. Black carbon that is left in, or discharged to, the environment is commonly referred to as environmental black carbon. It is an important part of the global carbon cycle and plays roles in various geo- and bio-geochemical processes. Soot is a significant component of atmospheric particles (Schmidt and Noack 2000). It contributes to the pulmonary toxicity of atmospheric aerosols (Lighty et al. 2000; Ramanathan and Carmichael 2008). Black carbon emissions from indoor cooking may cause the premature deaths of as many as 400,000 people per year (Smith 2005). Atmospheric black carbon is a major contributor to current global warming, being second only to carbon dioxide in importance. Furthermore, the atmospheric deposition of black carbon on snow and ice promotes melting, which may contribute significantly to the retreat of Arctic sea ice and glaciers (McMullen and Jabbour 2009; Ramanathan and Carmichael 2008). Environmental black carbon appears to be fairly widespread in undisturbed soils and sediments, where it typically occurs at low levels relative to total organic carbon. It can be more prevalent in areas of high industrial activity, on lands experiencing recent or frequent fires, and in deep sea and marine shelf sediments receiving inputs of black carbon from land fires over geological time (Masiello and Druffel 1998; Skjemstad et al. 1999).

Defined more broadly, black carbon includes engineered carbons such as activated carbon and "biochar." Activated carbon is charcoal from selected feedstock that is carbonized at relatively high temperatures (above 700 °C) and in many cases treated to tailor surface reactivity. Activated carbon has long been used as an adsorbent for water purification and has recently been deployed in the same role to stabilize contaminated sediments (Cho et al. 2009). Biochar is produced from waste biomass in reactors that operate under conditions of low oxygen and relatively low temperatures (~350 to ~700 °C). Biochar has attracted attention as a soil amendment of potential benefit in agriculture, as well as a sink for carbon due to the inherent stability of black carbon to biomineralization (Lehmann and Joseph 2009).

This chapter deals with a property of black carbon that is essential to its function in natural and anthropogenic systems—its surface activity, specifically its adsorptive properties toward organic compounds. Black carbon in soils and sediments can adsorb chemicals that have been discharged into them. The role of BC as adsorbent of environmental pollutants seems to have been first noted by Gustafsson and Gschwend, (1997) and Gustafsson et al. (1997). Black carbon may also carry chemicals formed during their synthesis that can subsequently be released into the environment or taken up by biological tissues. Polyaromatic hydrocarbons (PAHs), for example, are precursors of soot and can be found at significant levels in raw soot samples. Micron-size soot particles may be carried deep into the lung where associated PAHs may desorb into lung fluids. Charcoal present in or added to soils may affect the efficacy of soil pesticides and may interfere with signaling interactions among microorganisms, insects and plants mediated by natural allelochemicals and semiochemicals (Elmer and Pignatello 2011). This chapter will discuss the interactions of non-ionic compounds only, as little information is available on sorption of ionic and ionizable compounds by black carbon.

12.2 Structure and Relevant Properties of Black Carbon

12.2.1 Evolution of Structure During Pyrolysis

Black carbon is not a single substance, but a continuum of substances displaying a wide range of properties. These properties of biomass-derived sources (charcoal) depend strongly on the maximum heat treatment temperature, duration of maximum heat treatment temperature, source material, oxygen concentration, mineral content, and the extent to which tarry vapors evolved during pyrolysis are swept away The variation in black carbon structure makes it difficult to generalize about its sorptive properties.

Pyrolysis of lignocellulosic biomass under nitrogen yields H_2O, CO_2, CO, CH_4, and H_2. The initial heating stage (up to ~400 °C) releases mainly physically and chemically sorbed H_2O and CO_2. As temperature increases H_2O, CO_2, CO, and CH_4 continue to be evolved, peaking between 500 and ~600 °C. Ashes containing K, P, Mg and Na appear to catalyze the evolution of CO, which may continue even up to 1,000 °C (Mészáros et al. 2007). Hydrogen evolution commences at 600 °C and peaks at 800 °C. Pyrolysis of biomass also releases smaller amounts organic compounds including ethane, ethylene, toluene, benzene, phenol, formaldehyde, furan, and furfuraldehyde (Mészáros et al. 2007). Tarry liquids condensed from the evolved gases contain remnant molecules derived from sugars, anhydrosugars and lignin (Antal and Grønli 2003).

Charcoal produced from lignocellulosic biomass largely retains the macroscopic morphology of the feedstock (Antal and Grønli 2003). However, the microscopic structure changes with temperature, time and conditions. Keiluweit et al. (2010) studied the charring of wood and grass from 100 to 700 °C in an oxygen-limited atmosphere. (Biomass fires ordinarily do not exceed 700 °C.) They proposed four categories of char: *transition char*, in which transformations have occurred but the crystalline nature of the biopolymers is essentially preserved; *amorphous char*, in which altered biomolecules and incipient polyaromatic condensates are randomly mixed; *composite char* consisting of poorly-ordered graphene stacks embedded in amorphous phases; and *turbostratic char* composed of disordered graphitic microcrystallites. Carbonized charcoals are those enduring heating above 800 °C. Bourke et al. (2007) proposed a model for carbonized charcoals that posits the formation of graphitic microcrystallites of, at most, three platelets, intermingled with nongraphitic single-sheet polyaromatic ring systems.

Soot condenses out of the gas phase through a cascade of free radical reactions in which PAH molecules are important precursors (Xi and Zhong 2006). This process is summarized as follows: formation of precursors from fuel molecules and their fragments; nucleation (inception) of particles from precursors; mass growth of particles by addition of gas-phase molecules or radicals; coagulation of particles; further carbonization of particles; and oxidation of both precursors and particles. The primary particles are 15–50 nm diameter spheres that form "grape cluster" or "neckless" shaped aggregates up to a micrometer in dimension. Figure 12.1 shows that soot and char particles are essentially similar in structure at the nanometer scale.

The defining unit of the black carbon body is the polyaromatic ("graphene") platelet (Fig. 12.2). The platelets are functionalized along the rims with ketone, ether, hydroxyl, quinoid, carboxyl, and other polar functional groups that influence sorption of organic compounds (see below). Platelet size appears to increase with heat treatment temperature. Studying biomass charcoals prepared at 500 or 730 °C by DP-MAS ^{13}CNMR spectroscopy, Brewer et al. (2009) found that the graphene platelets average 7–8 and 17 fused rings, respectively. Platelets may exist singly or form short stacks up to five (Bourke et al. 2007; Xi and Zhong 2006). The stacks arrange in a disordered manner, creating a pore network. The graphene stacks are held together by both covalent and noncovalent bonds.

12.2.2 Pore and Surface Characteristics

The pore nanostructure of black carbon accounts for its strong adsorption capacity with respect to organic compounds. Partitioning of a molecule into any liquid or solid phase requires the formation of a cavity to accommodate the incoming molecule. For solids, this "cavity penalty" strongly depends on the visco-elastic properties and free volume distribution of the solid (Pignatello 2011) and can be substantial. For example, the free energy of cavitation to accommodate molecules the size of 1,2,4-trichlorobenzene or naphthalene in the dissolution domain of soil

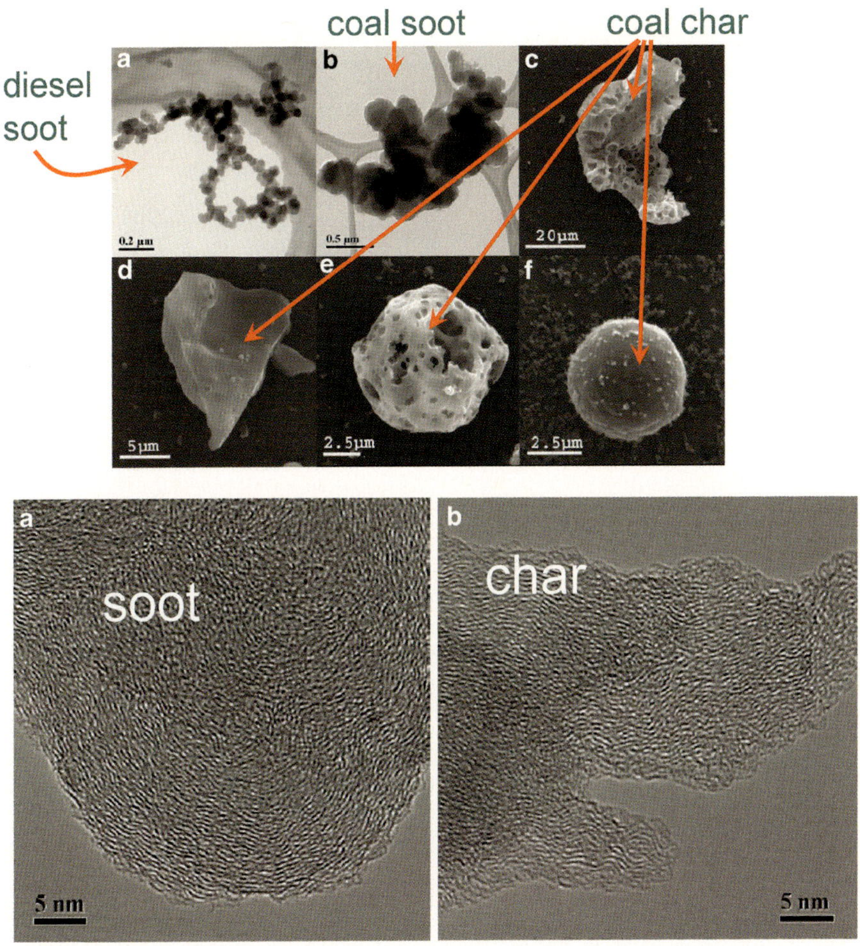

Fig. 12.1 (**a**) Morphologies of diesel soot, coal soot and coal char. (**b**) High-resolution transmission electron micrographs of a primary soot particle and a char particle (Chen et al. 2005)

organic matter has been estimated to be 15–20 kJ mol^{-1} (Lu and Pignatello 2004). The pre-existing small pores in black carbon enable solute molecules to sorb without the "cavity penalty" intrinsic to sorption in more flexible ("soft") solids, such as many polymers and natural organic matter. Some forms of black carbon may be able to undergo additional cavitation in response to penetration by certain adsorbates.

The porosity and surface area of black carbon depend greatly on source material and formation conditions. Surface area, normally determined from the adsorption isotherm of N_2 at 77 K using the B.E.T. equation, can range from a few to several hundred square meters per gram. For charcoals made from wood shavings in a stream of nitrogen, N_2-B.E.T. surface area tends to increase exponentially with heat treatment temperature from ~300 to ~450 °C, to plateau in the range 450–600 °C, and then to decline somewhat between 600 and 1,000 °C.

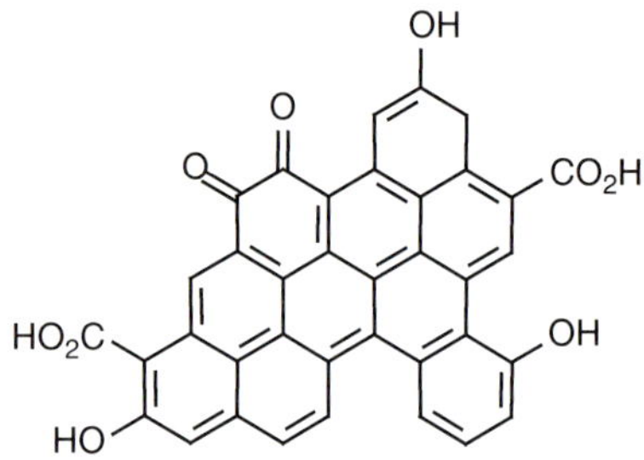

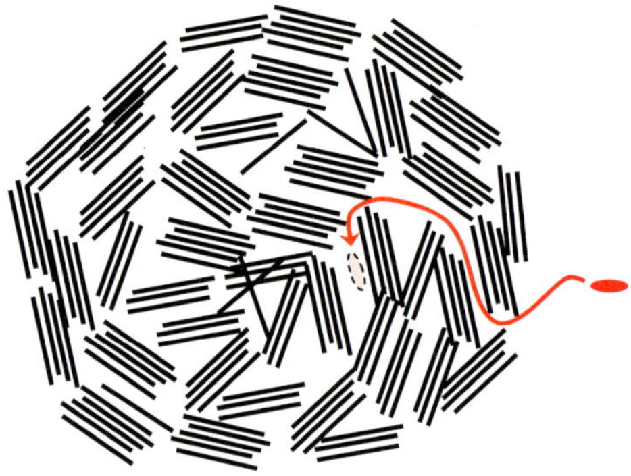

Fig. 12.2 Hypothetical structure of a black carbon graphene platelet and random packing of short stacks in the particle creating a network of micropores into which the solute diffuses

The majority of black carbon surface area is in the micropore (<2 nm) and small mesopore (2–∼20 nm) ranges. Since the space between platelets is too confined, the porosity of interest exists mainly in the spaces between the highly-disordered graphitic microcrystallites (Boehm 1964; Bourke et al. 2007; Goldberg 1985; Palotás et al. 1996). Pore size distribution may be determined by combined analysis of the CO_2 and N_2 adsorption isotherms using Monte Carlo and Non-local Density Functional Theory models. Figure 12.3 shows the pore size distributions of some charcoals. Most of the microporosity lies in the regions 0.4–0.7 nm and 0.8–1.0 nm. Some pores may be inaccessible to solute molecules due to, (*a*) an intrinsically more condensed structure that leaves pores isolated ("closed"); (*b*) the plugging of pore throats by uncombusted organic matter or fuels; (*c*) the plugging of pore throats by organic or inorganic substances that adsorb or condense on the surface once the particle is exposed to the natural environment; or (*d*) the steric size of the adsorbate molecule. In some diesel soots the ratio of N_2 BET surface area to the calculated smooth-sphere external surface area based on electron microscope images was close to unity (*10, 28,50*), suggesting that accessible micropores of soot might extend only a short distance into the solid. The high variation in adsorptivity among black carbon samples may be a consequence of differences in the interconnectivity of the pore networks. Furthermore, nitrogen may not be the best probe for surface area because, at 77 K, N_2 is kinetically restricted from entering very small pores or pores that may be blocked at their throats by the substances mentioned above (Braida et al. 2003; Kwon and Pignatello 2005; Pignatello et al. 2006a). At that temperature the pore walls or blocking substances may not flex sufficiently to let the N_2 molecules pass. Carbon dioxide is much less kinetically restricted at the temperature where its isotherm is constructed (273 K), and consistently gives higher surface areas than does N_2 (Braida et al. 2003; Kwon and Pignatello 2005; Pignatello et al. 2006a; Zhu et al. 2005).

Sorption generally increases with increasing specific surface area, although this correlation is complicated by pore size distribution, surface chemistry (Jonker and Koelmans 2002b) and the above-mentioned limitations of N_2 as the probe gas for measuring surface area.

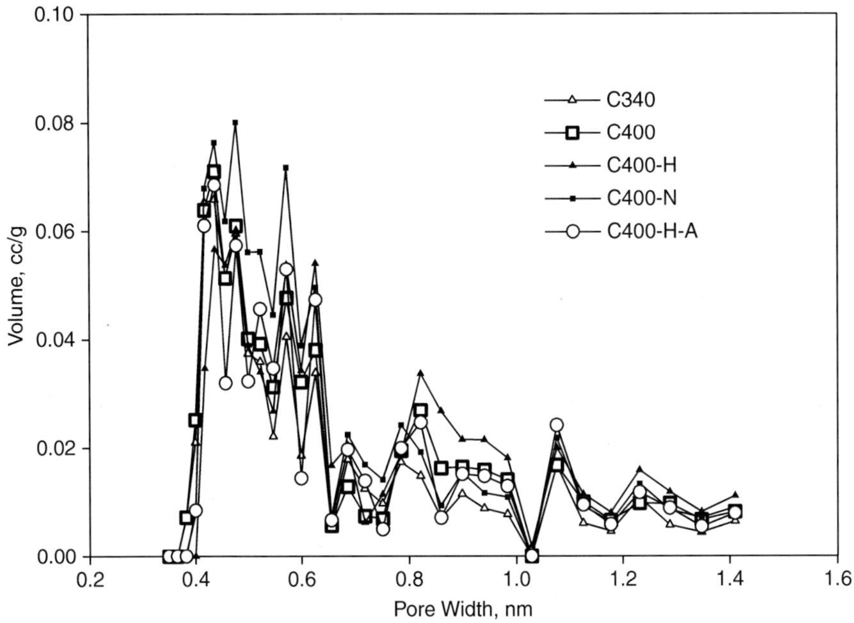

Fig. 12.3 Pore size distributions of different charcoals measured by analysis of CO_2 adsorption isotherms. The charcoals were produced from maple wood shavings under nitrogen flow at 340 or 400 °C (C340 and C400). C400-H is C400 reheated to 500 °C under H_2 in the presence of Pt/C catalyst; C400-N is C400 reheated to 500 °C under N_2; C400-H-A C400-H re-oxidized in air at 340 °C (From Zhu et al. (2005))

A large percentage of black carbon surface area is hydrophobic, so one might suppose that hydrophobicity is an important property responsible for the strong adsorptive capability of black carbon. Yet hydrophobicity *per se* seems to play a minor role. Figure 12.4 shows the isotherm of benzene in a charcoal having a surface area of 400 m^2 g^{-1}, and a porosity of 0.15 cm^3 g^{-1}, of which 80% is microporosity (Braida et al. 2003). Table 12.1 gives the charcoal-water distribution ratio, K_{BC}—the ratio of the sorbed to solution concentrations (units of L kg^{-1})—for benzene at select concentrations, compared to the *n*-octanol-water, *n*-hexadecane-water, and benzene-water partition coefficients for benzene solute (K_{OW}, K_{HD} and K_B, respectively, in units of L/L). Such solvents contain a very large "internal surface area" that is hydrophobic, or predominantly so. Also, benzene mimics, in some respects, the chemistry of the graphene surface. At low benzene solute concentration, the log K_{BC}, is around 4.0, a value that is more than 1.8 log units greater than K_{OW} or K_{HD}, and about 1.3 log units greater than K_B. At high concentrations, K_{BC} is at least as large as K_B. A more valid comparison requires knowledge of the volume fraction of charcoal accessible to benzene; it is reasonable to assume that the available volume fraction is equal to or less than the porosity, 0.15, determined by N_2 adsorption at a relative pressure of 0.995. Thus, the values

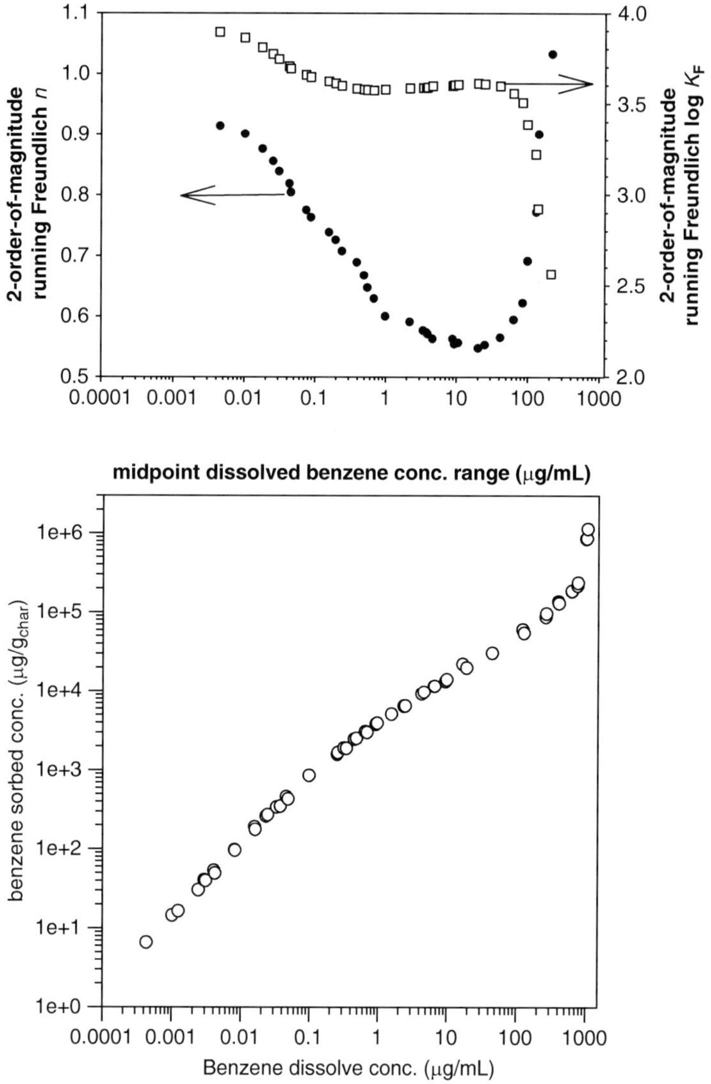

Fig. 12.4 The wood charcoal-water sorption isotherm of benzene (*the lower figure*) (Braida et al. 2003). The *upper figure* is the running slope and intercept of the isotherm taken at intervals of 2 log units, which represents the parameter values K_F and n, respectively, of the Freundlich equation $q = K_F C^n$, where q is the sorbed and C the dissolved concentration

of log K_{BC}, if known in volumetric units, would be roughly 1 log unit greater than the values listed in Table 12.1. Accordingly, black carbon is a far more potent sorbent of benzene than is expected based solely on its "hydrophobicity" relative to octanol, hexadecane, and benzene solvents. Phenanthrene is another example supporting this conclusion. Phenanthrene has a log K_{OW} of 4.46 and a log K_{HD} of 4.74 (Abraham

Table 12.1 Comparison of charcoal-water, benzene-water, n-hexadecane-water and n-octanol-water partition coefficients for benzene

System	Value
black carbon-water @ 3 μg/L, log K_{BC} [L/kg]	4.0[a]
black carbon-water @ 50 mg/L, log K_{BC} [L/kg]	2.7[a]
benzene-water, log K_B [L/L]	2.70[b]
n-octanol-water, log K_{OW} [L/L]	2.13[c]
n-hexadecane-water, log K_{HD} [L/L]	2.15[c]

[a]From Braida et al. (2003)
[b]From water solubility of benzene (Schwarzenbach et al. 2002)
[c]From Abraham et al. (1994)

et al. 1994). The distribution ratio log K_{BC} for phenanthrene sorption on the same charcoal ranged from 8.6 at 0.01 nmol L^{-1} to 6.8 at 10 nmol L^{-1}. The log K_{BC} for phenanthrene-d_{10} in various soots and a charcoal determined in a separate study ranged from 4.91 to 7.35 (4.34 for a coal soot) (Jonker and Koelmans 2002b). Again, these K_{BC} values would probably be about one log unit larger if expressed on the basis of available volume. Regardless, they are much larger than the K_{OW} and K_{HD} values. To conclude, the hydrophobic nature of the surface appears to be of secondary importance in the sorptive ability of black carbon; the primary factor driving sorption is the microporosity of the material. And a likely reason for this is the absence of a cavity penalty.

12.3 Adsorption Behavior of Nonionic Organic Compounds—Positive and Negative Driving Forces

12.3.1 Appropriate Isotherm Model

Models are useful for assigning mechanism, for comparing different sorbates or sorbents, for predicting the sorption strength of new materials, and as a source of input parameters for fate and transport models. One should be cautious, however, about assigning mechanism based on fit to any particular model. Consider the sorption isotherms of benzene, toluene and nitrobenzene on a charcoal (Braida et al. 2003; Sander and Pignatello 2005a). The isotherms of these three compounds gave reverse S-shaped curves (see Fig. 12.4 for the isotherm of benzene). Several models were tested. The number of fitting parameters and the central features of the models are as follows: Langmuir (2 parameters; sites of uniform energy limited in population), dual Langmuir (4; two Langmuir-type sites of high and low energy); Freundlich (2; derived from the Langmuir model assuming a continuous energy distribution of sites); dual mode (3; a partition phase plus a Langmuir type site), and Polanyi-Manes (3; sorption occurs by condensation as a liquid in pores, starting from the smallest).

$$\text{Langmuir}: q = \frac{Q_L K_L C}{1 + K_L C}$$

$$\text{Freundlich}: q = K_F C^n$$

$$\text{Dual} - \text{Mode}: q = K_D C + \frac{Q_H K_H C}{1 + K_H C}$$

$$\text{Polanyi} - \text{Manes}: \log q_{\bar{V}} = \log Q_{\bar{V}} + a\left[\varepsilon / \bar{V}\right]^b \text{ where } \varepsilon = RT \ln\left(C_{max}/C\right)$$

where q and C are the equilibrium sorbed and solute concentrations; Q_L and K_L are the Langmuir maximum capacity and affinity parameters; K_F and n are the Freundlich affinity and exponent parameters; K_D is the partition coefficient in a "dissolution" domain and Q_H and K_H are the capacity and affinity coefficients of a "hole-filling" domain; and for, the Polanyi-Manes equation, $q_{\bar{V}}$, $Q_{\bar{V}}$ are the sorbed and maximum sorbed concentrations normalized to the molar volume $\bar{V}$, C_{max} is the aqueous solubility, R is the gas constant, T is temperature, and a and b are regression parameters.

None of these models was entirely satisfactory. Based on the standard error of estimate, the order in goodness of fit was: dual Langmuir $\sim$ Polanyi-Manes $>$ dual mode $\sim$ Freundlich $\gg$ Langmuir. However, it is difficult to accurately compare these models as the number of fitting parameters is not the same. Only the dual mode and dual Langmuir models captured the reverse S-shape curvature. The dual Langmuir model, which has the most fitting parameters, gave unrealistic parameters, in that the maximum capacity of toluene was greater than benzene, even though toluene is larger in size, and even though competitive experiments indicated that the sorption domains of benzene and toluene were identical. The Polanyi-Manes model not only failed to capture the S-shape, but, contrary to underlying theory (Allen-King et al. 2002; Manes 1998; Xia and Ball 1999), the curves substantially deviated from one another using molar volume as the abscissa scaling factor. This deviation implies that the solute does not "condense" in the micropores of the charcoal in a form resembling the bulk liquid.

Of the above models, the Freundlich is the most frequently applied by researchers. But in most if not all cases, the parameters of the Freundlich equation depend on the range of concentration over which it is applied. For example, the 2 order-of-magnitude "running" Freundlich slope of the benzene isotherm in Fig. 12.4 results in a value of n that varies from nearly 1 on either end of the isotherm, to 0.55 in the "middle." The K_F value also changes considerably, especially at the high-concentration end. This limitation of the Freundlich model applied to black carbon should be kept in mind.

Assigning mechanism based on model fit is further made complicated by the fact that the nature of the sorbent itself appears capable of changing with adsorbate loading. Braida et al. (2003) attributed the tendency toward convex curvature of the benzene isotherm at high concentrations (see Fig. 12.4) to swelling by the adsorbate. They used two different methods to measure swelling—the method of specific gravity, which gives the displacement of water in an aqueous suspension of charcoal upon sorption, and a sedimentation technique, which measures the height of a charcoal column after sorption—that closely agreed. At concentrations above 10 mg g^{-1}, benzene swells the charcoal by an amount almost equal to its liquid volume. The ability of some solvents to extract PAHs and other compounds from black carbon has also been attributed to swelling (Akhter et al. 1985b; Jonker and Koelmans 2002a). Deformability as a function of black carbon synthesis conditions and as a function of pore size remain to be established.

As mentioned, sorption to raw black carbon is highly nonlinear. Nonlinearity can be so severe that the distribution coefficient can vary by several orders of magnitude along the isotherm. For example, adsorption of the antibiotic sulfamethazine to a charcoal at pH 5 (where the neutral molecule predominates in solution) is so nonlinear (a Freundlich n of 0.27), that the K_d varies by a factor of 1,000 over the range of 1 μg to 30 mg L^{-1} (Teixidó et al. 2011). The Freundlich n of benzene on charcoal illustrated in Fig. 12.4 over the entire isotherm is 0.706 overall, corresponding to a range in K_d of 20. Nonlinearity is due to a combination of surface site heterogeneity and pore size heterogeneity. Even the unhindered graphite surface has a surprisingly wide distribution of adsorption potentials; for example, the Freundlich n values for sorption of benzene and several other non-polar aromatic compounds from water on nonporous graphite (99.999% C; N_2-B.E.T specific surface area, 4.5 m^2 g^{-1}) are all below 0.60 (Zhu and Pignatello 2005). Graphite has at least three different types of sites: basal plane, defect sites along the basal plane (pits), and edge sites along step elevations. At higher concentrations, adsorbate-adsorbate interactions are possible as confirmed by atomic force microscopy for substituted benzoic acids (Martin 2003).

12.3.2 The Hydrophobic Effect

A major driving force for transfer of nonionic compounds from the aqueous phase to any solid, including black carbon, is the hydrophobic effect—the sum of the forces that limit the solubility of apolar molecules or parts thereof in water. The hydrophobic effect is not the result of any direct repulsion between apolar entities and water molecules—indeed, those entities interact attractively through London-van der Waals forces (dispersion, induction and dipolar forces). Nor is it the result of any special attraction among apolar molecules in the condensed phase. Rather, the hydrophobic effect results from the disruption of the cohesive energy of water as a result of bringing the apolar entity into solution. The disruption is due to the greater ordering of water molecules, and the reduced number of water-water hydrogen bonds, in the hydration shell of the apolar entity compared to bulk water itself (Chandler 2005; Lazaridis 2001; Southall et al. 2002), a situation that raises the entropy and/or enthalpy penalty for dissolution of the apolar entity in water.

The hydrophobic effect is clearly not the only driving force for adsorption to black carbon. Figure 12.5 shows the isotherms of selected aromatic compounds on non-porous microcrystalline graphite, which was chosen to represent

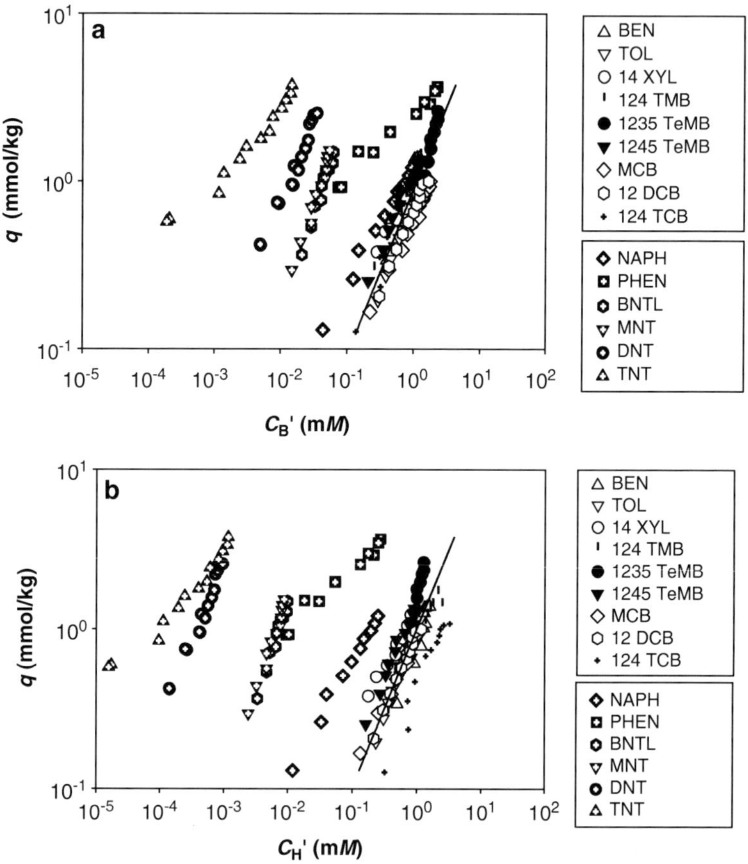

Fig. 12.5 Isotherms of polar and nonpolar aromatic compounds on graphite corrected for hydrophobic effects by benzene-water or *n*-hexadecane-water partition coefficient. C_S' is the "adjusted" aqueous concentration corresponding to the reference organic solvent S (benzene or *n*-hexadecane). The solid line represents the mean value of C_S' for all the compounds of the hydrophobic set (*top box*). TOL = toluene; XYL = 1,4-dimethylbenzene; 124 TMB = 1,2,4-trimethylbenzene; 124 TCB = 1,2,4-trichlorobenzene; 1235 TeMB = 1,2,3,5-tetramethylbenzene; 1245 TeMB = 1,2,4,5-tetramethylbenzene; MCB = chlorobenzene; 12 DCB = 1,2-dichlorobenzene; 124 TCB = 1,2,4-trichlorobenzene; BNTL = benzonitrile; MNT = 4-nitrotoluene; DNT = 2,4-dinitrotoluene; TNT = 2,4,6-trinitrotoluene; NAPH = naphthalene; PHEN = phenanthrene (From Zhu and Pignatello 2005 with permission)

the unhindered, unfunctionalized graphene surface area of black carbon. The isotherms are normalized for the hydrophobic effect based on the hexadecane-water or benzene-water partition coefficient (see Zhu and Pignatello (2005) for details). Whether benzene or hexadecane is used as the reference solvent, the isotherms of the non-polar single-ring compounds, having methyl, chlorine, or no substituents, on graphite are tightly clustered. This suggests a strong correlation between the free

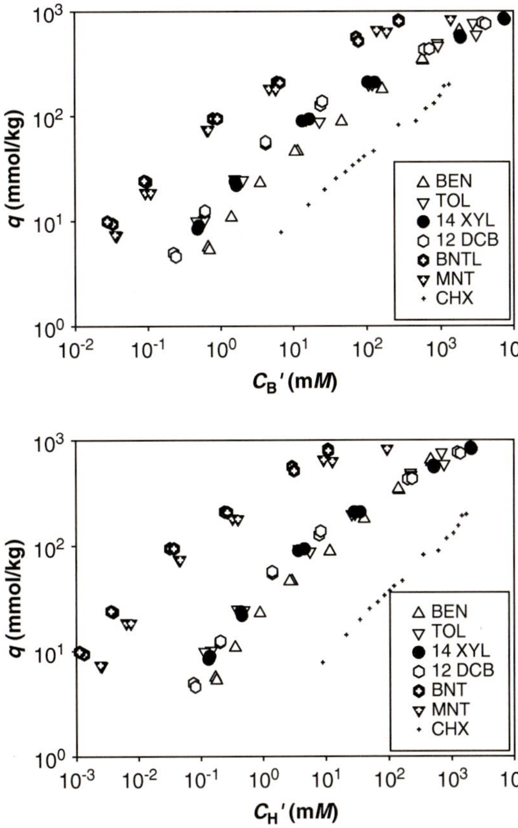

Fig. 12.6 Isotherms of polar and nonpolar aromatic compounds on charcoal black carbon corrected for hydrophobic effects by benzene-water or *n*-hexadecane-water partition coefficient. C_S' is the "adjusted" aqueous concentration corresponding to the reference organic solvent S. CHX is cyclohexane; other compound acronyms as in Fig. 12.5 (Data taken from Zhu and Pignatello 2005.)

energy of adsorption on graphite and the free energy of solvent-water partitioning. This correlation underscores the importance of molecular exclusion from water as a driving force for sorption. However, the isotherms of the polyaromatic hydrocarbons (naphthalene, phenanthrene), and the polar aromatic compounds (benzonitrile, 4-nitrotoluene, 2,4-dinitrotoluene, and 2,4,6-trintrotoluene) are displaced upward of the apolar, single-ring cluster of isotherms. This indicates that these compounds interact more strongly with the graphene surface than predicted by the hydrophobic effect based on the apolar, single-ring aromatic series, and suggest that other forces are involved, even on the relatively homogeneous, non-polar graphite surface.

Figure 12.6 shows the isotherms of selected compounds on black carbon, normalized for hydrophobic effects using benzene or hexadecane in the same manner as for graphite. It was further necessary to exclude compounds with more than two substituents in order to minimize steric effects, as will be discussed in a later section. As for graphite, the isotherms of the non-polar single-ring aromatic compounds (benzene, toluene, 1,4-xylene and 1,2-dichlorobenzene) cluster tightly, regardless of which reference solvent is used. The isotherms of the two polar compounds (benzonitrile and 4-nitrotoluene) are displaced upward of the non-polar cluster, while the isotherm of cyclohexane is displaced downward of the cluster. Therefore, besides the hydrophobic effect, intrinsic interactions of the adsorbate with the surface are important. In a later section we will discuss the reason for the stronger affinity of the polar solutes. Cyclohexane apparently has a weaker affinity for the surface than do the apolar aromatic compounds. This could be due to the greater polarizability of the aromatic compared to the aliphatic ring, or to the closer approach that the aromatic ring can achieve compared to the puckered cyclohexane ring. Both possibilities would result in greater London-van der Waals forces between the entities. Interestingly, given the same vapor pressure, alkyl-substituted benzene vapors adsorb more favorably on quartz than do *n*-alkane vapors (Goss 2004). This suggests that the effect originates from differences in the adsorbate, not the adsorbent.

12.3.3 Dipolar Interactions and the Influence of Water Molecules

The black carbon surface contains substantial O and to a lesser extent N functionality. What roles do these groups play in adsorption of organic compounds? Do dipole-dipole and H-bonding interactions between these groups and compatible groups on the solute contribute to adsorption? To answer these questions we must consider the multiple roles that are possible by water molecules. The enthalpy of adsorption is the net enthalpy of all interactions of solute occurring at the surface and in water. Water is both a polar and an H-bonding solvent. Therefore, bulk water will compete with surface groups for dipolar and H-bonding interactions with solute molecules. Moreover, H_2O and solute will compete for polar/H-bonding interactions with surface groups. Since H_2O is present at 55.5 M—many orders of magnitude greater than typical solute concentrations—the solute is at a hugely competitive disadvantage against H_2O. To illustrate, for a solute present at 0.001 M, the free energy advantage for water, assuming 1:1 competition, is about 27 kJmole^{-1} at 298 K due to the concentration effect alone! Solutes with phenolic, aromatic amine, amide, or carboxylate functional groups may compete more effectively with H_2O than other H-bonding compounds, since such groups form especially strong H-bonds with carboxylate and phenolate groups on the sorbent (29–67 kJ mol^{-1}), compared to H_2O (4–17 kJ mol^{-1}) (Gilli et al. 2009).

For activated carbons it is commonly observed that introducing O, during either synthesis or post-synthesis aging, reduces adsorption. One hypothesis is that O groups withdraw π electron density from the graphene surface, reducing the ability of the surface to undergo dispersion with the adsorbate (Coughlin and Ezra 1968; Radovic et al. 1997). Another hypothesis is that polar groups attract clusters of water molecules that suppress adsorption of organic solutes by thermodynamically blocking adsorption sites (Franz et al. 2000; Li et al. 2002; Pendleton et al.

1997; Snoeyink and Weber 1967)—i.e., a "crowding-out" effect—or by kinetically restricting penetration of adsorbate into micropore space (Kaneko et al. 1989). Molecular simulations (Müller and Gubbins 1998; Müller et al. 2000) show three-dimensional clustering of water molecules around H-bonding sites having the effect of crowding out CH_4 molecules.

We found that the sorption behavior of polar aromatic compounds (phenols, nitrobenzenes), nonpolar aromatic compounds, and cyclohexane for a series of charcoals (Zhu et al. 2005) was consistent with the crowding out hypothesis. We compared a char with 22.3% O content to the same char reduced to 4.2% O with H_2 catalyzed by Pt-on-C, and then re-oxidized to 18.6% O in air. Regardless of solute polarity, hydrogenation led to increased sorption, while subsequent re-oxidation led to decreased sorption. This result is consistent with the results of studies on activated carbons (Coughlin and Ezra 1968; Kaneko et al. 1989; Li et al. 2002). Hydrogenation enhanced the sorption of polar and non-polar compounds about equally, a result expected if electrostatic interactions of the polar compounds with surface functional groups are of minor importance compared to the available surface area. Adsorption enhancement after hydrogenation was greater at lower than at higher solute concentration, consistent with computational studies of the crowding-out effect of water on CH_4 adsorption in functionalized pores, which was greater in smaller (higher energy) pores than in larger (lower energy) pores (Müller and Gubbins 1998; Müller et al. 2000). In another study (Sander and Pignatello 2005a) we found that H-bonding of nitrobenzene with surface functional groups was unimportant because adsorption was unaffected by raising the pH above the point where nearly all –OH groups would be dissociated.

12.3.4 π-π Interactions

The graphene surface of black carbons is capable of interacting with certain aromatic adsorbates via π-π Electron Donor-Acceptor (π-π EDA) interactions. The π-π EDA interactions are weak interactions between arene monomers with opposing π-electron densities. Background and reviews on the potential role of π-π EDA forces in environmental partitioning processes are available (Keiluweit and Kleber 2009; Pignatello 2011). The nature of the π-π EDA bond is incompletely understood. Aromatic and related cyclic conjugated systems generally possess a quadrupole moment perpendicular to the plane of the ring. The predominant force of the π-π EDA bond appears to be attraction between quadrupole moments of the opposing ring systems, resulting in a sandwich or parallel-displaced alignment:

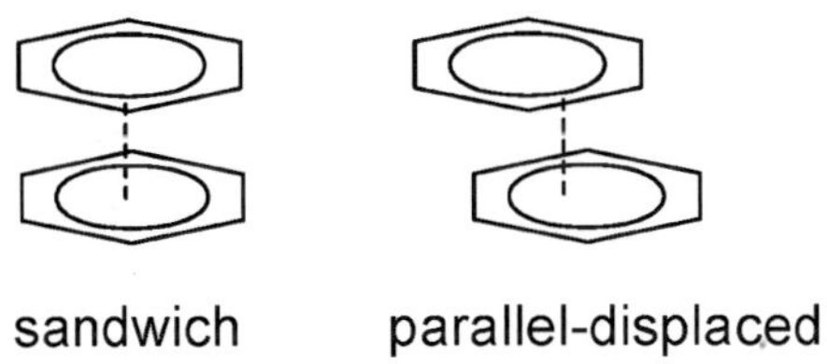

The enthalpy of the π-π EDA force, which may approach that of a strong H-bond, scales with the donor and acceptor ability of the opposing ring systems, as controlled by substituents, and the π cloud polarizability of the donor. Polarizability increases with fused ring number; thus, PAHs behave as strong π donors. The graphene surface may be π-donating or π-accepting depending on proximity to the interior or rim of the platelet. Interiors possess the polarizable π-donor property of PAHs, whereas rims may be lined with electron-withdrawing groups. The ability to form a π-π EDA bond can only enhance the driving force for sorption from water, since water lacks a π system.

We demonstrated π-π EDA interactions between π-accepting nitroaromatic compounds or π-donating PAHs with microcrystalline graphite and black carbon adsorbents (Sander and Pignatello 2005a; Zhu and Pignatello 2005). Adsorption of 4-nitrotoluene, 2,4-dinitrotoluene, and 2,4,6-trinitrotoluene on both adsorbents was much greater than predicted by hydrophobic effects based on a calibration set of compounds that included weak donors and acceptors. Their adsorption also increased in relation to π-acceptor strength (i.e., number of nitro substituents). Hydrogen bonding of the nitro groups was ruled out (Sander and Pignatello 2005a). Adsorption of PAHs on graphite and black carbon was, likewise, greater than predicted by the hydrophobic effect and followed the π-donor strength of the adsorbate (benzene < naphthalene < phenanthrene). Complexation of the nitroaromatics with model PAH donors (naphthalene, phenanthrene, pyrene) was observed spectrophotometrically in chloroform. The ^{1}H-NMR spectrum revealed ring-current induced upfield chemical shifts that result when the monomers associate closely in a face-to-face orientation. The formation constant for the complexed pair increased with the respective π-donor or π-acceptor character of the monomer. The absorption spectrum showed charge-transfer bands in the visible or near-UV region that are characteristic of many π-π EDA complexes. In support of a π-donor role for the graphene surface, we found a strong relationship to exist between the free energy of molecular complexation in chloroform and the excess free energy of nitroaromatic adsorption on graphite after normalization for hydrophobic effects. Using a similar approach, Chen et al. (2007b) detected π-π EDA interactions between polynitroaromatic compounds and the graphene-like surface of carbon nanotubes. Taken together, these results indicate that the graphene surface may be amphoteric with respect to π-interactive adsorbates i.e., electron rich regions of the surface attract strongly π-accepting molecules, while electron-poor regions attract strongly π-donating molecules.

12.3.5 *Steric Effects*

Steric hindrance is an important factor controlling adsorption to black carbons. There are at least two possible sources of steric effects in solids like black carbon: (*i*) conformational restrictions in the solute molecule that reduce its molecular contact with the surface, hence reduce London-van der Waal forces, which are inversely proportional to distance to the sixth power; and (*ii*) size exclusion at the throats of narrow pores that restrict penetration of solutes into pore networks.

The contact area hypothesis has not been firmly validated. The ratio of sorption coefficients for gas phase transfer to nonporous inorganic surfaces of normal- versus cyclo-alkanes—the former ostensibly better able to contact the surface due to their greater flexibility—is not much different than one (0.83–1.62), and the ratio *declines* with the number of carbon atoms, the opposite of the predicted effect (Endo et al. 2008). Researchers have shown that polychlorinated biphenyl (PCB) congeners with restricted co-planarity due to ortho chlorine substitution are less strongly adsorbed and more bioavailable in black carbons or sediments enriched in black carbon than congeners with the same number of Cl atoms that are less restricted (Barring et al. 2002; Bucheli and Gustafsson 2001; Jonker and Smedes 2000; Jonker et al. 2004; van Noort et al. 2002). However, PCBs with restricted coplanarity have an inherently weaker tendency to interact with neighboring molecules than do less restricted PCBs even in the liquid phase, as indicated by their higher subcooled liquid vapor pressures and lower octanol-water partition coefficients (Schwarzenbach et al. 2002).

By contrast, size exclusion (also known as molecular-sieving) is an established factor restricting adsorption to microporous inorganic solids. Diffusivity of compounds in zeolite micropores decreases exponentially as the minimum critical molecular diameter approaches the pore diameter (Kärger and Ruthven 1992). Strong evidence now exists for size exclusion in charcoal (Pignatello et al. 2006a; Zhu and Pignatello 2005). To demonstrate this effect, nonporous microcrystalline graphite was used as a reference material to represent the completely accessible graphene surface and to normalize for hydrophobic effects. In addition, the study focused on planar aromatic compounds with no conformational restrictions. Charcoal-graphite distribution isotherms were constructed from the experimental charcoal-water and graphite-water isotherms. Figure 12.7 shows that the charcoal-graphite distribution coefficient ($K_{\text{ch-gr}}$, m^2/m^2) as a function of molecular size. There was no significant difference in $K_{\text{ch-gr}}$ between benzene and monosubstituted benzenes (toluene, benzonitrile) (Fig. 12.7a). This is consistent with Sander and Pignatello (2005a) who found that benzene, toluene and nitrobenzene have identical sorption domains on charcoal. The closeness of $K_{\text{ch-gr}}$ to 1 suggests that graphite is a reasonable model for the black carbon surface. Placement of additional substituents on the benzene ring, however, reduces $K_{\text{ch-gr},}$ regardless of substituent polarity (Cl–, CH_3–, NO_2–). Tetra-substituted benzenes are fully an order of magnitude less intensely adsorbed than benzene and monosubstituted benzenes on charcoal relative to graphite. Increasing the fused ring size from 1 (benzene) to 3 (phenanthrene) decreased $K_{\text{ch-gr}}$ by an order of magnitude (Fig. 12.7b). These results are clearly consistent with a size-exclusion effect whereby the internal pore network surface area available for adsorption is restricted in relation to molecular size. Anthracene and phenanthrene are planer molecules with nearly identical K_{OW}s (Schwarzenbach et al. 2002), yet anthracene consistently adsorbs more strongly than phenanthrene to black carbon-rich sediments (Cornelissen et al. 2004; Jonker and Koelmans 2002b), presumably due to its smaller critical diameter.

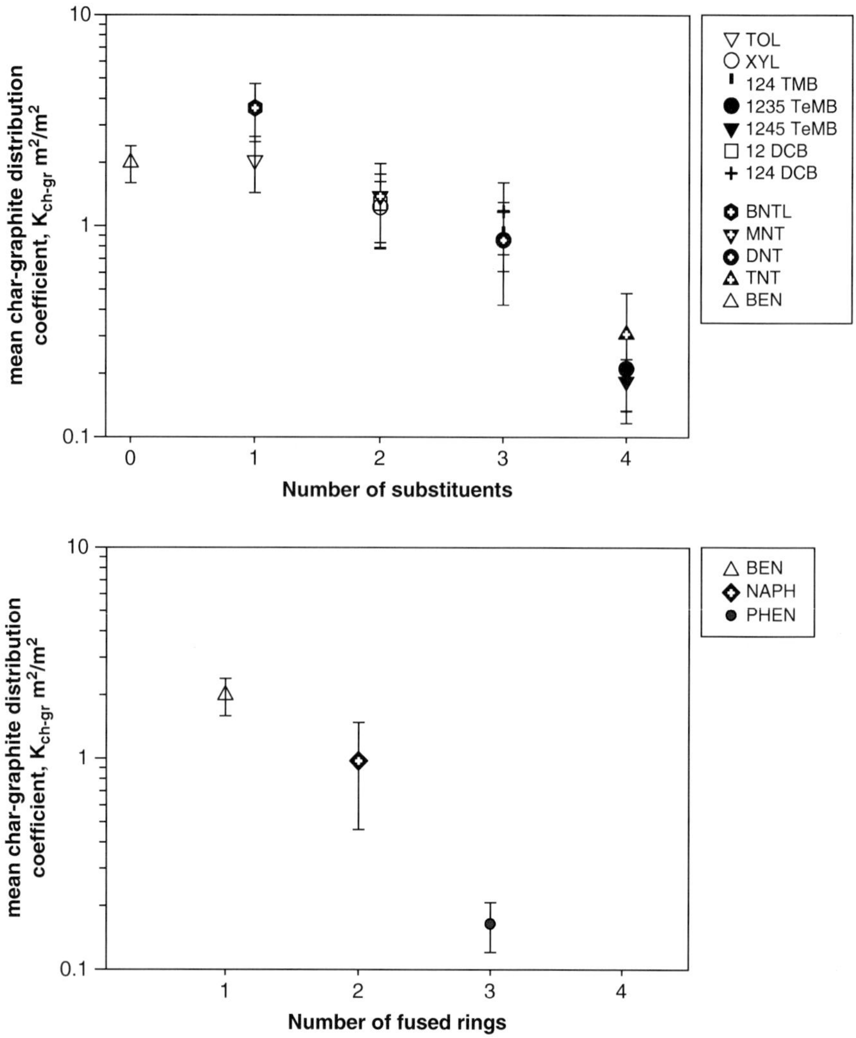

Fig. 12.7 Char-graphite distribution coefficients (N_2-B.E.T. surface area basis) as a function of molecular size, number of ring substituents (*upper figure*) or fused ring size (*lower figure*), for aromatic compounds. Each point represents the mean and standard deviation of 11–18 data measured over a range of concentrations (Reproduced from Pignatello (2011))

12.3.6 Competitive Effects and "Fouling" by Natural Substances

Competition between solutes for adsorption will be observed whenever sites on or within a solid have non-uniform energies and when sites of a given energy are limited in number—a situation that is certainly expected of black carbon. The more effective the competition of a co-solute, the more linear the isotherm of the principal solute will become (e.g., (McGinley et al. 1993; Sander and Pignatello 2005a, 2007)). This is because the co-solute suppresses the low-concentration end of the principal solute's isotherm more than the high-concentration end, rendering it more linear. Competition is important because contaminants are seldom present alone. Sander (Sander and Pignatello 2005a) showed that benzene, toluene and nitrobenzene not only suppress each other's sorption, but have identical sorption domains on a charcoal —that is, they compete for the same population of sites.

The surface activity of raw black carbon can be decreased by the presence of other sorbing substances the particle encounters, including unburned liquid fuel or fuel byproducts, natural organic matter, and metal ions or precipitates. Unburned liquid fuel or semivolatile fuel byproducts can condense in the pores of soot particles during synthesis and/or upon cooling (Akhter et al. 1985b). These condensates can suppress the sorption of contaminants. For example, phenanthrene sorption was weaker and more linear in a reference diesel soot having the greater amount of native extractable oils (20 vs 2% by weight) (Nguyen and Ball 2006). Atmospheric aerosol black carbon particles can accumulate secondary photolysis products of semivolatile organic compounds in the atmosphere that subsequently affect sorption (Dachs and Eisenreich 2000; Kamens et al. 1995; Strommen and Kamens 1997).

Upon weathering in natural environments, the ability of black carbon particles to sorb organic compounds may be attenuated by humic substances, metal ions and possibly metal sesquioxides that may adsorb or coat their surfaces. These substance suppress sorption by competing for surface area and/or by blocking pore throats. Dissolved humic substances are known to foul activated carbon adsorbents used in the water treatment industry (Kilduff and Wigton 1999; Li et al. 2003; Newcombe et al. 1997). Sorption of PAHs and PCBs in sediment believed to be rich in black carbon was up to nine times weaker than expected based on sorption coefficients for raw black carbon (Cornelissen and Gustafsson 2004; Jonker et al. 2004). Direct evidence for (Koelmans et al. 2009; Kwon and Pignatello 2005; Pignatello et al. 2006b; Wen et al. 2009) and against (Cornelissen and Gustafsson 2006) attenuation of sorption by natural substances has been reported. In our studies (Pignatello et al. 2006a), a charcoal-soil mixture suspended in water after a few weeks showed a 13-fold decline in the N_2-B.E.T. specific surface area and a 3-fold decline in the benzene sorption coefficient. Loading charcoal with humic acid, fulvic acid or triglycerides (as surrogates for the humic lipid fraction) from the aqueous phase decreased the specific surface area of the charcoal by up to 10^2 and caused a modest suppression of naphthalene and phenanthrene sorption (Pignatello et al. 2006b). Suppression of sorption became greater with increasing solute molecule size, consistent with competition between humic substances and the solute for external surfaces. The greatest suppression (an order of magnitude) was observed for phenanthrene when the charcoal was coated with either humic acid or a humic acid-aluminum ion flocculate. Tannic acid, a model for humic substances, decreased the surface area of black carbon and the sorption of hydrophobic compounds (Qiu et al. 2009). Metal ions compete with nonionic compounds on the surfaces of black carbon (Chen et al.

2007a), as well as carbon nanotubes (Chen et al. 2009). The cause of suppression was postulated to be "crowding out" of the organic adsorbate by the hydration sphere of the metal ion.

12.4 Sorption Reversibility

Physi-sorption is often considered reversible. There are, however, two meanings of the word "reversible" in the literature. One has a phenomenological connotation and can be expressed as "capable of being quantitatively retrieved, untransformed." Another is rooted in thermodynamics and means a process that follows the same pathway in the forward and reverse directions. Irreversibility is manifested by hysteresis wherein desorption points are displaced to higher sorbed concentrations relative to the original sorption (uptake) curve. There are artificial causes of hysteresis and true, or thermodynamic, causes of hysteresis. The latter is also known as irreversible sorption. Two types of thermodynamically irreversible sorption mechanisms have been identified: capillary condensation hysteresis and pore deformation hysteresis. Both result from achieving a metastable state. Capillary condensation hysteresis occurs in solids with fixed mesopores and is due to the formation of a metastable film during pore filling that eventually collapses to the thermodynamic meniscus plug state (*23*), from which desorption is less favorable. Pore deformation hysteresis applies to semi-flexible (glassy) organic substances. It results when adsorbate molecules intercalate in narrow pore spaces and cause the pores or proto-pores to expand inelastically; that is, the pore does not contract completely when the molecule leaves (Sander et al. 2006). Thermodynamically irreversible sorption obviously has important implications in regard to contaminant transport and bioavailability.

Sorption to black carbon in laboratory studies is often found to be hysteretic. Studies directed toward an understanding of the underlying causes are few. Many experimental pitfalls exist due to the possibility of artifacts, such as degradation, diffusion rate limitation and changes in the liquid-phase composition from the sorption to the desorption step. One must be especially cautious in interpreting hysteresis when desorption occurs from a solid that had initially sorbed a high percentage of the added solute (Sander and Pignatello 2005b, 2007).

Sander and Pignatello (2007) studied the sorption reversibility of benzene, toluene, and nitrobenzene, both singly and in pairs, by the charcoal mentioned earlier in which these compounds had completely overlapping sorption domains. Single-solute sorption was weakly hysteretic at high concentrations, which was taken as evidence for pore inelasticity based on analogy to previous studies with activated carbon (Bailey et al. 1971) and natural organic matter (Sander et al. 2005, 2006; Sander and Pignatello 2005b, 2009). The pore deformation hypothesis is supported by the aforementioned experiments (Section 12.3.1) that conclusively demonstrated swelling by the adsorbate. Swelling means that the internal matrix expands in response to the presence of the adsorbate.

The presence of a competing co-solute has a profound effect on hysteresis. Hysteresis is weak when the initial co-solute concentration is low, but grows as the initial co-solute concentration increases (Sander and Pignatello 2007). This trend is due to what is termed the "competitor dilution effect". The competitor dilution effect operates when the co-solute concentration is lowered during an experiment, whether by dilution or degradation. As a consequence, the solute then re-equilibrates to a new point where competitive pressure has been reduced relative to what it was at the initial point. For example, consider the situation where, after sorption of a solute and a co-solute is allowed to come to equilibrium, the aqueous phase is diluted to initiate desorption. Both the solute and co-solute are simultaneously diluted, and so the solute re-equilibrates to a new point at which it experiences less competition from the co-solute. A similar phenomenon can occur if the co-solute is degraded. Thus, the competitor dilution effect can have an impact on pollutant behavior under conditions where natural or anthropogenic competing substances become diluted or degraded. It can make the pollutant appear to be less thermodynamically accessible with time.

Lastly, we consider the phenomenon of "entrapment". The formation of highly desorption-resistant fractions that increase with aging up to a point is well documented for natural solids including black carbon (Pignatello and Xing 1996). Entrapment, however, must be distinguished from merely slow diffusion due to ordinary causes—the retarded movement of molecules through pores and phases due to interactions and steric hindrance. Entrapment occurs when a molecule reaching a site is subsequently inhibited or prevented from exiting that site due to changes in the physical-chemical structure of the solid, all the while maintaining its chemical identity. Establishing causation of entrapment is difficult experimentally. There are at least two hypothetical mechanisms for entrapment. Molecules may become trapped in closed pores during particle synthesis. This may explain high PAH desorption resistance in soot particles (Akhter et al. 1985a; Jonker and Koelmans 2002a, b; Jonker et al. 2005) because PAHs are among the molecular precursors of soot condensation from the gas phase (Akhter et al. 1985b; Lahaye 1990; Smedley et al. 1992). A second mechanism is physical rearrangement of the matrix induced by some force to convert an open pore (or cavity with flexible walls) to a closed pore (or cavity with rigid walls). Braida et al. (2003) suggested that entrapment could arise from pore deformation during a sorption-desorption cycle, where the sorbing molecule itself provides the force. According to this hypothesis, the original black carbon particle may contain both open sectors of highly interconnected pores that are in thermodynamic contact with bulk solution, and closed sectors of poorly interconnected pores not in thermodynamic contact with bulk solution. Molecules may leak into closed sections at higher concentrations by the process of swelling and plasticization. The reverse may occur during desorption through the process of shrinkage and deplasticization, but some molecules may end up becoming trapped in a developing closed sector as the matrix around them collapses and stiffens. Subsequent release may require cooperative flexing by the matrix, a process that may be accompanied by a high activation energy. It should be noted that this hypothesis has not been rigorously tested. The possibility of pore dilation caused by the packing effects of adsorbate molecules has also been discussed in regard to activated carbons (Olivier 1998). It is likely that the ability of a solute to deform black carbon depends on the flexibility of the carbon as well as solute molecular size and properties.

12.5 Summary and Concluding Remarks

Black carbon in the environment surely contributes to sorption of organic compounds, but its contribution depends greatly on its properties at the time of formation and the degree to which the surface activity is attenuated by uncarbonized fuels and weathering processes. Currently there are no generally accepted methods for quantifying black carbon in soils and sediments, nor for quantifying the contribution of black carbon to total sorption in a given sample. Clearly, further research in this area is required. The adsorbent ability of black carbon can be attributed to its porous nature which allows sorption without a cavity penalty. The available porosity of black carbon depends on synthesis conditions and the abundance of substances encountered by the particle that can block pores or compete with contaminants for pore space. Sorption is highly nonlinear and can result in orders of magnitude variation in the sorption distribution ratio over the whole isotherm. When used for mechanistic purposes all thermodynamic models for describing sorption have limitations. The hydrophobic effect is a major driving force for sorption of apolar compounds but interactions with the surface are clearly important. For example, benzene is preferred over cyclohexane after correcting for the hydrophobic effect by solvent-water partitioning, regardless of whether benzene or hexadecane is the reference solvent. We have demonstrated the importance of π-π EDA interactions in the adsorption of electron rich and (especially) electron poor aromatic compounds. Polar functional groups on the surface are inhibited from interacting strongly with polar molecules due to competition by water molecules, which can be overwhelming in the case of polar molecules that form H-bonds of ordinary strength. Instead, polar functional groups seem to suppress adsorption of organic solutes, regardless of polarity, by attracting water molecules that sterically crowd out solute molecules. Exclusion to intraparticle pore networks on the basis of steric size of the adsorbate is clearly an important consideration, but requires further quantification. Much less is known about sorption of ionic and ionizable organic compounds than neutral organic compounds by black carbon. Sorption to black carbon is often found to be hysteretic but the underlying causes of hysteresis require further attention. Clearly, this is a critical fate and transport issue. Finally, further research is needed to understand the reasons for highly resistant desorption.

References

Abraham MH, Chadha HS, Whiting GS, Mitchell RC (1994) Hydrogen bonding. 32. An analysis of water-octanol and water-alkane partitioning and the delta log P parameter of Seiler. J Pharm Sci 83:1085–1100

Akhter M, Chughtai A, Smith D (1985a) The structure of hexane soot I: spectroscopic studies. Appl Spectrosc 39:143–153

Akhter MS, Chughtai AR, Smith DM (1985b) The structure of hexane soot II: extraction studies. Appl Spectrosc 39(1):154–167

Allen-King RM, Grathwohl P, Ball WP (2002) New modeling paradigms for the sorption of hydrophobic organic chemicals to heterogeneous carbonaceous matter in soils, sediments, and rocks. Adv Water Resour 25:985–1016

Antal MJ Jr, Grønli M (2003) The art, science, and technology of charcoal production. Ind Eng Chem Res 42:1619–1640

Bailey A, Cadenhead DA, Davies DH, Everett DH, Miles AJ (1971) Low pressure hysteresis in the adsorption of organic vapours by porous carbons. Trans Faraday Soc 67:231–243

Barring H, Bucheli TD, Broman D, Gustafsson O (2002) Soot-water distribution coefficients for polychlorinated dibenzo-p-dioxins, polychlorinated dibenzofurans and polybrominated diphenylethers determined with the soot cosolvency-column method. Chemosphere 49:515–523

Boehm HP (1964) Some aspects of the surface-chemistry of carbon-blacks and other carbons. Carbon 32:759–769

Bourke J, Manley-Harris M, Fushimi C, Dowaki K, Nunoura T, Antal MJ (2007) Do all carbonized charcoals have the same chemical structure? 2. A model of the chemical structure of carbonized charcoal. Ind Eng Chem Res 46:5954–5967

Braida W, Pignatello JJ, Lu Y, Ravikovitch PI, Neimark AV, Xing B (2003) Sorption hysteresis of benzene in charcoal particles. Environ Sci Technol 37:409–417

Brewer CE, Schmidt-Rohr K, Satrio JA, Brown RD (2009) Characterization of biochar from fast pyrolysis and gasification systems. Environ Prog Sustain Energy 28:386–396

Bucheli TD, Gustafsson O (2001) Ubiquitous observations of enhanced solid affinities for aromatic organochlorines in field situations: are in situ dissolved exposures overestimated by existing partitioning models? Environ Toxicol Chem 20:1450–1456

Chandler D (2005) Interfaces and the driving force of hydrophobic assembly. Nature 437:640–647

Chen Y et al (2005) Electron microscopy investigation of carbonaceous particulate matter generated by combustion of fossil fuels. Energy Fuels 19:1644–1651

Chen J, Zhu D, Sun C (2007a) Effect of heavy metals on the sorption of hydrophobic organic compounds to wood charcoal. Environ Sci Technol 41:2536–2541

Chen W, Duan L, Zhu DQ (2007b) Adsorption of polar and nonpolar organic chemicals to carbon nanotubes. Environ Sci Technol 41:8295–8300

Chen G, Shan X, Wang Y, Wen B, Pei Z, Xie Y, Liu T, Pignatello JJ (2009) Adsorption of 2,4,6-trichlorophenol by multi-walled carbon nanotubes as affected by Cu(II). Water Res 43:2409–2418

Cho Y-M, Ghosh U, Kennedy AJ, Grossman A, Ray G, Tomaszewski JE, Smithenry DW, Bridges TS, Luthy RG (2009) Field application of activated carbon amendment for in-situ stabilization of polychlorinated biphenyls in marine sediment. Environ Sci Technol 43:3815–3823

Cornelissen G, Gustafsson O (2004) Sorption of phenanthrene to environmental black carbon in sediment with and without organic matter and native sorbates. Environ Sci Technol 38:148–155

Cornelissen G, Gustafsson O (2006) Effects of added PAHs and precipitated humic acid coatings on phenanthrene sorption to environmental black carbon. Environ Pollut 141:526–531

Cornelissen G, Elmquist M, Groth I, Gustafson Ö (2004) Effect of sorbate planarity on environmental black carbon sorption. Environ Sci Technol 38:3574–3580

Coughlin RW, Ezra FS (1968) Role of surface acidity in the adsorption of organic pollutants on the surface of carbon. Environ Sci Technol 2:291–297

Dachs J, Eisenreich SJ (2000) Adsorption onto aerosol soot carbon dominates gas-particle partitioning of polycyclic aromatic hydrocarbons. Environ Sci Technol 34:3690–3697
Elmer W, Pignatello JJ (2011) Effect of biochar amendments on mycorrhizal associations and fusarium crown and root rot of asparagus in replant soils. Plant Dis 95:960–966
Endo S, Grathwohl P, Schmidt TC (2008) Absorption or adsorption? Insights from molecular probes n-alkanes and cycloalkanes into modes of sorption by environmental solid matrices. Environ Sci Technol 42:3989–3995
Franz M, Arafat HA, Pinto NG (2000) Effect of chemical surface heterogeneity on the adsorption mechanism of dissolved aromatics on activated carbon. Carbon 38:1807–1819
Gilli P, Pretto L, Bertolasi V, Gilli G (2009) Predicting hydrogen-bond strengths from acid-base molecular properties. The pK(a) slide rule: toward the solution of a long-lasting problem. Acc Chem Res 42:33–44
Goldberg ED (1985) Black carbon in the environment. Wiley, New York, NY
Goss KU (2004) The air/surface adsorption equilibrium of organic compounds under ambient conditions. Crit Rev Environ Sci Technol 34:339–389
Gustafsson Ö, Gschwend PM (1997) Soot as a strong partition medium for polycyclic aromatic hydrocarbons in aquatic systems. In: Eganhouse RP (ed) Molecular markers in environmental geochemistry. American Chemical Society, Washington, D.C., pp 365–381
Gustafsson Ö, Haghseta F, Chan C, MacFarlane J, Gschwend PM (1997) Quantification of the dilute sedimentary soot phase: implications for PAH speciation and bioavailability. Environ Sci Technol 31:203–209
Jonker MT, Koelmans AA (2002a) Extraction of polycyclic aromatic hydrocarbons from soot and sediment: solvent evaluation and implications for sorption mechanism. Environ Sci Technol 36:4107–4113
Jonker MTO, Koelmans AA (2002b) Sorption of polycyclic aromatic hydrocarbons and polychlorinated biphenyls to soot and soot-like materials in the aqueous environment: mechanistic considerations. Environ Sci Technol 36:3725–3734
Jonker M, Smedes F (2000) Preferential sorption of planar contaminants in sediments from Lake Ketelmeer, the Netherlands. Environ Sci Technol 34:1620–1626
Jonker MTO, Hoenderboom AM, Koelmans AA (2004) Effects of sedimentary sootlike materials on bioaccumulation and sorption of polychlorinated biphenyls. Environ Toxicol Chem 23:2563–2570
Jonker MTO, Hawthorne SB, Koelmans AA (2005) Extremely slowly desorbing polycyclic aromatic hydrocarbons from soot and soot-like materials: evidence by supercritical fluid extraction. Environ Sci Technol 39:7885–7895
Kamens RM, Odum JR, Fan Z-H (1995) Some observations on times to equilibrium for semivolatile polycyclic aromatic hydrocarbons. Environ Sci Technol 29:43–50
Kaneko K, Abe M, Ogino K (1989) Adsorption characteristics of organic compounds dissolved in water on surface-improved activated carbon fibres. Colloids Surf 37:211–222
Kärger J, Ruthven DM (1992) Diffusion in zeolites and other microporous solids. Wiley, New York, NY
Keiluweit M, Kleber M (2009) Molecular-level interactions in soils and sediments: the role of aromatic pi interactions. Environ Sci Technol 43:3421–3429
Keiluweit M, Nico PS, Johnson MG, Kleber M (2010) Dynamic molecular structure of plant biomass-derived black carbon (biochar). Environ Sci Technol 44:1247–1253
Kilduff JE, Wigton A (1999) Sorption of TCE by humic-preloaded activated carbon: incorporating size-exclusion and pore blockage phenomena in a competitive adsorption model. Environ Sci Technol 33:250–256
Koelmans AA, Meulman B, Meijer T, Jonker MTO (2009) Attenuation of polychlorinated biphenyl sorption to charcoal by humic acids. Environ Sci Technol 43:736–742
Kwon S, Pignatello JJ (2005) Effect of natural organic substances on the surface and adsorptive properties of environmental black carbon (char): pseudo pore blockage by model lipid components and its implications for N2-probed surface properties of natural sorbents. Environ Sci Technol 39:7932–7939

Lahaye J (1990) Mechanisms of soot formation. Polym Degrad Stabil 30:111–121

Lazaridis T (2001) Solvent size vs cohesive energy as the origin of hydrophobicity. Acc Chem Res 34:931–937

Lehmann J, Joseph S (eds) (2009) Biochar for environmental management: science and technology. Earthscan Publications Ltd., London, UK

Li L, Quinlivan PA, Knappe DRU (2002) Effects of activated carbon surface chemistry and pore structure on the adsorption of organic contaminants from aqueous solution. Carbon 40:2085–2100

Li Q, Snoeyink VL, Mariãas BJ, Campos C (2003) Elucidating competitive adsorption mechanisms of atrazine and NOM using model compounds. Water Res 37:773–784

Lighty SJ, Veranth JM, Sarofim AF (2000) Combustion aerosols: factors governing their size and composition and implications to human health. J Air Waste Manag Assoc 50:1565–1618

Lu Y, Pignatello JJ (2004) History-dependent sorption in humic acids and a lignite in the context of a polymer model for natural organic matter. Environ Sci Technol 38:5853–5862

Manes M (1998) Activated carbon adsorption fundamentals. In: Meyers RA (ed) Encyclopedia of environmental analysis and remediation. Wiley, New York, pp 26–68

Martin DS (2003) The adsorption of aromatic acids onto the graphite basal surface. Surf Sci 536:15–23

Masiello CA, Druffel ERM (1998) Black carbon in deep-sea sediments. Science 280:1911–1913

McGinley PM, Katz LE, Weber WJ Jr (1993) A distributed reactivity model for sorption by soils and sediments. 2. Multicomponent systems and competitive effects. Environ Sci Technol 27:1524–1531

McMullen CP, Jabbour J (2009) Climate Change Science Compendium 2009. United Nations Environment Programme. http://www.unep.org/compendium2009/

Mészáros E, Jakab E, Varhegyi G, Bourke J, Manley-Harris M, Nunoura T, Antal MJ (2007) Do all carbonized charcoals have the same chemical structure? 1. Implications of thermogravimetry – mass spectrometry measurements. Ind Eng Chem Res 46:5943–5953

Müller EA, Gubbins KE (1998) Molecular simulation study of hydrophilic and hydrophobic behavior of activated carbon surfaces. Carbon 36:1433–1438

Müller EA, Hung FR, Gubbins KE (2000) Adsorption of water vapor-methane mixtures on activated carbons. Langmuir 16:5418–5424

Newcombe G, Drikas M, Hayes R (1997) Influence of characterised natural organic material on activated carbon adsorption: II. Effect on pore volume distribution and adsorption of 2-methylisoborneol. Water Res 31:1065–1073

Nguyen TH, Ball WP (2006) Absorption and adsorption of hydrophobic organic contaminants to diesel and hexane soot. Environ Sci Technol 40:2958–2964

Olivier JP (1998) Improving the models used for calculating the size distribution of micropore volume of activated carbons from adsorption data. Carbon 36:1469–1472

Palotás ÁB, Rainey LC, Feldermann CJ, Sarofim AF, Vander Sande JB (1996) Soot morphology: an application of image analysis in high-resolution transmission electron microscopy. Microsc Res Tech 33:266–278

Pendleton P, Wong SH, Schumann R, Levay G, Denoyel R, Rouquerol J (1997) Properties of activated carbon controlling 2-methylisoborneol adsorption. Carbon 8:1141–1149

Pignatello JJ (2011) Interactions of anthropogenic organic chemicals with natural organic matter and black carbon in environmental particles. In: Huang PM (ed) Biophysico-chemical processes of anthropogenic organic compounds in environmental systems, vol 3, IUPAC series on biophysico-chemical processes in environmental systems. John Wiley & Sons, Hoboken, New Jersey, U.S.A. pp 3–50

Pignatello JJ, Xing B (1996) Mechanisms of slow sorption of organic chemicals to natural particles. Environ Sci Technol 30:1–11

Pignatello JJ, Kwon S, Lu Y (2006a) Effect of natural organic substances on the surface and adsorptive properties of environmental black carbon (char): attenuation of surface activity by humic and fulvic acids. Environ Sci Technol 40:7757–7763

Pignatello JJ, Lu Y, LeBoeuf EJ, Huang W, Song J, Xing B (2006b) Nonlinear and competitive sorption of apolar compounds in black carbon-free natural organic materials. J Environ Qual 35(4):1049–1059
Qiu Y, Xiao X, Cheng H, Zhou Z, Sheng GD (2009) Influence of environmental factors on pesticide adsorption by black carbon: pH and model dissolved organic matter. Environ Sci Technol 43:4973–4978
Radovic LR, Silva IF, Ume JI, Menéndez JA, Leon LY, Scaroni AW (1997) An experimental and theoretical study of the adsorption of aromatics possessing electron-withdrawing and electron-donating functional groups by chemically modified activated carbons. Carbon 35:1339–1348
Ramanathan V, Carmichael G (2008) Global and regional climate changes due to black carbon. Nat Geosci 1:221–227
Sander M, Pignatello JJ (2005a) Characterization of charcoal adsorption sites for aromatic compounds: insights drawn from single-solute and bi-solute competitive experiments. Environ Sci Technol 39:1606–1615
Sander M, Pignatello JJ (2005b) An isotope exchange technique to assess mechanisms of sorption hysteresis applied to naphthalene in kerogenous organic matter. Environ Sci Technol 39:7476–7484
Sander M, Pignatello JJ (2007) On the reversibility of sorption to black carbon: distinguishing true hysteresis from artificial hysteresis caused by dilution of a competing adsorbate. Environ Sci Technol 41:843–849
Sander M, Pignatello JJ (2009) Sorption irreversibility of 1,4-dichlorobenzene in two natural organic matter-rich geosorbents. Environ Toxicol Chem 28:447–457
Sander M, Lu Y, Pignatello JJ (2005) A thermodynamically based method to quantify sorption hysteresis. J Environ Qual 34:1063–1072
Sander M, Lu Y, Pignatello JJ (2006) Conditioning annealing studies of natural organic matter solids linking irreversible sorption to irreversible structural expansion. Environ Sci Technol 40:170–178
Schmidt MWI, Noack AG (2000) Black carbon in soils and sediments: analysis, distribution, implications, and current challenges. Global Biogeochem Cycles 14:777–793
Schwarzenbach RP, Gschwend PM, Imboden DM (2002) Environmental organic chemistry, 2nd edn. Wiley, New York, NY
Skjemstad JO, Taylor JA, Smernik RJ (1999) Estimation of charcoal (char) in soils. Commun Soil Sci Plant Anal 30:2283–2298
Smedley JM, Williams A, Bartle KD (1992) A mechanism for the formation of soot particles and soot deposits. Combust Flame 91:71–82
Smith KR (2005) National burden of disease in India from indoor air pollution. Proc Natl Acad Sci USA 97:13286–13293
Snoeyink VL, Weber WJ Jr (1967) The surface chemistry of active carbon. Environ Sci Technol 1:228–234
Southall NT, Dill KA, Haymet ADJ (2002) A view of the hydrophobic effect. J Phys Chem B 106:521–533
Strommen MR, Kamens RM (1997) Development and application of a dual-impedance radial diffusion model to simulate the partitioning of semivolatile organic compounds in combustion aerosols. Environ Sci Technol 31:2983–2990
Teixidó M, Pignatello JJ, Beltran JL, Grenados M, Peccia J (2011) Speciation of the ionizable antibiotic sulfamethazine on black carbon (biochar). Environ Sci Technol 45:10020–10027
van Noort PCM, Cornelissen G, Ten Hulscher TEM, Belfoid A (2002) Influence of sorbate planarity on the magnitude of rapidly desorbing fractions of organic compounds in sediment. Environ Toxicol Chem 21:2326–2330
Wen B, Huang RX, Li RJ, Gong P, Zhang S, Pei ZG, Fang J, Shan XQ, Khan SU (2009) Effects of humic acid and lipid on the sorption of phenanthrene on char. Geoderma 150:202–208
Xi J, Zhong B-J (2006) Soot in diesel combustion systems. Chem Eng Technol 29:665–673

Xia G, Ball WP (1999) Adsorption-partitioning uptake of nine low-polarity organic chemicals on a natural sorbent. Environ Sci Technol 33:262–269
Zhu D, Pignatello JJ (2005) Characterization of aromatic compound sorptive interactions with black carbon (charcoal) assisted by graphite as a model. Environ Sci Technol 39:2033–2041
Zhu D, Kwon S, Pignatello JJ (2005) Adsorption of single-ring organic compounds to wood charcoals prepared under different thermochemical conditions. Environ Sci Technol 39:3990–3998

Part IV
Soil Management

Chapter 13
Subsurface Soil Acidification in Farming Systems: Its Possible Causes and Management Options

Caixian Tang, Chandrakumara Weligama, and Peter Sale

Abstract Subsurface soil acidity is widespread and its amelioration is costly and often practically infeasible. This paper summarizes recent research on the causes and management of subsurface soil acidity in farming systems. The development of subsurface soil acidity depends largely on acid production by plant roots due to excess cation uptake, particularly under legume-based agriculture. The deposition and decomposition of plant residues do not cause subsurface soil acidification but contribute to the development of subsurface soil acidity profiles through their liming effect on the topsoil. Nitrification from ammonium-based fertilizers or from organic N in plant residues, and subsequent leaching of nitrate, contribute mainly to topsoil acidification. In contrast, the uptake of leached nitrate and denitrification in deeper layers may decrease subsurface soil acidity. Various application methods and liming materials have been studied for their effectiveness in ameliorating subsurface soil acidity. These include application of lime, gypsum and organic materials, and growing acid-tolerant crops. Biological amelioration through managing excess anion (nitrate) uptake has been tested in the field and is a promising method to ameliorate subsurface soil acidity. The major challenge for the biological amelioration method is to synchronize nitrate movement and root capture in the subsurface soil for maximal alkalization and minimal nitrate leaching loss.

Keywords Biological amelioration • C cycle • Cation-anion uptake • N cycle • Crop residues • Subsoil acidity

---

C. Tang (✉) • C. Weligama • P. Sale
Department of Agricultural Sciences, La Trobe University, Melbourne Campus, Bundoora, VIC 3086, Australia
e-mail: c.tang@latrobe.edu.au; ckumara@email.com; p.sale@latrobe.edu.au

J. Xu and D.L. Sparks (eds.), *Molecular Environmental Soil Science*, Progress in Soil Science, DOI 10.1007/978-94-007-4177-5_13,

13.1 Introduction

Approximately 50% of the world's potentially arable land area consists of acid soils (pH ≤ 5.5) (Kochian et al. 2004) and many soils have acidity below the plough layer (Sumner 1995). Topsoil acidity is commonly ameliorated with liming. However, lime movement down into the subsurface soil is slow, especially in variable charge soils and hence surface application of lime is often inefficient at ameliorating subsurface soil acidity. Although heavy top dressings with lime may have an ameliorating effect on subsurface soil acidity (see below), such high rates have adverse effects on some crops in the rotation (McLay et al. 1994a) or cause lime-induced deficiencies of nutrients such as Mn and Zn. Deep placement of lime into acid subsurface soil through injection and/or ripping is generally considered to be uneconomic.

Unfortunately, subsurface soil acidification (below the plough layer) is a continuing process under many farming systems, especially legume-based systems. Soil acidification can occur in soil profiles to a depth of more than 80 cm (Williams 1980; Coventry and Slattery 1991; Dolling and Porter 1994; Conyers et al. 1996; Moody and Aitken 1997; Noble et al. 1997). In many soil profiles, acidification rates are greater in the subsurface soil than in the topsoil. For example, in the central wheatbelt of Western Australia, soil pH increased with increasing soil depth at uncleared (native vegetation) sites, whereas at cleared (farmed) sites the lowest pH occurred in the 10–30 cm layers. In southern New South Wales, even fertile red earth soils are showing the development of an acidic layer below the normal plough depths (Conyers et al. 1996).

Subsurface soil acidity is associated with toxicities of aluminium and manganese, and deficiencies of calcium and phosphorus. It primarily decreases root elongation, resulting in a poor root system and hence a decrease in the uptake of nutrients and water, particularly for acid-sensitive crops (Tang et al. 2002). Moisture in the topsoil is often depleted late in the growing season as temperatures and growth rates increase. Plants are therefore forced to rely on the supply of water and nutrients from deeper soil layers. However, the poor root systems of crops grown with subsurface soil acidity will limit reproductive growth and final grain yields.

There are several excellent reviews on the general topic of soil acidification and the management of soil acidity (e.g. Helyar and Porter 1989; Bolan and Hedley 2003; Coventry et al. 2003) which do not cover much on subsurface soil acidity and will not be the focus of this chapter. Instead, the chapter will review current knowledge on possible causes of subsurface soil acidification, especially under legume-based farming systems. It will then discuss management options for subsurface soil acidity with an emphasis on the biological amelioration through managing cation/anion uptake. This chapter does not attempt to review causes of subsurface soil acidification as a result of natural weathering processes (see Sumner 1995) or acid precipitation (see Alewell 2003).

13.2 Causes of Subsurface Soil Acidification

Soil acidification in farming systems is attributed to a number of processes. The major causes in farming systems are the removal of alkalinity in farm products in the carbon cycle, and release of protons during N_2 fixation and nitrate leaching in the nitrogen cycle. Subsurface soil acidification, on the other hand, appears to be mainly caused by spatial separation of acid and alkali produced in these cycles (Fig. 13.1). Tang and Rengel (2003) described the processes which contribute to this separation and to the development of acidity in deeper soil layers. In the carbon and nitrogen cycles, the acid produced by the roots due to excess uptake of cations over anions is distributed through the rooting zone, with the amount and distribution of the acid being proportional to the root length distribution. Plant residues in the carbon cycle however, are mainly oxidised in the surface soil. This oxidation of organic anions present in the residues is an acid-consuming process (increases pH), and thus the acidity generated by roots is neutralised in the topsoil, but still remains in the subsurface soil. In the nitrogen cycle, mineralization and nitrification mainly occur in the topsoil. When these processes are accompanied by nitrate uptake in the topsoil, then there is no change in topsoil pH. However, if nitrate is leached from the topsoil, then there will be a surplus of protons in the topsoil (Figs. 13.1 and 13.2) resulting in topsoil acidification. The leaching of nitrate originating from the topsoil is unlikely to cause subsurface soil acidification. In contrast, uptake of nitrate by roots and denitrification in subsurface soil layers may reduce the net production of acid by the roots. These processes are discussed in detail below.

13.2.1 Excess Uptake of Cations Over Anions

Plant roots take up nutrients from soil solutions to satisfy their growth requirements. The uptake of cations across the plasma membrane is associated with H^+ extrusion while uptake of anions is associated with OH^-/HCO_3^- release (or H^+ consumption).

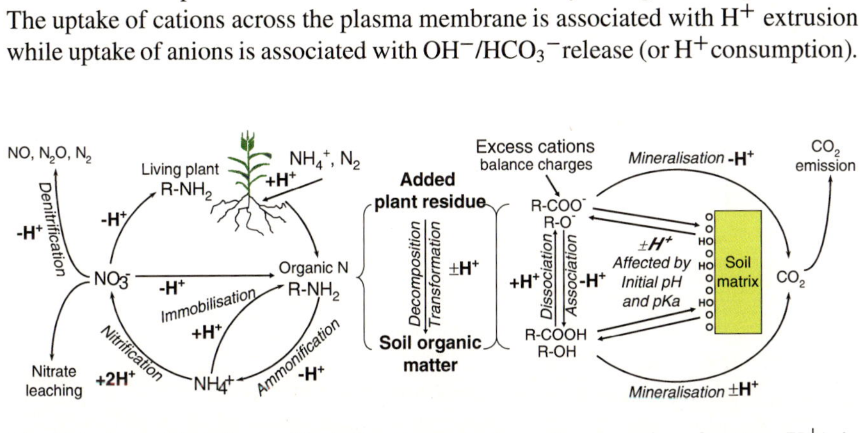

Fig. 13.1 A simplified diagram illustrating the production and consumption of protons (H^+), in the nitrogen (*left*) and carbon (*right*) cycles. $+H^+$ indicates acidification (decrease in pH); **-H**$^+$, alkalization (increase in pH) and $\pm H^+$, unknown effects

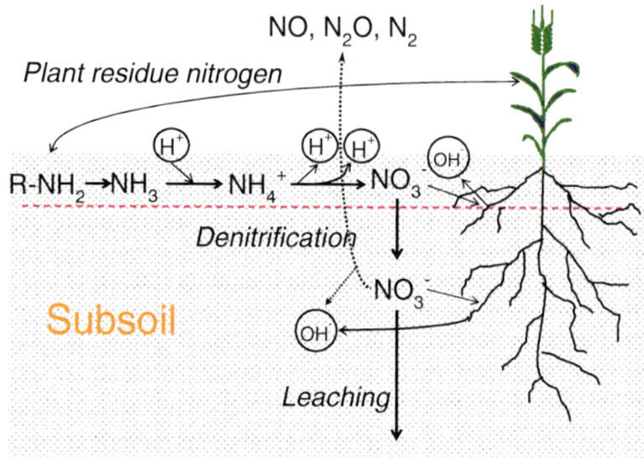

Fig. 13.2 Schematic illustration of nitrogen transformations, and nitrate leaching and uptake, in relation to acidification in the soil profile

If plants take up more cations than anions, then there is an overall net extrusion of H^+ around the roots. If plants take up more anions than cations, there is a net alkalization along the roots (Marschner 1995; Paul et al. 2003; Tang and Rengel 2003; references therein).

The form of nitrogen plays a prominent role in the cation-anion balance (see detailed discussion below). NH_4^+ nutrition leads to an excess uptake of cations over anions and to strong acidification, whereas NO_3^- nutrition results in a rhizosphere alkalization. By comparison, legumes that are actively fixing N_2, take up more cations than anions, and hence release H^+ and acidify their rhizosphere (van Beusichem 1981; Jarvis and Hatch 1985; Loss et al. 1993b; Tang and Rengel 2003). A close linear relationship was found between the amount of H^+ released and the excess of cations over anions taken up by N_2-fixing plants from various legume species (Tang et al. 1997).

Plant species differ substantially in their acid production and their capacity to acidify soils (Tang and Rengel 2003; references therein). Legumes generally cause more soil acidification than non-leguminous species. This may be attributed to the following likely causes. The first is that legumes excrete large amounts of protons due to excess cation uptake during N_2 fixation. Second, legume residues contain high N, resulting in large amounts of nitrate produced during residue decomposition. Finally, legumes generally have a restricted rooting system, with a poor ability to take up soil nitrate during growth, and this can lead to nitrate leaching from the root zone and to acidification due to a net proton surplus (Fig. 13.1).

Tropical legumes generally have lower excess uptake of cations over anions than temperate legumes, and thus have a lower acidifying effect on the rhizosphere and the bulk soil (Andrew et al. 1973). In addition, the products of N assimilation in tropical legumes appear to be ureides (allantoin and allantoic acid, which have high pK_a values) and are unlikely to be dissociated to donate protons at physiological pH values in the cytoplasm and the xylem (Bolan et al. 1991).

The release of acid produced by excess cation uptake from the soil depends on the distribution of the roots and nutrients, the pattern of nutrient uptake and H^+ extrusion along the roots. In an experiment using 1-m soil columns (Tang et al. 2000), where nutrients were applied uniformly throughout the column, root length density of NH_4^+-fed lupin and subterranean clover plants tended to increase with depth, and relatively uniform decreases of soil pH occurred in the soil profile. Where nutrients were applied only in the top 10-cm layer, then 50–70% of the roots of N_2-fixing plants were distributed in this layer and the soil pH decreased most in this top layer. The decrease of pH in different soil layers was proportional to the root length density (Tang et al. 2000). In the field, deeper-rooted legumes appear to cause subsurface soil acidification to greater depths than shallower-rooted legumes. For example, Loss et al. (1993a) observed more soil acidification in deeper layers of soil profiles under a deep-rooted lupin-wheat rotation than under shallow-rooted clover-based pastures. The results do suggest that excess cation uptake by plant roots plays an important role in subsurface soil acidification.

13.2.2 Role of Plant Residues

The accumulation of organic matter has been suggested as one of the causes of soil acidification. Two reasons for this explanation have been proposed: (i) organic matter forms soluble complexes with cations such as Ca and Mg and thus facilitates the leaching loss of these cations, and (ii) organic matter contains functional groups which can be a source of H^+ ions after dissociation (Brady and Weil 2002; Bolan et al. 2003). Organic matter is also thought to lower soil pH by releasing H^+ ions that are associated with organic anions, by the mineralisation of its N and the subsequent nitrification or by an increased cation exchange capacity and corresponding increase in exchangeable acidity (Williams 1980).

However, in many soil profiles the most acidified layers were below 10 cm, even though organic matter generally accumulated in the top 10 cm of the soil (Williams 1980; Dolling and Porter 1994). Indeed, in some farmed soils, the pH increased in the topsoil compared to the nearby uncultivated bushland soils (e.g. Dolling and Porter 1994). Similarly, in a field survey, Griffin et al. (2003) examined the impact of land use on soil organic carbon and pH across various agro-ecosystems in Western Australia. It was found that soil pH was consistently lower at soil depths of 10–100 cm at sites that had been farmed for many years, compared to the corresponding uncleared 'native vegetation' sites. It was also shown that the organic carbon in the topsoils from uncleared bush sites tended to be less than those from farmed sites, but the reverse occurred in the subsurface soil, where the organic carbon in uncleared sites was generally greater than that in farmed sites. This indicates that organic matter accumulation may not be always associated with soil acidification.

The addition of organic matter to soil has been found to increase, decrease or have no effect on soil pH (Ritchie and Dolling 1985; Yan et al. 1996; Tang and Yu 1999; Tang et al. 1999; Marx et al. 2002; Paul et al. 2003; Xu et al. 2006b). These conflicting accounts of the effect of organic matter on soil acidification have probably resulted from differences in (i) composition of organic matter present in or added to soil, (ii) soil characteristics and (iii) decomposition resulting from different experimental conditions, such as temperature, water regime and leaching, that were used in the different experiments.

Our recent studies have shown that addition of plant residues increased the pH of acidic soils. Furthermore, the amounts of alkalinity produced in the soil correlated positively with concentrations of excess cations and total nitrogen in the added residues, and negatively with the initial pH of the soil (Tang and Yu 1999; Xu et al. 2006b). Thus, the amount of alkalinity produced in soil decreased with an increase in the initial pH of the soil. The studies also demonstrated that the effects of the plant residue on alkalinity production were less under sterile conditions than under non-sterile conditions, and changes in soil pH were related to CO_2 release, and the dynamics of NH_4^+ and NO_3^- in the soil. Xu et al. (2006b) suggested that there were at least four components responsible for soil pH changes after plant residues were added to the soil: (i) mineralisation of organic anions (balanced by excess cations) from added plant residues, (ii) ammonification (organic $N \rightarrow NH_4^+$) of the residue N, (iii) nitrification ($NH_4^+ \rightarrow NO_3^-$) of mineralised residue N, and (iv) association/dissociation of protons from organic compounds. Processes (i) and (ii) result in an alkalization of the soil while Process (iii) acidifies the soil. The influence of Process (iv) on soil pH change depends on the initial pH of the soil and the pKa of the organic compounds present. The overall effect of plant residue addition on soil pH therefore depends on the extent of each of these processes.

From above studies and those published earlier, it is suggested that the deposition and subsequent decomposition of plant residues do not necessarily cause subsurface soil acidification. Rather they contribute to the development of the pH variation in soil profiles through their effect on the topsoil. The addition of plant materials to acidic soils generally increases soil pH. While the extent of proton excretion from roots during plant growth is proportional to the root distribution, the oxidation of organic anions during decomposition of the shoot residue mainly occurs in the topsoil since residues are generally not incorporated into deeper layers. Thus, the decomposition of shoot residues neutralises the acid created during plant growth in the topsoil. In some soils where subsurface soil acidification has occurred, an increased soil pH with time of farming has been observed in the topsoil (e.g. Dolling and Porter 1994). Root residues usually have lower concentrations of excess cations than shoot residues and have a lesser "liming" effect (Tang and Yu 1999; Tang et al. 1999). Therefore, root residue decomposition will not fully neutralize the acidity created along the roots in the soil profile and this acidity, produced during growth, will persist in the subsurface soil layers.

13.2.3 Nitrification and Nitrate Leaching

Nitrification and the subsequent leaching of nitrate have often been suggested as a major cause of soil acidification. However, such processes do not appear to be a cause of subsurface soil acidification. Mineralization and nitrification of plant residue N have been shown to occur mainly in the topsoil due to the relative immobility of NH_4^+ and deposition of plant residues on the soil surface. For example, 64% of N mineralisation and 50% of nitrification in the soil profile of a loamy sand, and 78% of N mineralisation and 41% of nitrification in a sandy clay loam occurred in the top 5-cm soil layer (Murphy et al. 1998). More recently, Dear et al. (2009) showed that while NH_4^+ was mainly concentrated in the topsoil of two Red Kandosols over the growing season, the concentration of NO_3^- decreased in the topsoil and increased below 30 cm, with annual pasture legumes. Thus, acidification caused by nitrification, and the subsequent leaching of nitrate, occurs mostly in the topsoil. There will be no net acidification in the topsoil if plant roots utilize all of the nitrate generated *in situ*. Nitrate uptake enhances excess anion uptake, and thus decreases extrusion of H^+ or increases extrusion of OH^-/HCO_3^-. However, if the nitrate is leached, then the topsoil will acidify because of the net increase in H^+ ions (Fig. 13.2) and because the leaching of nitrate is accompanied by the leaching of cations other than H^+ ions (Di and Cameron 2005). If plant roots take up the leached nitrate below the topsoil layer, it is expected that alkalization (or less acidification) will then occur in the subsurface soil. These processes are illustrated in Fig. 13.2.

The hypothesis that nitrification in the topsoil, and the subsequent leaching of nitrate into the subsurface soil, can decrease subsurface soil acidification, was tested in two column experiments with a sandy soil by Tang et al. (2000). These experiments examined the effect of nitrate and ammonium addition to the topsoil on subsurface soil acidification under wheat and the N_2-fixing legumes, lupin and subterranean clover. Nitrate leaching was achieved by adding excess water to the surface of the columns 18 days after sowing. In the first experiment, the addition of $Ca(NO_3)_2$ to the topsoil layer increased nitrate concentration in soil profiles but did not affect the growth of shoot and roots, or the N concentration in shoots of either species. It caused alkalization or reduced acidification at all depths, compared to the treatment without $Ca(NO_3)_2$.

In the second experiment, the addition of $(NH_4)_2SO_4$ to the top 10 cm significantly increased NO_3^- concentration in all layers with NH_4^+ being mainly retained in the top 20-cm layer. Compared to the plants receiving no $(NH_4)_2SO_4$, lupin grown with $(NH_4)_2SO_4$ in 0–10 cm of the column, caused more acidification in the top 10 cm but less acidification in 10–40 cm of the column. Similarly, subterranean clover grown with added $(NH_4)_2SO_4$ in the 0–10 cm layer caused more acidification in the top 10 cm and less acidification in the 20–50 cm layer. Growing wheat with $(NH_4)_2SO_4$ in the topsoil decreased soil pH in top 20 cm but slightly increased soil pH in the deeper layers (Tang et al. 2000). Clearly, leaching of nitrate from topsoil is unlikely to cause subsurface soil acidification. In contrast, the uptake of nitrate by the roots reduces net acid production in subsurface soil layers. Further research is needed to validate the results under field conditions.

13.2.4 Acid Movement

The downward movement of H^+ and soluble Al, and NH_4^+ may contribute to subsurface soil acidification (Hue and Licudine 1999) but the magnitude of the downward movement is unknown. It is expected that such acid movement would be small because of the high affinity of H^+ and Al^{3+} ions for negatively charged surfaces of soil colloids. In many soil profiles, pH is also higher in topsoil than in the subsurface soil so that little acid is able to move downward via mass flow or diffusion. Another consideration is that although the highest concentration of acids (lowest pH) is generally present in the subsurface soil layer, there is no evidence that this high concentration of acid moves lower over time. For example, the lowest soil pH in a deep yellow sand always remained between 10 and 30 cm irrespective of the years of farming (Dolling and Porter 1994). If the acid had moved deeper in the profile, this low soil pH should have moved into the lower layers with time. Further studies are required to verify this view.

13.3 Current Methods for Ameliorating Subsurface Soil Acidity

Various application methods and liming materials have been studied for their effectiveness in ameliorating subsurface soil acidity. Some of these methods will be briefly discussed below. The first method that should be mentioned is surface liming. It may partially ameliorate subsurface soil acidity while direct liming into acidic subsurface soil has generally produced a quicker and greater yield response. Field trials have demonstrated yield advantages in growing acid-tolerant wheat varieties in soil with subsurface soil acidity (Scott et al. 2001). Modelling work has indicated that growing acid-tolerant wheat partially eliminates the adverse effects of acidity on crop yields in soils with severe subsurface soil acidity and almost fully eliminates the effect in soils with shallow/moderate subsurface soil acidity (Tang et al. 2003a).

13.3.1 Application of Ground Limestone

The surface application or otherwise shallow incorporation (0–10 cm) of lime is the most common practice to combat soil acidity in agriculture. Short-term yield responses to surface liming of soils with acid subsurface soils are often lacking or vary greatly, depending on application rate, lime quality, crop species, rainfall, acidity profile, soil type and the severity of acidity (Conyers and Scott 1989; Conyers et al. 1991; Scott et al. 1997; Whitten et al. 2000; Tang et al. 2003a, b). Normally, the response to lime application by plants starts slowly and diminishes over the time (Mahoney and Uren 1982). The impact of liming is mainly restricted to the depth of its application because lime is a relatively insoluble material. The alkalinity-producing products in the form of OH^-, $HCO_3{}^-$ and $CO_3{}^{2-}$, move slowly in soil profiles (Farina 1997; Scott et al. 1997; Conyers et al. 2003). Furthermore, the increase in the surface soil pH by liming is likely to increase the negative charge

(CEC) thereby increasing the retention of Ca, especially in variable-charge soils. Therefore, the effect of surface-applied lime on increasing subsurface soil pH can be limited with realistic lime application rates (2 $t\,ha^{-1}$) (Conyers and Scott 1989; Scott et al. 1997).

Many reports have therefore stated that surface liming gives no yield increase under different crops and soil types with subsurface soil acidity (Helyar 1991; Coventry 1992; Farina 1997). This problem can be partially overcome by applying high rates on the soil surface or incorporating lime into deep soil layers (Pinkerton and Simpson 1986; Conyers et al. 2003). Application of large doses of lime cannot be justified due to problems associated with over-liming including trace element deficiencies (Porter 1984; Bolland et al. 2001; Tang et al. 2003b). On the other hand, lime application is becoming a costly endeavour in many countries due to the transportation cost (Conyers et al. 1996). In addition, the cost of liming can be much higher compared to the returns from liming in low rainfall areas (Helyar et al. 1988). Nevertheless, a limited number of studies have showed that surface liming may partially ameliorate subsurface soil acidity in the long term (e.g. Sumner 1995; Whitten et al. 2000; Tang et al. 2003a).

Direct liming into the acidic subsurface soil layer through injection and/or ripping results in quicker and greater yield responses than surface lime applications (e.g. Coventry 1991; Sumner 1995; Scott et al. 1997). However, deep incorporation of lime is impractical due to lack of suitable machinery and high energy cost (Sumner et al. 1986; Scott et al. 1997).

Liming increases mineralisation and nitrification in acidic soils and hence can increase the likelihood of NO_3^- losses from leaching (Ridley et al. 2001). This in turn reduces the effect of the lime application on increasing the pH in the topsoil because NO_3^- leaching contributes to topsoil acidification. However, if the leached NO_3^- ions are taken up by the plants in the subsurface soil, then it is likely that the plant roots in the deeper soil layers will release OH^- and therefore counteract subsurface soil acidification and reduce subsurface soil acidity (as shown in Fig. 13.2).

13.3.2 Application of Gypsum

Surface-applied gypsum has been recognised as an economically viable, alternative method for combating subsurface soil acidity (McLay et al. 1994a). Gypsum ($CaSO_4{\cdot}2H_2O$, Ksp 2.1×10^{-5}) is more soluble than lime ($CaCO_3$, Ksp 8.7×10^{-9}) and thus moves faster in soil profiles. Additionally, the increase in pH in the surface layer increases the retention of Ca (see above). Gypsum was previously used in agriculture because of its ameliorating qualities in sodic soils, its ability to act as a soil conditioner, and its high content of Ca and S enabling it to act as a source of nutrients for plants (Shainberg et al. 1989). Yield responses to gypsum addition to acid soils, in most cases, were due to either improvement of Ca status or to a decrease in bioavailable Al in subsurface soil, which in turn stimulated root growth (Sumner 1993; Farina et al. 2000). As a result, plants can have better access to water and nutrients in acid soils. The effect of gypsum lasts longer in heavy-textured than light-textured soils. In a loamy sand profile, surface-applied gypsum can be leached out from the profile in 5 years (Sumner 1990).

There are a number of explanations for the ameliorating effect of gypsum in acid soils. The first explanation is that the sulphate ions displace hydroxide ions from soil surfaces into the soil solution, which in turn react with free Al^{3+} (Reeve and Sumner 1972). The second suggestion is that sulphate can form non-toxic $AlSO_4^+$ ion pairs or non-soluble aluminium hydroxy sulphate (Alva et al. 1988). Thirdly, leaching of gypsum can increase ionic strength in the subsurface soil, which in turn reduces Al^{3+} activity (Ritchie 1989). Displacement of Al from soil colloids and the subsequent leaching of Al from soil is another suggestion (Oates and Caldwell 1985).

There is however conflicting evidence relating to the effect of gypsum on soil pH. Smith et al. (1994) found that the gypsum treatment (2.5 t ha^{-1}) reduced pH by 0.2 units in the 5–10 and 10–15 cm soil layers compared to the control. Probable displacement of Al^{3+} and H^+ from cation exchange sites by soluble Ca^{2+} appears to be a possible reason for the pH decrease. In contrast, the application of gypsum increased soil pH in other experiments (Sumner 1995). In these situations, the pH increase was attributed to the ligand exchange process in which the SO_4^{2-}ions replaced OH^- ions (Smith et al. 1994). The above two processes are likely to happen simultaneously and so the final pH in a gypsum-amended soil would be the net effect from both processes.

There are some implications for gypsum use as an ameliorant of acid soils. Gypsum application, especially at high rates can have adverse effects on the yield of some crops such as lupins (McLay et al. 1994a) and can cause Mg and K to leach from the upper layers in sandy soils (McLay et al. 1994b). The application of gypsum can also be costly as more than 5 t ha^{-1} may be required to get a significant benefit (Farina 1997).

13.3.3 Application of Plant Residue and Animal Manure

The application of plant residues and animal manures to soil is considered to be a good agricultural management practice. This is because of the positive impacts on plant growth by providing plant nutrients (Bessho and Bell 1992; Naramabuye and Haynes 2006) and by improving soil structure. Organic acids are released during the decomposition process of organic materials. These acid ligands are known to be effective in binding toxic Al^{3+} (Hue et al. 1986). The organically-complexed Al species are less toxic to plant roots (Wong and Swift 2003). For example, in an experiment with highly acidic mine-spoil material (pH below 3.5) high rates of compost provided sufficient Al sorption capacity to keep the Al^{3+} activity in soil solution low enough for plant growth (von Willert and Stehouwer 2003).

Apart from the effect of $CaCO_3$ in some animal manures, the addition of organic materials can also increase soil pH at least in the short term (Haynes and Mokolobate 2001; Wong and Swift 2003). This pH increase can be mainly due to decarboxylation of water-soluble organic anions in the materials (Yan et al. 1996; Rukshana et al. 2011). Other mechanisms that have been suggested to account for an increase in soil pH include: (i) adsorption of H^+ onto the surface of the organic materials (Hoyt and Turner 1975), (ii) production of OH^- ions by ligand exchange between the terminal OH^- on the sesquioxide surfaces and organic anions (Hue and Amien 1989), (iii) accumulation of $NH_4{}^+$ and hydrolysis of NH_3 during the decomposition of organic N (Hoyt and Turner 1975; Naramabuye and Haynes 2006), (iv) increase of soil base saturation through the replacement of protons and aluminium from exchange sites by cations added with organic matter (Bessho and Bell 1992), and (v) development of reducing conditions and associated denitrificaiton due to stimulated microbial activity during the rapid decomposition of the added organic matter (Hue 1992).

The composition and the degree of decomposition of organic materials may affect the amount of alkalinity released during decomposition (Noble et al. 1996; Tang et al. 1999; Tang and Yu 1999; Marschner and Noble 2000; Haynes and Mokolobate 2001; Xu and Coventry 2003; Xu et al. 2006a, b). For example, Noble et al. (1996) showed that the application of leaf litter neutralised soil acidity, and the extent of neutralisation was related to the content of their ash alkalinity. Similarly, Xu et al. (2006b) found that the amounts of alkalinity produced in soil correlated positively with concentrations of excess cations and total N in the added materials. The N concentration in organic materials influences whether the organic N is mineralized or immobilised which is crucial towards its effect on soil pH (Bolan et al. 1991).

Relatively little information is available on the movement down the soil profile of the alkalinity, released from the addition of organic materials. Lund and Doss (1980) showed that repeated applications of dairy manure at 135 $t\,ha^{-1}$ $year^{-1}$ for 3 years increased the pH and concentrations of exchangeable Mg and K in the subsurface soil of a Plinthic Paleudult. Similarly, Wright et al. (1985) found that the combined application of cow manure (44.8 $t\,ha^{-1}$) and dolimite (8.4 $t\,ha^{-1}$) to the topsoil increased subsurface soil pH and decreased exchangeable Al in a Typic Paleudult. However, in a soil column experiment, the presence of plant residues had little effect on the mobilization of the reaction products of surface-applied lime in the profile of an acidic Red Latosol (Moraes et al. 2007). Recently, Butterly et al. (2011) showed that the soluble fraction of residues is important for alkalinity release during the initial stages of decomposition. It is expected that the organic compounds in the soluble fraction can move down in the soil profile before they are decomposed by microbes. Depending on residue and soil type, significant amounts of the alkalinity moved below the soil layer in which the crop residue was applied, within 3 months of the application (Butterly et al. 2012).

An alternative strategy that should have potential in alleviating subsurface soil acidity is the practice known as subsoil manuring. This involves the deep incorporation of high rates of organic manures in subsurface layers to improve the physical properties of dense clay subsurface soils that restrict root growth, water uptake and crop yields in dry years (Gill et al. 2009). The marked improvement in macroporosity and hydraulic conductivity in subsurface soil layers with subsoil manuring, were attributed to the increased microbial activity associated with the breakdown of the organic amendment (Clark et al. 2007) and the breakdown of exudates from the increased mass of roots in the organic-amended subsurface soil layers. Not surprisingly, there were large increases in crop yields following this subsurface soil intervention. Given that the organic acids released from the decomposing organic amendments should alleviate any toxic Al species that might be present in acidic subsurface soils (Wong and Swift 2003) then one would expect that subsoil manuring would have an additional benefit of ameliorating subsurface soil acidity, if the subsurface soils being treated with the deep-placed manures, are also constrained by subsurface soil acidity.

13.3.4 Addition of Coal-Derived Organic Materials

The humic materials produced during oxidation of coal have been used for ameliorating soil acidity in general, and subsurface soil acidity in particular, in recent years (van der Watt et al. 1991; Noble and Randall 1998; Inoue et al. 2001; Imbufe et al. 2004; 2005). These materials are soluble in water and can leach down the soil profile in contrast to lime. The humic products are rich in carboxylic and phenolic groups, which provide reactive sites for cation exchange, bind phytotoxic elements, increase pH buffering and promote the penetration and retention of Ca in soils (Imbufe et al. 2004). Potassium humate is one of humic materials and is produced by the alkaline treatment of Victorian brown coal. It can effectively increase pH and levels of K, Mg and Ca in soil (Imbufe et al. 2004, 2005). In a column experiment, calcium fulvate increased soil pH to a depth of 100 cm in soils with a clay content of 6–32%, after columns were leached by two pore volumes of water (van der Watt et al. 1991). Use of coal-derived organic materials as an ameliorant of acid subsurface soils is promising. Further research is required to validate their effectiveness and economics in the broad-acre agriculture.

13.4 Biological Amelioration Through Managing Cation-Anion Uptake by Crops

13.4.1 Impact of the Form and Uptake of N on Rhizosphere pH

Nitrogen uptake has a marked influence on the cation–anion uptake balance and rhizosphere pH due to the high demand for N by plants (Marschner 1995; Mengel et al. 2001). This is because total N uptake can be responsible for 70–80% of the total ion uptake by plants (van Beusichem et al. 1988; Neumann and

Römheld 2002). Moreover, plants can take up the N either in the cation (NH_4^+) or anion (NO_3^-) form. In addition, legumes assimilate N in its neutral gaseous N_2 form. Some plants even take up organic forms of N such as amino acids (Jones and Darrah 1994) and urea (Harper 1984). Therefore, the role of N in cation-anion balance depends on the form of the available N in the soil as well as the crop or pasture species. It follows that a predominant supply of nitrate would lead to a greater anion uptake than cations while a predominant ammonium and uncharged N (i.e. biological N_2 fixation) supply would result in greater cation uptake than anions (Marschner 1995; Tang and Rengel 2003; references therein).

The metabolic assimilation of the N ions following the uptake into root cells produces H^+ or OH^- ions which in turn contribute to pH changes in the rhizosphere. For example, the assimilation of NO_3^- results in the production of OH^-, which is released into the rhizosphere during NO_3^- reduction in root tissue, in order to stabilize intercellular pH (Neumann and Römheld 2002). In contrast, assimilation of NH_4^+ by roots produces H^+ ions which can result in lower pHs in soil (Marschner 1995).

13.4.2 Factors Affecting Nitrate Uptake

Nitrate uptake occurs via H^+ co-transport system which requires energy (Taiz and Zeiger 2006). This process depends on the trans-membrane H^+ gradient generated by the H^+-ATPase in the membrane. The pH decrease in the growth medium increases the H^+ gradient between the cell and external environment favouring NO_3^- uptake. However, the uptake of NO_3^- is regulated by the availability of NO_3^-. Indeed, a key factor for generating enzymes that are required for NO_3^- transportation and assimilation is the presence of NO_3^- in the environment (Taiz and Zeiger 2006).

Plants prefer to take up NH_4^+ compared to NO_3^- when they have access to NH_4^+ ions (Colmer and Bloom 1998). However, the diffusion coefficient of NO_3^- is much higher than that of NH_4^+ (Miller and Cramer 2004). In addition, NO_3^- ions are not adsorbed onto soil colloidal surfaces with negative charges and thus freely move in the soil profile. These two characteristics have contrasting consequences on nitrate availability and access to the root system. Nitrate can be more available to roots via diffusion or mass flow than ammonium, and this increases the possibility of its interception by roots. On the other hand, nitrate can be easily lost from the soil profile by leaching (Black 1992). As a result, the timing of root proliferation is important for interception and take up nitrate from the soil (Diggle et al. 1990).

Plant species differ greatly in their demand for N and in root morphology and root distribution in the soil profile. They therefore differ in their ability to intercept and to take up NO_3^- in the soil profile (see Tang and Rengel 2003). Generally, perennial species can take up more NO_3^- than annual species, and non-leguminous species take up more than legumes. In the field, NO_3^- leaching was much greater under the grain legume *Lupinus angustifolius* than under wheat (Anderson et al. 1998). Also, the concentrations of NO_3^- in soil profiles were higher under subterranean clover and yellow serradella than under the mixed culture of subterranean clover and perennial grass species at a low and medium rainfall site of south-eastern Australia (Dear et al. 2009). Genotypes within the same species may also differ in utilizing soil NO_3^- . For example, less NO_3^- was leached under deep-rooted than shallow-rooted genotypes of bentgrass (Bowman et al. 1998). Furthermore, growing acid-tolerant species and genotypes, compared to acid-sensitive ones, on acidic soils usually leads to increased root growth in these soils, particularly in acidic subsurface soils, and this will increase the capacity of roots to absorb NO_3^-. Selecting genotypes that intercept and take up more soil NO_3^- may provide an option for minimizing subsurface soil acidification and ameliorating subsurface soil acidity.

Nitrate availability in soil will influence root growth, which in turn affects NO_3^- uptake. Localised sources of NO_3^- are known to increase root proliferation in the soil where they occur (Robinson 1994). The enhanced root proliferation results from increased root elongation and initiation of lateral roots. The localised stimulatory effect of root growth by NO_3^- supply occurs due to specific signalling property of NO_3^- ions themselves. It has also been suggested that the concentration of NO_3^- in the plant affects the expression of the NO_3^--inducible MADS-box gene (ANR1). This gene encodes "a component of the signal transduction pathway linking external NO_3^- supply to increased rate of lateral root elongation" (Zhang and Forde 2000). Selection of varieties that can over-express the NO_3^--inducible MADS-box gene would benefit rhizosphere alkalization through improved root proliferation and NO_3^- capture and uptake in acid subsurface soils.

13.4.3 Excess Nitrate Uptake and Subsurface Soil Alkalization

As discussed earlier, differences in anion/cation uptake by roots can lead to the release of protons or hydroxyl ions, resulting in acidification or alkalization in the rhizosphere. There needs to be an excess of anion uptake over cation uptake in order to increase rhizosphere pH. Therefore, a large amount of anions should be supplied to roots to encourage excess anion uptake in order to combat the acidity. Given that the magnitude of N uptake by plants, it follows that a large supply of nitrate anions, that will result in excess anion uptake (Mengel et al. 2001) and subsequent rhizosphere alkalization, will be required in the acidic soil layer. Root-induced rhizosphere alkalization can be extended to the bulk soil depending on soil pH buffer capacity (Hinsinger et al. 2003).

There is evidence that subsurface soil alkalization can occur following N fertilizer applications to the soil surface. This has occurred both in glasshouse experiments (Black 1992; Tang et al. 2000) and under field conditions (Adams and Pearson 1969; Noble et al. 2008; Tang et al. 2011). In soil-column experiments, Tang et al. (2000) showed that the columns receiving $Ca(NO_3)_2$ in the topsoil

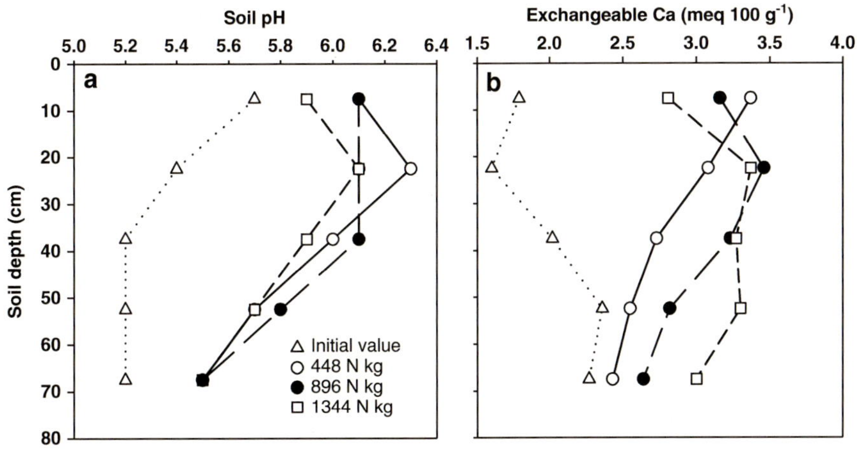

Fig. 13.3 Effect of application rates of $Ca(NO_3)_2$ (kg ha^{-1} year^{-1}) to the surface on pH (**a**) and on exchangeable Ca (**b**) changes with soil depth after 4 years of application in a sandy loam soil (Adopted from Adams and Pearson (1969), with kind permission of Soil Science Society of America)

layer had pH 0.1–0.2 units higher in deeper soil layers than columns receiving no $Ca(NO_3)_2$, after growing N_2-fixing lupin and subterranean clover plants for 105 days. Although N_2 fixation itself causes soil acidification, a net alkalization occurred below 20 cm with lupins and below 30 cm with clover plants.

Similar findings were observed in a 4-year field experiment where coastal bermudagrass was grown in an irrigated sandy loam soil. In this study, Adams and Pearson (1969) applied NH_4NO_3 at an annual rate of 896 kg N ha^{-1} year^{-1}. This resulted in a decline in soil pH of around 1.3–1.4 units in the surface 0–15 cm, and a decline of 0.3–0.5 of a unit in the 15–30 cm soil layers, compared to the initial soil pH. However, there was either no pH decline or an increase of about 0.3 units in 30–45, 45–60 and 60–75 layers. The effect of $Ca(NO_3)_2$ was also investigated (Fig. 13.3), with the $Ca(NO_3)_2$ being supplied at rates of 448, 896 and 1,344 kg N ha^{-1} to the surface 0–20 cm layer. Changes in pH and exchangeable Ca in the soil profile were measured after 4 years of N application. The alkalizing effect of $Ca(NO_3)_2$ was evident in all soil layers compared to the initial soil pH (Fig. 13.3). Exchangeable Ca data indicate an increased Ca movement in the deeper layers with increasing N rates. Calcium movement is generally accompanied by an anion, which most likely would be the NO_3^- anion. Therefore, it appears that NO_3^- leaching increased with increasing N rates which in turn led to increased NO_3^- uptake by bermudagrass roots in deeper layers, resulting in alkalization in these soil layers. The increased movement NO_3^- ions facilitated the movement of Ca to subsurface soils, thereby helping in mitigating subsurface soil acidity.

In another field study with Gamba grass (*Andropogon gayanus*), a significant pH increase was observed in 0–30 cm soil layer when N was supplied in the NO_3^-

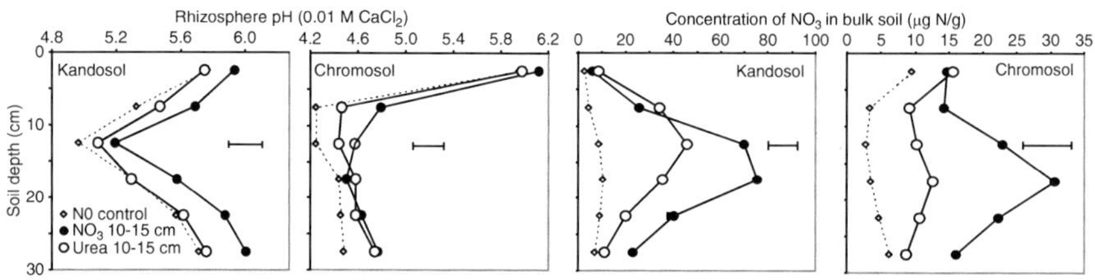

Fig. 13.4 Changes in rhizoshere pH and NO_3 concentration in profiles of Kandosol and Chromosol soil under a wheat crop supplied with no N fertilizer, $Ca(NO_3)_2$ or urea at 10–15 cm. Bars are LSD values at $P = 0.05$ (Adopted from Tang et al. 2011)

form compared to the control and the NH_4^+ treatments (Noble et al. 2008). The study was conducted in an acid sandy soil and the grass was given three different N treatments; these were no N, KNO_3 or $(NH_4)_2SO_4$ with the N applications totalling 287 kg N ha^{-1} over a 3-year period. The supply of KNO_3 increased the pH by 0.2–0.3 units in the 0–30 cm soil layers after 3 years. This alkalization was a direct result of excess anion uptake by plants as there was a little possibility for denitrification in this lightly textured soil. In contrast, the supply of $(NH_4)_2SO_4$ decreased pH by 0.4 units, the acidity likely resulting from plant uptake and nitrification of ammonium.

Biological amelioration of subsurface soil acidity by managing the supply and uptake of nitrate from the subsurface soil, has recently been examined in more detail at two field sites using acid-tolerant and acid-sensitive wheat varieties. The results collated from the 2006 trials show the potential for ameliorating subsurface soil acidity using nitrate fertilizers (Tang et al. 2011). Similar yields and N-use efficiency were achieved with the supply of nitrate, compared to the supply of urea. Application of nitrate increased rhizosphere pH up to 0.5 units and the bulk soil pH to 0.3 units, to a depth exceeding 30 cm at both sites (Fig. 13.4). The placement of nitrate fertilizer in the 10–15 cm soil layer increased subsurface soil pH more than the surface application. The treatment effects on the bulk soil pH became more significant in the second and third season. In contrast, the application of urea (as the most common N fertilizer) acidified the soil in the 5–10 cm soil layer by up to 0.4 units compared to the control, and up to 0.7 units in the 10–15 cm layer, compared to the nitrate treatment, after 3 years of treatment (Conyers et al. 2011). Nitrate application markedly increased nitrate concentration in soil profiles, particularly at the 10–20 cm depth (Fig. 13.4), whereas urea application increased the NH_4^+ concentration. The majority of the nitrate was retained in the top 30-cm layers in the medium rainfall region. In the soil-induced process, the net effect of urea (ammonification followed by nitrification) is a decrease in pH, whereas nitrate source does not contribute to this process at all. More field work is required to assess the impact of nitrate leaching, and uptake from deeper soil layers, in different soils under different climatic conditions.

Root proliferation in acid subsurface soils is important if the biological amelioration of subsurface soil acidity is to occur. Localised application of N and P can play a vital role in the stimulation of root growth either in the surface or subsurface soil layer, whereas localized applications of K had relatively little effect on root growth (Robinson 1994). A later study demonstrated that localised supply of nitrate and P together was required to maximise alkalization in an acid subsurface soil (Weligama et al. 2008). This alkalization resulted from an increased uptake of NO_3^- and stimulated root proliferation in the fertilizer-applied layer, which in turn enhanced shoot growth due to efficient exploration of resources in the acid subsurface soil. While nitrate supply alone increased root growth, P placement alone did not stimulate root proliferation in the acid soil (Weligama et al. 2008). The study suggests that P fertilizer should be applied together with nitrate to achieve maximal benefits in extra root growth and pH increase.

An alternative fertilizer strategy for ameliorating subsurface soil acidity was proposed by Noble et al. (1997). This involved providing the ingredients of $Ca(NO_3)_2$ in the soil surface layer by applying lime and an NH_4^+-based fertilizer (apart from nitrate fertilizers) at the surface. The subsequent nitrate leaching into the subsurface soil, and the uptake of nitrate from the subsurface soil layers, were thought to offer a feasible and practical method to address subsurface soil acidity. This proposal has been verified in a number of studies. Poss et al. (1995) reported an alkalization that occurred in subsurface soil of a red Kandosol in a semi-arid climate with mean annual rainfall of 539 mm. The soil was limed (2.5 t ha^{-1}) 2 years before the start of the experiment and urea was applied to the surface at 140 kg N ha^{-1}. The study showed that a net acidification of 1.8 kmol H^+ ha^{-1} occurred in the topsoil whereas a net alkalization of 0.9 kmol OH^- ha^{-1} occurred in the 25–90 cm layer. Nitrate uptake by wheat roots and denitrification were postulated as possible reasons for the alkalization in the subsurface soil. Tang et al. (2000) showed that the addition of $(NH_4)_2SO_4$ in the top 10 cm of 1-m soil columns significantly increased NO_3^- concentration in all layers with NH_4^+ being mainly retained in the top 20-cm layer, and that growing wheat decreased soil pH in top 20 cm but increased soil pH in the deeper layers. Similarly, Black (1992) observed that the application of urine or a urea solution resulted in acidification in the 2–8 cm deep layer and significant alkalization in the 8–10 cm layer. Uptake of NO_3^- by plants was the most likely reason for the alkalization that occurred in the deeper soil layer. However, in another experiment, the supply of urea with surface liming was not successful in rhizosphere alkalization in the subsurface layer. One explanation was that the severe acidity of the soil reduced the population of nitrifying microorganisms resulting in limited nitrification (Weligama et al. 2010b).

There are several key requirements for optimizing the use of the biological amelioration in addressing subsurface soil acidity. These are that the crop should be tolerant of soil acidity, that a large amount of NO_3^- needs to be supplied to the surface soil, and that sufficient water or rainfall is supplied to the soil surface to leach the NO_3^- into the subsurface soil layers. However, there is a delicate balance required between the NO_3^- supply and the rain events that supply water to the surface. Insufficient water supply with high NO_3^- will restrict deep root growth and plants may not be able to capture any of the NO_3^- that is subsequently leached. Similarly, insufficient NO_3^- supply and adequate water may also fail to produce sufficient deep roots to capture the leached NO_3^- (Weligama et al. 2010a). There also needs to be synchrony in timing, such that the root system has time to develop to capture NO_3^- from subsequent leaching events. The major challenge for the biological amelioration method is to minimize NO_3^- leaching loss, especially in sandy soils and in high rainfall environments. Modeling work is needed to study the feasibility of the biological method and how to maximize alkalization effects while at the same time minimizing NO_3^- leaching losses under different scenarios (e.g. Dunbabin et al. 2003).

13.5 Conclusions

There is strong evidence that root activity due to excess cation uptake is a major cause of subsurface soil acidification in many farming systems. Nitrification of ammonium, from the ammonification of organic N, and the subsequent leaching of the nitrate, appear to be the major causes of topsoil acidification. However, they are not likely to cause subsurface soil acidification. Subsurface soil acidification appears to be mainly caused by the spatial separation of acid and alkali produced by the C and N cycles. While much emphasis has been given to the amelioration of soil acidity, minimizing or preventing subsurface soil acidification should receive more attention. This may be achieved through selection of lower acid-producing species with less excess cations in their tissue, and managing cation/anion uptake in soil profiles.

Surface liming is currently the most common strategy to combat soil acidity, and may partially ameliorate subsurface soil acidity in long-term (Sumner 1995; Whitten et al. 2000; Tang et al. 2003b). Direct liming into the acidic subsurface soil produced a quicker and greater yield response, but is generally considered to be uneconomic. Field trials have demonstrated yield advantages of growing Al-tolerant wheat cultivars in soil profiles with subsurface soil acidity (Scott et al. 2001). The use of Al-tolerant cultivars now forms part of a different strategy to deep liming, known as biological amelioration, for combating subsurface soil acidity.

Biological amelioration operates by enabling roots in the subsurface soil to take up nitrate ions. The approach has been tested in the field and appears to be a promising method to counteract subsurface soil acidification and to ameliorate subsurface soil acidity. This is certainly the case where rainfall is low and profit margins from farming are small. Based on current N application rates of 30–150 kg ha^{-1} $year^{-1}$, with an N-use efficiency of 40%, the application of the N in the nitrate form would generate alkalinity equivalent to 42–214 kg lime ha^{-1} $year^{-1}$. This should be able to stop and reverse the acidification process (Porter et al. 1995). The major challenge for the biological amelioration method is to synchronize NO_3^- movement and root capture, in order to minimize NO_3^- leaching loss and maximise alkalization in the subsurface soil.

Acknowledgments We thank Professor Nanthi Bolan (University of South Australia) for his constructive comments and Dr Mark Conyers for enjoyable discussions. Financial support from the Australian Research Council (LP0562504, DP0877882 and DP120104100) and the Grains Research, Development Corporation (Australia) are greatly acknowledged.

References

Adams F, Pearson RW (1969) Neutralizing soil acidity under Bermudagrass sod. Soil Sci Soc Am Proc 33:737–742

Alewell C (2003) Acid inputs into the soils from acid rain. In: Rengel Z (ed) Handbook of soil acidity. Marcel Dekker, Inc, New York

Alva AK, Sumner EM, Noble AD (1988) Alleviation of aluminium toxicity by phospho-gypsum. Commun Soil Sci Plant Anal 19:385–403

Anderson GC, Fillery IRP, Dunin FX, Dolling PJ, Asseng S (1998) Nitrogen and water flows under pasture-wheat and lupin-wheat rotations in deep sands in Western Australia. II. Drainage and nitrate leaching. Aust J Agric Res 49:345–361

Andrew CS, Johnson AD, Sandland RL (1973) Effect of aluminium on the growth and chemical composition of some tropical and temperate pasture legumes. Aust J Agric Res 24:325–339

Bessho T, Bell LC (1992) Soil solid and solution phase-changes and mung bean response during amelioration of aluminium toxicity with organic-matter. Plant Soil 140:183–196

Black AS (1992) Soil acidification in urine and urea affected soil. Aust J Soil Res 30:989–999

Bolan NS, Hedley MJ (2003) Role of carbon, nitrogen and sulfur cycles in soil acidification. In: Rengel Z (ed) Handbook of soil acidity. Marcel Dekker, Inc, New York

Bolan NS, Hedley MJ, White RE (1991) Process of soil acidification during nitrogen cycling with emphasis on legume based pastures. Plant Soil 134:53–63

Bolan NS, Adriano DC, Curtin D (2003) Soil acidification and liming interactions with nutrient and heavy metal transformation and bioavailability. Adv Agron 78:216–272

Bolland MDA, Rengel Z, Paszkudzka-a-Baizert L, Osborne LD (2001) Responses of subterranean clover and Italian ryegrass to application of lime. Aust J Exp Agric 41:177–185

Bowman DC, Devitt DA, Engelke MC, Rufty TW (1998) Root architecture affects nitrate leaching from bentgrass turf. Crop Sci 38:1633–1639

Brady NC, Weil RR (2002) The nature and properties of soils, 13th edn. Prentice Hall, New Jersey

Butterly CR, Baldock JA, Tang C (2012) The contribution of crop residues to changes in soil pH under field conditions. Plant and Soil. doi: 10.1007/s11104-012-1422-1

Butterly CR, Bhatta Kaudal B, Baldock JA, Tang C (2011) Contribution of soluble and insouble fractions of agricultural residues to short-term pH changes. Eur J Soil Sci 62:718–727

Clark GJ, Dodgshun N, Sale PWG, Tang C (2007) Changes in chemical and biological properties of a sodic clay subsurface soil with addition of organic amendments. Soil Biol Biochem 39: 2806–2817

Colmer TD, Bloom AJ (1998) A comparison of NH_4^+ and NO_3^- net fluxes along root of rice and maize. Plant Cell Environ 21:240–246

Conyers MK, Scott BJ (1989) The influence of surface incorporated lime on sub surface acidity. Aust J Exp Agric 29:201–207

Conyers MK, Poile GJ, Cullis BR (1991) Lime responses by barley as related to available soil aluminium and manganese. Aust J Agric Res 42:379–390

Conyers MK, Heenan DP, Poile GJ, Cullis BR, Helyar KR (1996) Influence of dryland agricultural management practices on the acidification of soil profile. Soil Till Res 37:127–141

Conyers MK, Mullen CL, Scott BJ, Poile GJ, Braysher BD (2003) Long term benefits of limestone application to soil profiles and to cereal crop yields in southern and central NSW. Aust J Exp Agric 43:71–78

Conyers MK, Tang C, Poile GJ, Liu DL, Chen D, Nuruzzaman M (2011) A combination of biological activity and the nitrate form of nitrogen can be used to ameliorate subsurface soil acidity under dryland wheat farming. Plant Soil 348:155–166. doi:10.1007/s11104-011-0827-6

Coventry DR (1991) The injection of slurries of lime, associated with deep tillage, to increase wheat production on soils with subsurface soil acidity. In: Wright R, Baligar V, Murrman R (eds) Plant-soil interaction at low pH. Kluwer Academic, Dordrecht

Coventry DR (1992) Acidification problems of duplex soils used for crop-pasture rotations. Aust J Exp Agric 32:901–914

Coventry DR, Slattery WJ (1991) Acidification of soil associated with Lupins grown in a crop rotation in North-Eastern Victoria. Aust J Agric Res 42:391–397

Coventry DR, Farhoodi A, Xu RK (2003) Managing soil acidification through crop rotations in southern Australia. In: Rengel Z (ed) Handbook of soil acidity. Marcel Dekker, Inc, New York

Dear BS, Virgona JM, Sandral GA, Swan AD, Morris B (2009) Changes in soil mineral nitrogen, nitrogen leached, and surface pH under annual and perennial pasture species. Crop Pasture Sci 69:975–986

Di HJ, Cameron KC (2005) Reducing environmental impacts of agriculture by using a fine particle suspension nitrification inhibitor to decrease nitrate leaching from grazed pastures. Agric Ecosyst Environ 109:202–212

Diggle AJ, Bowden JW, D'Antuono MF (1990) A comparison of the effects of mineral and organic nitrogen sources on the distribution of wheat roots in a leaching environment. Aust J Soil Res 28:963–971

Dolling PJ, Porter WM (1994) Acidification rates in the central wheatbelt of Western Australia. I. On a deep yellow sand. Aust J Exp Agric 34:1155–1164

Dunbabin V, Diggle A, Rengel Z (2003) Is there an optimal root architecture for nitrate capture in leaching environments? Plant Cell Environ 26:835–844

Farina MBW (1997) Management of subsurface soil acidity in environments outside the humid tropics. In: Moniz A (ed) Plant-soil interaction at low pH: sustainable agriculture and forestry production. Campains, Brazil

Farina MPW, Channon P, Thibaud GR (2000) A comparison of strategies for ameliorating subsurface soil acidity: II. Long-term soil effects. Soil Sci Soc Am J 64:652–658

Gill JS, Sale PWG, Peries RR, Tang C (2009) Changes in soil physical properties and crop root growth in dense sodic subsurface soil following incorporation of organic amendments. Field Crop Res 114:137–146

Griffin EA, Verboom WH, Allen DG (2003) National Carbon Accounting System Technical Report no. 38, Australian Government

Harper JF (1984) Uptake of organic nitrogen forms by roots and leaves. American Society of Agronomy, Madison

Haynes RJ, Mokolobate MS (2001) Amelioration of Al toxicity and P deficiency in acid soils by additions of organic residues: a critical review of the phenomenon and the mechanisms involved. Nutr Cycl Agroecosyst 59:47–63

Helyar KR (1991) The management of acid soils. In: Wright R, Baligar V, Murrman R (eds) Plant-soil interaction at low pH. Kluwer Academic, Dordrecht

Helyar KR, Porter WM (1989) Soil acidification, its measurement and process involved. In: Robson AD (ed) Soil acidity and plant growth. Academic, Sydney

Helyar KR, Hochman Z, Brennan JP (1988) The problem of acidity in temperate area soils and its management. In: Loveday J (ed) National soil conference review papers. Australian Society Soil Science Inc., University of Western Australia, Perth

Hinsinger P, Plassard C, Tang C, Jaillard B (2003) Origins of root mediated pH changes in rhizosphere and their responses to environmental constrains: a review. Plant Soil 248:43–59

Hoyt PB, Turner RC (1975) Effects of organic materials added to a very acid soil on pH, aluminium, exchangeable NH_4^+, and crop yield. Soil Sci 119:227–237

Hue NV (1992) Correcting soil acidity of a highly weathered ultisol with chicken manure and sewage-sludge. Commun Soil Sci Plant Anal 23:241–264

Hue NV, Amien I (1989) Aluminium detoxification with green manures. Commun Soil Sci Plant Anal 20:1499–1511

Hue NV, Licudine D (1999) Amelioration of subsurface acidity through surface application of organic manures. J Environ Qual 28:623–632

Hue NV, Craddock GR, Adams F (1986) Effect of organic acids on aluminium toxicity in subsurface soils. Soil Sci Soc Am J 50:28–34

Imbufe IU, Patti AF, Surapaneni A, Jackson R, Webb JA (2004) Effects of brown coal derived materials on pH and electrical conductivity of an acidic vineyard soil. In: Singh B (ed) SuperSoil 2004: 3rd Australian New Zealand soils conference, University of Sydney, Australia, 5–9 Dec 2004

Imbufe IU, Patti AF, Burrow D, Surapaneni A, Jackson WR, Milner A (2005) Effects of potassium humate on aggregate stability of two soils from Victoria, Australia. Geoderma 125:321–330

Inoue K, Kondo S, Tamano Y, Yokota H (2001) Amelioration of subsurface soil acidity in a nonallophanic Andosol by surface application of organic calcium salts. Soil Sci Plant Nutr 47:113–122

Jarvis SC, Hatch DJ (1985) Rates of hydrogen efflux by nodulated legumes grown in solution culture with continuous pH monitoring and adjustment. Ann Bot 55:41–51

Jones DL, Darrah PR (1994) Amino-acid influx and efflux at the soil-root interface of *Zea Mays* L. and its implications in the rhizosphere. Plant Soil 163:1–12

Kochian LV, Hoekenga OA, Pineros MA (2004) How do crop plants tolerate acid soils? – mechanisms of aluminum tolerance and phosphorous efficiency. Annu Rev Plant Biol 55: 459–493

Loss SP, Ritchie GSP, Robson AD (1993a) Effect of lupins and pasture on soil acidification and fertility in Western Australia. Aust J Exp Agric 33:457–464

Loss SP, Robson AD, Ritchie GSP (1993b) H^+/OH^- Excretion and nutrient uptake in upper and lower parts of lupin (*Lupinus angustifolius* L.) root systems. Ann Bot 72:315–320

Lund ZF, Doss BD (1980) Coastal bermudagrass yield and soil properties as affected by surface-applied dairy manure and its residue. J Environ Qual 9:157–162

Mahoney GP, Uren NC (1982) The long term effect of lime on soil pH. In: Norman MJT (ed) Australian Agronomy Conference, Wagga Wagga, Australia

Marschner H (1995) Mineral nutrition of higher plants, 2nd edn. Academic, London

Marschner B, Noble AD (2000) Chemical and biological processes leading to the neutralization of acidity in soil incubated with litter materials. Soil Biol Biochem 32:805–813

Marx M, Marschner B, Nelson P (2002) Short-term effects of incubated legume and grass materials on soil acidity and C and N mineralisation in a soil of north-east Australia. Aust J Soil Res 40:1231–1241

McLay CDA, Ritchie GSP, Porter WM (1994a) Amelioration of subsurface acidity in sandy soils in low rainfall regions. I. Responses of wheat and lupins to surface-applied gypsum and lime. Aust J Soil Res 32:835–846

McLay CDA, Ritchie GSP, Porter WM, Cruse A (1994b) Amelioration of subsurface acidity in sandy soils in low rainfall regions. II. Changes to soil solution composition following the surface application of gypsum and lime. Aust J Agric Res 32:847–865

Mengel K, Kirkby EA, Kosegrten H, Appel T (2001) Principles of plant nutrition. Kluwer Academic, Dordrecht

Miller AJ, Cramer MD (2004) Root nitrogen acquisition and assimilation. Plant Soil 274:1–36

Moody PW, Aitken RL (1997) Soil acidification under some tropical agricultural systems.1. Rates of acidification and contributing factors. Aust J Soil Res 35:163–173

Moraes MF, Cantarella H, Quaggio JA, Coscione AR (2007) Ion mobility in acid soils with surface application of lime, organic acid and crop residues. Rev Bras Ciênc Solo 31:673–684

Murphy DV, Sparling GP, Fillery IRP (1998) Stratification of microbial biomass C and gross N mineralisation with soil depth in two contrasting Western Australian agricultural soils. Aust J Soil Res 36:45–55

Naramabuye FX, Haynes RJ (2006) Effect of organic amendments on soil pH and Al toxicity and use of laboratory indices to predict their liming effect. Soil Sci 171:754–763

Neumann G, Römheld V (2002) Root induced changes in the availability of nutrients in the rhizosphere. In: Waisel Y, Eshel A, Kafkafi U (eds) Plant Roots: the hidden half. Marcel Dekker, New York

Noble AD, Randall PJ (1998) Ameliorating acidic soils with organic materials: effect of amendments derived from coal on yield and composition of young wheat plants grown on an acid red podzol. Commun Soil Sci Plant Anal 29:3023–3043

Noble AD, Zenneck I, Randall PJ (1996) Leaf litter ash alkalinity and neutralisation of soil acidity. Plant Soil 179:293–302

Noble AD, Bramley RGV, Wood PM, Hurney AP (1997) Sugarcane and soil acidity- why should we be worried? In: Proceedings of Australian Society of Sugarcane Technologists. pp 187–199

Noble AD, Suzuki S, Soda S, Ruaysoongnern S, Berthelsen S (2008) Soil acidification and carbon storage in fertilized pastures of Northeast Thailand. Geoderma 144:248–255

Oates KM, Caldwell AG (1985) Use of by-product gypsum to alleviate soil acidity. Soil Sci Soc Am J 49:915–918

Paul KI, Black AS, Conyers MK (2003) Development of acidic subsurface layers of soil under various management systems. Adv Agron 78:187–214

Pinkerton A, Simpson JR (1986) Interactions of surface drying and subsurface nutrients affecting plant growth on acidic soil profiles from an old pasture. Aust J Exp Agric 26:681–689

Porter WM (1984) Causes of acidity. J Agric West Aust 25:119–120

Porter WM, McLay CDA, Dolling PJ (1995) Rates and sources of acidification in agricultural systems of southern Australia. In: Date RA et al (eds) Plant soil interactions at low pH. Kluwer Academic, Dordrecht

Poss R, Smith CJ, Dunin FX, Angus JF (1995) Rate of soil acidification under wheat in a semi-arid environment. Plant Soil 177:85–100

Reeve NG, Sumner EM (1972) Amelioration of subsurface soil acidity in Natal Oxisols by leaching of surface applied amendments. Agrochemophysica 4:1–6

Ridley AM, White RE, Helyar KR, Morrison GR, Heng LK, Fisher R (2001) Nitrate leaching loss under annual and perennial pastures with and without lime on a duplex (texture contrast) soil in humid southeastern Australia. Eur J Soil Sci 52:237–252

Ritchie GSP (1989) The chemical behavior of aluminium, hydrogen and manganese in acid soils. In: Sobson AD (ed) Soil acidity and plant growth. Academic, London

Ritchie GSP, Dolling PJ (1985) The role of organic matter in soil acidification. Aust J Soil Res 23:569–576

Robinson D (1994) The responses of plants to non-uniform supplies of nutrients. New Phytol 127:635–674

Rukshana F, Butterly CR, Baldock JA, Tang C (2011) Model organic compounds differ in their effects on pH changes of two soil differing in initial pH. Biol Fert Soil 47:51–62. doi:10.1007/s00374-010-0498-0

Scott BJ, Conyers MK, Poile GJ, Cullis BR (1997) Subsurface acidity and liming affect yield of cereals. Aust J Agric Res 48:843–854

Scott BJ, Fisher JA, Cullis BR (2001) Aluminium tolerance and lime increase wheat yield on the acidic soils of central and southern New South Wales. Aust J Exp Agric 41:523–532

Shainberg I, Sumner ME, Miller WP, Farina MPW, Pavan MA, Fey MV (1989) Use of gypsum on soils: a review. In: Stewart BA (ed) Advances in soil science. Springer, New York

Smith CJ, Peoples MB, Keerthisinghe G, James TR, Garden DL, Tuomi SS (1994) Effect of surface application of lime, gypsum and phosphogypsum on the alleviating of surface and subsurface acidity in a soil under pasture. Aust J Soil Res 32:995–1008

Sumner ME (1990) Gypsum as an ameliorant for the subsoil acidity syndrome. Floride Institute of Phosphate Research, Bartow, Florida

Sumner ME (1993) Gypsum and acid soils: the world scene. Adv Agron 51:1–32

Sumner ME (1995) Amelioration of subsurface soil acidity with minimum disturbance. In: Jayawardane NS, Stewart BA (eds) Subsurface soil management techniques, Advances in soil science series. CRC Press, Florida

Sumner ME, Shahandesh H, Bonton J, Hammel J (1986) Amelioration of acid soil profile through deep liming and surface application of gypsum. Soil Sci Soc Am J 50:1254–1258

Taiz L, Zeiger E (2006) Plant physiology, 4th edn. Sinauer Associates Inc, Sunderland

Tang C, Rengel Z (2003) Role of plant cation/anion uptake ratio in soil acidification. In: Rengel Z (ed) Handbook of soil acidity. Marcel Dekker, Inc, New York

Tang C, Yu Q (1999) Impact of chemical composition of legume residues and initial soil pH on pH change of a soil after residue incorporation. Plant Soil 215:29–38

Tang C, McLay CDA, Barton L (1997) A comparison of proton excretion of twelve pasture legumes grown in nutrient solutions. Aust J Exp Agric 37:563–570

Tang C, Sparling GP, McLay RC (1999) Effect of short-term legume residue decomposition on soil acidity. Aust J Soil Res 37:561–573

Tang C, Raphael C, Rengel Z, Bowden JW (2000) Understanding subsurface soil acidification: effect of nitrogen transformation and nitrate leaching. Aust J Soil Res 38:837–849

Tang C, Rengel Z, Abrecht D, Tennant D (2002) Aluminium-tolerant wheat uses more water and yields higher than aluminium-sensitive one on a sandy soil with subsurface acidity. Field Crop Res 78:93–103

Tang C, Asseng S, Diatloff E, Rengel Z (2003a) Modelling yield losses of aluminium resistant and aluminium-sensitive wheat due to subsurface soil acidity: effects of rainfall, liming and nitrogen application. Plant Soil 254:349–360

Tang C, Rengel Z, Diatloff E, Gazey C (2003b) Response of wheat and barley to liming sandy soil with subsurface soil acidity. Field Crop Res 80:235–244

Tang C, Conyers MK, Nuruzzaman M, Poile GJ, Liu DL (2011) Biological amelioration of subsurface soil acidity through managing nitrate uptake by wheat crops. Plant Soil 338: 383–397

van Beusichem ML (1981) Nutrient absorption by pea plants during dinitrogen fixation. 1. Comparison with nitrate nutrition. Netherl J Agric Sci 29:259–273

van Beusichem ML, Kirkby EA, Bass R (1988) Influence of nitrate and ammonium nutrition and the uptake, assimilation and distribution of nutrients in *Ricinus communis*. Plant Physiol 86:914–921

van der Watt HVH, Barnard RO, Conje IJ, Dekker J, Croft GJB, van der Walt MM (1991) Amelioration of subsurface soil acidity by application of coal derived calcium fulvate to the soil surface. Nature 350:146–148

von Willert FJ, Stehouwer RC (2003) Compost and calcium surface treatment effects on subsurface soil chemistry in acidic minespoil columns. J Environ Qual 32:781–790

Weligama C, Tang C, Sale PWG, Conyers MK, Liu DL (2008) Localised nitrate application together with phosphorus enhances root proliferation of wheat and maximises rhizosphere alkalization in acid subsurface soil. Plant Soil 312:101–115

Weligama C, Sale P, Conyers MK, Liu DL, Tang C (2010a) Nitrate leaching stimulates subsurface root growth of wheat and increases rhizosphere alkalization in a highly acidic soil. Plant Soil 328:119–132

Weligama C, Tang C, Sale P, Conyers MK, Liu DL (2010b) Application of nitrogen in NO_3^- form increases rhizosphere alkalization in the subsurface soil layers in an acid soil. Plant Soil 333:403–416

Whitten MG, Wong MTF, Rate AW (2000) Amelioration of subsurface acidity in the south-west of Western Australia: downward movement and mass balance of surface-incorporated lime after 2–15 years. Aust J Soil Res 38:711–728

Williams C (1980) Soil acidification under clover pasture. Aust J Exp Agric 20:561–567

Wong MTF, Swift RS (2003) Role of organic matter in alleviating soil acidity. In: Rengel Z (ed) Handbook of soil acidity. Marcel Dekker, Inc, New York

Wright RJ, Hern JL, Baligar VC, Bennett OL (1985) The effect of surface applied soil amendments on barley root-growth in an acid subsurface soil. Comm Soil Sci Plant Anal 16:179–192

Xu RK, Coventry D (2003) Soil pH changes associated with lupin and wheat plant materials incorporated in a red-brown earth soil. Plant Soil 250:113–119

Xu JM, Tang C, Chen Z (2006a) Chemical composition controls residue decomposition in soils differing in initial pH. Soil Biol Biochem 38:544–552
Xu JM, Tang C, Chen Z (2006b) The role of plant residues in pH changes of acid soils differing in initial pH. Soil Biol Biochem 38:709–719
Yan F, Schubert S, Mengel K (1996) Soil pH changes during legume growth and application of plant matter. Boil Fertil Soil 23:236–242
Zhang H, Forde BG (2000) Regulation of *Arabidopsis* root development by nitrate availability. J Exp Bot 51:51–59

Chapter 14
Microbial Indicators of Soil Quality in Upland Soils

Philip C. Brookes, Juan C. Aciego Pietri, Yuping Wu, and Jianming Xu

Abstract The concept of soil quality is very difficult to satisfactorily define. Possibly a definition based upon the suitability of a soil to perform the purpose for which it is required, i.e. function, while having limitations, is suitable in most circumstances. Certainly, some of the most important determinants of soil quality include the ability of a soil to decompose plant and animal residues, maintain adequate nutrient and soil organic matter pools and to act as a filtering system to supply pure water to our rivers, lakes and ground waters. The activities of the soil micro-organisms (i.e. the fungi, bacteria, protozoa, yeasts etc.) drive these processes. Any factors which inhibit their activities therefore have large effects on soil fertility, and therefore soil quality. Linking the activities of one, or at best a few, species of soil micro-organism to soil quality is normally an impractical proposition. Another problem is that soil microbial activity, resulting for example, in carbon dioxide evolution from soil, is the net result of complex substrate-microbial interactions, again involving many different microbial populations. To overcome this, we have developed a number of methods to measure the soil micro-organisms as a single, undifferentiated, unit, or 'black box' – termed the microbial biomass. Other methods can indicate the activity of this biomass. While having obvious limitations, this approach has enabled us to detect the direction of change

P.C. Brookes (✉) • J.C.A. Pietri
Soil Science Department, Rothamsted Research, Harpenden, Hertfordshire, AL5 2JQ, UK
e-mail: philip.brookes@bbsrc.ac.uk

J.C.A. Pietri
Agriculture Faculty, Central University of Venezuela, Maracay, Aragua State, Ap. 4579, Venezuela
e-mail: juanaciego@gmail.com

Y. Wu • J. Xu
Institute of Soil and Water Resources and Environmental Science, Zhejiang University, Hangzhou 310058, China
e-mail: a-angelfish@163.com; jmxu@zju.edu.cn

J. Xu and D.L. Sparks (eds.), *Molecular Environmental Soil Science*,
Progress in Soil Science, DOI 10.1007/978-94-007-4177-5_14,

(i.e. decrease or increases) in soil organic matter and in soil ecosystem functioning due, for example to increases in heavy metal concentrations, changes in pH or to inputs of crop residues. The value of the microbial biomass as an indicator, or 'early warning' of changing soil conditions, or changing soil quality, will be illustrated and discussed.

Keywords Soil quality • Soil pH • Soil microbial biomass • Microbial activity

14.1 The Concept of Soil Quality

Chemical, physical and biological characteristics of air and water can mainly be precisely and accurately defined because of their comparatively simple structure and composition. In contrast, soil is a much more complex medium, encompassing air, water and the living and non-living fractions of the soil itself. Its very complexity and variability, within and between both space and time, still largely defy anything like a complete analysis of chemical, biological and physical properties, despite many attempts to do so. It therefore follows that a single, universal, definition of 'soil quality' is doomed to failure from the start.

Certainly, observing a decline in soil quality may be easier, in many cases, than actually defining what comprises good soil quality. Thus, since the early 1980s, there has been a decline in soil productive capacity in more than 10% of the world's cultivated land. This is attributed to a number of different environmental impacts e.g. erosion, atmospheric pollution, enhanced grazing pressure, salinisation and desertification (e.g. Van-Camp et al. 1987). However, despite several international attempts, there is still not a definitive set of agreed basic indicators for the evaluation This is directly attributed to the ever continuing difficulty in defining soil quality and lack of agreement on analytical methods and their standardisation so that it may be assessed on an international, or even national (Bloem et al. 2005).

Despite this, the search goes on. Possibly the definition that still best represents the concept of soil quality was that given by Doran and Parkin (1994):

> ··· the capacity of the soil to interact with the ecosystem in order to maintain biological productivity, environmental quality and to promote animal and plant *health.*

The Rodale Institute (1991) introduced a similar concept for soil quality, based upon three essential criteria, i.e.

1. *Productivity* — the soil's capacity to increase plant biological productivity.
2. *Environmental quality* — the soil's capacity to attenuate environmental contamination, pathogens and external damage.
3. *Health of living organisms* — the inter-relationship between soil quality and animal, plant and human health (from Bloem et al. 2005).

14.2 Parameters to Assess Soil Quality

These can be divided into physical, chemical and biological. Physical and chemical properties are less variable within a site than biological ones which may change by orders of magnitude even as a result of only comparatively small environmental fluctuations.

This paper is concerned with biological parameters and then microbial ones only. Because they are in intimate contact with the soil and have both mass and activity, soil micro-organisms are, in many ways, ideal indicators of soil functioning. However, any soil property, be it chemical, physical or biological, operates within a range which can be considered 'normal' and some stresses, e.g. air-drying-rewetting of soil, may cause activity to fluctuate over several orders of magnitude. This can make it very difficult to decide if an observed parameter is indicating normal functioning in a particular soil or is responding to an external stress, so that 'soil quality' may be considered to be impaired. This problem was first observed and a solution offered by Domsch (1980) and Domsch et al. (1983) when considering the effects of pollutants, especially pesticides, on the soil micro-organisms. They first considered the effects of naturally occurring stresses on soil microbial populations and their activities, including fluctuations in temperature, extremes of water potential, pH extremes, physical disturbances to soil (e.g. ploughing or minimum till), changes in gaseous exchange or nutrient supply and changes in predator-prey balances. Clearly, any of these phenomena, singly or jointly, may be readily detectable by current methodologies and could, quite wrongly, be considered to indicate significant changes in soil functioning and so, according to the definitions above, indicate changes in soil quality. The question is, what measurable changes should be treated with concern and which should not, when considering natural soil functioning? In an attempt to resolve this, Domsch et al. (1983) analysed 55 published fluctuations in bacterial populations under field conditions and found that up to 90% of fluctuations occurred naturally. On this basis they suggested that any measured parameter(s) that returned to a 'natural' baseline within 30 days of being measured was normal, those involving delays of 60 days were tolerable while those that took longer than 60 days were taken to indicate that further investigation was required (Fig. 14.1). While this study was undertaken to provide some sort of standardisation for determining possible damage to the microbial population in soil due to pesticide applications it illustrates the crucial point that fluctuations in microbial populations within a very wide range are completely normal. This therefore indicates, in turn, that soil quality indicators, be they based upon microbial population, activity measurements or both, must be interpreted in similarly robust ways.

This was also clearly demonstrated by Cook and Greaves (1983) who provided an excellent example of the natural variation in biological activity which may occur over a 12-month period. They directly measured CO_2 evolution from a field plot for 12 months. They found that CO_2 evolution was at a maximum (around 950 $\mu g\ cm^{-3}$ soil day^{-1}) under the warm wet conditions of summer but had declined to less than

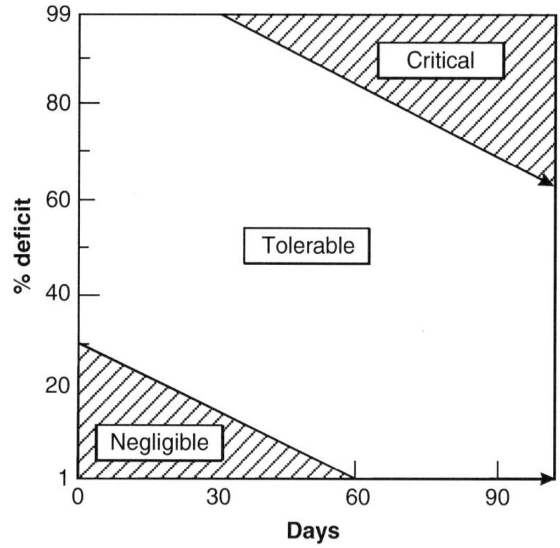

Fig. 14.1 Relationship between monitoring period and changes in biological activity (deficit) in three categories describing the ecological significance of injuries (From Domsch et al. 1983. Reproduced with permission from the publisher)

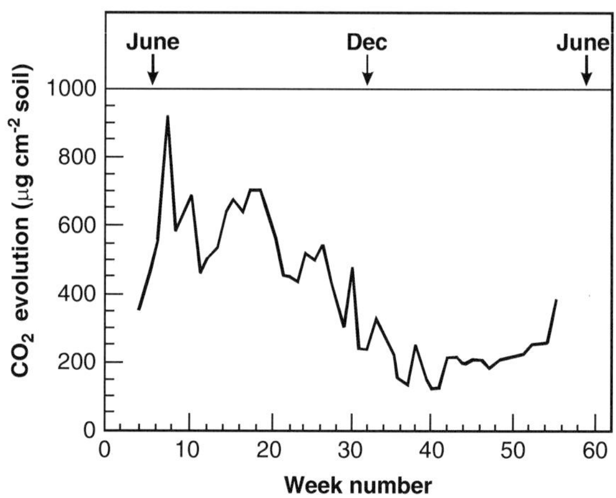

Fig. 14.2 Carbon dioxide evolved from a UK field over 12 months (From Cook and Greaves 1983. Reproduced with permission of the publisher)

50% of this by December of the same year (Fig. 14.1). Nitrogen mineralisation followed a similar pattern. Both properties showed a long stationary phase of 15–20 weeks into the next spring, then activity recommenced. Clearly, both C and N mineralisation, the products of natural biological activity, are changing markedly due to changing moisture and temperature. However there is no indication that these changes are induced by changes in soil quality. Therefore, indicators of soil quality have to be sufficiently robust to deal with natural variations in ecosystem functioning of this magnitude. This is precisely why the concept introduced by Domsch (1980) is so important and why its philosophy should be incorporated into any attempts to adopt and standardise methods to quantify soil quality.

14.3 Criteria to Be Used in Selection of Possible Microbiological Properties as Indicators of Soil Quality

There are a number of basic criteria that a microbiological property might be expected to fulfill as an indicator of soil quality (modified from Brookes 1995).

1. The property needs to be accurately and precisely measureable across a wide range of soil types and soil conditions.
2. Because a large number of samples usually have to be analysed it is preferable that the property can be easily and economically measured.
3. The property needs to be one that also permits a control, or background, measurement to be made so that changes in soil quality can be precisely determined.
4. The property needs to be sensitive enough to indicate genuine changes in soil quality, however defined, but also sufficiently robust not to give false signals.
5. The property needs scientific validity based upon reliable and contemporary scientific knowledge.
6. Reliance upon a single property may be unsafe. Two or more, preferably independent, should be selected. However, their interrelationships in a range of environments should be previously established.

14.4 Possible Microbial Indicators of Soil Quality

Microbial parameters as indicators of soil quality fall into three main groups. The first measures the microbial population at the single organism level, at the functional group level or at the whole population level The second type measures the activity of the entire population e.g. soil organic N or C mineralisation and the third type measures a combination of both activity and population, giving specific activities. Examples are given in Table 14.1.

Table 14.1 Possible physicochemical, biological and biochemical indicators of soil quality

Physicochemical properties	*Microbial population*
Organic C	Microbial biomass C
Total N	Microbial Ninhydrin-N
Inorganic N	ATP
pH	
Microbial activity	*Microbial community structure*
CO_2 evolution; N mineralisation	PLFAs
Microbial-specific respiration (qCO_2)	FAMEs
Substrate-induced respiration (SIR)	PHB
	Ergosterol
	DNA

The use of micro-organisms following growth in culture media is rejected completely. Pure cultures may be atypical in form or physiology when grown in artificial media as they are isolated from their normal ecological environment. In view of this, interpretation of results and extrapolation to field conditions will be largely meaningless. Similarly the use of soil enzymes may be questionable as the total enzymic enzyme capacity of a soil may be a function of a number of fractions e.g. living microbial cells, recently dead microbes and other organisms, extracellular stabilised enzymes from a number of sources etc.

14.5 Materials and Methods

14.5.1 Soils, Sampling and Preparation

The soil of Hoosfield, at Rothamsted Research, U.K. is classified as Typic Paleudalf (U.S.D.A., 1992) or Chromic Luvisol (F.A.O., 1989). Such soils were originally acidic, well-drained to moderately well-drained and developed in a relatively silty (loess-containing) superficial deposit overlaying, and mixed with, clay-with-flints. The topsoil is a flinty, silty clay loam (18–27% clay).

Hoosfield has been under arable management since before the nineteenth century. It probably received only a single heavy (150–250 $t\,ha^{-1}$) dressing of chalk initially. The soil from the site which we studied (the Hoosfield acid strip) has not received any other amendment including chemical or organic fertilizer since then (A. E. Johnston and P. R. Poulton, Personal Communication). At that time, chalk dressings were dug from infield 'bell-pits' or from 'dell-holes' on bordering slopes and spread by hand to improve the fertility and workability of the originally acid soils.

By the 1950s, reserves of $CaCO_3$ remaining from earlier application had become exhausted by leaching at further distances from the original chalk pits, where the original dressings became increasingly smaller as the effort to carry the chalk increased. Here the soil had become acidic (Avery and Catt 1995).

In our work, soils were sampled along a 200 m soil transect of the Hoosfield acid strip. At each core position, 20×5 cm diameter cores (0–23 cm depth) were taken across the strip and bulked in the field. Initially, a bulk sample was taken every 10 m, increasing to 5 m at between 50 and 100 m, giving a total of 27 bulked samples. The 5 m samples were taken over the distance where it was anticipated that pH changes would be greatest. All the soil samples were then sieved separately (<2 mm) in the laboratory and dry matter contents determined (105 °C, 24 h). The soils were then adjusted to 40% water-holding-capacity (WHC) and pre-incubated for 7 days at 25 °C to allow microbial activity to decline after the initial disturbance (Jenkinson 1988). Other <2 mm sieved small sub-samples were air-dried and ground (<180 μm) for chemical analysis.

14.5.2 Soil Chemical Analyses

Soil pH was measured at a soil:water ratio of 1:2.5 (weight/weight). Air-dry soil (10 g, <2 mm) and 25 ml of deionised water were shaken together for 1 min, left to settle for 30 min, repeating this procedure once more, then pH determined with a pH electrode. The total C and total N measurements were done by dry combustion using a Leco CNS-2000 autoanalyser. The $CaCO_3$ concentrations in the soils were determined with an Eijkelkamp Calcimeter (Williams 1948). Exchangeable Al and Mn concentrations in 0.5 M K_2SO_4 extracts of unfumigated soil (previously diluted in water, 1:100) were determined by inductively coupled plasma-optical emission spectrometry (ICP, Applied Research Laboratories Accuris, Fisons Instruments).

14.5.3 Soil Microbial Analyses

14.5.3.1 Microbial Biomass

Soil microbial biomass C measurements were made by the fumigation-extraction method (Vance et al. 1987). Organic C in the soil extracts was measured by an automated procedure (Wu et al. 1990), using a Total Carbon Analyser (TOC 200). Carbon evolved as carbon dioxide (CO_2-C) was measured by automated infrared analysis. Soil microbial biomass C (Bc) was calculated from: Bc = 2.22 Ec, where Ec = [(organic C extracted from fumigated soil) minus (organic C extracted from non-fumigated soil)] and expressed as $\mu g\ C\ g^{-1}$ soil.

K_2SO_4-extractable ninhydrin nitrogen was measured by the method of Amato and Ladd (1988) as modified by Joergensen and Brookes (1990). Soil microbial biomass ninhydrin-nitrogen (B_{NIN}) was calculated from: B_{NIN} = [(ninhydrin-N extracted from fumigated soil) minus (ninhydrin-N extracted from non-fumigated soil)] and expressed as μg ninhydrin-N g^{-1} soil.

14.5.3.2 Microbial Activity

Carbon evolved as CO_2-C was determined by trapping CO_2 in 1 M NaOH (25 ml) and titrating with standardised 1 M HCl (Tinsley et al. 1951). Each moist soil portion (50 g on an oven-dry soil basis) was put into a glass jar (100 ml) which was then placed in a 1 l brown glass jar with 25 ml 1 M NaOH in a vial and 10 ml free water. The jar was sealed with a rubber bung. Three blank replicate treatments of jars without soil were also prepared. After the desired period of incubation, each NaOH sample (5 ml) was mixed with water (10 ml) and titrated from pH 8.3 to 3.7 with standardised 0.5 M HCl using a Radiometer Copenhagen TTT80 Titrator. The qCO_2 quotient was calculated from: $qCO_2 = (CO_2\text{-C respired time}^{-1})/(\text{microbial biomass C})$, and expressed as mg CO_2-C evolved g^{-1} Bc day^{-1}.

14.5.4 Statistical Analyses

All microbial and chemical variables measured along the Hoosfield acid strip were graphically fitted against soil pH using the best curve model. Correlation analysis between biological and chemical variables was done. For these analyses the statistical package GenStat® 2000 for Windows (5th Edition) was used. All measurements are the mean of triplicate determinations.

14.6 Results and Discussion

14.6.1 The Hoosfield Acid Strip Soil pH Gradient

Here we summarise recent data from the Hoosfield acid strip at Rothamsted Research (Aciego Pietri and Brookes 2008), in relation to soil quality. This experiment has a continuous range of soil pHs from about pH 8 to 4 on a single soil type (Chromic Luvisol), a rare occurrence as most soil pH gradients cover several soil types. The large and continuous pH range arises as a result of uneven calcium carbonate applications in the nineteenth century. This pH gradient appears to be one of the best published so far. The changing soil pH caused considerable changes in concentrations of soil organic C and total N, with the largest concentrations and C/N ratio around pH 7. Soil $CaCO_3$ was greatest at pH 7.5, declined to pH 7 then remained at negligible concentrations below this until pH 3.7 (results not given) (Figs. 14.3 and 14.4).

14.6.2 Microbial Properties Along the Hoosfield Acid Strip Soil pH Gradient

Microbial biomass C and ninhydrin N concentrations were maximal at above pH 7. Both biomass C and ninhydrin N were positively correlated with soil pH. However,

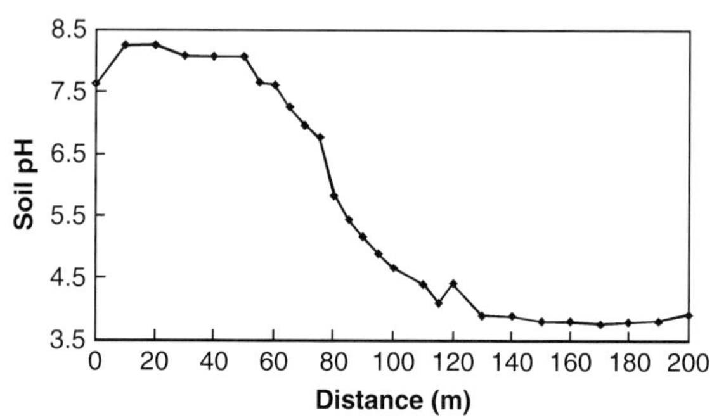

Fig. 14.3 pH in soils from Hoosfield acid strip

Fig. 14.4 Effect of soil pH on: (**a**) organic C, (**b**) total N, and (**c**) organic C/total N ratio in soils from Hoosfield acid strip

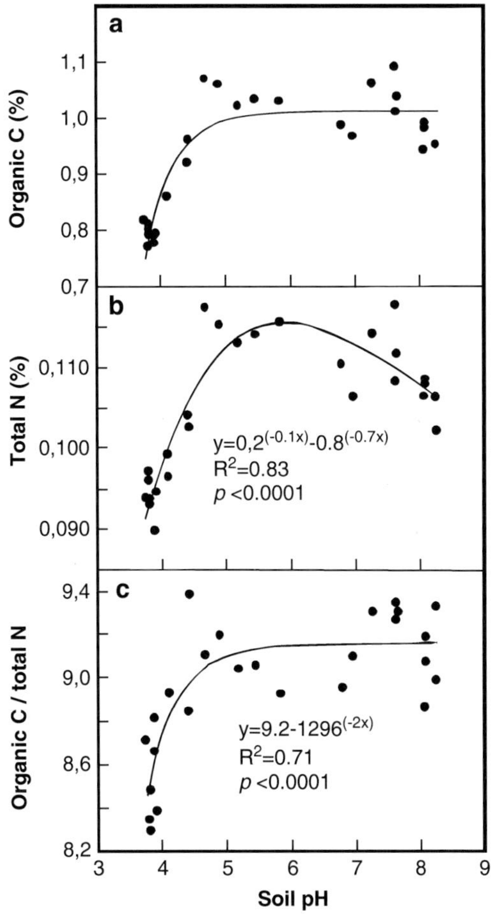

biomass C data was best fitted to a exponential equation ($R^2 = 0.80$) while the relationship with ninhydrin N was more complicated (i.e. a positive polynomial: $R^2 = 0.90$) (Figs. 14.5 and 14.6).

The biomass specific respiration quotient (qCO_2) ranged from about 15 to 66 mg CO_2-C g^{-1} microbial C d^{-1} (Fig. 14.4d). A polynomial regression ($R^2 = 0.33$, $p < 0.001$, curve not shown) fitted the relationship between soil pH and biomass specific respiration rate. However, in this case the specific respiration rates were maximal at both low and high extremes of pH.

Biomass C was closely correlated with pH (r = 0.81), organic C (r = 0.84) and total N (r = 0.80), while biomass ninhydrin-N was better correlated with pH (r = 0.89) than with organic C (r = 0.79) and total N (r = 0.71). The CO_2 evolution was slightly more correlated with organic C (r = 0.90) than with pH (r = 0.84). Above about pH 5.7, K_2SO_4-extractable Al was very small but below this pH a rapid linear increase in extractable Al occurred.

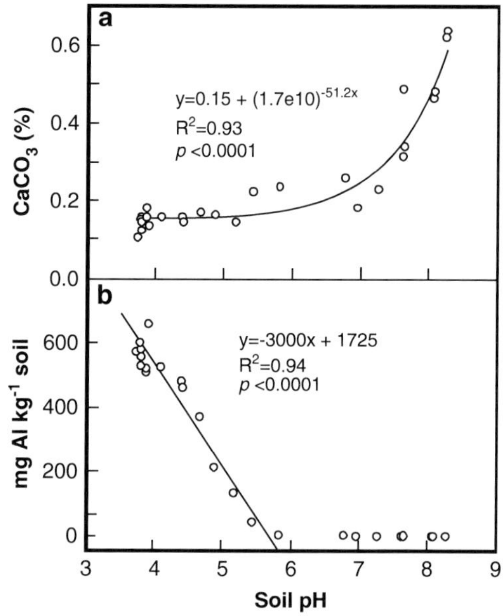

Fig. 14.5 Relationship between soil pH and: (**a**) $CaCO_3$ concentration, (**b**) Al concentration in soils from Hoosfield acid strip

14.7 Phospholipid Fatty Acids

Microbial community structure was evaluated by phospholipids fatty acids (PLFAs) in unamended soil at 0, 5, 25 and 50 days incubation at 25°C and 40% Water Holding Capacity, and the CO_2 evolution measurement was extended up to 100 days. The fatty acids 16:1ω5, 16:1ω7c, 18:1ω7c&9 t and i17:0 had significant and positive linear relationships with soil pH. In contrast, the fatty acids i15:0, a15:0, i16:0 and br17:0, 16:02OH, 18:2ω6,9, 17:0, 19:0, 17:0c9,10 and 19:0c9,10 were greatest at the lowest pHs. The PLFA profiles indicated that Gram-negative bacteria tended to increase and Gram-positive bacteria tended to decrease with increasing pH. The ratio fungal PLFA 18:2ω6,9/bacterial PLFAs indicated that fungal biomass was largest at the lowest pH.

14.8 An Approach to Evaluate Soil Quality Indicators Under Different Soil Uses

Soils have many uses. Producing food and energy crops are obvious ones but they are also critical in many other situations, e.g. buildings and roads, parks, sport fields and other recreational areas, providing drainage and supporting natural habitats for wildlife. Soils also vary naturally in pH, as in most or all other physical,

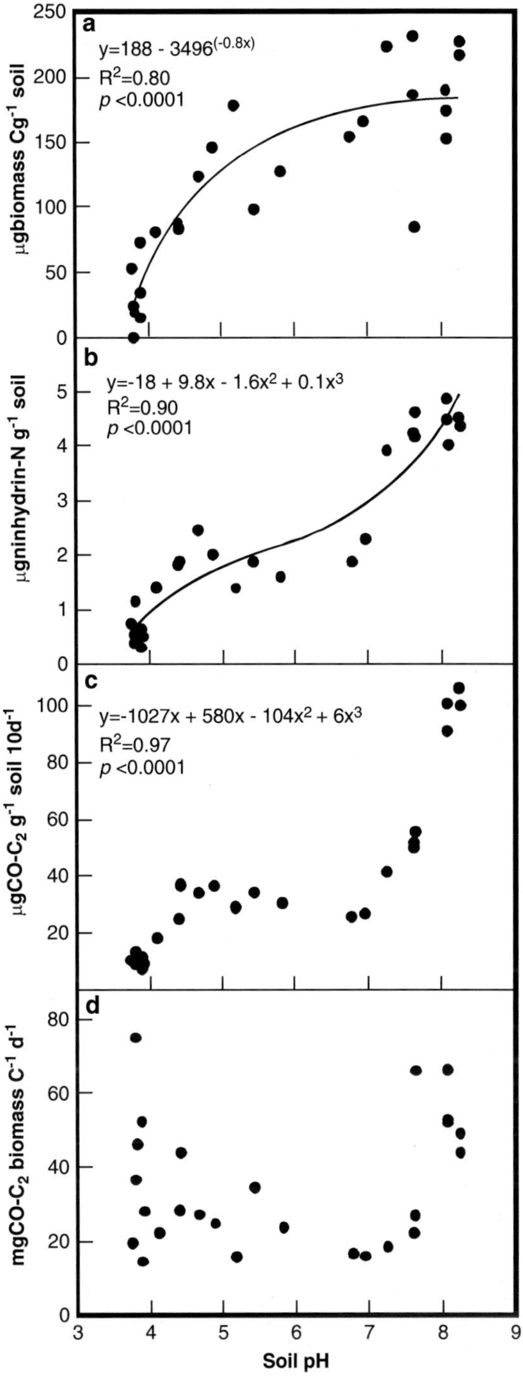

Fig. 14.6 Effect of soil pH on: (**a**) microbial biomass C, (**b**) microbial biomass ninhydrin-N, (**c**) evolved CO_2-C, (**d**) microbial specific respiration in soils from Hoosfield acid strip

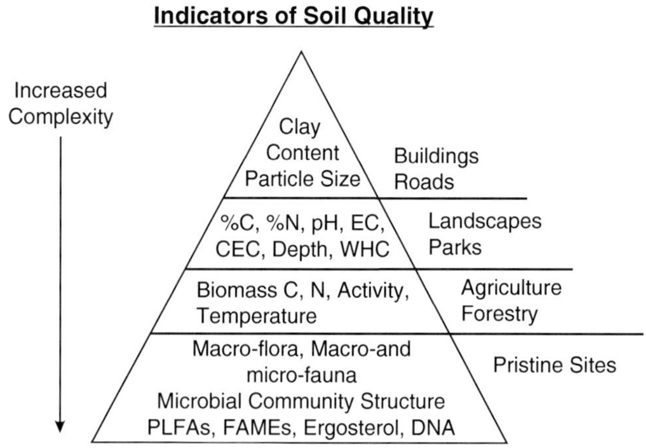

Fig. 14.7 Selecting soils as 'best fitted for purpose'

chemical and biological characteristics. A given soil may range from calcareous to highly acidic, each supporting a plant-animal microbe community which may have adapted to these specific conditions over many thousands, or even tens of thousands, of years. For example, if a given soil is to support a road, an ability to withstand compaction (i.e. aggregate stability) is likely to be of much more value than ensuring that the soil is exhibiting a high degree of microbial biodiversity. However, soil physical or chemical conditions may be vital considerations when growing different crops. Consider, for example, the tea plant, *Camellia senensis*, grown widely throughout Asia. It thrives at pH 4.5, requiring large concentrations of available Al. In contrast the grass we know as wheat, *Triticum* sp., has a soil pH optimum around 6.5–7.5. If it survived at all at pH 4 .5 the yield would be much impaired if it occurred at all. Which soil, therefore, is of better quality?

We depend upon soil for so many different things, many of which are certainly not interchangeable. For example, to produce food and energy crops requires a very different soil quality to one required to support buildings and roads, parks, sportfields and other recreational areas, or a soil supporting natural habitats for wildlife, or another providing drainage. To apply a universal indicator of soil quality to these diverse situations would be fairly meaningless. This brings us to the concept of 'best fitted for purpose'. We contend that there is not, nor ever will be, a single soil property that can be considered as a universal indicator of soil quality. We therefore consider that to apply a universal indicator of soil quality to cover the diverse requirements we have of soil is fairly meaningless. A more realistic approach is to decide parameters which indicate 'the best fit for purpose'. Thus it is possible to design a 'hierarchical tree' of soil quality, with increasing complexity of soil quality indicators as the demands for better soil quality increases as the requirements for more complexity of the ecosystem increase (Fig. 14.7).

14.9 Microbial Properties as Indicators of Soil Quality

The use of soil micro-organisms, in terms of biomass, community structure or activity, as shown above, certainly has much to offer in this sort of study. Firstly, soil micro-organisms are literally bathed in soil solution, so provide a good sampling device for detection of pollutants or changes in substrate supply. Secondly, most micro-organisms are largely immobile in soils, being attached to soil particles by a variety of means (e.g. polysaccharide gums or invading hyphae). For example, increases in Al concentration caused by decreasing soil pH (Fig. 14.5) will also certainly directly interact with the biomass due to increased Al concentrations in the soil diffusion stream. Giller et al. (1998) reported that a decrease in soil microbial biomass under chronic metal stress has been observed in many field and laboratory studies but it is likely to be preceded by changes in community structure.

It is very likely that the decrease in microbial biomass in soils from Hoosfield at lower pH values is caused by aluminium toxicity (Blake et al. 1999 and Fig. 14.3) so that the mechanisms operating in the microbial cells are similar to those caused by heavy metal toxicity. The large concentrations of Al at low pHs may decrease inputs of C from plants to the soils and decrease efficiency of conversion of this C into biomass C. These mechanisms were also suggested by Chander and Brookes (1991a, b, c, 1992) in order to explain the toxic effects of heavy metals: (a) directly on microbial biomass and, (b) indirectly, by decreasing inputs of plant-derived substrates (e.g. roots, root exudates). Aluminium has been previously reported to be toxic to micro-organisms and microbial processes, but compared with other metals, less work has been done on this subject.

Biomass C and biomass ninhydrin-N also decreased in UK soils that were metal-contaminated from past applications of contaminated sewage sludge or sludge-containing compost (Brookes and McGrath 1984; Chander and Brookes 1991a, b, c, 1992) or following the recent addition of metal-enriched sewage sludge in laboratory incubations (Chander and Brookes 1993). Similarly, the biomass specific respiration quotient (qCO_2) had the greatest values at high (above 7.5) and very low (below 4) pHs indicating a possible stress condition, i.e. more CO_2-C was respired per unit of microbial biomass C (Fig. 14.4). The qCO_2 quotient indicates the efficiency of the soil microbial biomass in utilising organic substrates during biosynthesis. The diversion of substrate C from biosynthesis to activity under stress conditions, caused, for example, by heavy metals, has been observed previously (e.g. Brookes and McGrath 1984; Chander and Brookes 1991a, b, c). It is believed to be due, for example, to the expenditure of energy to remove heavy metals (or protons in this case) against a concentration gradient, the production of protective chelates etc., and a lower efficiency of enzyme activity.

Similarly, the biomass specific respiration quotient (qCO_2) had the greatest values at high (above 7.5) and very low (below 4) pHs indicating a possible stress condition, i.e. more CO_2-C was respired per unit of microbial biomass C (Fig. 14.4). The qCO_2 quotient indicates the efficiency of the soil microbial biomass in utilising organic substrates during biosynthesis. The diversion of substrate C from biosynthesis to activity under stress conditions, caused, for example, by heavy metals, has been observed previously (e.g. Brookes and McGrath 1984; Chander and Brookes 1991a, b, c). It is believed to be due, for example, to the expenditure of energy to remove heavy metals (or protons in this case) against a concentration gradient, the production of protective chelates etc., and a lower efficiency of enzyme activity. A decrease in soil microbial biomass under chronic metal stress has been observed in other field and laboratory studies but it is likely to be preceded by changes in community structure (Giller et al. 1998). This was also observed in the soils of the Hoosfield acid strip experiment. Thus, the PLFA profiles indicated that Gram-negative bacteria tended to increase and Gram-positive bacteria tended to decrease with increasing pH (results not given).

14.10 The Soil Quality Paradox

The above discussion illustrates the difficulties with which we are confronted when dealing with the issue of soil quality. The lowestl pH of the Hoosfield soil is around 3.7 (Aciego Pietri and Brookes 2008). At this pH, Al is toxic, both to the soil microbial biomass and to the wheat crop which has been planted. However, the acidification is, at least partly, naturally produced. Therefore the increased toxic Al activity is a natural product of the interactions between soil chemistry and pH. Indeed, many calcifuge-based natural soil ecosystems depend upon this reaction for their continued existence. In contrast, the heavy metals Cu, Ni, Cr and Zn were artificially introduced to the soils with industrially produced sewage sludge. They similarly decreased the soil microbial biomass concentration and altered biomass specific respiration. Most people, including the authors, would therefore consider that the quality of these soils have been impaired. However, in terms of soil microbial ecosystem functioning, this might seem to be a difficult case to wholly support when one also considers the situation with Al. Here, quite similar responses by the soil microbial biomass to both Al and to heavy metals were observed at low soil pHs.

14.11 The Meaning of Soil Quality

Despite all of the problems in defining soil quality, there can be little doubt that it is here to stay. The term implies that we do in fact have reliable measurements of soil quality, which may be used to rank one soil against another. As can be seen from the discussion above, this approach falls apart without much effort. In part, soil scientists themselves are responsible for this situation. Grants are offered to cash-strapped research scientists to evaluate parameters, or schemes, for indicating, measuring or predicting soil quality and we are (mainly) happy to place our heads upon the block! Each soil is different, even if it has never been faced with pollutants or grown a single agricultural crop. Therefore there can be no absolute soil quality measurement and each soil needs to be evaluated separately in terms of its natural differences in soil type, texture, chemical, physical and biological properties and land use. Therefore, to assign any 'quality' to a soil is a value judgment rather than a decision based upon rigorous scientific process. Again, the least unsatisfactory definition is one that takes a low key approach along the lines that 'good soil quality' is best defined as the soil being 'well fitted for purpose'. It is best to leave it there.

Acknowledgments Rothamsted Research is an Institute of the UK Biotechnological and Biological Sciences Research Council. We thank P. R. Poulton for helpful discussion. We also thank the Council of Scientific and Humanistic Development (CDCH) of the Central University of Venezuela for financial support.

References

Aciego Pietri JC, Brookes PC (2008) Relationships between soil pH and microbial properties in a UK arable soil. Soil Biol Biochem 40:1856–1861

Amato M, Ladd JN (1988) Assay for microbial biomass based on ninhydrin-reactive nitrogen in extracts of fumigated soils. Soil Biol Biochem 20:107–114

Avery BW, Catt JA (1995) The soil at Rothamsted. Lawes Agricultural Trust (IACR-Rothamsted), Harpenden, 44 pp

Blake L, Goulding KWT, Mott CJB, Johnston AE (1999) Changes in soil chemistry accompanying acidification over more than 100 years under woodland and grass at Rothamsted experimental station. Eur J Soil Sci 50:401–412

Bloem J, Schouten AJ, Soren J, Sorensen MR, Van der Werf A, Bruere M (2005) Monitoring and evaluating soil quality. In: Bloem J, Hopkins DW, Benedetti A (eds) Microbiological methods for evaluating soil quality. CABI Publishing, Wallingford, pp 23–49

Brookes PC (1995) The use of microbial parameters in monitoring soil pollution by heavy metals. Biol Fertil Soils 19:269–279

Brookes PC, Mcgrath SP (1984) The effects of metal toxicity on the soil microbial biomass. J Soil Sci 35:341–346

Chander K, Brookes PC (1991a) Microbial biomass dynamics during the decomposition of glucose and maize in metal-contaminated and non-contaminated soils. Soil Biol Biochem 23:917–925

Chander K, Brookes PC (1991b) Effects of heavy metals from past applications of sewage sludge on microbial biomass and organic matter accumulation in a sandy loam and silty loam UK soil. Soil Biol Biochem 23:927–932

Chander K, Brookes PC (1991c) Plant inputs of carbon to metal-contaminated and non-contaminated soil and effects on the synthesis of soil microbial biomass. Soil Biol Biochem 23:1169–1177

Chander K, Brookes PC (1992) Synthesis of microbial biomass from added glucose in metal-contaminated and non-contaminated soils following repeated fumigation. Soil Biol Biochem 24:613–614

Chander K, Brookes PC (1993) Residual effects of zinc, copper and nickel in sewage sludge on microbial biomass in a sandy loam. Soil Biol Biochem 25:1231–1239

Cook KA, Greaves MP (1983) Natural variability in microbial activities. In: Somerville L, Greaves MP (eds) Pesticide effects on soil microflora. Taylor and Francis, London, pp 15–43

Domsch KH (1980) Interpretation and evaluation of data. In: Recommended tests for assessing the side effects of pesticides on the soil microflora. Technical report no. 59, Weed Research Organisation, Oxford, pp 6–8

Domsch KH, Jagnow G, Anderson TH (1983) An ecological concept for the assessment of side effects of agrochemicals on soil micro-organisms. Residue Rev 86:185–235

Doran JW, Parkin TB (1994) Defining and assessing soil quality. In: Doran JW, Coleman DC, Bezdicek DF, Stewart BA (eds) Defining soil quality for a special environment, Special Publication 35. American Society of Agronomy, Madison, pp 3–21

Giller KE, Witter E, McGrath SP (1998) Toxicity of heavy metals to microorganisms and microbial processes in agricultural soils: a review. Soil Biol Biochem 30:1389–1414

Jenkinson DS (1988) The determination of microbial biomass carbon and nitrogen in soil. In: Wilson JR (ed) Advances in nitrogen cycling in agricultural ecosystems. CAB International, Wallingford, pp 368–386

Joergensen RG, Brookes PC (1990) Ninhydrin-reactive nitrogen measurements of microbial biomass in 0.5 M K_2SO_4 soil extracts. Soil Biol Biochem 22:1023–1027

Rodale Institute (1991) Conference report and abstract. International conference on the assessment and monitoring of soil quality. Pensylvania, USA

Tinsley J, Taylor TJ, Moore JH (1951) The determination of carbon dioxide derived from carbonates in agricultural and biological materials. Analyst 76:300–310

Van-Camp L, Bujarrabal B, Gentle AR, Jones RJA, Montanarella L, Olazabal C, Selvaradjou Vance ED, Brookes PC, Jenkinson DS (1987) An extraction method for measuring microbial biomass C. Soil Biol Biochem 19:703–707

Vance ED, Brookes PC, Jenkinson DS (1987) An extraction method for measuring microbial biomass C. Soil Biol Biochem 19:703–707

Williams DE (1948) A rapid manometric method for the determination of carbonate in soils. Proc Soil Sci Soc Am 13:127–129

Wu J, Joergensen RG, Pommerening B, Chaussod R, Brookes PC (1990) Measurement of soil microbial biomass C by fumigation-extraction – an automated procedure. Soil Biol Biochem 22:1167–1169

Chapter 15
Impacts of Agricultural Management Practices on Soil Quality

Bocar Ahamadou and Qiaoyun Huang

Abstract Soil quality includes assessment of a variety of soil properties and processes as they relate to the ability of soil to function effectively as a component of a healthy ecosystem. A comprehensive understanding of the impact of agricultural management practices on soil physical, chemical and biological properties is of paramount importance since they are key indicators and components of soil quality/health. The current article gives an extensive review on the recent studies of these soil key properties as impacted by agricultural practices such as cultivation, crop rotation, fertilization and pesticide application following the presentation of a sound concept of soil quality. Future research on this extremely challenging and important area should be stimulated to sustain and enhance ecosystem productivity, services, and integrity. To face the challenges of the inherent alteration of human and environmental interactions, several scientific questions for future investigations on soil quality/health are presented.

Keywords Agricultural management • Soil quality • Soil health • Soil biological property • Soil physical property • Soil chemical property

Nowadays, the need for assessing soil physical, chemical and biological properties has expanded because of growing public interest in determining the consequences of agricultural management practices on the quality of soil relative to sustainability of

B. Ahamadou • Q. Huang (✉)
State Key Laboratory of Agricultural Microbiology, Huazhong Agricultural University, Wuhan 430070, China

Q. Huang
Key Laboratory of Subtropical Agriculture and Environment, Ministry of Agriculture, Huazhong Agricultural University, Wuhan 430070, China
e-mail: qyhuang@mail.hzau.edu.cn

J. Xu and D.L. Sparks (eds.), *Molecular Environmental Soil Science*, Progress in Soil Science, DOI 10.1007/978-94-007-4177-5_15,

agroecosystem functions and productivity. In fact, there is increasing concern about some important global issues, which include the increases in human population, energy and food demands, atmospheric CO_2, the decline in per capita availability of cultivable land and renewable fresh water resources. However, the increase in the extent and severity of inappropriate human-induced agricultural management practices with adverse impacts on soil quality/health, ecosystem services and environment are of particular interest.

The concept of soil quality includes assessment of soil properties and processes as they relate to the ability of soil to function effectively as a component of a healthy ecosystem. Specific functions and subsequent values provided by agroecosystems are variable and rely on numerous soil physical, chemical, and biological properties and processes, which can differ across spatial and temporal scales. Choice of a standard set of specific properties as indicators of soil quality can be complex and will vary among agroecosystems and management objectives. Indices of soil quality which incorporate soil chemical, physical, and biological properties will be most readily adopted if they are sensitive to management-induced changes, easily measured, relevant across sites or over time, inexpensive, closely linked to measurement of desired values, and adaptable for specific ecosystems. Globally, the challenge consists firstly in understanding how soil physical, chemical and biological properties are affected by intensive agriculture, and then, in developing new strategies that take advantage of ecological interactions within agricultural systems. While the issues are serious and diverse, common links among them could be soil microbes and soil organic matter as influenced by interactive effects of anthropogenic and natural perturbations. Strategies to address these issues involve a comprehensive understanding of the impact of agricultural management practices on soil physical, chemical and biological properties since they are key indicators and components of soil quality/health. This paper traces the development of the concept of soil quality, explores the use of soil chemical, physical and biological properties as determinants of soil quality as well as the impact of agricultural management practices on these properties. It further elucidates the significance of humic substances, enzymes and microorganisms in soils. The review also presents challenges and opportunities for soil and environmental scientists to play a relevant role in the assessment and advancement of sustainable agroecosystem management by developing the concept of soil quality as an indicator of sustainability.

15.1 Soil Quality/Health

15.1.1 Concepts and Indicators

Growing concern about soil degradation and the need for sustainable soil management in agroecosystems, coupled with societal concern for environmental quality, has given new impetus to research on soil quality characterization (Fig. 15.1). Many

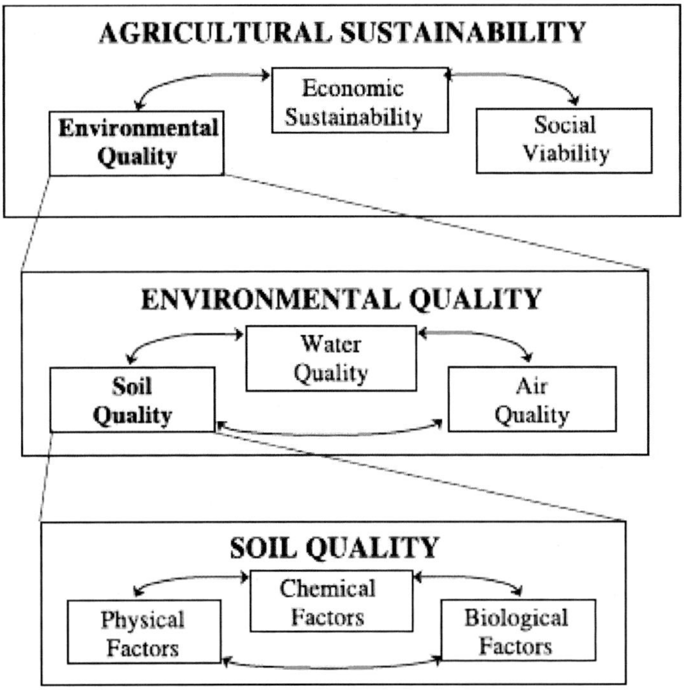

Fig. 15.1 Hierarchical relationship of soil quality to agricultural sustainability (After Karlen et al. 2003)

factors influence the complex chemical, physical, and biological processes which govern soil quality and crop productivity. For instance, changes in soil quality caused by imbalanced fertilizer use, acidification, salinity, alkalinity, and decline in soil organic matter may take several years to appear. Long-term experiments provide the best means of studying changes in soil properties and processes over time, and these experiments are important for obtaining information on long-term sustainability of agricultural systems to formulate future strategies for maintaining soil health (Hati et al. 2006).

Although the concept of soil quality has been embraced since ancient times, and a variety of definitions have been proposed for the term soil quality, agreement on assessing and interpreting soil quality remains elusive (Zvomuya et al. 2008; Bastida et al. 2008). This is probably related to the innate difficulty in defining soil itself and to the multifaceted nature (i.e., scientific, personal, and social) of environmental concerns (Carter 2002). Many definitions have been suggested in recent years, but one that best represents the concept was given by Doran and Parkin (1994): 'The capacity of the soil to interact with the ecosystem in order to maintain biological productivity, environmental quality and to promote animal and plant health.' Karlen et al. (1997) also have suggested a widely accepted definition stipulated as follows: "soil quality is the capacity of a specific kind of soil to function, within natural or

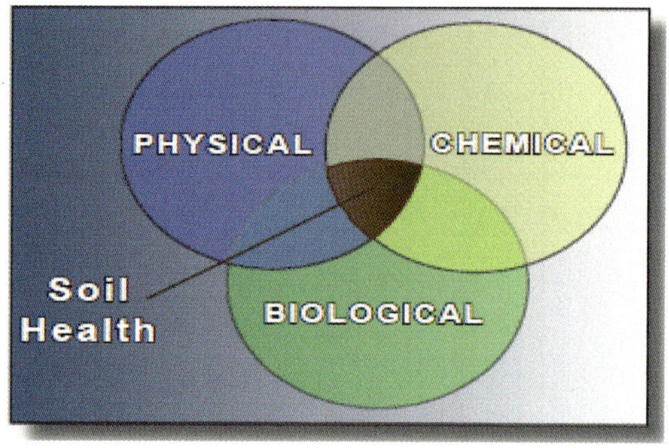

Fig. 15.2 The concept of soil health deals with integrating the physical, chemical and biological components of the soil (Adapted from the Rodale Institute)

managed ecosystem boundaries, to sustain plant and animal productivity, maintain or enhance water and air quality, and support human health and habitation". Both definitions include an inherent and a dynamic component of soil characteristics. The former is an expression of the soil forming factors, whereas the dynamic soil quality generally refers to the condition of soil that is changeable in a short period of time by human impact, including agricultural management practices (Idowu et al. 2008).

Biological and chemical processes influence pore size distribution, density and stability of soil structure. In turn, the physical structure of a soil also plays an integral role in controlling chemical and biological processes (Dexter 2004). Agricultural management practices such as tillage, traffic patterns, crop rotation practices, cover crops and organic matter additions strongly influence the components of soil quality and thus crop performance (Doran and Parkin 1994).

Attempts have been made to separate the concept of soil quality from that of soil health (Figs. 15.1 and 15.2). The limits of the two concepts are not particularly clear, but it is currently accepted that the term quality refers to the aptitude of the soil for carrying out a specific function, while health refers to its overall condition (Gil-Sotres et al. 2005).

The importance of developing the concept of soil quality was enhanced because of the need to apply soil science to address the problems of nonagricultural uses of soil (e.g., mineland restoration, urban uses and disposal of urban wastes, soil contamination and pollution by industrial activities, athletic and recreational uses of soil, and environmental regulatory functions with particular reference to water quality and the greenhouse effect). A strong need, therefore, arose to develop appropriate indicators of soil quality in relation to specific soil function (Fig. 15.3 and Tables 15.1 and 15.2).

Several physical, chemical and biological attributes have been used by different authors as soil quality indices in agro-ecosystems (Tables 15.3 and 15.4). In fact, defining soil quality is, usually attempted using somewhat arbitrarily chosen chemical, biological and physical indicators which represent particular constituents, processes or conditions (Burns et al. 2006). While, the physical and physico-chemical parameters are of little use as they alter only when the soil undergoes a really drastic change, the biological and biochemical parameters are sensitive to the

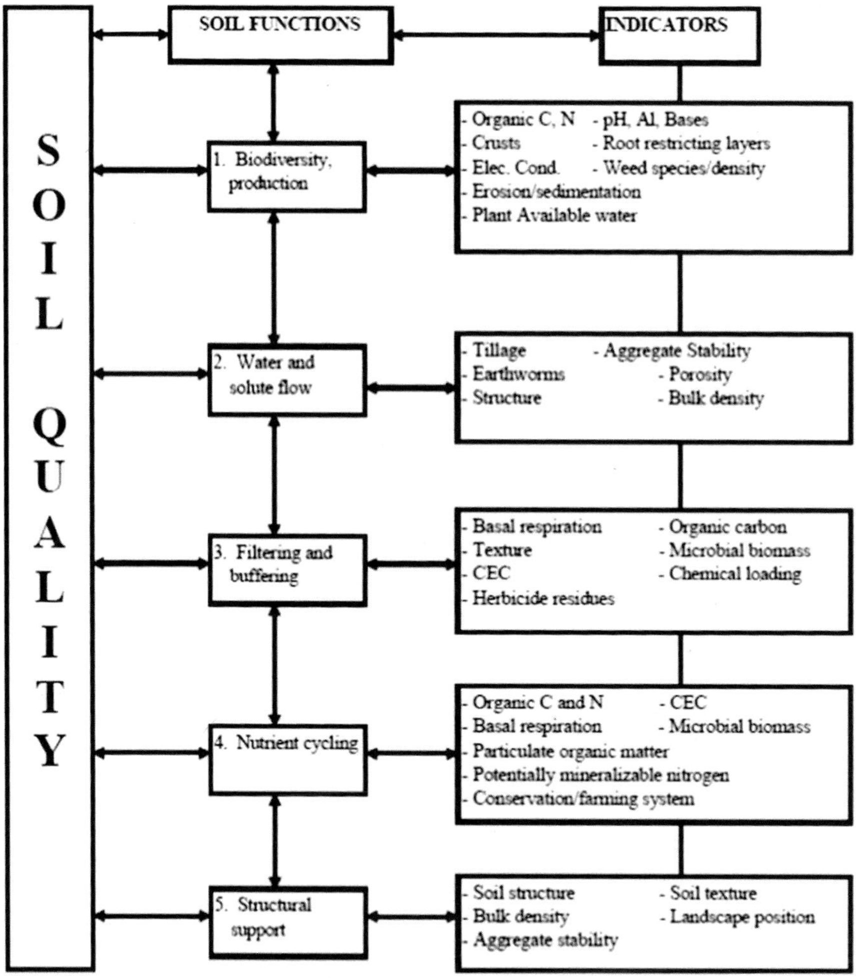

Fig. 15.3 Graphical representation of the concept of soil quality using soil function and indicators of soil quality (After Mausbach and Tugel 1997)

slight modifications that the soil can undergo in the presence of any degrading agent (Gil-Sotres et al. 2005). Soil quality is based on a series of thresholds defined by limiting factors and user needs. Its assessment depends on indicators, defined as those soil properties and processes that have greatest sensitivity to changes in soil function. Indicators should correlate well with ecosystem processes, integrate soil properties and processes, be accessible to many users and sensitive to management and climate, and if possible, be components of existing databases (Doran and Parkin 1996). A minimum data set (MDS) was proposed (Appendix F) to measure soil quality and its changes due to management practices through selection of key indicators such as soil texture, organic matter, pH, nutrient status, bulk density,

Table 15.1 Brief descriptions of the selected soil health assessment indicators

Physical	**Aggregate Stability**: is a measure of the extent to which soil aggregates resist falling apart when wetted and hit by rain drops. It is measured using a rain simulation sprinkler that steadily rains on a sieve containing a known weight of soil aggregates between 0.5 and 2.0 mm. The unstable aggregates slake (fall apart) and pass through the sieve. The fraction of soil that remains on the sieve determines the percent aggregate stability
	Available Water Capacity: reflects the quantity of water that a disturbed sample of soil can store for plant use. It is the difference between water stored at field capacity and the wilting point, and is measured using pressure chambers
	Surface Hardness: is calculated from the maximum soil surface (0–6 in. depth) penetration resistance (psi) determined using a field penetrometer
	Subsurface Hardness: is a measure of the maximum resistance (in psi) encountered in the soil at the 6–18 in. depth using a field penetrometer
Biological	**Organic Matter**: is any material that is derived from living organisms, including plants and soil fauna. Total soil organic matter consists of both living and dead material, including well decomposed humus. The percent OM is determined by loss on ignition, based on the change in weight after a soil is exposed to approximately 850 °F in a furnance
	Active Carbon: is a measure of the fraction of soil organic matter that is readily available as a carbon and energy source for the soil microbial community (the fuel of the soil food web). Active carbon is a "leading indicator" of soil health response to changes in crop and soil management, usually responding much sooner than total organic matter content. The soil sample is mixed with potassium permanganate (deep purple in color) and as it oxidizes the active carbon, the color (absorbance) is measured using a spectrophotometer
	Potentially Mineralizable Nitrogen: is the amount of nitrogen that is converted (mineralized) from an organic form to a plant-available inorganic form by the soil microbial community over 7 days in an incubator. It is a measure of soil biological activity and an indicator of the amount of nitrogen that is rapidly available to the plant
	Root Health Rating: is a measure of the quality and function of the roots as indicated by size, color, texture and absence of symptoms and damage by root pathogens such as *Fusarium*, *Pythium*, *Rhizoctonia*, and *Thielaviopsis*. Been seeds are grown in a portion of the soil sample in the greenhouse for 4 weeks. Low ratings (1–3) suggest healthy roots because pathogens are not present at damaging level and /or are being suppressed by the beneficial microorganisms in the soil
Chemical	**Soil Chemical Composition**: a standard soil test analysis package measures levels of pH, plant nutrients and toxic element. Measured levels are interpreted in the framework of sufficiency and excess but are not crop specific

electrical conductivity and rooting depth (Larson and Pierce 1994). The MDS must be sufficiently diverse to represent biological, chemical, and physical properties and processes of complex systems. Currently, the soil management assessment framework (SMAF) has 11 scored indicators with several (>20) being proposed, evaluated, or having preliminary scoring curves (Appendix G). Karlen et al. 2006 reported that the SMAF uses quantitative laboratory analyses, provides site-specific interpretations, and can lead to a better understanding of how management affects a

Table 15.2 Soil quality indicators and example scoring functions (Reported by Mausbach and Tugel 1997)

Indicators	Soil functions	Range/limits for scoring function	Scoring function
Physical			
% Stable aggregate (*more is* better)	Regulating and partitioning water and nutrient flow	30–60%	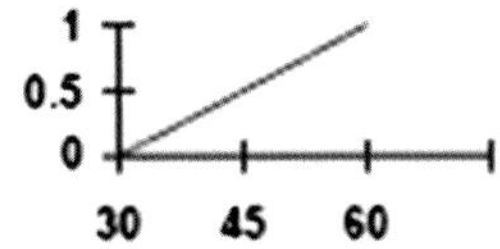
Porosity – surface 75 mm (*optimum*)	Regulating and partitioning water and nutrient flow	20–80%	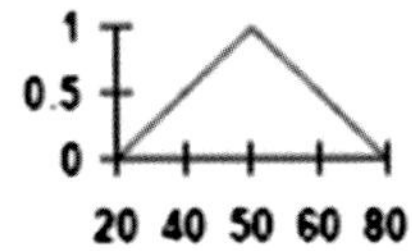
Bulk density (*less is better*)	Regulating and partitioning water and nutrient flow	1.3–2.1 mg/m^3	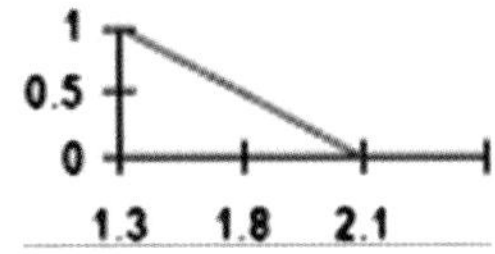
Rooting depth (*more is better*)	Biological activity	60–250 cm	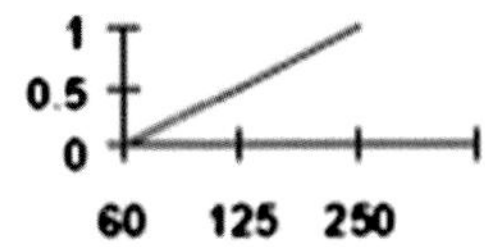
Biological			
Microbial biomass (*more is better*)	Biological activity, nutrient cycling, and filtering: and buffering	75–700 mg C/kg	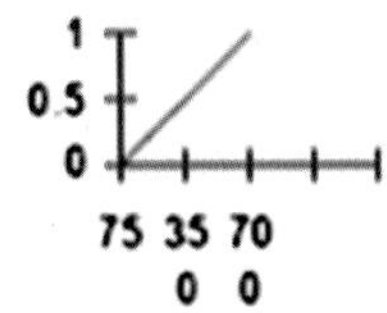
Respiration (*more is better)*	Biological activity, nutrient cycling	0.5–8.0 mg C/kg	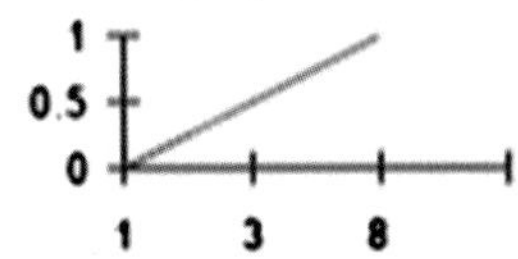
Total nitrogen (*more is better*)	Biological activity, nutrient cycling	1.5–5.0 mg/cm^3	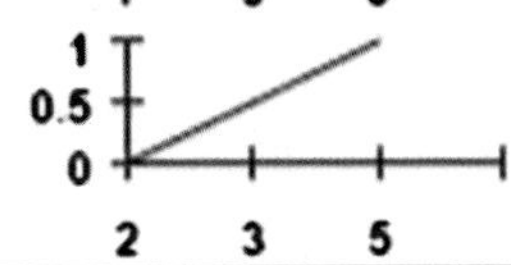

specific soil resource with respect to multiple endpoints that are outcomes driven by landowner/operator or societal goals (e.g., productivity and environmental quality).

Recently it was reported that soil quality indexing involves the choice of appropriate indicators for minimum data set, the transformation of indicator to scores; and combination of the scores into index (Masto et al. 2008). These researchers further reported that scoring and combining the indicators into indices can be done by linear and non-linear scoring techniques. Statistical techniques like

Table 15.3 Indicators used by different authors in soil quality indexes in agro-ecosystems (After Bastida et al. 2008)

Authors	Objective	Indicators used
Karlen et al. (1994a)	Soil Quality Index: Evaluation of effects of crop residue management on soil quality under corn culture	Aggregate stability, porosity, worms, microbial biomass, respiration, ergosterol, total C, total N, bulk density, available water, pH, electrical conductivity
Wang and Gong (1998)	Relative Soil Quality Index: Evaluation of changes in soil quality in natural and agriculture systems	Soil depth, texture, slope, organic matter, total and bioavailable N, P and K, cation exchange capacity and pH
Hussain et al. (1999)	Adaptation of indices to evaluate effect of three cultivation systems on soil quality	Aggregate stability, organic C crop residues, porosity, exchangeable K, pH
Glover et al. (2000)	Soil Quality Index: Evaluation of effects of different apple production systems: conventional, organic and integrated	Aggregate stability, porosity, worms, porosity, organic C, microbial biomass C and N, cationic exchange capacity, pH and N
Liebig et al. (2001)	Land Quality Index: Agroecosystem performance: effects of conventional and alternative agricultural systems	Seed yield, N content of seed, pH, organic C, nitrates
Andrews et al. (2002b)	Soil Quality Index: Evaluation of tomato and cotton crop quality in conventional and organic cultivation	Organic matter, electrical conductivity, pH, water water-stable aggregates, real density and Zn
Koper and Piotrowska (2003)	Biochemical Soil Fertility Index: Comparison of long term effect of organic and mineral fertilisation in sugar beet	Organic C, total N, dehydrogenase activity, alkaline phosphatase activity, protease activity, amylase activity
Kang et al. (2005)	Sustainability Index: Comparison of long term effect of organic amendments in systems for cultivating maize and rice	Organic C, total N, extractable K, extractable nitrates and ammonium content, microbial biomass C and N, mineralizable N, respiration, bacterial counts, mycchorhizal infection, dehydrogenase activity
Sharma et al. (2005)	Soil Quality Index: Selection of adequate managements in drylands comparing between conventional and minimal cultivation	Available N, K and S, microbial biomass C and saturated hydraulic conductivity
Lee et al. (2006)	Soil Quality Index: Effects of swine manure compost application on soil quality under different vegetable and rice systems	Bulk density, aggregates, organic C, pH, available K and P, extractible Cu and Zn, microbial biomass, C, mineralizable N

Puglisi et al. (2005)	Alteration index: Effects on the quality of agricultural soils contaminated with industrial and municipal wastes, organic fertilisation or irrigation with poor quality water under different crops: Ficus carica, maize, tomato, etc.	Arylsulphatase enzymatic activities, β-glucosiadase, phosphatase, urease, invertase, dehydrogenase and phenoloxidase
Puglisi et al. (2006)	Soil Alteration Index: Effects on the quality of agricultural soils contaminated with industrial and municipal wastes, organic fertilisation or irrigation with poor quality water under different crops: F. carica, maize, tomato, etc.	PLFAs (Phospholipid fatty acid)
Mohanty et al. (2007)	Soil Quality Index: Effects of cultivation practice (conventional and without ploughing) in rice–wheat systems, and maintaining vegetal residues on soil quality	Bulk density, aggregate stability, resistance to penetration, organic matter
Masto et al. (2007)	Soil Quality Index: Evaluation of agricultural soils fertilized with inorganic and/or farm yard manure	Bulk density, water retention, pH, electrical conductivity, bioavailable nutrients, organic matter, microbial biomass and crop yield
Erkossa et al. (2007)	Soil Quality Index: Compare the effect of land preparation methods (broad bed and furrows, green manure, ridge and furrows, reduced tillage) on soil quality	Bulk density, aggregate stability, organic C, microbial biomass C, pH, available water capacity

Table 15.4 Thirty-nine potential indicators evaluated for use in the soil health assessment protocol

Physical	Biological	Chemical
1. Bulk density	17. Root health assessment	28. Phosphorus
2. Macro-porosity	18. Beneficial nematode population	29. Nitrate nitrogen
3. Meso-porosity	19. Parasitic nematode population	30. Potassium
4. Micro-porosity	20. Potential mineralizable nitrogen	31. pH
5. Available water capacity	21. Decomposition rate	32. Magnesium
6. Residual porosity	22. Particulate organic matter	33. Calcium
7. Penetration resistance at 10 kPa	23. Active carbon test	34. Iron
8. Saturated hydraulic conductivity	24. Weed seed bank	35. Aluminum
9. Dry aggregate size (<0.25 mm)	25. Microbial respiration rate	36. Manganese
10. Dry aggregate size (0.25–2 mm)	26. Glomalin	37. Zinc
11. Dry aggregate size (2–8 mm)	27. Organic matter content	38. Copper
12. Wet aggregate stability (<0.25–2 mm)		39. Exchangeable acidity
13. Wet aggregate stability (2–8 mm)		
14. Surface hardness with penetrometer		
15. Subsurface hardness with penetrometer		
16. Field infiltrability		

principal component analysis (Andrews et al. 2002; Sharma et al. 2005) and factor analysis (Brejda et al. 2000a, b) were also employed to identify the critical soil parameters.

The identification of biological indicators of soil quality was also reported as critically important because soil quality is strongly influenced by microbiological mediated processes such as nutrient cycling, nutrient capacity, aggregate stability (Doran and Parkin 1994). Biological indicators of soil quality that are commonly measured include soil organic matter, respiration, microbial biomass (total bacteria and fungi,) and mineralizable nitrogen.

Microbial biomass may be a useful indication for soil quality since it is a susceptible soil component and sensitive to management practices or pollution. However, it must be realized that between different soil samples different biomass may occur without direct correlation to soil quality (Schloter et al. 2003). Microbial diversity parameters are often used for the assessment of the quality of contaminated soils. Muller et al. (2001), for example, investigated the long-term exposure to mercury on the microbial community at a contaminated site in Assens, Denmark. They found that bacterial diversity was reduced in the contaminated soils, whereas there was no difference in fungal biomass. However, in such investigations, it is well known that the huge amount of diversity (e.g. 1 g soil contents about 10^9 bacteria belonging to about 10,000 different species) makes it often difficult to handle the microbial community as an indicator for soil quality. Therefore, due to the mentioned complicity of the whole microbial community it might be useful to look at indicator organisms only, which are correlated to soil quality, for example, beneficial microbes like *Rhizobium* or arbuscular mycorrhiza (Schloter et al. 2003; Bending et al. 2004).

Soil microbial activity leads to the liberation of nutrients available for plants but also to the mineralization and mobilization of pollutants and xenobiotics. The measurements of microbial activity include enzymatic assays that catalyze substrate-specific transformations and may be helpful to ascertain effects of soil management, land use and specific environmental conditions. In considering soil enzymes as an indicator of soil quality, the question that remains is which enzymes should be measured for this purpose?

The turnover of nitrogen is also considered as an indicator for soil quality, since bioavailable nitrogen is one of the keys for plant growth in agriculture. At the same time, nitrate, nitrite or N_2O play an important role in environmental pollution. Therefore, it is of great interest to understand the key processes in the nitrogen cycle (e.g. nitrogen fixation, mineralization, nitrification and denitrification) in more detail, to define ways to produce a high productive agriculture which protects environment.

By using faunal groups as indicators for soil quality the choice of organisms should meet some requirements or conditions such as the formation of a dominant group and the occurrence in all soil types, the existence of high abundance and biodiversity and the importance of their role in soil functioning, e.g. in food webs. Nematodes (Table 15.4) fulfill these conditions and seem to be, at our present state of knowledge, the most promising group, also because different tests in ecotoxicology, realized for single species as well as for communities, shows the suitability (Schloter et al. 2003). These authors further stated that the use of soils fauna as indicators offers different possibilities. For example they argued that single species bioassays are important to assess effects of single stressors and bioconcentration studies. However, they remarked that these tests are often realized in laboratory experiments, with soil samples transferred in experimental systems and spiked with contaminants. Therefore, they stressed that the experiments on community level are ecologically more relevant, since they integrate interactions of all soil factors including management and pollutants effects.

In summary, the aggregation of indicators for evaluating soil quality should consider the complexity of microbial life in soil. Multiple indicators can be considered to refer to the 'driving forces' for C and N cycling in soils. As a minimum data set, microbial biomass content and microbial activity rates including enzyme activities were often estimated together with measures on some basic soil components, i.e. organic C content (Carter et al. 1997). Moreover, it is important to note that the soil quality concept encompasses both inherent and dynamic soil properties and processes. The assessment of soil quality must reflect biological, chemical, and physical properties, processes and their interactions. It should also be stressed that there is no ideal or magic index value, but soil quality assessments can be made using a framework that prioritizes management goals, identifies critical soil functions necessary for achieving those goals, and selecting indicators that provide

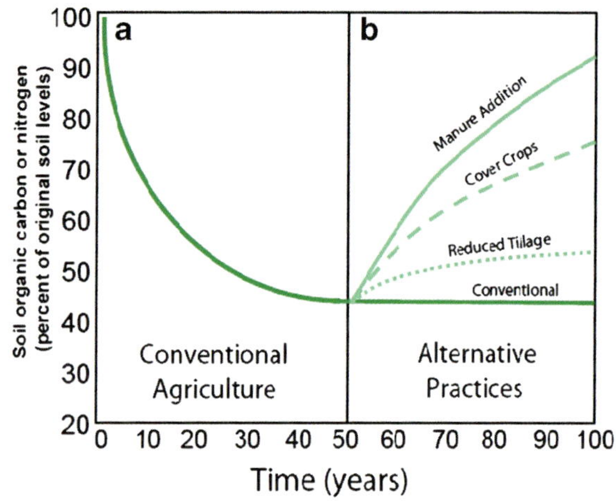

Fig. 15.4 Generalized effects of different agricultural practices on total soil organic C or N content of soil (From Ann et al. 2004)

useful information regarding how a specific soil is functioning. For researchers, there is need to conduct sufficiently detailed tests while controlling for variation in order to develop meaningful assessments of soil status, often expressed as an index of soil quality (Kang et al. 2005).

15.1.2 Effects of Agricultural Management Practices on Soil Quality

Monitoring soil quality in agricultural soils is important for assessing the long-term sustainability of agricultural operations. In general, conventional agricultural practices (i.e., high input, high intensity systems) have caused soil quality to decline over the last century (Ann et al. 2004). For example, following 50 years of tillage and commercial fertilization, fields can lose about half of their original organic C or N content. However, alternative practices can help to recover these losses (Fig. 15.4)

Expansion and intensification of cultivation are among the predominant global changes of this century. Matson et al. (1997) reported that the intensification of agriculture by use of high-yielding crop varieties, fertilization, pesticides, irrigation and mechanization has contributed substantially to the tremendous increases in food production over the past 50 years. Land conversion and intensification, however, also alter the biotic interactions and patterns of resource availability in ecosystems and can have serious local, regional, and global environmental consequences through, erosion, compaction, reduction of organic matter, non-point source pollution, reduced biodiversity and negative global consequences, including impacts on atmospheric constituents and climate. Agricultural management practices such as tillage, crop rotation, and fertilization affect the levels of soil organic matter (SOM) by influencing the amount of plant residues returned to the soil, and the rate of

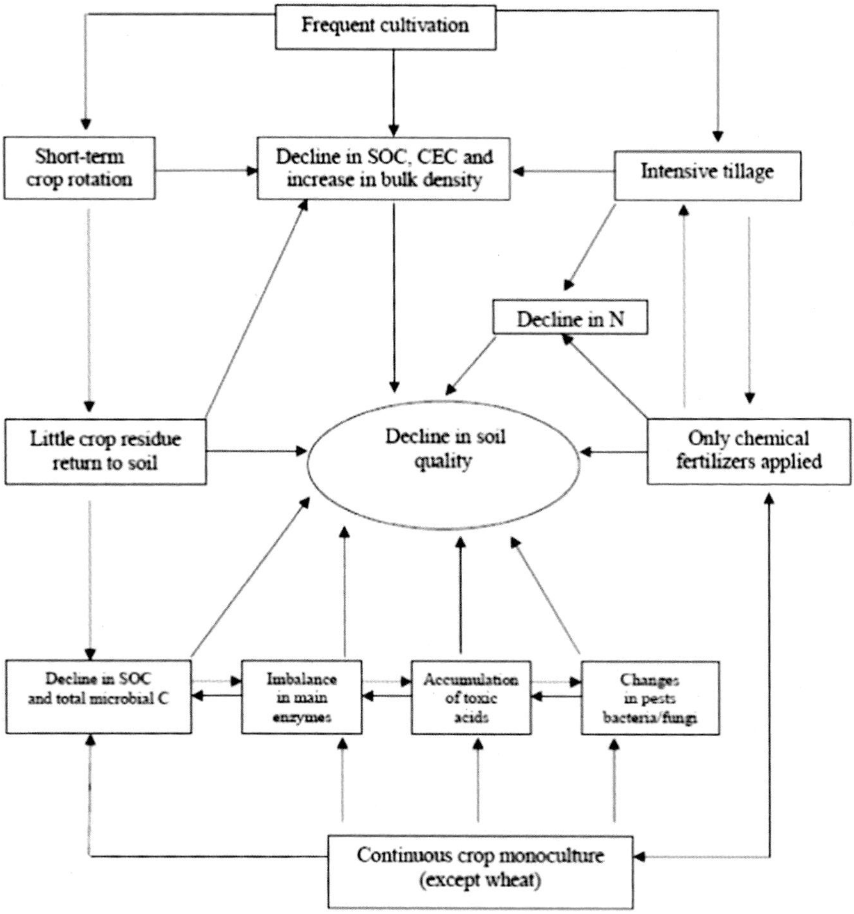

Fig. 15.5 Diagram of interactions and negative impacts of agricultural management on soil quality (After Liu et al. 2006)

residues and SOM decomposition (Antil et al. 2005). This influence is not only by levels of organic matter input, but also by the related biological activity, overall nutrient status of the soil, vegetation, climate, and soil minerals. The interactions and negative impacts of agricultural management on soil quality are illustrated in Fig. 15.5.

15.1.2.1 Cultivation/Tillage

Tillage was reported to be a major factor dictating loss of soil organic matter, and in order to maintain a high level of soil organic matter to enhance soil tilth, fertility, and productivity, there has been a growing awareness among researchers to identify suitable soil management practices depending upon climatic and edaphic conditions (Sharma et al. 2008). If the soil has not been tilled before, the existing equilibrium of loosing and compressing forces, of texturing and decay of structures is interrupted. By using several physical, chemical and biological attributes, Karlen et al. (1994) have studied the long-term effects of tillage on the quality of Orthic Luvisol and silt loam (fine-silty, mixed, mesic Typic Hapludalfs) soils from Rozetta and Palsgrove, respectively (USA). They found that plots managed using no-till practices for 12 years before the collection of samples had surface soil aggregates that were more stable in water and had higher total carbon, microbial activity, ergosterol concentrations, and earthworm populations than either the chisel or plow treatments. It was also reported that in many cases tillage enhances the breakdown of organic material in soil and that minimum tillage can lead to an increase of SOM contents (Tatzber et al. 2008). Previous studies on the changes in soil organic carbon (SOC) levels in numerous small tillage plots of Norfolk loamy sand, South Carolina, demonstrated that, after 9 years of conservation tillage, the SOC content in the top few centimeters was significantly higher than the soil under conventional tillage management (Hunt et al. 1996). In a clayey soil from Ibadan (Nigeria), there occurred a significant decrease in soil organic carbon (SOC) and total soil nitrogen (TSN), and an increase in C: N ratio with cultivation duration in the 0–5-cm depth (Lal 1997).

It is well known that the manner by which crop residues are introduced to the soil matrix differs dramatically between no-till (NT) and tillage management (TM). Under NT management, crop residues are left on the surface of the soil after harvest, whereas residues are mechanically incorporated into the soil during tillage. Stable isotopes have shown that the majority of C from corn residues is concentrated in the upper 5 cm under NT management, but a more uniform vertical distribution throughout the plow depth is achieved during tillage (Olchin et al. 2008). It has also been reported that soil tillage, for example, strongly influences the placement and distribution of crop residues, resulting in differences in the composition and activity of microbial and faunal communities (Giller and Cadisch 1997). This can markedly impact rates of residue decomposition and carbon sequestration as well as dynamics of nutrient mineralization/immobilization and the availability of suitable refuges for plant pathogens. The intensity of soil tillage may also indirectly impact physical processes in soils (e.g. bioturbation, soil aggregation) through changes in the diversity and composition of biological communities in addition to direct mechanical alterations in structure that result from cultivation. Soil tillage can also directly disrupt earthworm populations and render them susceptible to predation by birds, and can destroy termite galleries (Giller and Cadisch 1997).

Zalidis et al. (2002) reported that soil compaction has been caused by the repetitive and cumulative effect of heavy machinery in Chromic Cambisols and the Chromic Luvisols from the Mediterranean region (Greece), resulting in a decrease of soil porosity, on the farm scale and an increase of the surface runoff on the watershed scale. The 0–5 cm layer under non tillage system showed increases up to 68% for amylase, 90% for cellulase, 219% for arylsulfatase, 46% for acid phosphatase, and 61% for alkaline phosphatase in an oxisol (Typic Haplorthox) soil from the State of Paraná, Brazil (Balota et al. 2004). At the 0–10-cm layer and

10–20 cm of a humus-rich volcanic ash (Umbric Andosol) soil from Fukushima (Japan), Miura et al. (2008) found that, the effects of tillage on soil organisms significantly differed, and negative impacts of tillage were smaller in the deeper layer. In addition, bacterial and fungal substrate-induced respiration (SIR) and the population density of enchytraeids were higher under no tillage than under conventional tillage, but the population densities of protozoa, mites, and collembolans did not differ significantly.

Following 4 years of a cotton-tomato rotation on the west side of the San Joaquin Valley in California, USA, conservation tillage and cover crops altered physical and chemical properties of a sandy clay loam soil (Veenstra et al. 2006). For example the conservation tillage increased available phosphorus but redistributed potassium from the subsurface to the surface by accumulating organic matter at the soil surface and not remixing it with tillage. Nitrate accumulated at the surface as well. With regard to the long-term soil fertility and quality, the authors did not see any increases in the conservation tillage systems, but cover crops increased soil carbon significantly regardless of the tillage treatment. Cover cropping also increased total soil nitrogen, phosphorus and potassium, and in the conservation tillage treatments mitigated the increases in salt concentration.

15.1.2.2 Crop Rotation

Crop rotation is a "system of growing different kinds of crops in recurrent succession on the same land". However, in the strictest sense, crop rotation is more than just changing crops from year to year based on current economic situations. Rather, it is a long-term plan for soil and farm management (Singer and Bauer 2009). The effects of extended crop rotations on soil quality have been discussed in general terms and specifically with regard to aggregate size distribution and stability and other indicators, but not with regard to soil quality per se (Karlen et al. 2006). It has been reported that the off-site environmental problems caused by cropping systems are often linked to a compromised soil function resulting from poor soil management (Liebig et al. 2006). Consequently, information on how cropping systems influence soil quality will allow agriculturists to design systems that are more environmentally sustainable. For instance, Karlen et al. (2006) found that the quality of fine loamy soils in the Northern Corn and Soybean Belt of the Midwestern USA was the highest under extended crop rotations.

Another study was conducted by Masto et al. (2008) on the long-term effects of land use and soil management practices in New Delhi (India) at a Typic Haplustept soil. They noticed that the adoption of maize–wheat system proved to be better than traditional rice–wheat; and in general, agricultural crops have a significant advantage than vegetable crops, in terms of maintaining soil quality. Additionally, soil physical quality parameters were significantly affected only in the sewage-irrigated fields system when compared with the uncultivated soils, particularly for the bulk density. Among chemical parameters, electrical conductivity was high in the sewage-irrigated fields (SF), and soil nutrients (N, P, K, S, Zn, Fe, Cu and Mn) were well above the critical limits of deficiency in all the systems. In several long-term on-farm and on-station field experiments of Australian Vertisols, Hulugalle and Scott (2008) also reported that, due to the greater intensity of wet/dry cycles of wheat, better soil structure resulted by growing wheat after cotton than by growing legumes such as dolichos, chickpea, faba bean, and field pea. They further stated that the increasing complexity of crop rotations ensures improvements in soil quality and agricultural sustainability.

At the State of Sonora (Mexico), Limon-Ortega et al. (2009) have examined the effects of crop rotation, wheat straw management, and chicken manure on the quality of a Chromic Haplotorrert (Vertisol Calcaric Chromic) soil. They observed that the improvement of the rotation complexity and the reduction of the fallow by including summer maize or sesbania increased the soil organic matter. Moreover, they highlighted that planting sesbania as a summer crop after harvesting wheat in the spring enhanced both the amounts of carbon and nitrogen mineralized from the soil microbial biomass (C and N).

In integrated crop–livestock systems at a Virden series silty clay loam soil from Pana, IL (USA), Maughan et al. (2009) demonstrated that cover crops and perennial pastures greatly increased the contents of total N and C, and had larger soil aggregates as compared to the continuous corn monoculture. Coulter et al. (2009) also realized that cropping system in poorly-drained fine- and coarse-silty soils at three diverse locations in Illinois (USA) influenced soil C and N more than N fertilization, and that this influence was greater in the labile fraction of soil organic matter (SOM) than in total SOM. With regard to the total soil organic carbon (SOC) stocks in both rice–wheat and maize–wheat cropping systems from a sandy loam soil of Ludhiana (India), Kukal et al. (2009), found higher contents of these stocks in farmyard manure (31.3 and 23.3 Mg ha^{-1}, respectively) followed by balanced fertilization (29.6 and 21.3 Mg ha^{-1}), whereas the lowest contents were recorded in the control (21.4 and 18.7 Mg ha^{-1}).

15.1.2.3 Inorganic and Organic Amendments

According to Wu et al. (2005), fertilization can enhance the growth of shoots and roots of crops and thus promote more carbon input into soil, resulting in higher soil organic carbon content eventually. However, due to the continuous use of chemical fertilizers, there is increased emphasis on their impacts on the environmental quality, depending on the type and rate of fertilizers, crop rotation systems, soil properties, and even the nature of the ecozones in which the soil developed (Kaur et al. 2005; Wu et al. 2005). For instance, a study over 20 years with farmyard manure (FYM) and inorganic fertilizers in a Yellow fluvo-acquic soil (Jiangsu province, China), showed an increase of soil organic matter of 80% compared to only 10% with NPK (Dong et al. 2006). These researchers further observed that total nutrient input with the combined FYM and inorganic fertilizer treatments was much higher than with only the inorganic fertilizer treatments. Similar findings were obtained by Yang et al. (2007) who carried out a long-term field experiment on a calcareous desert soil in Guansu Province (NW China). They found that integrated application of N, P, and K fertilizers and manure is an important strategy to maintain or increase soil organic C and N, improve soil fertility, maintain nutrients balance, and minimize damage to the environment. In a paddy soil derived from Quaternary Red Clay in Hunan Province (China), Xu et al. (2008) also noticed the increase of soil organic matter by 18.5% in the chemical fertilizers combined with swine manure (NPKM) treatment, which were significantly higher than those in the chemical fertilizers only (NPK). From another long-term fertilizer experiment at Ranchi (India), the application of balanced fertilizer along with manure (NPKM) or lime (NPKL) improved aggregation, water retention, microporosity and available water capacity of an acidic Alfisol (Typic Haplustalf) and reduced its bulk density in 0–30 cm depth over the control plot (Hati et al. 2008). In contrast, soil aggregate stability, microporosity and available water capacity were significantly lower in the nitrogen plots than that in the control.

Zhong and Cai (2007) found that all enzyme activities except invertase further increased after 13 years of application of inorganic fertilizers in a rice-planting soil derived from Quaternary Red earth of Yingtan (China). However, at the same site, Zhong and Cai (2007) remarked that, the application of P and N did not directly affect microbial parameters in the soil, but did so indirectly by increasing crop yields, thus promoting the accumulation of soil organic matter. Similar results were found by Kong et al. (2008) who investigated the effects of long-term application of chemical fertilizers on microbial biomass and functional diversity in a Back clay loam soil of Hailun, Northeast China. They found that the soil microbial biomass, and the microbial functional diversity and evenness did not show any significant differences among the different fertilizer treatments including control.

Nemergut et al. (2008) used a variety of techniques to examine the effects of chronic N amendments on SOM chemistry and microbial community structure and function in an alpine tundra soil at the Niwot Ridge Longterm Ecological Research (LTER) site. They found that the ratio of fungi to bacteria did not change in the N-amended soils, but that of archaea to bacteria dropped from 20% to less than 1% in these plots. No between treatment differences were further revealed in the comparisons of aliphatic and aromatic carbon compounds. However, G-lignins were found in higher relative abundance in the fertilized soils, while proteins were detected in lower relative abundance. Additionally, the activity of leucine amino peptidase declined, while that of urease increased. Thus, they concluded that chronic N fertilization induces significant shifts in the soil carbon dynamics due to shifts in microbial community structure and function.

Results of the study carried out by Zhong and Cai (2007) in a rice-planting Red soil of Yingtan, Jiangxi Province (China) suggested that the long-term application of different inorganic fertilizers promoted certain changes in microbial community structure. These findings are in some extent contradictory to those found by Wang et al. (2008) in a Black soil of Dehui region (Jilin province, China). The results of these researchers showed that the substrate richness and catabolic diversities of bacterial communities were the highest in the treatment of farm yard manure, and the lowest in the chemical fertilizer treatment. The DGGE fingerprint further revealed that the majority of bands were similar among all treatments. However, the investigation demonstrated that in general, chemical fertilizer decreased the diversity of the fast growth bacteria or eutrophic bacteria. Recently Gong et al. (2009) also advocated that in a fluvo-aquic soil after 18 years of fertilization in Fengqiu county (China), the application of half organic manure with mineral fertilizer NPK produced higher culturable microbial counts than application of mineral fertilizers alone, and the NPK treatment gave higher culturable microbial counts than other mineral fertilizer treatments (NP, PK or NK).

Long-term applications of organic and inorganic fertilizers influence organic matter as well as other soil-quality parameters, such as the biomass, diversity and activity of soil microbes. However, the magnitude of changes in these characteristics depends on soil climatic conditions, the type, quantity and quality of the incorporated amendments. The literature clearly shows that organic amendments enhance soil quality and microbial characteristics, whereas the effects of inorganic fertilizers remain controversial and tend to decrease the diversity of soil microbial community. In fact as stated by Dick (1992), the additions of inorganic fertilizers can increase soil biological activity because of increased plant biomass production which upon incorporation stimulates soil biological activity. Conversely, limited evidence suggests that repeated applications of inorganic fertilizers can also suppress the production of certain enzymes that are involved in cycling of a given nutrient. The studies reviewed have been useful in understanding the long-term effects in terms of how agricultural management practices change soil quality. In addition it has been shown that the development of "universal" soil quality indices still remains challenging for soil and environmental scientists.

15.1.2.4 Pesticides

The extensive use of pesticides in agricultural land influences both the biotic and abiotic processes within the soil. As a result, several soil functions are degraded including the food web support, the retention and transformation of toxicants and nutrients, soil resilience, and the ability of soil to protect surface and ground water (Zalidis et al. 2002). On a farm scale, pesticides cause the destruction of part of the soil flora and fauna, which in turn causes both physical and chemical deterioration. Effects on non-target organisms in soil occur frequently. On watershed scale, the main problem derives from the leaching and drainage of pesticides into the surface and ground water. Additionally, the reduction of the soil's ability to remove other pollutants, due to the alteration of soil properties by pesticides and the degradation of soil's toxicant retention and transformation function, allows the movement of those pollutants to adjacent water bodies.

As soil quality is composed of a series of soil characteristics, the following sections present in detail the roles of soil physical, chemical and biological properties in soil quality/health and the impacts of agricultural practices on these soil natures.

15.2 Soil Biological Properties

The soil environment is teeming with biological life and is one of the most abundant and diverse ecosystems on earth. Soil biota, including flora (plants), fauna (animals) and microorganisms (Fig. 15.6), perform functions that contribute to the soil's development, structure and productivity. For example, plants act on the soil environment by aiding in structure and porosity, and in supplying SOM via shoot and root residue. Root channels can remain open for some time after the root decomposes, allowing an avenue for water and air movement. Roots also act to stabilize soil through aggregation and intact root systems can decrease soil

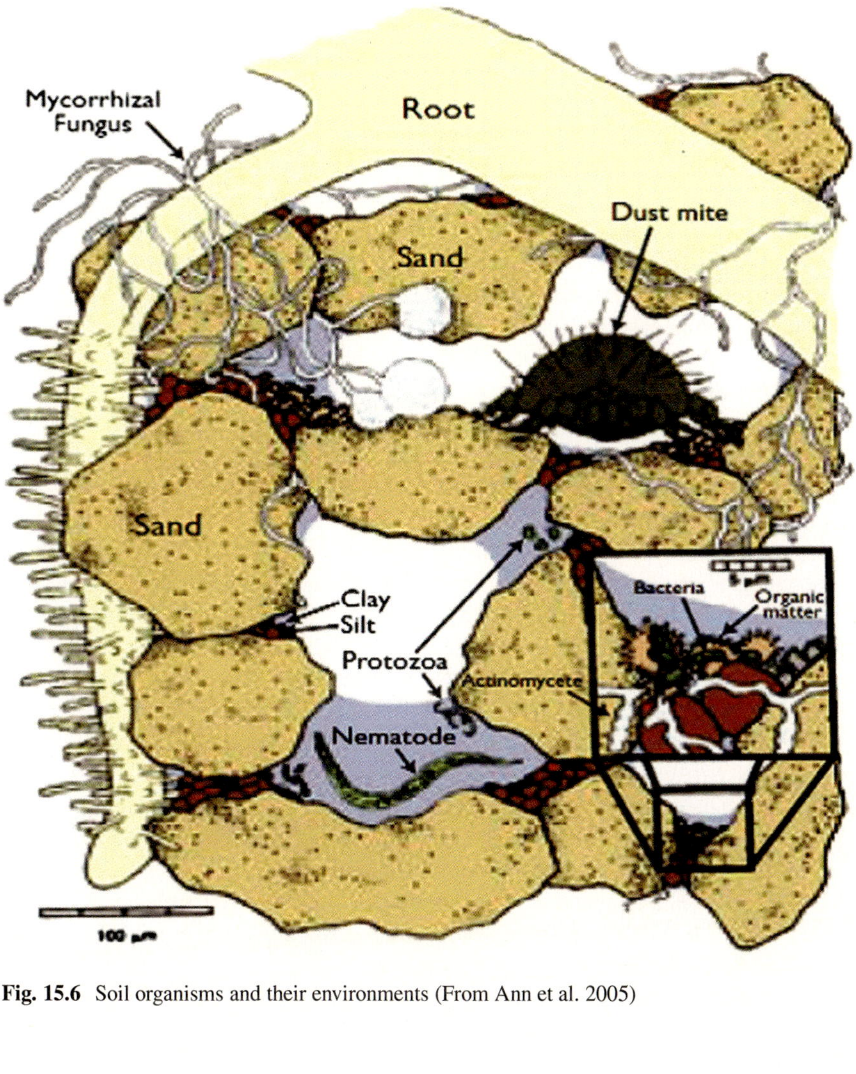

Fig. 15.6 Soil organisms and their environments (From Ann et al. 2005)

loss. The ‘rhizosphere,’ the narrow zone of soil directly surrounding plant roots, is the most biologically active region of the soil. It contains sloughed root cells and secreted chemicals (i.e., sugars, organic acids) that provide organisms with food.

15.2.1 Microorganisms in Soils

Microorganisms represent the largest and most diverse biotic group in soil, with an estimated one million to one billion microorganisms per 1 gm of agricultural top soil (Tugel and Lewandowski 1999). Soil microbes mainly include bacteria, protozoa, algae, fungi and actinomycetes (Fig. 15.6).

15.2.1.1 Significance of Soil Microorganisms

Microbes aid soil structure by physically surrounding particles and ‘gluing’ them together through the secretion of organic compounds, mainly sugars. This contributes to the formation of granular structure in the A horizon where microbial populations are greatest. They influence the availability of nutrients for crop production via a range of activities such as the decomposition of crop residues, immobilization of nutrients, mineralization, biological nitrogen fixation, and bioturbation.

The soil fauna are crucial for the initial comminution and mixing of residues into the soil, whilst the microflora have a greater suite of enzymes for chemical breakdown of organic material (Bunemann et al. 2006). Many immature and mature insects, other arthropods, earthworms, nematodes and larger macro-organisms live in the soil and have an important influence on soil structure. They ingest and egest soil material, relocate plant material and form burrows (Amezketa 1999). Biogenic structures formed by soil organisms such as termite mounds and casts are often protected from mineralization. The effects of these activities are variable. Macro-organisms improve aeration, porosity, infiltration, aggregate stability, litter mixing, improved N and C stabilization, C turnover and carbonate reduction and N mineralization, nutrient availability and metal mobility.

Bacteria and fungi are often considered as a labile pool of nutrients (C, N, P, and S) called the soil microbial biomass that has a pivotal role in nutrient immobilization and mineralization. The release of nutrients from the microbial biomass is partly regulated through grazing by the soil fauna. Roots, fungi and bacteria enhance aggregation by enmeshing soil particles and providing extracellular compounds that bind particles together (Bronick and Lal 2005). Bacteria are often associated with clay and polysaccharides in microaggregates, resulting in lower microbial biomass in microaggregates than macroaggregates (Lupwayi et al. 2001). Lower bacteria/ fungi ratio in macroaggregates than microaggregates suggests that bacterial activity may dominate in microaggregation while fungal activity dominates in macroaggregate formation (Schutter and Dick 2002). Increases in macroaggregates are associated with increases in fungal activities and fresh residues (Denef et al. 2001). Microbial influence is most pronounced in sandy soils where soil microorganisms produce readily available C-source for the rapid stabilization of aggregates (Kiem and Kandeler 1997). Fungal hyphae improve aggregate stability by reorientation of clay particles, binding particles with extracellular polysaccharides, and enmeshing particles (Ternan et al. 1996). Hyphae also enmesh microaggregates to form macroaggregates suggesting that aggregation increases with hyphal density (Haynes and Beare 1997). Microorganisms are also of great importance in the accumulation of heavy metals. Studying microorganisms as metal sorbents, Ledin et al. (1999) from Linkoping (Sweden) have used five common soil components (quartz, feldspar plagioclase, kaolin, goethite and peat) with a bacterium and a fungus to make an artificial soil system. They found that the two microorganisms T. harzianum and B. subtilis accumulated a considerable part (up to 26, 16 and 38%) of the total inventory of Zn, Cd, and Hg, respectively despite the fact that they constituted only a minor fraction (0.4 or 1.7%) of the total solid mass. In contrast, quartz and feldspar, which together formed 80% of the soil components, accumulated less than 10%.

In contrast to other species, microorganisms have the ability to physiologically adapt quickly which allows them to survive and remain active in the face of differing environmental stresses (Schimel et al. 2007). Based on these properties, differences in the structures and functional processes of microbes could emerge simultaneously if circumstances changed, so microbes can be used as a sensitive indicator of soil environmental quality.

15.2.1.2 Impacts of Agricultural Practices on Soil Microorganisms

Management practices can affect the activity of soil organism through changes in aeration and structure, cropping systems, and inputs.

Tillage typically accelerates short-term bacteria and protozoa activity by increasing aeration and breaking up residue into smaller particles that are more exposed to microbial attack (Vigil and Sparks 2003). Conversely, fungal biomass has been shown to increase in conservation tillage systems, possibly as a result of less tillage disrupting fungal hyphal networks and/or increases in SOM levels (Frey et al. 1999).

Crop rotation systems may support more organism diversity and activity than monoculture systems due to increased and more diverse residues and specific interactions occurring between certain plants and organisms (Olfert et al. 2002).

External agricultural inputs such as mineral fertilizers, organic amendments, microbial inoculants, and pesticides are applied with the ultimate goal of maximizing productivity and economic returns, while side effects on soil organisms are often neglected. In this respect, Bunemann et al. (2006) reported that mineral fertilizers have limited direct effects, but their application can enhance soil biological activity via increases in system productivity, crop residue return, and soil organic matter. Another important indirect effect especially of N fertilization is soil acidification, with considerable negative effects on soil organisms. Organic amendments such as manure, compost, biosolids, and humic substances provide a direct source of

C for soil organisms as well as an indirect C source via increased plant growth and plant residue returns. Organic fertilizers usually increase soil microbial biomass, CO_2 evolution and enzyme activities while inorganic fertilizers have relatively less effect on these soil properties (Chu et al. 2007). Effects of long-term cropping, fertilization, manuring and their integration on microbial community were studied in soil samples from five long-term fertilizer experiments under various rainfed production systems in the semi-arid tropics (SAT) of India (Vineela et al. 2008). Higher microbial numbers were obtained in the combinations of organic and inorganic fertilizers treatments compared to control. However, these authors noticed that fungal population was higher in acidic soils and in treatments under continuous inorganic fertilization treatments, whereas a high numbers of bacteria were found in integrated use of organic and inorganic fertilizers.

Non-target effects of microbial inoculants appear to be small and transient. Among the pesticides, few significant effects of herbicides on soil organisms have been documented, whereas negative effects of insecticides and fungicides are more common. Copper fungicides are among the most toxic and most persistent fungicides, and their application warrants strict regulation. Fox et al. (2007) showed that a subset of organochlorine pesticides, agrichemicals, and environmental contaminants induced a symbiotic phenotype of inhibited or delayed recruitment of rhizobia bacteria to host plant roots, fewer root nodules production, lower rates of nitrogenase activity, and a reduction in overall plant yield at time of harvest. These researchers further observed significant reductions in the number of nodules per plant in all chemical treatment groups compared with + Rhiz-positive control at 4 and 6 weeks after inoculation. Additionally, nitrogenase activity was significantly reduced in all chemical treatment groups compared with the + Rhiz-positive control at 3, 4, and 5 weeks after inoculation, whereas at all time, the -Rhiz uninoculated control group exhibited essentially no nitrogenase activity. Therefore, the authors proposed that one explanation of the disparity in N fixation in treated vs. untreated legumes was that some pesticides disrupted the natural phytochemical signaling of legumes which resulted in disconnecting the chemical communication between the plant and Rhizhobia that was necessary for optimal symbiotic nitrogen fixation.

It is well accepted that heavy metal contamination has a negative effect on bacterial communities (Moffett et al. 2003; Antolín et al. 2005; Gremion et al. 2004). Recently, Oliveira and Pampulha (2006) observed a marked decrease in population size of aerobic heterotrophic and asymbiotic nitrogen fixing bacteria (about 47 and 77% in 2003, and 67 and 77% in 2004, respectively) in heavy metal contaminated sandy loam soils from the Estarreja region in Portugal. They further noticed that differences in the viable count of fungal and actinomycetes were also significant; however, these two microbial groups seemed to be less sensitive to the presence of heavy metals. The authors reported various similar studies in which fungi have been found to be more resistant than bacteria to long-term heavy metal contamination (Fliessbach et al. 1994; Frostegård et al. 1996). In contrast to their findings, Hayat et al. (2002) found that the actinomycetes had lower tolerance to heavy metals than other bacteria. Propitious effects of heavy metals on microbial community and soil quality were observed by Zhang et al. (2008) in soils irrigated with sewage for over 30 years in the Boxing County (Shandong province, China). Moreover, the authors argued that heavy metals in these irrigated soils have not reached hazardous concentrations, even though a twofold increase in their concentrations was recorded in the agricultural soils as compared to the control area.

Many studies have shown a suppressive effect of heavy metals on microbes (Giller et al. 1998; Liao and Xie 2007). However, a higher concentration of heavy metals did not always reduce microbial activities. It has been reported that certain microbial species, such as α-subdivision of Proteobacteria and the genus Burkholderia, include species that may prosper even under heavy metal-polluted conditions (Wang et al. 2010). Therefore, the role of microorganisms in risk assessment of heavy metal polluted soils should be carefully considered according to practical situations and most of the focus should be on further probing between microbial parameters and environmental characteristics.

The population dynamics of soil microorganisms is influenced not only by agricultural external inputs, but also by tillage and crop rotation practices. It has been reported that cultivation may reduce soil biological activity through the decrease of macroaggregates, whereas crop rotations compared to monoculture improve the microbial biomass and the activities of soil enzymes by suppressing the deleterious rhizobacteria (Dick 1992). By testing whether communities of arbuscular mycorrhizal fungi (AMF) colonizing the roots of maize (*Zea mays* L.) were affected by soil tillage practices (plowing, chiseling, and no-till) in Tänikon (Switzerland), Jansa et al. (2003) found that the presence of the genus Scutellospora in maize roots was strongly reduced in plowed and chiseled soils. In addition, they observed that fungi from the suborder Glomineae were more prevalent colonizers of maize roots growing in plowed soils, but were also present in the roots from other tillage treatments. According to these authors, the observed changes in AMF community in response to different soil tillage practices might be due to the disturbance of AMF hyphae integrity, the changes in nutrient content, microbial activity, or in weed populations. Within this context, Oehl et al. (2003) also have investigated the impact of land use intensity on the diversity of arbuscular mycorrhizal fungi (AMF) in agroecosystems of Central Europe and found that for the field samples, the numbers of AMF spores and species were highest in the grasslands, lower in the low- and moderate-input arable lands, and lowest in the lands with intensive continuous maize monocropping. Their findings also clearly indicated that agricultural intensification, severely affects AMF abundance and community structure. Moreover, the AMF communities remarkably differed not only in diversity but also in functional aspects (rates of root colonization and spore formation in the trap cultures). Recently, at a typical Hapludoll, sandy loam soil from Cordoba (Argentina), Vargas Gil et al. (2009) found that soil populations of Actinomycetes, Trichoderma spp., and Gliocladium spp. were 49% higher under conservation tillage systems, in soil amended with diammonium phosphate (DAP) and not previously grazed. It has been also reported that tillage destroyed the mycelia of saprotrophic basidiomycetes, resulting in the relative enrichment of bacteria and after stopping agricultural practices, fungal biomass in soils increased within 2 years (Štursová and Baldrian 2010).

15.2.2 Enzymes Activities in Soils

15.2.2.1 Importance of Soil Enzymes

Soil enzymes are important for catalyzing innumerable reactions necessary for life processes of microorganisms in soils, decomposition of organic residues, cycling of nutrients, and formation of organic matter and soil structure (Balota et al. 2004). According to Acosta-Martínez and Klose (2008), enzymes (E) catalyze reactions without undergoing permanent alteration by reducing the activation energy via the formation of an enzyme–substrate complex (ES), thus the product (P) of the reaction is released at faster rates than if the enzymes were not present.

Enzyme activities in soil can be associated with active cells (animal, plant, microbial), entire dead cell, and cell debris as well as being complexed with clay minerals and humic colloids (Taylor et al. 2002). In addition, it was reported that the microbial component, composed of protozoa, fungi, actinomycetes, and bacteria, is the main source of enzymes in soils (Acosta-Martínez and Klose 2008). These authors also stated that the total activity of any enzyme in soil is a composite of activities associated to ten different locations (Fig. 15.7) and distinguished several approaches to separate extracellular enzymatic activities from those associated with soil microorganisms. These approaches included exposure of soils to: (a) elevated temperatures, (b) fumigants, (c) plasmolytic and antiseptic agents, and (d) irradiation with gamma rays or electron beams. Soil enzymes can be stabilized in two locations (Fig. 15.4): adsorbed to internal or external clay surfaces; and complexed with humic colloids by adsorption and cross-linking, microencapsulation, ion exchange, entrapment or co-polymerization (Weetal 1975; Boyd and Mortland 1990). Clay and humic colloids have been shown to play an important role in the stability of abiontic enzymes to thermal stress, such as heating and repeated freezing-thawing cycles (Miller and Dick 1995) (Fig. 15.8).

Soil enzymes activities have been suggested as suitable indicators of soil quality because: (a) they are a measure of the soil microbial activity and therefore they are strictly related to the nutrient cycles and transformations; (b) they rapidly may respond to the changes caused by both natural and anthropogenic factors; (c) they are easy to measure (Gianfreda et al. 2005). Moreover, as claimed by several authors (Dick and Tabatabai 1993; Dick 1997; Trasar-Cepeda et al. 2000), soil enzymes activities may be considered early and sensitive indicators to measure the degree of soil degradation in both natural and agro-ecosystems, being thus well suited to measure the impact of pollution on the quality of soil.

Enzyme activity in soils has been studied for several decades. The most widely assayed enzymes are those involved in the degradation of the plant cell wall polysaccharides, lignin and the biopolymers forming microbial biomass-polysaccharide hydrolases and oxidative ligninolytic enzymes (Šturzová and Baldrian 2010). Due to the importance of N, P, and S for plant nutrition and microbial metabolism, enzymes involved in assimilation of these elements phosphatases, arylsulfatase, proteases, peptidases and others are also frequently studied.

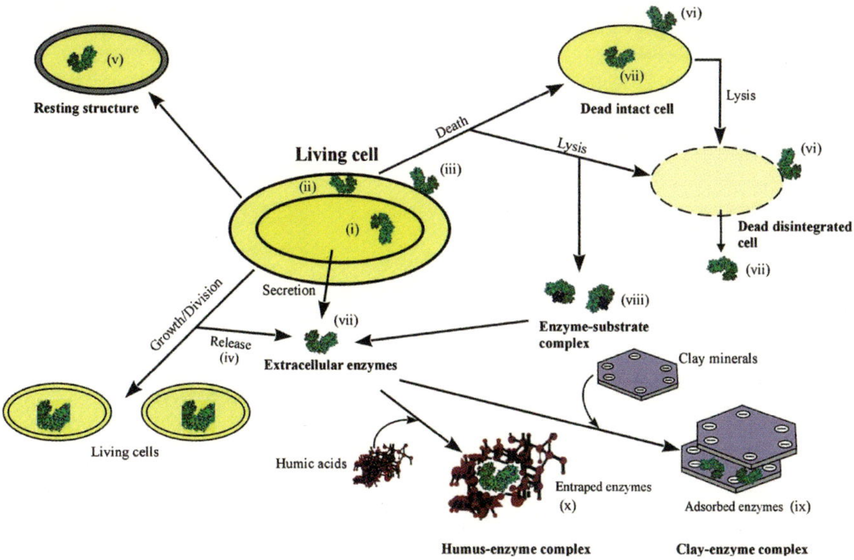

Fig. 15.7 Locations of enzymes in soil (after Acosta-Martínez and Klose 2008): (*i*) enzymes functioning within the cytoplasm of proliferating microbial, animal, and plant cells; (*ii*) enzymes restricted to the periplasmatic space of proliferating Gram-negative bacteria; (*iii*) enzymes attached to the outer surface of viable cells with active sites extending into the soil environment; (*iv*) enzymes released into the soil solution by living cells during cell growth and division; (*v*) enzymes within non-proliferating cells such as fungal spores, protozoa cysts, plant seeds, and bacteria endospores; (*vi*) enzymes attached to an entire dead cell and cell debris; (*vii*) enzymes leaking from intact cells or released from lysed cells, originally located on the cell membrane or within the cell, which may survive for a short period in the soil solution; (*viii*) enzymes temporarily associated with soluble or insoluble enzyme–substrate complexes; (*ix*) enzymes absorbed to the external or internal (i.e. within the lattices of 2:1 layer silicates) surfaces of clay minerals; (*x*) enzymes complexed with humic colloids by absorption, entrapment, or copolymerization during humification

There are several factors that affect the activity of extracellular enzymes in soils independently from their producers and ecosystem processes and the impacts of agricultural management practices seem to be of great importance among other factors.

15.2.2.2 Impact of Agricultural Management Practices on Soil Enzyme Activities

Soil degradation through loss of organic matter and structural integrity is a well known outcome of an intensive cropping regime. In fact, agricultural practices such as cultivation/tillage, fertilization, crop rotation, and soil amendment can be detrimental and/or propitious to soil enzymes. Dick (1984) observed that the activities of acid phosphatase, arylsulfatase, invertase, amidase and urease in the

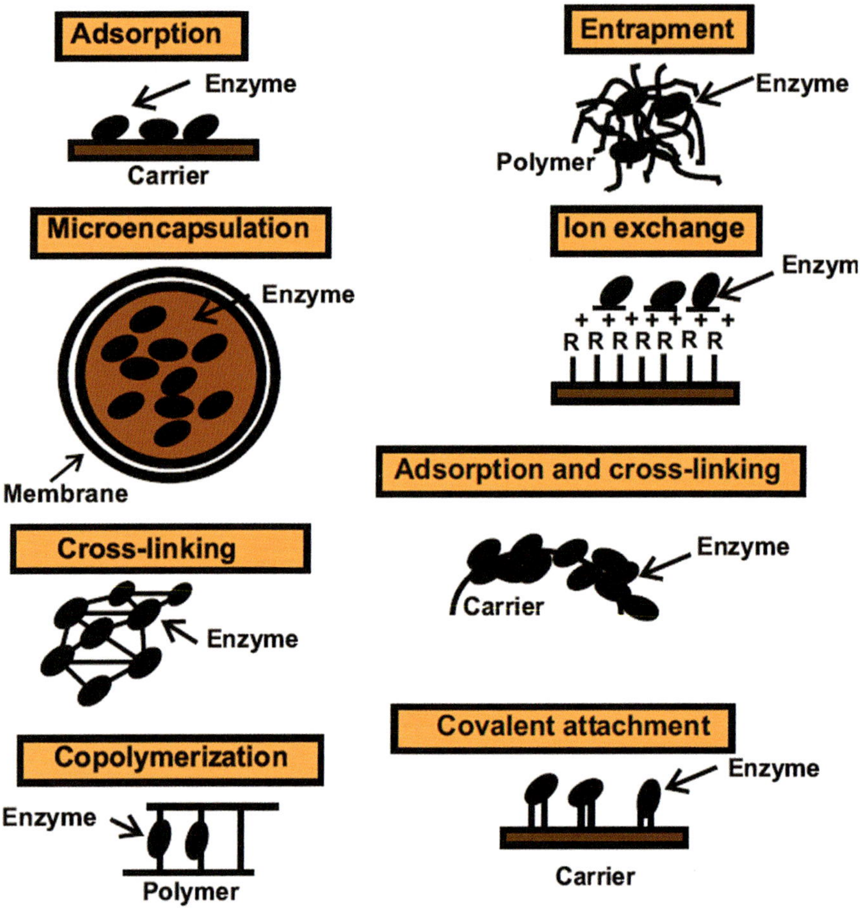

Fig. 15.8 Schematic representation of methods of immobilizing enzymes (Weetal 1975)

top 7.5 cm of soil were higher under no tillage than under ploughed treatments and concluded that changes in activity were not ascribed to long-term pesticide application. Similar results were also reported by Speir and Ross (2002) for urease, protease and phosphatase activities. At different sites in Carlentini (Italy), Riffaldi et al. (2002) studied the biochemical properties of a clay soil as affected by long-term crop management systems and found that dehydrogenase, amylase and ß-Glucosidase activities were considerably reduced in intensively cultivated soils compared to undisturbed grasslands. However, protease activity was inversely related to the intensity of cultivation while soil alkaline phosphatase activity was also depressed by this practice. Among the tested enzymes, the authors found that catalase activity best discriminated between the management systems. Ekenler and Tabatabai (2003) showed that ß-glucosaminidase activity, one of the major enzymes that play a key role in N mineralization in soils, was significantly affected by soil disturbance, tillage and residue management practices. Within this context, Green et al. (2007) examined tillage impacts on soil biological activity and aggregation in a Brazilian Cerrado Oxisol and found that ß-Glucosidase, arylamidase, and acid phosphatase activities were significantly greater under notill and disk harrow than under disk plow in the 0–5 cm depth. They also observed that no-till and disk harrow management practices changed only the vertical distribution of enzyme activity within the profile (0–30 cm). Similarly, in a silt loam (Inceptisol) soil of Luoyang city (China), Jin et al. (2009) found that the activities of catalase, urease, and invertase under no-till with mulch (NT) and subsoiling with mulch (SS) were generally significantly higher than that under conventional control (CT) and reduced tillage (RT). In addition, throughout the wheat-growing season, these three enzymes had higher activities in the stages with vigorous vegetative growth than in stages with productive growth. Štursová and Baldrian (2010) have identified the factors determining the activity and size of the mobile fraction of extracellular enzymes (laccase, Mn-peroxidase, endocellulase, cellobiohydrolase, β-glucosidase, endoxylanase, β-xylosidase, α-glucosidase, chitinase, arylsulfatase, phosphatase, phosphodiesterase, alanine and leucine aminopeptidase) in central Europe using a set of soils covering a wide range of physico-chemical properties. They found that with the exception of Mn peroxidase and β-glucosidase, the specific activities of all enzymes in arable fields under tillage were similar to those in grasslands.

The distribution of soil microorganisms and enzyme activities in particle-size fractions can be important in determining how tillage systems influence SOM accumulation and soil productivity (Salinas-Garcia et al. 1997). Kandeler et al. (1999b) investigated the response of microbial biomass, urease and xylanase within particle size fractions of a Haplic Phaeocem soil at Bad Lauchstädt (Germany). In the bulk soil, they observed that farmyard manure increased urease, arginine deaminase and alkaline phosphatase activity, whereas the effect of additional NPK fertilization depended on the enzyme assayed. For example, Xylanase activity was mainly influenced by the quality and quantity of the residues and the amount of below- ground plant biomass. With regard to the particle size fractions, they found that urease activity was essentially located in the 63–2 μm (56%) and 2–0.1 μm (33%) fractions, whereas xylanase was enriched in both sand fractions (2,000–250 and 250–63 μm). Matocha et al. (2004) also examined the distribution of phenol oxidase activity in the particle size fractions of a silt loam soil after 33 years of imposed tillage and N fertilization treatments. They found that phenol oxidase was located primarily in the silt fraction, followed by the clay and sand in the no-tillage plots. Several authors have shown that soil enzymes are heterogeneously distributed across particle size fractions (Saviozzi et al. 2007; Lagomarsino et al. 2009) or bimodally distributed between the sand and clay fractions (Marx et al. 2005). Some other researchers argued that in general, the carbohydrases (ß-glucosidase, ß-cellobiohydrolase, N-acetyl-ß-glucosaminidase and ß-xylosidase) were reduced from the coarse sand to the clay fraction (Kandeler et al. 1999a; Marx et al. 2005), while the enzymes involved in nitrogen (urease and protease), in phosphorus (acid and alkaline phosphatase) and in sulphur (arylsulfatase) transformations are predominant in both the silt and clay fractions (Kandeler et al. 1999b; Saviozzi et al. 2007).

The activity of ß-Glucosidase was found to be sensitive to soil management and relatively stable with season in a study conducted by Bandick and Dick (1999). Under various poultry litter application rates at the Houston Black clay soils (Riesel, USA), Acosta-Martinez and Harmel (2006) obtained significantly greater activities of both ß-Glucosidase and α-Galactosidase in pasture and cultivated sites as compared to the untreated soils. They concluded that litter rates of approximatively 6.7 Mg ha^{-1} and greater have potential to increase soil microbial properties and enzymatic activities after only four consecutive annual applications compared to untreated soils. However they also noticed that the rates of litter greater than 6.7 Mg ha^{-1} could result in phosphorus (P) levels exceeding crop needs, which leaves the potential for excess P loss in runoff. In a sandy clay loam soil at the experimental farm of the Central Rice Research Institute, Cuttack (India), Nayak et al. (2007) found that the addition of compost and inorganic fertilizer stimulated dehydrogenase, invertase, and urease activity as well as that of ß-Glucosidase and FDA hydrolysis. They further observed that throughout the cropping period, ß-Glucosidase activity increased with plant growth and was highest at grain filling stage. On the contrary, the activities of FDA hydrolysis and invertase were highest at the panicle initiation stage of the rice crop. The activities of Cellulase and amidase were at their peaks during maximum tillering stage and decreased thereafter till maturity to increase again at harvest.

Crecchio et al. (2004) stated that municipal solid waste compost and mineral N amendments significantly increased the activities of dehydrogenase (9.6%), ß-glucosidase (13.5%), urease (15.4%), nitrate reductase (21.4%) and phosphatase (9.7%) in a clay soil of Italy. They further observed a significant reduction in protease activity (from 3.6 to 2.8 U g^{-1} soil) when a double dose of compost was added to the cropped plots, whereas no dosage effect was detected for the other enzymes.

Inorganic fertilizer applications can affect enzyme activities through higher plant yields, crop residue amounts, and changes in soil pH and soil solution chemistry but, the addition of enzyme reaction products by inorganic fertilizers can also suppress enzyme synthesis (Gil-Sotres et al. 2005). These workers also reported that amendments with urban and industrial wastes, releasing various organic and inorganic pollutants, waste or high-salinity water, or atmospheric depositions may affect soil microorganisms and enzyme activities in various ways. For example, the addition of organic waste materials generally enhances soil enzyme activities owing to the increase in soil organic matter and nutrient contents, and improvement of soil physical properties unless these organic wastes also contain high levels of heavy metals and other toxic compounds. They demonstrated that the degree of enzyme activity inhibition by organic wastes depends on the enzyme, soil organic matter and clay contents, as well as the concentration and form of heavy metals in these waste materials.

Dick (1988) indicated that long-term application of organic manure increased soil enzymatic activity, and microbial biomass, but $NH4^{+}$–N fertilizer caused a decrease of amidase and urease activities related to N fertilizer cycle. In general, soil enzymatic activity in the treatment of organic manure combined with N, P and

K fertilizers is higher than that of the treatment of only manure or only chemical fertilizers. In Yingtan, China, Zhong and Cai (2007) found that the activities of acid phosphatase and urease in the original degraded soil were very low and the activities of dehydrogenase and invertase were not detectable. In addition, they noticed that all enzyme activities except invertase increased after 13 years of planting double-cropped rice under flooding conditions, even without fertilization. However, fertilization further increased these enzyme activities, although, generally speaking, they were still low and varied significantly with fertilizer application. These authors observed significantly higher dehydrogenase and acid phosphatase activities in the treatments with P (NPK, NP, and PK) as compared to those in the treatments without P, whereas, the urease activity was significantly higher in the treatments applied with N (NK, NP, and NPK) than in the treatments without N (CK and PK). Recently, phosphorus (P) and potassium (K) fertilizers significantly promoted phosphatase activity in a study conducted by Yang et al. (2008) in Meadow soil (Liaoning, China), which contrasted with the results obtained by Zhao (1998) who concluded that P and K fertilizers reduced the phosphatase activity of both brown fluvoaquic soil and grey cinnamonic soil. The former authors additionally found that catalase activity was reduced when organic manure was applied with mineral fertilizers, whereas invertase activity was not affected significantly by any of the treatments. With respect to urease, they observed that the MNPK treatment significantly increased its activity, but decreased importantly when only N fertilizer was applied with manure. Furthermore the results of their study revealed that the activities of all the studied enzymes were significantly low at the early growth stage of cucumber, but peaked at early-mid growth stage and a net decrease was observed towards the end of the growing season.

Stimulation of microbial biomass and enzyme activities in soil is usually greater in organically fertilized soils than in inorganically ones (Gianfreda and Bollag 1996). For example, it was reported that residue management in a 55 year-old straw and N fertilization experiment significantly influenced the activity of ß-glucosidase, urease, amidase, phosphatase and arylsulfatase in a winter wheat fallow system on semiarid soils of the Pacific Northwest, showing highest activities in manure-treated plots (Kandeler et al. 1999b). By reviewing the long-term effects of organic amendments on soil fertility, Diacono and Montemurro (2009) reported that enzymatic activity increased by 30% with sludge addition. Treatments of fungal residues return (100 and 150%) also had higher level of invertase, urease, neutral-phosphatase, catalase activities than chemical fertilizer treatments at different growth stages of rice and wheat in a paddy submergeric soil under a rice-wheat rotation in Chengdu Plain, China (Huang et al. 2010).

Phenol oxidase activity was found to be 1.7 times greater in no-tillage compared with moldboard plow in the control treatment (Matocha et al. 2004). These authors further observed that in plots under no-tillage, N fertilization (336 kg N ha^{-1}) had a marked negative effect on soil phenol oxidase activity, showing a 38% decrease ($P < 0.01$), whereas moldboard plow plots were relatively insensitive to applied N rate. Roldán et al. (2005) also observed that dehydrogenase and hydrolase (urease, protease-BAA, phosphatase and ß-glucosidase) activities decreased significantly with intensive (mouldboard) and even with reduced tillage (subsoilbedding and shred-bedding) in a Vertisol soil of Northern Tamaulipas in Mexico. However, the activities of alkaline, acid phosphatases and dehydrogenase were found to be significantly higher under non-tillage than those under minimum and conventional tillage in a study conducted by Omidi et al. (2008) at the Research Agricultural Center of Sari (Iran). In addition, under conventional tillage, the authors noticed that the activities of these enzymes were at the lowest. They further observed that non-tillage significantly affected the enzymatic activities at the 0–10 cm depth, but at the 10–20 cm depth minimum tillage resulted in the highest activity of dehydrogenase. Similarly, results obtained by Qin et al. (2010) in both bulk silt loam soil and in its four particle-size fractions at the piedmont of the Taihang Mountains (China) showed that no-till with residue retention as compared to moldboard plowing without residue retention significantly promoted the activities of ß glucosidase, N-acetyl- ß-glucosaminidase, protease, urease and alkaline phosphomonoesterase. They also noticed that the accumulated particular organic carbon in the sand fractions under no-till with residue retention significantly increased the enzyme activities and consequently accelerated the nutrient cycling in the sand fractions, SOC and enzymes within the silt and clay fractions were stabilized by mineral particles and hence had limited capacity to release plant nutrients.

Gil-Sotres and associates (2005) reported that liming of agricultural soils can stimulate most enzyme activities because of increases in soil pH, except for acid phosphatase activity, which is decreased by increasing soil pH. They also stated that organic soil amendments can lead to an initial increase in enzyme activities, but subsequent additions may fail to sustain high enzyme activities.

By studying soil the effects of anthropogenic alterations on enzyme activities in alluvial and calcareous soils of North and South Italy, Gianfreda et al. (2005) noticed that for agricultural soils, arylsulphatase activities increased by about 2.5–3.0 times in the hazel sites periodically flooded compared to the sites never flooded, whereas urease activity in these sites remained lower. They further observed that dehydrogenase activity reached high levels in both North Italy and South Italy soils, irrespective of soil alteration, while invertase activity did not differ between flooded and never flooded South Italy soils. The authors also stated that as compared to agricultural soils, non-cultivated polluted soils in general showed lower values of enzymatic activities. For example urease and dehydrogenase were totally absent, or at not detectable levels, in both the 3-year-high and low polycyclic aromatic hydrocarbons (PAH) polluted soils.

A long-term intensive monoculture usually supplies lower amounts and diversity of organic matter than crop rotation, thus suppressing microbial activities and consequently decreasing enzymatic ones (Klose and Tabatabai 2000). Soils under crop rotations generally show higher enzyme activities compared with monocropping systems owing to diversified organic inputs, improved soil structure, nearly year-round rhizosphere and plant cover and higher root density (Gil-Sotres et al. 2005). These workers also reported that changes in enzyme activities may even anticipate those in organic matter of soil as affected by soil management. However, the changes in enzyme activities due to crop rotations can depend on the soil type and the quantity and quality of plant residues and nutrients entering the soil.

Organic contaminants, heavy metals, and pesticides are pollutants of major environmental concern because of their persistence and broad biocidal activity against key soil processes affecting ecosystem stability and resilience. Each soil enzyme reacts differently to soil pollution depending on the type of pollution, pollutant concentration, and soil type. Heavy metals can react with active groups of enzymes resulting in inhibition or inactivation, or decreasing enzyme proliferation by changing the microbial community (Gil-Sotres et al. 2005). The effect of heavy metals on enzyme activities may vary considerably among the elements, enzymes and soils. Indeed, it is connected both with physical and chemical properties of the soil, especially soil organic matter and clay content, as well as to the kind of enzyme and metal involved. Some authors have proposed enzyme activities, especially dehydrogenase, urease and phosphatase, as indicators of soil contamination with heavy metals (Al-Khafaji and Tabatabai 1979; Welp 1999). For instance Oliveira and Pampulha (2006) observed that a sandy loam soil contaminated by heavy metals in the Estarreja region (Portugal) significantly decreased dehydrogenase activity and ATP content. Thus, they proposed dehydrogenase to be a sensitive assay for determining the effect of heavy metals on the physiologically active soil microbial biomass, being reduced by about 80 and 95% respectively in 2003 and 2004, in relation to the control soil samples. The toxicity of Cd was significantly enhanced by the addition of phenanthrene and the effect is more significant for invertase than that for urease in a study conducted by Shen et al. (2005) in a paddy soil of Qibao campus, Shanghai, China. The findings also revealed that in the case of dehydrogenase, this effect occurred in the latter experiment, whereas the phenanthrene addition did not significantly influence the phosphatase activity in soil with Cd. Lead (Pb) treatments had a stimulating effect on the activities of urease, acid phosphatase and dehydrogenase at low concentration and an inhibitory influence at higher concentration in a greenhouse pot experiment with rice crop, using silty loam and yellowish red soils collected respectively, from Kuajiachi campus and Deqing County, Zhejiang Province, China (Zeng et al. 2007). These workers further found that changes of the soil dehydrogenase activity in the two kinds of paddy soils were much greater than the other two enzymes. Moreover, they observed that when the level of Pb treatments increased to 500 mg/kg, ecological risk existed evidently to both soil enzymatic activities and plants for the soil–Pb–rice system. They concluded that 500 mg/kg Pb treatment could be considered as the critical concentration and that the degree of influence in enzymatic activities by Pb was related to the clay and organic matter contents of the soils.

Pesticide effects on soil enzyme activities are difficult to summarize because pesticides are subjected to different microbial processes, including biodegradation, cometabolism, polymerization or conjugation, and accumulation in microorganisms (Gil-Sotres et al. 2005). These authors have further reported that a small initial stimulatory effect on dehydrogenase activity and slight increases or no significant effect on the activities of ureases and phosphatases when pesticides are applied at recommended field rates. Recent studies carried out by Grenni et al. (2009) to investigate the degradation of the phenylurea herbicide linuron in soil amended with lignocellulosic wastes used as soil organic amendments showed that this amendment is a carbon sources for soil microbiota that increase soil dehydrogenase activity. Within this context, Romero et al. (2010) conducted a study on enzyme activities and diuron persistence in a soil amended with vermicompost and classified as Chromic Vertisol, in Jaén (Andalusia, Spain). Their findings revealed significant differences in diuron availability between Spent Grape marc vermicompost-amended soil (SVD) and urea-amended soil (SDU) with low dehydrogenase and urease activities recorded in unamended soil samples (S). The authors concluded that although it cannot be ruled out that certain bacterial groups could be negatively affected by diuron; there was no evidence for this on the basis of the dehydrogenase and urease enzymatic findings.

15.3 Soil Chemical Properties

According to Schoenholtz et al. (2000), soil chemical properties can be divided into two categories: static (i.e. point-in time) and dynamic (i.e. process-related) soil parameters. They can further be grouped into parameters related to soil carbon status, soil acidity, and measures of nutrient availability. They express to some extent, the dichotomy between the need for simplicity and practicability, which tends to favor static parameters that are easily and routinely measured, but are hierarchically several levels removed from soil function, and the desire to more accurately represent the dynamic processes that underlie site productivity, which tend to involve more laborious and/or costly assays.

15.3.1 Soil Humic Substances

15.3.1.1 Roles of Humic Substances (HS)

Humic substances (HS) are relatively stable and active fractions of soil organic matter pool. They are able to partly resist microbial decomposition (Dou et al. 2008) and are ubiquitous major components of soil organic matter (SOM). Humic substances are also extremely important attributes of soil quality since they profoundly influence the physical, chemical and biological properties and processes of soils. Humic substances are sources of energy and nutrients for soil biota (carbon-cycling and sequestration) and a sink of plant nutrients (N, P, and S via mineralization). They contribute to the charge characteristics of soils and have ability to complex with multivalent ions and organic compounds (control of pollutant toxicity, transport and fate). In addition they are involved in soil and humus-enzyme complexes formation and affect aggregate stability, trafficability, water retention and hydraulic properties. As a result, soil organic matter content and quality are now recognized as key factors in the evaluation of the sustainability of soil management practices (Haynes 2008). At the same time, they possess a considerable chemical reactivity through which they contribute to soil properties and productivity (Galantini and Rosell 2006; Zhang et al. 2009). All the three main components of HS, namely humic acids (HAs), fulvic acids (FAs) and humin (HU) have similar structures but they are different in molecular weights, elemental analysis and functional group contents.

15.3.1.2 Influence of Agricultural Practices on the Composition and Structure of HS

Agricultural management practices, namely tillage, crop rotation and fertilization are known to affect SOM strongly with respect to amount and composition at shorter as well as at longer time scales. Tillage was found to increase the relative aromaticity in humic acid (HA) as compared to that from no-till soils (Madari et al. 1998). These authors have commented upon the observed increase as a probable result of the more oxidative conditions in soils with continuous mechanical disturbance and consequently with faster decomposition rates of the SOM, mainly the aliphatic part of the humic acid (HA). Within this context, Ding et al. (2002) studied the changes of SOM quantity and quality under conventional and conservation tillage systems in a Norfolk loamy sand soil (South Carolina, USA). They found that the elemental composition of HA from the two tillage systems was similar but the ^{13}C-NMR results showed that the aliphatic-C content of HA was higher in the top soil (0–5 cm) under conservation than conventional tillage. Conversely, they observed that the aromatic C of HA was higher under conventional than conservation tillage. However aliphatic-C of HA declined with increase of soil depth under both tillage systems, whereas the aromatic C of HA increased with soil depth. Furthermore, their results revealed that HA contained more recalcitrant functional groups in conventional than in conservation tillage system. Szajdak et al. (2003) studied the concentrations of bound amino acids in humic (HA) and fulvic acids (FA) in samples collected from sandy clay loam soils under no-till (NT) and conventional till (CT) management in Georgia, USA. Their results revealed that the total amount of bound amino acids in HA was higher in NT than in CT samples. In contrast, the concentration of bound amino acids in FA was higher in CT than in NT samples. They further observed that in all samples, the dominant amino acid concentration in HA and FA was higher in NT than in CT, indicating a higher microbial biomass in NT soils. In addition, HA from NT management was characterized by higher molecular weight and a higher degree of condensation of aromatic constituents than HA from CT. After 11 years of different tillage operations in Chinese Mollisols, Liu et al. (2005) reported that integrated tillage, where the system varied with each crop in the rotation (i.e. moldboard plow for wheat, deep chisel for corn and rotatory plow for soybean), had the highest levels of soil organic C and N in the upper soil layer.

By evaluating the chemical changes of SOM in a sandy clay loam in organically and conventionally managed fields in north-east Viterbo city (Central Italy), Marinari et al. (2007) found that the humic C content was 55% greater under organic than conventional management in the upper layer (5–20 cm). They further noticed a significantly higher humification degree and rate as well as richer "young" pseudo-stable, less condensed humic substances under organic than conventional management.

In a Pullman soil of Texas (USA), Acosta-Martinez et al. (2004) found that organic C was higher in perennial pasture compared with continuous cotton, and microbial biomass C, N and the enzymes were also lower in the continuous cotton. Recently, Liu et al. (2006) reported that the inclusion of legume crops in a long-term rotation trial of a Black Earth (Pellic vertisol) and Red clay (Chromic vertisol) soils in Australia resulted in an increase in liable carbon concentrations compared with continuous wheat or a long fallow period. They also noticed that during a 9-year crop rotation experiment in the Chinese Mollisol, the soil organic carbon in the treatments of the wheat-sweet clover and wheat-soybean with addition of pig manure or wheat straw was significantly greater(22%) than that of the commonly used wheat-soybean alone (wheat straw removed).

Fertilizer effects on SOM quality were smaller in loamy than sandy soils (Ellerbrock et al. 1999) and this difference was attributed to clay stability effects on humic substances (HS) extraction. In a fluvi-calcaric cambisol of the long-term rotation experiment at Legnaro (Italy), Nardi et al. (2004) demonstrated that different use of organic and mineral fertilizations influenced the humic carbon molecular complexity. They further noticed that farmyard fertilizations improved the production of humus with a high degree of polycondensation, whereas the absence of organic fertilizer inputs led to a higher percentage of non-complex and light-weight humus.

Long-term effects of fertilizer and manure application as well as the potential impact of diversified cropping systems and soil types on the annual C-inputs and soil quality parameters have been evaluated in sub-humid and semi-arid tropical India (Manna et al. 2005). The findings revealed that the soil organic carbon (SOC) in the unfertilized plot (control) decreased by 41.5, 24.5, and 15.5% compared to initial values in the sites of Barrackpore, Ranchi and Akola, respectively, wherein the treatment receiving NPK and NPK + FYM either maintained or improved it over initial SOC content in these sites. Furthermore, in NPK + FYM treatments, they obtained annual C input values estimated to 4,392, 4,159 and 3,113 kg ha^{-1} $year^{-1}$ in rice–wheat–jute, sorghum–wheat and soybean–wheat system, respectively. They also noticed that the active fractions of SOC, viz., water-soluble carbon and hydrolysable carbohydrates, soil microbial biomass C and N, dehydrogenase and alkaline phosphatase activity, improved significantly with the application of NPK and NPK + FYM.

Due to management differences in a complex of Entic Haplustolls soils of the semiarid Pampean soils (Argentina), Galantini and Rosell (2006) found higher fulvic acid (FA) contents in fertilized systems, resulting in a decreased of the soil HA/FA ratio, whereas, that of humin was lower under the wheat-grazing (WG)-fertilized, probably due to lower tillage intensity. The authors further observed that the elemental composition of humic acid (HA) was modified by fertilization with a decrease in C content and with variable changes in N, H and S contents. Their findings suggested that fertilizer application produce HA with different degree of maturity, with higher transformation in continuous wheat (WW) than in wheat-grazing, in agreement with HA and FA accumulation, respectively. In addition, fertilization produced an increase of aliphatic groups as compared to aromatic ones, probably due to the higher residue inputs in fertilized than in non-fertilized treatments.

Results of a conducted study by Reyes-Solís et al. (2009) in calcaric pheozemes soils irrigated with wastewater in the Mezquital Valley (Mexico) showed that metals such as Zn and Cu are mainly associated with the humic acids. These researchers further observed that Fulvic acids retained mainly Cr while Pb was distributed among humic and fulvic acids. They also observed that in general, metal retention by humic substances increases with irrigation time. However, the authors reported that although the total concentration of the metals was increased with the irrigation time, the metal associated with the organic fraction was not increased, and was accumulated on the top soil, which was the general view because in calcareous soils, the top soil has higher clays and organic matter content. It has been also shown in Southwestern Spain that Cu (II) complexing capacity was higher in leonardite humic acid (LE-HA) and unamended soil humic acids than that in biosolids compost humic acid (BI-HA), whereas the opposite was true for the conditional stability constant of Cu (II)–HA complexes (Soler-Rovira et al. 2010). In addition, these authors observed that Cu (II) complexing capacity increased as the content of acidic ligands, especially COOH groups, aromaticity, and humification degree increased. The results of these researchers further suggested that the pH of the soil amendment system was the most important chemical property governing Cu(II) solubility and bioavailability in metal-contaminated soils remediated with biosolids compost (BI) and leonardite (LE), although soil organic matter and the HA fraction may also be important factors. In particular, the authors stated that binding sites formed by N-, S-, and O-containing acidic functional moieties in HAs may play an important role in the Cu (II) behavior.

Pesticides may affect soil quality but the use of organic amendments, such as sewage sludge, peat or surfactants, may increase their retention and degradation in the upper soil layers, and thus diminish their environmental fate (Hernández-Soriano et al. 2007). Within this context, in a calcareous silt loam soil (Southeast of Spain), these authors studied the adsorption and desorption of several organophosphorous insecticides (diazinon, dimethoate, malathion and methidathion) and found that humic acids induce an adsorption increment of the pesticides, while peat and sewage sludge do not significantly affect pesticide adsorption at the applied dosage (the usual land use dosage for sewage sludge is 40–45 t ha^{-1}). They further reported that the use of a cationic surfactant considerably enhances the insecticide retention and the increase was highest for the combined application of the surfactant and the humic acids. Filipe et al. (2010) investigated the influence of different organic amendments on the adsorption of thiram onto a luvisol soil derived from loess (Portugal) when compared to mineral fertilizer. Their findings revealed that organic amendments greatly affect thiram adsorption–desorption processes, since the magnitude of adsorption, characterized by the KD values, was found to be higher in the organic household compost (COM), followed by the sewage sludge from municipal water treatment facilities (SLU), farmyard manure (FYM) and the lowest value of KD was recorded in the mineral fertilizer application (MIN).

15.3.2 Soil pH, Plant Nutrients and Toxic Elements

15.3.2.1 Importance of Soil pH and Nutrients

The pH of the soil solution is very important because soil solution carries in it nutrients such as nitrogen (N), potassium (K), and phosphorus (P) that plants need in specific amounts to grow, thrive, and resist diseases. No other single chemical soil characteristic is more important in determining the chemical environment of higher plants and soil microbes than the pH (Bolan et al. 2003). There are few reactions involving any component of the soil or of its biological inhabitants that are not sensitive to soil pH. This sensitivity must be recognized in any soil-management system. For example if the pH of the soil solution is increased above 5.5, nitrogen (in the form of nitrate) is made available to plants. Phosphorus, on the other hand, is available to plants when soil pH is between 6.0 and 7.0. Certain bacteria help plants obtain N by converting atmospheric nitrogen into a form of N that plants can use. These bacteria live in root nodules of legumes (like alfalfa and soybeans) and function best when the pH of the plant they live in is growing in soil within an acceptable pH range. For instance, alfalfa grows best in soils of pH 6.2–7.8, while soybean grows best in soils of pH 6.0–7.0. Peanuts grow best in soils of pH 5.3–6.6. Many other crops, vegetables, flowers and shrubs, trees, weeds and fruit are pH dependent and rely on the soil solution to obtain nutrients.

If the soil solution is too acidic plants cannot utilize N, P, K and other nutrients they need. In acidic soils, plants are more likely to take up toxic metals and some plants eventually die of toxicity (poisoning). Herbicides, pesticides, fungicides and other chemicals are used on and around plants to fight off plant diseases and get rid of pathogens that feed on and kill plants. Knowing whether the soil pH is acidic or basic is important because if the soil is too acidic the applied pesticides, herbicides, and fungicides will not be absorbed and they will end up in garden water and rain water runoff, where they eventually become pollutants in our streams, rivers, lakes, and ground water.

15.3.2.2 Impact of Agricultural Management Practices on Soil pH, Plant Nutrients and Toxic Elements

Long-term application of chemical fertilizers without FYM did not significantly increase soil nutrient content, whereas the combination of FYM with inorganic fertilizers resulted in a substantial increase of nitrogen, phosphorus, and potassium in a silt loam soil from Gansu Province, northwestern China (Su et al. 2006). They further observed that imbalanced fertilization significantly decreased soil phosphorus and potassium. This demonstrated that nutrient depletion and deterioration of soil fertility is inevitable when only chemical fertilizers are added. Diepeningen et al. (2006) have studied the effects of organic versus conventional management on chemical and biological parameters in 14 conventional fields of Netherlands and found that the soil type, clayey or sandy, had a more pronounced effect on most of the determined soil characteristics than its management types.

The Olsen phosphorus, exchangeable cations and salinity have been investigated by Morari et al. (2008) in two long-term experiments in north-eastern Italy under various management practices and soil types. They found that in the crop rotations (CR) experiment, the mineral fertilizer rate (MFR) affected the pH value, reducing it proportionally to the rate applied. With regard to this observed soil acidification, the authors argued that ammoniac and ureic fertilizers, in particular, would have a marked effect on the pH, due to the absorption of ammonium or nitrate by plants, since these processes are known to produce hydrogen ions. In general, they further observed that the use of organic fertilizer resulted in higher available P, exchangeable K and Mg concentrations than the mineral treatments at the same levels, whereas no differences for exchangeable Na and Ca were observed.

From an analysis of the effects of tillage, stubble and fertilizer management on the distribution of organic matter and nutrients in the topsoil of a Luvisol in southern Queensland, Australia, Thomas et al. (2007) concluded that organic matter and exchangeable K concentrations were greater under no-till than under conventional tillage and reduced till, whereas exchangeable Na was lower under no-till. They also noticed that the decrease in soil pH with N fertilizer application may be a cause for concern, especially if the soil is already acidic and N fertilizer use continues.

15.4 Soil Physical Properties

Soil is comprised of minerals, soil organic matter (SOM), water, and air (Fig. 15.9). The composition and proportion of these components greatly influence soil physical properties, including texture, structure, and porosity, the fraction of pore space in a soil. In turn, these properties affect air and water movement in the soil, and thus the soil's ability to function.

15.4.1 Importance of Soil Physical Properties

Productive soils have attributes that (1) promote root growth; (2) accept, hold, and supply water; (3) hold, supply, and cycle mineral nutrients; (4) promote optimum gas exchange; (5) promote biological activity; and (6) accept, hold, and release carbon (Burger and Kelting 1999). All of these attributes are, in part, a function of soil physical properties and processes. Some of these soil physical properties are static in time, and some are dynamic over varying time scales. Some are resistant to change by agricultural management practices, while some are changed easily in positive and negative ways. If changed, some properties and processes will recover

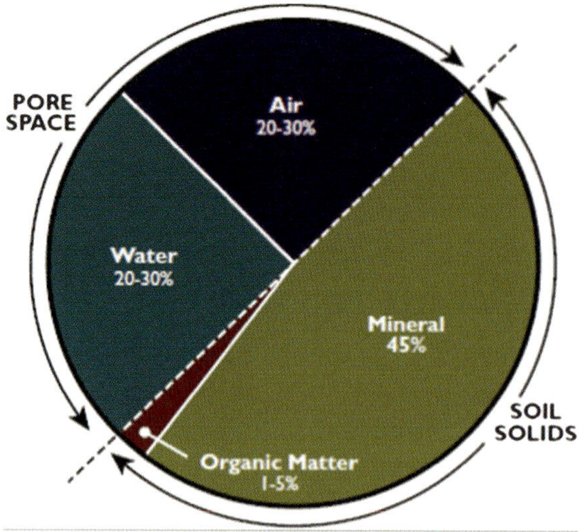

Fig. 15.9 The four components of soil: Minerals and SOM make up the solid fraction, whereas air and water comprise the pore space fraction. A typical agricultural soil is usually around 50% solid particles and 50% pores (Adapted from Brady and Weil 2002)

at varying rates while others are irreversible. All of these factors will determine the extent to which each soil property or process is useful for measuring soil quality and monitoring the maintenance of soil quality through time.

Soil texture is the most fundamental qualitative soil physical property controlling water, nutrient, and oxygen exchange, retention, and uptake (Schoenholtz et al. 2000). However, soil texture and depth are also soil properties that would change little through time for a given soil, and so they would not be very useful for assessing management effects. Soil bulk density varies among soils of different textures, structures, and organic matter content, but within a given soil type, it can be used to monitor degree of soil compaction and puddling.

Soil structure refers to the size and shape of soil aggregates held together by organic matter and other chemical precipitates (Fig. 15.10). Aggregate stability describes the ability of the soil to retain its arrangement of solid and void space when exposed to different stresses (Kay 1990). Stability characteristics are generally specific for a structural form and the type of stress being applied. A measure of aggregate stability could serve as a surrogate for soil structure, which is critical for development of root systems (Kay and Grant 1996).

15.4.2 Soil Physical Properties as Influenced by Agricultural Management Practices

Soils undergo intensive changes in their physical, chemical and biological properties during natural soil development and as a result of anthropogenic processes such as plowing, sealing, erosion by wind and water, amelioration, excavation and

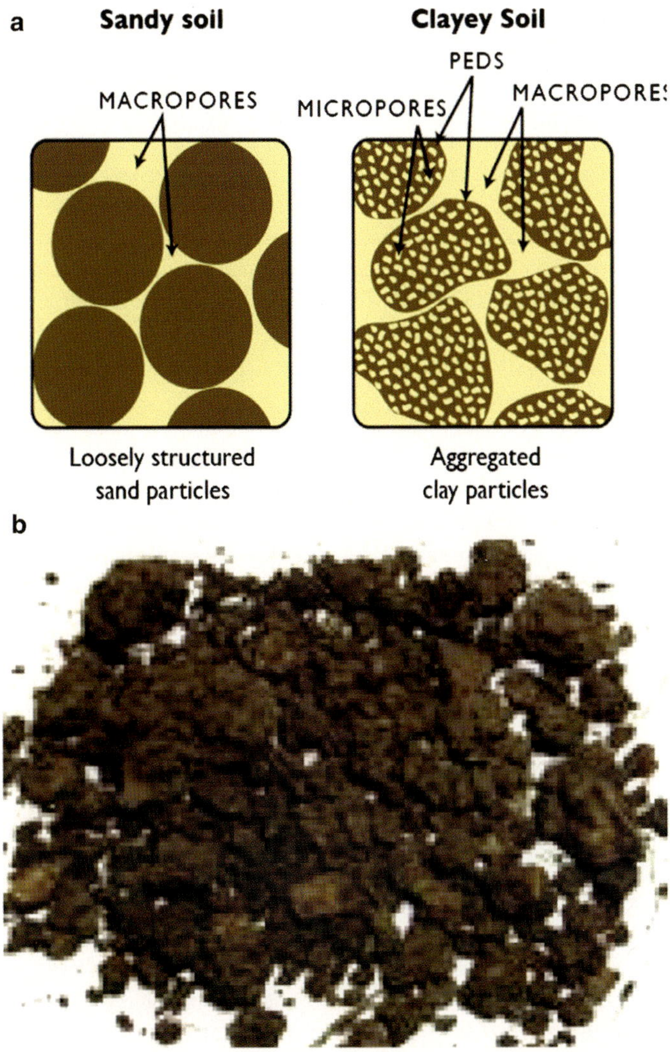

Fig. 15.10 Generalized porosity in sandy and clayey soils (**a**) and top soil exhibiting granular structure (**b**) (From Ann et al. 2005)

reclamation of devastated land (Horn and Baumgartl 2001). With respect to soil physical properties, soil structure is fundamental. Its stability is expressed by the stability of soil aggregates, it directly or indirectly influences other physical and chemical properties of the soil and it can be used as an indicator of soil degradation (Rohošková and Valla 2004). Soil structure exerts important influences on the edaphic conditions and the environment since it is often expressed as the degree of stability of aggregates (Bronick and Lal 2005). Aggregation results from the rearrangement, flocculation and cementation of particles. It is mediated by soil organic carbon (SOC), biota, ionic bridging, clay and carbonates. The measurement of soil aggregate stability becomes important because it can give general information about soil conditions. The aggregate stability is the ability of the bonds of the aggregates to resist when exposed to stresses causing their disintegration (e.g. tillage, swelling and shrinking processes, kinetic energy of raindrops, etc.). According to Kay and Angers (2001), soil structure has a major influence on the ability of soil to support plant growth, cycle C and nutrients, receive, store and transmit water, and to resist soil erosion and the dispersal of chemicals of anthropogenic origin. These authors further stressed that particular attention must be paid to soil structure in managed ecosystems where human activities can cause both short- and long-term changes that may have positive or detrimental impacts on the functions that soil fulfills.

The loss of organic matter and consequently soil fertility is often driven by unsustainable practices such as deep plowing on fragile soils and cultivation of erosion-facilitating crops and the continuous use of heavy machinery which destroys soil structure through compaction. However, it was also reported that tillage and manure application did not influence soil aggregates distribution in a silt-loam soil at the MAPAQ experimental farm in St. Lambert, Canada (Kheyrodin and Antoun 2007). Soil organic matter content has a direct relationship with soil erodibility. It was reported that the stability of soil aggregates is enhanced where organic material is combined with clay particles and where it contributes to chemical bonding (Emadodin et al. 2009).

It is important to note that physical controls on the storage and loss of organic matter can not be viewed in isolation from biological influences. Soil biota is clearly important in mediating physical changes in soil structure that may alter the storage and transformations of SOM. Biological constituents ranging from roots and fungi to microarthropods and earthworms can influence the formation and stabilization of soil aggregates. For example, several studies (e.g. Martin 1992; Lavelle and Martin 1992) have shown that earthworm casts store and protect about 20% more organic C than non-ingested soils. Other organisms may also contribute to the physical protection of organic matter through their influence on soil aggregation. For example, Beare et al. (1997) indicated that fungal hyphae were responsible for about 40% of the macroaggregation (>2,000 μm) and significantly greater retention of soil organic matter in soils under no-tillage management but a much lesser role in conventionally tilled soils. Rillig and Mummey (2006) reviewed the contribution of mycorrhizas, especially arbuscular mycorrhizal fungi (AMF), to soil structure at various hierarchical levels: plant community; individual root; and the soil mycelium. They additionally provided an overview of the various, in part hypothetical, mechanisms by which the mycorrhizal fungal mycelium may influence soil aggregation (Fig. 15.11). These mechanisms have been loosely divided into biophysical, biochemical and biological processes, but they are strongly interrelated. Roots and hyphae enmesh and release organic compounds that act as glue to hold particles together. Consequently, fungal hyphae improve aggregate

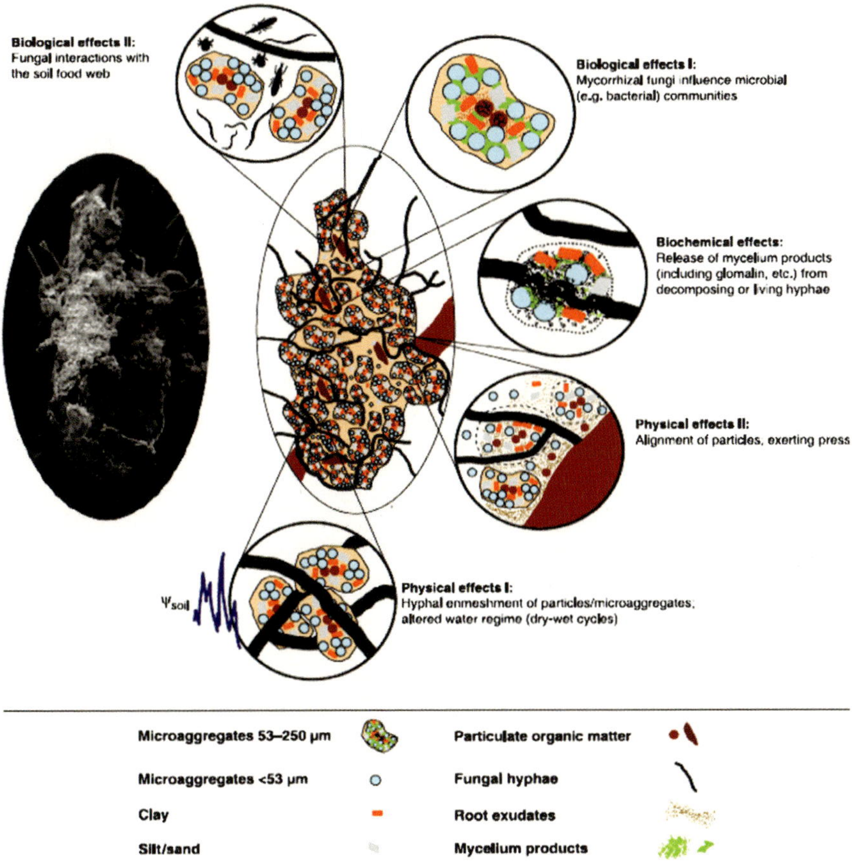

Fig. 15.11 Overviews of various AMF hyphal mediated mechanisms in the formation or stabilization of soil at macroaggregate and microaggregate scales (After Rillig and Mummey 2006)

stability (Bronick and Lal 2005) by reorientation of clay particles, binding particles with extracellular polysaccharides, and enmeshing particles. Hyphae also enmesh microaggregates to form macroaggregates suggesting that aggregation increases with hyphal density (Haynes and Beare 1997). Particles can be rearranged during enmeshment, while wet–dry cycles help to stabilize the aggregates.

Soil disturbance from tillage is a major cause of organic matter depletion and reduction in stability of soil aggregates when native ecosystems are converted to agriculture (Six et al. 2000a). Several authors have reported that tillage enhances the decomposition of SOM, significantly lowers the water stability of aggregate >2 mm, disrupts soil aggregates, exposes physically protected organic material and disturbs plant and animal communities that contribute to aggregation (Chan et al. 2002; Bronick and Lal 2005). According to Pagliai et al. (2004), aggregates are less stable in ploughed soils and these result in a more pronounced tendency to form surface crust compared with soils under minimum tillage and ripper subsoiling. It has been reported that no-tilled soil has increased water-soluble C, dehydrogenase, urease and acid phosphatase activities, aggregate stability and glomalin compared to tilled soils, especially in the shallowest (0–5 cm) layer (Roldán et al. 2005). No-tillage management also increases soil aggregation compared with conventional tillage which destroys the original soil structure by breaking up the macroaggregates into microaggregates and this results in an increase in microporosity and may increase soil bulk density values (Wei et al. 2006). These authors further reported that compared with conventional tillage, no-tillage contributes to the formation and stabilization of macroaggregates by promoting fungal growths, substantially increasing the mass of macroaggregates and by decreasing that of microaggregate.

Soil organic matter (SOM) is a major resource that links the chemical, physical and biological properties of soils, and is considered a major binding agent that stabilizes soil aggregates (Haynes and Beare 1997). Soil aggregation is severely influenced by human activities, including land utilization, tillage disturbance, manure application, and cropping (Bronick and Lal 2005). Aggregates physically protect SOM but also influence soil tilth regulate water flow, determine microbial biomass and mineral nutrient reserves (e.g., N, P, S) and reduce run-off and erosion (Wei et al. 2006).

The addition of maize root powder and glucose played an important role in protecting the water stability of aggregate of diameter >1 mm, but could not protect the water stability of aggregate smaller than 1 mm in diameter (Chen et al. 1977). Manure application can contribute to the accumulation of macroaggregate-protected C and N, improve the nutrient status of the soil and increase SOC levels (Schjonning et al. 2002; García-Orenes et al. 2005). Soil porosity, aggregation and the percentage of large sized water stable aggregates (>5 mm) were improved due to the application of manure and compost but that of smaller sized aggregates were reduced (Pagliai et al. 2004). It was reported that the addition of composted residues is the most effective measure for increasing rhizosphere aggregate stability while the application of biosolids (sewage sludge) significantly increases the organic C, carbohydrates and aggregate stability, resulting in a decrease of the bulk density of soils (Wei et al. 2006).

Different crops have different effects on soil aggregation and C accumulation: perennial grasses due to their extensive root systems are more efficient in soil aggregation than annual crops (Arshad et al. 2004). All the aggregates of pasture soil are significantly richer in monosaccharides than those in the cultivated soil, the aggregates 200–2,000 μm accounting for most of the soil carbohydrates. It appears that xylose which is highly depleted within macroaggregates after cultivation might be responsible for the management-induced changes in aggregate stability (Larre-Larrouy et al. 2004). Guo and associates (2010) have investigated the changes of soil pH, soil bulk density, soil aggregate distribution and stability, and soil organic carbon, total nitrogen, total phosphorus, and available phosphorus associated with soil aggregates when cropland was converted to grassland on the semiarid Loess Plateau of China. Overall, they found that soil physical properties significantly improved after the cropland was converted to grassland. This was especially true for the number of soil aggregates in the 2 to 1 mm and 1 to 0.5 mm categories and the stability of aggregates which significantly increased ($P < 0.05$) in all treatments.

15.5 Conclusions and Prospects

Soil physical, chemical and biological properties affect many processes in the soil that make it suitable for agricultural practices and other purposes. Texture, structure, and porosity influence the movement and retention of water, air and solutes in the soil, which subsequently affect plant growth and organism activity. Most soil chemical properties are associated with the colloid fraction and affect nutrient availability, biota growing conditions, and, in some cases, soil physical properties. Biological properties in soil contribute to soil aggregation, structure and porosity, as well as SOM decomposition and mineralization. Organism activity is controlled by various soil conditions and may be altered by management practices. Since many soil properties are interrelated with one another, it is difficult to draw distinct lines of division where one type of property dominates the behavior of the soil. Therefore, understanding and recognizing soil properties and their connections with one another is important for making sound decisions regarding soil use and management.

Soil physical, chemical, and biological interfacial interactions in the rhizosphere play a vital role in carbon cycling and climate change, the formation and transformation of environmental nanoparticles, the fate of nutrients, metals, metalloids, and xenobiotics, ecotoxicological problems, biodiversity, and geomedicine (Huang 2008). Fundamental understanding of soil physical, chemical, and biological interfacial interactions in the rhizosphere at the molecular level is essential for developing innovative strategies for land resource management to sustain food security and ecosystem integrity. Future research on this extremely challenging and important area of science should be stimulated to sustain and enhance ecosystem productivity, services, and integrity.

To face the challenges of the inherent alteration of human and environmental interactions, a few scientific questions for future research on soil quality/health are listed as follows: (1) How does a shift from agricultural economy to industrial economy alter human and environmental interactions? How does the globalization of our life style alter biogeochemical cycles? Can we find practical agricultural management strategies to stabilize soil organic carbon and avoid eutrophication in aquatic systems? (2) Does global climate change have any direct effect on soil organisms? How do soil organisms adapt or acclimatise to global climate change? (3) What is the relative contribution of each functional group or even specific species to ecosystem? (4) How much can different soil organisms contribute to the remediation of soil pollutants?

References

Acosta-Martínez V, Harmel RD (2006) Soil microbial communities and enzyme activities under various poultry litter application rates. J Environ Qual 35:1309–1318

Acosta-Martínez V, Klose S (2008) Soil enzymes. Encyclopedia Soil Sci 1:1–5. doi:10.1081/E-ESS-120044019

Acosta-Martínez V, Zobeck TM, Allen V (2004) Soil microbial, chemical and physical properties in continuous cotton and integrated crop – livestock systems. Soil Sci Soc Am J 68:1875–1884

Al-Khafaji AA, Tabatabai MA (1979) Effect of trace elements on arylsulphatase activity in soils. Soil Sci 127:129–133

Amezketa E (1999) Soil aggregate stability: a review. J Sus Agric 14:83–151

Andrews SS, Karlen DL, Mitchell JPA (2002) A comparison of soil quality indexing methods for vegetable production systems in Northern California. Agr Ecosyst Environ 90:25–45

Ann MC, Clain J, Jeff J (2004) Nutrient management module no. 15: sustainable agriculture. MSU Extension Serv 4449:1–12

Ann MC, Clain J, Jeff J (2005) Soil and water management module 1: basic soil properties. MSU Extension Serv 4481:1–12

Antil RS, Gerzabek MH, Haberhauer G, Eder G (2005) Long-term effects of cropped vs. fallow and fertilizer amendments on soil organic matter I. Organic carbon. J Plant Nutr Soil Sci 168: 108–116

Antolín MC, Pascual I, García C, Polo A, Sánchez-Díaz M (2005) Growth, yield and solute content of barley in soils treated with sewage sludge under semiarid Mediterranean conditions. Field Crop Res 94:224–237

Arshad MA, Franzluebbers AJ, Azooz RH (2004) Surface soil structural properties under grass and cereal production on a Mollic Cyroboralf in Canada. Soil Till Res 77:15–23

Balota EL, Kanashiro M, Filho AC, Andrade DS, Dick RP (2004) Soil enzyme activities under long-term tillage and crop rotation systems in subtropical agro-ecosystems. Braz J Microbiol 35:300–306

Bandick AK, Dick RP (1999) Field management effects on soil enzyme activities. Soil Biol Biochem 31:1471–1479

Bastida F, Zsolnay A, Hernández T, García C (2008) Past, present and future of soil quality indices: a biological perspective. Geoderma 147:159–171

Beare MH, Hu S, Coleman DC, Hendrix PF (1997) Influences of mycelial fungi on soil aggregation and organic matter storage in conventional and no-tillage soils. Appl Soil Ecol 5:211–219

Bending GD, Turner MK, Rayns F, Marx M-C, Wood M (2004) Microbial and biochemical soil quality indicators and their potential for differentiating areas under contrasting agricultural management regimes. Soil Biol Biochem 36:1785–1792

Bolan NS, Adriano DC, Curtin D (2003) Soil acidification and liming interactions with nutrient and heavy metal transformation and bioavailability. Adv Agron 78:215–272

Boyd SA, Mortland MM (1990) Enzymes interactions with clays and clay-organic matter complexes. In: Bollag JM, Stozsky G (eds) Soil biochemistry. Marcel Dekker, New York

Brady N, Weil R (2002) The nature and properties of soils, 13th edn. Prentice Hall, Upper Saddle River, p 960

Brejda JJ, Karlen DL, Smith JL, Allan DL (2000a) Identification of regional soil quality factors and indicators: II. Northern Mississippi Loess Hills and Palouse Prairie. Soil Sci Soc Am J 64:2125–2135

Brejda JJ, Moorman TB, Karlen DL, Dao TH (2000b) Identification of regional soil quality factors and indicators: I. Central and southern high plains. Soil Sci Soc Am J 64:2115–2124

Bronick CJ, Lal R (2005) Soil structure and management: a review. Geoderma 124:3–22

Bunemann EK, Schwenke GD, Van Zwieten L (2006) Impact of agricultural inputs on soil organisms – a review. Aust J Soil Res 44:379–406

Burger JA, Kelting DL (1999) Using soil quality indicators to assess forest stand management. For Ecol Manage 122:155–156

Burns RG, Nannipieri P, Benedetti A, Hopkins DW (2006) Defining soil quality. In: Bloem J, Hopkins DW, Benedetti A (eds) Microbiological methods for assessing soil quality. CABI Publishing, United Kingdom

Carter MR (2002) Soil quality for sustainable land management: organic matter and aggregation interactions that maintain soil functions. Agron J 94:38–47

Carter MR, Gregorich EG, Anderson DW, Doran JW, Janzen HH, Pierce FJ (1997) Concepts of soil, quality and their significance. In: Gregorich EG, Carter MR (eds) Soil quality for crop production and ecosystem health. Elsevier, Amsterdam

Chan KY, Heenan DP, Oates A (2002) Soil carbon fractions and relationship to soil quality under different tillage and stubble management. Soil Till Res 63:133–139

Chen Y, Senesi N, Schnitzer M (1977) Information provided on humic substances by E_4/E_6 ratio. Soil Sci Soc Am J 41:352–358

Chu H, Lin X, Fujii T, Morimoto S, Yag K, Hu J, Zhang J (2007) Soil microbial biomass, dehydrogenase activity, bacterial community structure in response to long-term fertilizer management. Soil Biol Biochem 39:2971–2976

Coulter JA, Nafziger ED, Wander MM (2009) Soil organic matter response to cropping system and nitrogen fertilization. Agron J 101:592–599

Crecchio C, Curci M, Pizzigallo MDR, Ricciuti P, Ruggiero P (2004) Effects of municipal solid waste compost amendments on soil enzyme activities and bacterial genetic diversity. Soil Biol Biochem 36:1595–1605

Denef K, Six J, Bossuyt H, Frey SD, Elliott ET, Merckx R, Paustian K (2001) Influence of dry-wet cycles on the interrelationship between aggregate, particulate organic matter, and microbial community dynamics. Soil Biol Biochem 33:1599–1611

Dexter AR (2004) Soil physical quality Part I. Theory, effects of soil texture, density, and organic matter, and effects on root growth. Geoderma 120:201–214

Diacono M, Montemurro F (2009) Long-term effects of organic amendments on soil fertility. A review. Agron Sus Dev 30:401–422

Dick WA (1984) Influence of long-term tillage and crop combinations on soil enzyme activities. Soil Sci Soc Am J 48:569–574

Dick RP (1988) Influence of long-term residue management on soil enzyme activity in relation to soil chemical properties of a wheat–fallow system. Soil Biol Biochem 6:159–164

Dick RP (1992) A review: long-term effects of agricultural systems on soil biochemical and microbial parameters. Agr Ecosyst Environ 40:25–36

Dick RP (1997) Soil enzyme activities as integrative indicators of soil health. In: Pankhurst CE, Doube BM, Gupta VVSR (eds) Biological indicators of soil health. CAB International, Wallingford, pp 121–156

Dick WA, Tabatabai MA (1993) Significance and potential uses of soil enzymes. In: Metting FB (ed) Soil microbial ecology: application in agricultural and environmental management. Marcel Dekker, New York, pp 95–125

Ding G, Novak JM, Amarasiriwardena D, Hunt PG, Xing B (2002) Soil organic matter characteristics as affected by tillage management. Soil Sci Soc Am J 66:421–429

Dong LH, Yuan Q, Yuan HL (2006) Changes of chemical properties of humic acids from crude and fugal transformed lignite. Fuel 85:2402–2407

Doran JW, Parkin TB (1994) Defining and assessing soil quality. In: Doran JW, Coleman DC, Bezdicek DF, Stewart BA (eds) Defining soil quality for a sustainable environment, vol 35. Soil Sci Soc Am, Madison, pp 3–21

Doran JW, Parkin TB (1996) Quantitative indicators of soil quality: a minimum data set. In: Doran JW, Jones AJ (eds) Methods for assessing soil quality, vol 49. Soil Sci Soc Am, Madison, pp 25–37

Dou S, Zhang JJ, Li K (2008) Effect of organic matter applications on $^{13\circ}$C-NMR spectra of humic acids of soil. Eur J Soil Sci 59:532–539

Ekenler M, Tabatabai MA (2003) Tillage and residue management effects on ß-glucosaminidase activity in soils. Soil Biol Biochem 35:871–874

Ellerbrock R, Hohn A, Gerke HH (1999) Characterization of soil organic matter from a sandy soil in relation to management practice using FT-IR spectroscopy. Plant Soil 213:55–61

Emadodin I, Reiss S, Bork HR (2009) A study of the relationship between land management and soil aggregate stability (case study near albersdorf, northern-germany). ARPN J Agr Biol Sci 4:48–53

Erkossa T, Itanna F, Stahr K (2007) Indexing soil quality: a new paradigm in soil science research. Aust J Soil Res 45:129–137

Filipe OMS, Vidal MM, Scherer HW, Schneider RJ, Duarte AC, Esteves VI, Santos EBH (2010) Effect of long term organic amendments on adsorption–desorption of thiram onto a luvisol soil derived from loess. Chemosphere. doi:10.1016/j.chemosphere.2010.04.003

Fliessbach A, Martens R, Reber HH (1994) Soil microbial biomass and activity in soils treated with heavy metal contaminated sewage sludge. Soil Biol Biochem 26:1201–1205

Fox JE, Gulledge J, Engelhaupt E, Burow ME, McLachlan JA (2007) Pesticides reduce symbiotic efficiency of nitrogen-fixing rhizobia and host plants. P Natl Acad Sci U S A 104:10282–10287

Frey SD, Elliot ET, Paustian K (1999) Bacterial and fungal abundance and biomass in conventional and no-tillage agroecosystems along two climatic gradients. Soil Biol Biochem 31:573–585

Frostegård Å, Tunlid A, Bååth E (1996) Changes in microbial community structure during long term incubation in two soils experimentally contaminated with metals. Soil Biol Biochem 28:55–63

Galantini J, Rosell R (2006) Long-term fertilization effects on soil organic matter quality and dynamics under different production systems in semiarid pampean soils. Soil Till Res 87: 72–79

García-Orenes F, Guerrero C, Mataix-Solera J, Navarro-Pedreño J, Gómez I, Mataix-Beneyt J (2005) Factors controlling the aggregate stability and bulk density in two different degraded soils amended with biosolids. Soil Till Res 82:65–76

Gianfreda L, Bollag JM (1996) Influence of natural and anthropogenic factors on enzyme activity in soil. In: Stotzky G, Bollag JM (eds) Soil biochemistry. Marcel Dekker, New York

Gianfreda L, Rao MA, Piotrowska A, Palumbo G, Colombo C (2005) Soil enzyme activities as affected by anthropogenic alterations: intensive agricultural practices and organic pollution. Sci Total Environ 341:265–279

Giller KE, Cadisch G (1997) Driven by nature: a sense of arrival or departure. In: Cadisch G, Giller KE (eds) Driven by nature: plant litter quality and decomposition. CAB International, Wallingford

Giller KE, Witter E, McGrath S (1998) Toxicity of heavy metals to microorganisms and microbial processes in agricultural soils: a review. Soil Biol Biochem 30:1389–1414

Gil-Sotres F, Trasar-Cepeda C, Leiros MC, Seoane S (2005) Different approaches to evaluating soil quality using biochemical properties. Soil Biol Biochem 37:877–887

Glover JD, Reganold JP, Andrews PK (2000) Systematic method for rating soil quality of conventional, organic, and integrated apple orchards in Washington State. Agric Ecosyst Environ 80:29–45

Gong W, Yan X, Wang J, Hu T, Gong Y (2009) Long-term manure and fertilizer effects on soil organic matter fractions and microbes under a wheat–maize cropping system in Northern China. Geoderma 149:318–324

Green VS, Stott DE, Cruz JC, Curi N (2007) Tillage impacts on soil biological activity and aggregation in a Brazilian Cerrado Oxisol. Soil Till Res 92:114–121

Gremion F, Chatzinotas A, Kaufmann K, Sigler WV, Harms H (2004) Impacts of heavy metal contamination and phytoremediation on a microbial community during a twelve month microcosm experiment. FEMS Microbiol Ecol 48:273–283

Grenni P, Barra Caracciolo A, Rodriguez-Cruz MS, Sanchez-Martin MJ (2009) Changes in the microbial activity in a soil amended with oak and pine residues and treated with linuron herbicide. Appl Soil Ecol 41:2–7

Guo ZB, Yan G, Zhang RH, Li FM, Zeng ZX, Liu H (2010) Improvement of soil physical properties and aggregate-associated C, N, and P after cropland was converted to grassland in semiarid loess plateau. Soil Sci 175:99–104. doi:10.1097/SS.0b013e3181cda54a

Hati KM, Mandal KG, Misra AK, Ghosh PK, Bandyopadhyay KK (2006) Effect of inorganic fertilizer and farmyard manure on soil physical properties, root distribution, and water use efficiency of soybean in Vertisols of central India. Bioresour Technol 97:2182–2188

Hati KM, Swarup A, Mishra B, Manna MC, Wanjari RH, Mandal KG, Misra AK (2008) Impact of long-term application of fertilizer, manure and lime under intensive cropping on physical properties and organic carbon content of an Alfisol. Geoderma 148:173–179

Hayat S, Iqbal A, Azam ZM, Ahmad A, Inam A, Samiullah A (2002) Effect of long-term application of oil refinery wastewater on soil health with special reference to microbiological characteristics. Bioresour Technol 84:159–163

Haynes RJ (2008) Soil organic matter quality and the size and activity of the microbial biomass: their significance to the quality of agricultural soils. In: Huang Q, Huang PM, Violante A (eds) Soil mineral–microbe-organic interactions theories and applications. Springer-Verlag, Berlin/Heidelberg

Haynes RJ, Beare MH (1997) Influence of six crop species on aggregate stability and some labile organic matter fractions. Soil Biol Biochem 29:1647–1653

Hernández-Soriano MC, Peña A, Mingorance MD (2007) Retention of organophosphorous insecticides on a calcareous soil modified by organic amendments and a surfactant. Sci Total Environ 378:109–113

Horn R, Baumgartl T (2001) Dynamic properties of soils. In: Warrick AW (ed) Soil physics companion. CRC Press, Washington

Huang PM (2008) Soil physicochemical and biological interfacial interactions in the rhizosphere: impacts on food security and ecosystem integrity. Plenary 5th international symposium, ISMOM, Pucón, Chile, 24–28 Nov, pp 1–5

Huang C, Deng L, Gao X, Zhang S, Luo T, Ren Q (2010) Effects of fungal residues return on soil enzymatic activities and fertility dynamics in a paddy soil under a rice-wheat rotation in Chengdu Plain. Soil Till Res 108:16–23

Hulugalle NR, Scott F (2008) A review of the changes in soil quality and profitability accomplished by sowing rotation crops after cotton in Australian Vertosols from 1970 to 2006. Aust J Soil Res 46:173–190

Hunt PG, Karlen DL, Matheny TA, Quisenberry VL (1996) Changes in carbon content of Norfolk loamy sand after 14 years of conservation or conventional tillage. J Soil Water Conserv 51:255–258

Hussain I, Olson KR, Wander MM, Karlen DL (1999) Adaptation of soil quality indices and application to three tillage systems in southern Illinois. Soil Tillage Res 50:237–249

Idowu OJ, van Es HM, Abawi GS, Wolfe DW, Ball JI, Gugino BK, Moebius BN, Schindelbeck RR, Bilgili AV (2008) Farmer-oriented assessment of soil quality using field, laboratory, and VNIR spectroscopy methods. Plant Soil 307:243–253

Jansa J, Mozafar A, Kuhn G, Anken T, Ruh R, Sanders IR, Frossard E (2003) Soil tillage affects the community structure of mycorrhizal fungi in maize roots. Ecol Appl 13:1164–1176

Jin K, Sleutel S, Buchan D, Neve SD, Cai DX, Gabriels D, Jin JY (2009) Changes of soil enzyme activities under different tillage practices in the Chinese Loess Plateau. Soil Till Res 104: 115–120

Kandeler E, Palli S, Stemmer M, Gerzabek MH (1999a) Tillage changes microbial biomass and enzyme activities in particle size fractions of a Haplic Chernozem. Soil Biol Biochem 31:1253–1264

Kandeler E, Stemmer M, Klimanek EM (1999b) Response of soil microbial biomass, urease and xylanase within particle size fractions to long-term soil management. Soil Biol Biochem 31:261–273

Kang GS, Beri V, Sidhu BS, Rupela OP (2005) A new index to assess soil quality and sustainability of wheat-based cropping systems. Biol Fertil Soil 41:389–398

Karlen DL, Wollenhaupt NC, Erbach DC, Berry EC, Swan JB, Eash NS, Jordhal JL (1994) Long-term tillage effects on soil quality. Soil Till Res 32:313–327

Karlen DL, Mausbach MJ, Doran JW, Cline RG, Harris RF, Schuman GE (1997) Soil quality: a concept, definition, and framework for evaluation. Soil Sci Soc Am J 61:4–10

Karlen DL, Ditzler CA, Andrews SS (2003) Soil quality: why and how? Geoderma 114:145–156
Karlen DL, Hurley EG, Andrews SS, Cambardella CA, Meek DW, Duffy MD, Mallarino AP (2006) Crop rotation effects on soil quality at three Northern Corn/Soybean Belt locations. Agron J 98:1–12
Kaur K, Kapoor KK, Gupta AP (2005) Impact of organic manures with and without mineral fertilizers on soil chemical and biological properties under tropical conditions. J Plant Nutr Soil Sci 168:117–122
Kay BD (1990) Rates of change of soil structure under different cropping systems. Adv Soil Sci 12:1–52
Kay BD, Grant CD (1996) Structural aspects of soil quality. In: MacEwan RJ, Carter MR (eds) Soil quality is in the hands of the land manager. Centre for environmental management. University of Ballarat, Victoria, pp 37–41
Kay BD, Angers DA, Structure S (2001) In: Warrick AW (ed) Soil physics companion. CRC Press, Washington
Kheyrodin H, Antoun H (2007) Long–term tillage and manure effect on soil physical and chemical properties and carbon and nitrogen mineralization potentials. Iran Agric Res 25:1–14
Kiem R, Kandeler E (1997) Stabilization of aggregates by the microbial biomass as affected by soil texture and type. Appl Soil Ecol 5:221–230
Klose S, Tabatabai MA (2000) Urease activity of microbial biomass in soil. Biol Fertil Soil 6:68–72
Kong WD, Zhu YG, Fu BJ, Han XZ, Zhang L, He JZ (2008) Effect of long-term application of chemical fertilizers on microbial biomass and functional diversity of a black soil. Pedosphere 18:801–808
Koper J, Piotrowska A (2003) Application of biochemical index to define soil fertility depending on varied organic and mineral fertilization. Electron J Pol Agric Univ 6
Kukal SS, Rasool R, Benbi DK (2009) Soil organic carbon sequestration in relation to organic and inorganic fertilization in rice–wheat and maize–wheat systems. Soil Till Res 102:87–92
Lagomarsino A, Grego S, Marhan S, Moscatelli MC, Kandeler E (2009) Soil management modifies micro-scale abundance and function of soil microorganisms in a Mediterranean ecosystem. Eur J Soil Sci 60:2–12
Lal R (1997) Long-term tillage and maize monoculture effects on a tropical Alfisol in western Nigeria. II. Soil chemical properties. Soil Till Res 42:161–174
Larre-Larrouy MC, Blanchart E, Albrecht A, Feller C (2004) Carbon and monosaccharides of a tropical Vertisol under pasture and market gardening: distribution in secondary organo-mineral separates. Geoderma 119:163–178
Larson WE, Pierce FJ (1994) The dynamics of soil quality as a measure of sustainable management. In: Doran JW, Coleman DC, Bezdicek DF, Stewart BA (eds) Defining soil quality for a sustainable environment, vol 35. Soil Sci Soc Am J, Madison, pp 37–51
Lavelle P, Martin A (1992) Small-scale and large scale effects of endogeic earthworms on soil organic matter dynamics in soils of the humid tropics. Soil Biol Biochem 24:1491–1498
Ledin M, Krantz-Rülcker C, Allard B (1999) Microorganisms as metal sorbents: comparison with other soil constituents in multi-compartment systems. Soil Biol Biochem 31:1639–1648
Lee CH, Wu MY, Asio VB, Zueng-Sang C (2006) Using a soil quality index to assess the effects of applying swine manure compost on soil quality under a crop rotation system in Taiwan. Soil Sci 171:210–222
Liao M, Xie XM (2007) Effect of heavy metals on substrate utilization pattern, biomass, and activity of microbial communities in a reclaimed mining wasteland of red soil area. Ecotoxicol Environ Saf 66:217–223
Liebig M, Carpenter-Boggs L, Johnson JMF, Wright S, Barbour N (2006) Cropping system effects on soil biological characteristics in the Great Plains. Renewable Agric Food Syst 21:36–48
Liebig MA, Varvel G, Doran J (2001) A simple performance-based index for assessing multiple agroecosystem functions. Agron J 93:313–318
Limon-Ortega A, Govaerts B, Sayre KD (2009) Crop rotations, wheat straw management and chicken manure effects on soil quality. Agron J 101:600–606

Liu X, Liu J, Xing B, Herbert SJ, Meng K, Han X, Zhang X (2005) Effects of long-term continuous cropping, tillage, and fertilization on soil organic carbon and nitrogen of black soils in China. Commun Soil Sci Plant Anal 36:1229–1239

Liu X, Herbert SJ, Hashemi AM, Zhang X, Ding G (2006) Effects of agricultural management on soil organic matter and carbon transformation – a review. Plant Soil Environ 52:531–543

Lupwayi NZ, Arshad MA, Rice WA, Clayton GW (2001) Bacterial diversity in water-stable aggregates of soils under conventional and zero tillage management. Appl Soil Ecol 16:251–261

Madari B, Micheli E, Czinkota I, Johnston CT, Graveel JG (1998) Soil organic matter as indicator of changes in the environment. Anthropogenic influences: tillage. Agrokem Talajtan 47:1–4

Manna MC, Swarup A, Wanjari RH, Ravankar HN, Mishra B, Saha MN, Singh YV, Sahi DK, Sarap PA (2005) Long-term effect of fertilizer and manure application on soil organic carbon storage, soil quality and yield sustainability under sub-humid and semi-arid tropical India. Field Crop Res 93:264–280

Marinari S, Liburdi K, Masciandaro G, Ceccanti B, Grego S (2007) Humification mineralization pyrolytic indices and carbon fractions of soil under organic and conventional management in central Italy. Soil Till Res 92:10–17

Martin A (1992) Short-term and long-term effect of the endogeic earthworm Millsonia anomala (Omodeo Megascolecidae, Oligochaeta) of a tropical savanna, on soil organic matter. Biol Fertil Soil 11:234–238

Marx MC, Kandeler E, Wood M, Wermbter N, Jarvis SC (2005) Exploring the enzymatic landscape: distribution and kinetics of hydrolytic enzymes in soil particle-size fractions. Soil Biol Biochem 37:35–48

Masto RE, Chhonkar PK, Purakayastha TJ, Patra AK, Singh D (2008) Soil quality indices for evaluation of long-term land use and soil management practices in semi-arid sub tropical India. Land Degrad Dev 19:516–529

Masto RE, Chhonkar PK, Singh D, Patra AK (2007) Soil quality response to long-term nutrient and crop management on a semi-arid Inceptisol. Agric Ecosyst Environ 18:130–142

Matocha CJ, Haszler GR, Grove JH (2004) Nitrogen fertilization suppresses soil phenol oxidase enzyme activity in no-tillage systems. Soil Sci 169:708–714

Matson PA, Parton WJ, Power AG, Swift MJ (1997) Agricultural intensification and ecosystem properties. Science 277:504–509

Maughan MW, Flores JPC, Anghinoni I, Bollero G, Fernández FG, Tracy BF (2009) Soil quality and corn yield under crop–livestock integration in Illinois. Agron J 101:1503–1510

Mausbach MJ, Tugel A (1997) Soil quality – a multitude of approaches. In: Berkeley, California soil quality: from critical research to sustainable management. California: Keynote address Kearney Foundation Symposium, March 25, pp 1–13

Miller M, Dick RP (1995) Thermal stability and activities of soil enzymes as influenced by crop rotations. Soil Biol Biochem 27:1161–1166

Miura F, Nakamoto T, Kaneda S, Okano S, Nakajima M, Murakami T (2008) Dynamics of soil biota at different depths under two contrasting tillage practices. Soil Biol Biochem 40:406–414

Moffett BF, Nicholson FA, Uwakwe NC, Chambers BJ, Harris JA, Hill TCJ (2003) Zinc contamination decreases the bacterial diversity of agricultural soil. FEMS Microbiol Ecol 43:13–19

Mohanty M, Painuli DK, Misra AK, Ghosh PK (2007) Soil quality effects of tillage under rice -wheat cropping on a Vertisol in India. Soil Tillage Res 92:243–250

Morari F, Lugato E, Giardini L (2008) Olsen phosphorus, exchangeable cations and salinity in two long-term experiments of north-eastern Italy and assessment of soil quality evolution. Agr Ecosyst Environ 124:85–96

Muller AK, Westergaard K, Christensen S, SÖrensen SJ (2001) The effect of long-term mercury pollution on the soil microbial community. FEMS Microbiol Ecol 36:11–19

Nardi S, Morari F, Berti A, Tosoni M, Giardini L (2004) Soil organic matter properties after 40 years of different use of organic and mineral fertilizers. Eur J Agron 21:357–367

Nayak DR, Babu YJ, Adhya TK (2007) Long-term application of compost influences microbial biomass and enzyme activities in a tropical Aeric Endoaquept planted to rice under flooded condition. Soil Biol Biochem 39:1897–1906

Nemergut DR, Townsend AR, Sattin SR, Freeman KR, Fierer N, Neff JC, Bowman WD, Schadt CW, Weintraub MN, Schmidt SK (2008) The effects of chronic nitrogen fertilization on alpine tundra soil microbial communities: implications for carbon and nitrogen cycling. Environ Microbiol 10:3093–3105

Oehl F, Sieverding E, Ineichen K, Mäder P, Boller T, Wiemken A (2003) Impact of land use intensity on the species diversity of arbuscular mycorrhizal fungi in agroecosystems of Central Europe. Appl Environ Microbiol 69:2816–2824

Olchin GP, Ogle S, Frey SD, Filley TR, Paustian K, Six J (2008) Residue carbon stabilization in soil aggregates of no-till and tillage management of dryland cropping systems. Soil Sci Soc Am J 72:507–513

Olfert O, Johnson GD, Brandt SA, Thomas AG (2002) Use of arthropod diversity and abundance to evaluate cropping systems. Agron J 94:210–216

Oliveira A, Pampulha ME (2006) Effects of long-term heavy metal contamination on soil microbial characteristics. J Biosci Bioeng 102:157–161

Omidi H, Tahmasebi Z, Torabi H, Miransari M (2008) Soil enzymatic activities and available P and Zn as affected by tillage practices, canola (*Brassica napus* L.) cultivars and planting dates. Eur J soil boil 44:443–450

Pagliai M, Vignozzi N, Pellegrini S (2004) Soil structure and the effect of management practices. Soil Till Res 79:131–143

Puglisi E, Del Re AAM, Rao MA, Gianfreda L (2006) Development and validation of numerical indices integrating enzyme activities of soils. Soil Biol Biochem 38:1673–1681

Puglisi E, Nicelli M, Capri E, Trevisan M, Del Re AAM (2005) A soil alteration index based on phospholipid fatty acids. Chemosphere 61:1548–1557

Qin S, Hu C, He X, Dong W, Cui J, Wang Y (2010) Soil organic carbon, nutrients and relevant enzyme activities in particle-size fractions under conservational versus traditional agricultural management. Appl Soil Ecol. doi:10.1016/j.apsoil. 2010.03.007

Reyes-Solís IE, Solís C, Isaac-Olive K, García NE, Andrade E (2009) Fractionation analysis of trace metals in humic substances of soils irrigated with wastewater in Central Mexico by particle induced X-ray emission. Microchem J 91:129–132

Riffaldi R, Saviozzi A, Levi-Minzi R, Cardelli R (2002) Biochemical properties of a Mediterranean soil as affected by long-term crop management systems. Soil Till Res 67:109–114

Rillig MC, Mummey DL (2006) Mycorrhizas and soil structure. New Phytol 171:41–53. doi:10.1111/j.1469-8137.2006.01750.x

Rohošková M, Valla M (2004) Comparison of two methods for aggregate stability measurement – a review. Plant Soil Environ 50:379–382

Roldán A, Salinas-García JR, Alguacila MM, Caravaca F (2005) Changes in soil enzyme activity, fertility, aggregation and C sequestration mediated by conservation tillage practices and water regime in a maize field. Appl Soil Ecol 30:11–20

Romero E, Fernández-Bayo J, Díaz JMC, Nogales R (2010) Enzyme activities and diuron persistence in soil amended with vermicompost derived from spent grape marc and treated with urea. Appl Soil Ecol 44:198–204

Salinas-Garcia JR, Hons FM, Matocha JE, Zuberer DA (1997) Soil carbon and nitrogen dynamics as affected by long-term tillage and nitrogen fertilization. Biol Fertil Soil 25:182–188

Saviozzi A, Cardelli R, Labbaci L, Levi-Minzi R, Riffaldi R (2007) Selected enzyme activities in particle-size fractions from an organically and conventionally managed soil. Fresenius Environ Bull 16:1195–1200

Schimel J, Balser TC, Wallenstein M (2007) Microbial stress-response physiology and its implications for ecosystem function. Ecology 88:1386–1394

Schjonning P, Elmholt S, Munkholm LJ, Debosz K (2002) Soil quality aspects of humid sandy loams as influenced by organic and conventional long-term management. Agric Ecosyst Environ 88:195–214

Schloter M, Dilly O, Munch JC (2003) Indicators for evaluating soil quality. Agric Ecosyst Environ 98:255–262

Schoenholtz SH, Miegroet HV, Burger JA (2000) A review of chemical and physical properties as indicators of forest soil quality: challenges and opportunities. For Ecol Manage 138:335–356

Schutter ME, Dick RP (2002) Microbial community profiles and activities among aggregates of winter fallow and cover-cropped soil. Soil Sci Soc Am J 66:142–153

Sharma KL, Mandal UK, Srnivas K, Vittal KPR, Mandal B, Grace JK, Ramesh V (2005) Long-term soil management effects on crop yields and soil quality in a dryland Alfisol. Soil Till Res 83:246–259

Sharma KL, Grace JK, Mandal UK, Gajbhiye PN, Srinivas K, Korwar GR, Bindu VH, Ramesh V, Ramachandran K, Yadav SK (2008) Evaluation of long-term soil management practices using key indicators and soil quality indices in a semi-arid tropical Alfisol. Aust J Soil Res 46:368–377

Shen G, Cao L, Lu Y, Hong J (2005) Influence of phenanthrene on cadmium toxicity to soil enzymes and microbial growth. Environ Sci Pollut Res Int 12:259–263

Singer JW, Bauer PJ (2009) Crop rotation in row crop production systems. Popular publication. Available: http://soilquality.org/practices/row_crop_rotations.html

Six J, Elliott ET, Paustian K (2000) Soil macroaggregate turnover and microaggregate formation: a mechanism for C sequestration under no-tillage agriculture. Soil Biol Biochem 32:2099–2103

Soler-Rovira P, Madejón E, Madejón P, Plaza C (2010) In situ remediation of metal contaminated soils with organic amendments: role of humic acids in copper bioavailability. Chemosphere 79:844–849

Speir TW, Ross DJ (2002) Hydrolytic enzyme activities to assess soil degradation and recovery. In: Burns RG, Dick RP (eds) Enzymes in the environment. Activity, ecology and applications. Marcel Dekker, New York

Štursová M, Baldrian P (2010) Effects of soil properties and management on the activity of soil organic matter transforming enzymes and the quantification of soil-bound and free activity. Plant Soil. doi:10.1007/s11104-010-0296-3

Su YZ, Wang F, Suo DR, Zhang ZH, Du MW (2006) Long-term effect of fertilizer and manure application on soil-carbon sequestration and soil fertility under the wheat–wheat–maize cropping system in northwest China. Nutr Cycl Agroecosyst 75:285–295. doi:10.1007/s10705-006-9034-x

Szajdak L, Jezierski A, Cabrera MT (2003) Impact of conventional and no-tillage management on soil amino acids, stable and transient radicals and properties of humic and fulvic acids. Org Geochem 34:693–700

Tatzber M, Stemmer M, Spiegel H, Katzlberger C, Haberhauer G, Gerzabek MH (2008) Impact of different tillage practices on molecular characteristics of humic acids in a long term field experiment—An application of three different spectroscopic methods. Sci Total Environ 406:256–268

Taylor JP, Wilson B, Mills MS, Burns RG (2002) Comparison of microbial numbers and enzymatic activities in surface soils and subsoils using various techniques. Soil Biol Biochem 34:387–410

Ternan JL, Williams AG, Elmes A, Hartley R (1996) Aggregate stability of soils in central Spain and the role of land management. Earth Surf Process Landforms 21:181–193

Thomas GA, Dalal RC, Standley J (2007) No-till effects on organic matter, pH, cation exchange capacity and nutrient distribution in a Luvisol in the semi-arid subtropics. Soil Tillage Res 94:295–304

Trasar-Cepeda C, Leirós MC, Seoane S, Gil-Sotres F (2000) Limitation of soil enzymes as indicators of soil pollution. Soil Biol Biochem 32:1867–1875

Tugel AJ, Lewandowski AM (1999) Soil biology primer. NRCS Soil Quality Institute, Ames, p 50

van Diepeningen AD, Vos OJ, Korthals GW, van Bruggen AHC (2006) Effects of organic versus conventional management on chemical and biological parameters in agricultural soils. Appl Soil Ecol 31:120–135

Vargas Gil S, Meriles J, Conforto C, Figoni G, Basanta M, Lovera E, March GJ (2009) Field assessment of soil biological and chemical quality in response to crop management practices. World J Microbiol Biotechnol 25:439–448

Veenstra JJ, Horwath WR, Mitchell JP, Munk DS (2006) Conservation tillage and cover cropping influence soil properties in San Joaquin Valley cotton-tomato crop. Calif Agric 60:146–153

Vigil MF, Sparks D (2003) Conservation tillage fact sheet. Central Great Plains Research Station, Akron

Vineela C, Wani SP, Srinivasarao Ch, Padmaja B, Vittal KPR (2008) Microbial properties of soils as affected by cropping and nutrient management practices in several long-term manurial experiments in the semi-arid tropics of India. Appl Soil Ecol 40:165–173

Wang G, Liu J, Qi X, Jin J, Wang Y, Liu X (2008) Effects of fertilization on bacterial community structure and function in a black soil of Dehui region estimated by Biolog and PCR-DGGE methods. Acta Ecol Sin 28:220–226

Wang Q, Dai J, Yu Y, Zhang Y, Shen T, Liu J, Wang R (2010) Efficiencies of different microbial parameters as indicator to assess slight metal pollutions in a farm field near a gold mining area. Environ Monit Assess 161:495–508

Wang X, Gong Z (1998) Assessment and analysis of soil quality changes after eleven years of reclamation in subtropical China. Geoderma 81:339–355

Weetal HH (1975) Immobilized enzymes and their application in the food and beverage industry. Process Biochem 10:3–24

Wei C, Gao M, Shao J, Xie D, Pan G (2006) Soil aggregate and its response to land management practices. China Particuology 4:211–219

Welp G (1999) Inhibitory effects of the total and water-soluble concentrations of nine different metals on the dehydrogenase activity of a loess soil. Biol Fertil Soil 30:132–139

Wu T, Schoenau JJ, Li F, Qian P, Malhi SS, Shi Y (2005) Influence of fertilization and organic amendments on organic-carbon fractions in Heilu soil on the loess plateau of China. J Plant Nutr Soil Sci 168:100–107

Xu MG, Li DC, Li JM, Qin DZ, Kazuyuki Y, Yasukazu H (2008) Effects of organic manure application with chemical fertilizers on nutrient absorption and yield of rice in Hunan of Southern China. Agric Sci China 7:1245–1252

Yang SM, Malhi SS, Li FM, Suo DR, Xu MG, Wang P, Xiao GJ, Jia Y, Guo TW, Wang JG (2007) Long-term effects of manure and fertilization on soil organic matter and quality parameters of a calcareous soil in NW China. J Plant Nutr Soil Sci 170:234–243

Yang L, Li T, Li F, Lemcoff JH, Cohen S (2008) Fertilization regulates soil enzymatic activity and fertility dynamics in a cucumber field. Sci Hortic (Amsterdam) 116:21–26

Zalidis G, Stamatiadis S, Takavakoglou V, Eskridge K, Misopolinos N (2002) Impacts of agricultural practices on soil and water quality in the Mediterranean region and proposed assessment methodology. Agric Ecosyst Environ 88:137–146

Zeng LS, Liao M, Chen CL, Huang CY (2007) Effects of lead contamination on soil enzymatic activities, microbial biomass, and rice physiological indices in soil–lead rice (*Oryza sativa* L.) system. Ecotoxicol Environ Saf 67:67–74

Zhang YL, Dai JL, Wang RQ, Zhang J (2008) Effects of long-term sewage irrigation on agricultural soil microbial structural and functional characterizations in Shandong, China. Eur J Soil Biol 44:84–91

Zhang J, Dai J, Wang R, Li F, Wang W (2009) Adsorption and desorption of divalent mercury (Hg^{2+}) on humic acids and fulvic acids extracted from typical soils in China. Colloid Surf A 335:194–201

Zhao ZZ (1998) Research on relationship between soil enzyme and soil fertility. J Qinghai Univ (Nat Sci) 16:24–29

Zhong WH, Cai ZC (2007) Long-term effects of inorganic fertilizers on microbial biomass and community functional diversity in a paddy soil derived from quaternary red clay. Appl Soil Ecol 36:84–91

Zvomuya F, Janzen HH, Larney FJ (2008) A long-term field bioassay of soil quality indicators in a semiarid environment. Soil Sci Soc Am J 72:683–692

Index

J. Xu and D.L. Sparks (eds.), *Molecular Environmental Soil Science*,
Progress in Soil Science, DOI 10.1007/978-94-007-4177-5,

Printed by Printforce, the Netherlands